Applied Mathematical Sciences

Volume 206

The mathematization of all sciences, the fading of traditional scientific boundaries, the impact of computer technology, the growing importance of computer modeling and the necessity of scientific planning all create the need both in education and research for books that are introductory to and abreast of these developments. The purpose of this series is to provide such books, suitable for the user of mathematics, the mathematician interested in applications, and the student scientist. In particular, this series will provide an outlet for topics of immediate interest because of the novelty of its treatment of an application or of mathematics being applied or lying close to applications. These books should be accessible to readers versed in mathematics or science and engineering, and will feature a lively tutorial style, a focus on topics of current interest, and present clear exposition of broad appeal. A compliment to the Applied Mathematical Sciences series is the Texts in Applied Mathematics series, which publishes textbooks suitable for advanced undergraduate and beginning graduate courses.

More information about this series at http://www.springer.com/series/34

Bangti Jin

Fractional Differential Equations

An Approach via Fractional Derivatives

 Springer

Bangti Jin
Department of Computer Science
University College London
London, UK

ISSN 0066-5452 ISSN 2196-968X (electronic)
Applied Mathematical Sciences
ISBN 978-3-030-76045-8 ISBN 978-3-030-76043-4 (eBook)
https://doi.org/10.1007/978-3-030-76043-4

Mathematics Subject Classification: 26A33, 33E12, 35R11, 34A08, 35R30

This Springer imprint is published by the registered company Springer Nature Switzerland AG
The registered company address is: Gewerbestrasse 11, 6330 Cham, Switzerland

Preface

Fractional differential equations (FDES), i.e., differential equations involving fractional-order derivatives, have received much recent attention in engineering, physics, biology and mathematics, due to their extraordinary modeling capability for describing certain anomalous transport phenomena observed in real world. However, the relevant mathematical theory of FDES is still far from complete when compared with the more established integer-order counterparts.

The main modeling tools, i.e., fractional integrals and fractional derivatives, have a history nearly as old as calculus itself. In 1695, in a letter to Leibniz, L'Hospital asked about the possible meaning of a half derivative. Leibniz in a letter dated September 30, 1695—now deemed as the birthday of fractional calculus—replied "*It will lead to a paradox, from which one day useful consequences will be drawn.*" The first definitions of general fractional derivatives, e.g., Riemann-Liouville fractional integral and derivatives, that supported rigorous analysis were proposed in the nineteenth century. By the middle of the twentieth century, fractional calculus was already an almost fully developed field. The practical applications of these operators, as conjectured by Leibniz, were realized only much later. Recent experimental observations of "fractional" diffusion have led to a flurry of studies in physics and engineering. By now FDES have been established as the lingua franca in the study of anomalous diffusion processes.

This motivates a mathematical study of differential equations containing one or more fractional derivatives. This is the primary purpose of this textbook: we give a self-contained introduction to basic mathematical theory of FDES, e.g., well-posedness (existence, uniqueness and regularity of the solutions) and other analytic properties, e.g., comparison principle, maximal principle and analyticity, and contrast them with the integer counterparts. The short answer is that there is much in common, but with some significant and important differences and additional technical challenges. One distinct feature is that the solution usually contains weak singularity, even if the problem data are smooth, due to the nonlocality of the operators and weak singularity of the associated kernels. These new features have raised many outstanding challenges for relevant fields, e.g., mathematical analysis, numerical analysis, inverse problems and optimal control, which have witnessed exciting developments in recent years. We will also briefly touch upon these topics to give a flavor, illustrating distinct influences of the nonlocality of fractional operators.

The term FDE is a big umbrella for any differential equations involving one or more fractional derivatives. This textbook only considers the models that involve only one fractional derivative in either spatial or temporal variable, mostly of the Djrbashian-Caputo type, and only briefly touches upon that of the Riemann-Liouville type. This class of FDEs has direct physical meaning and allows initial/boundary values as for the classical integer-order differential equations, and thus has been predominant in the mathematical modeling of many practical applications.

The book consists of seven chapters and one appendix, which are organized as follows. Chapter 1 briefly describes continuous time random walk, which shows that the probability density function of certain stochastic process satisfies a FDE. Chapters 2 and 3 describe, respectively, fractional calculus (various fractional integral and derivatives), and two prominent special functions, i.e., Mittag-Leffler function and Wright function. These represent the main mathematical tools for FDEs, and thus Chapters 2 and 3 build the foundation for the study of FDEs in Chapters 4–7. Chapters 4–5 are devoted to initial and boundary value problems for fractional ODEs, respectively. Chapters 6–7 describe mathematical theory for time-fractional diffusion (subdiffusion) in Hilbert spaces and Hölder spaces, respectively. In the appendix, we recall basic facts of function spaces, integral transforms and fixed point theorems that are extensively used in the book. Note that Chapters 4–7 are organized in a manner such that they only loosely depend on each other and the reader can directly proceed to the chapter of interest and study each chapter independently, after going through Chapters 2–3 (and the respective parts of the appendix).

The selected materials focus mainly on solution theory, especially the issues of existence, uniqueness and regularity, which are also important in other related areas, e.g., numerical analysis. Needless to say, these topics can represent only a few small tips of current exciting developments within the FDE community, and many important topics and results have to be left out due to page limitation. Also we do not aim for a comprehensive treatment and describing the results in the most general form, and instead we only state the results that are commonly used in order to avoid unnecessary technicalities. (Actually each chapter can be expanded into a book itself!) Whenever known, references for the stated results are provided, and the pointers to directly relevant further extensions are also given. Nonetheless, the reference list is nowhere close to complete and clearly biased by the author's knowledge.

There are many excellent books available, especially at the research level, focusing on various aspects of fractional calculus or FDEs, notably [SKM93] on Riemann-Liouville fractional integral/derivatives, [Pod99, KST06] on fractional derivatives and ODEs, [Die10] on ODEs with Djrbashian-Caputo fractional derivative. Our intention is to take a broader scope than the more focused research monographs and to provide a relatively self-contained gentle introduction to current mathematical theory, especially distinct features when compared with the integer-order counterparts. Besides, it contains discussions on numerical algorithms and inverse problems, which have so far been only scarcely discussed in textbooks on fdes, despite their growing interest in the community. Thus, it may serve as an introductory reading for young researchers entering the field. It can also be used as the textbook for a single semester course. Many exercises (of different degree of difficulty) are provided at the end of each chapter, and quite a few of them give further extensions to the results stated in the book. The typical audience would be senior undergraduate and graduate students with a solid background in real analysis and basic theory of ordinary and partial differential equations, especially Chapters 6 –7.

The book project was initiated in 2015, and but stopped in Summer 2016. The writing was resumed in later 2019, and the scope and breadth have grown significantly, in order to properly reflect several important developments in the last few years. The majority of the writing was done during the COVID 19 pandemic, and partly during an extended research stay in 2020 at Department of Mathematics, The Chinese University of Hong Kong, when the book was being finalized. The hospitality is greatly appreciated. The writing has benefited enormously from the collaborations with several researchers on the topic of FDES during the past decade, especially Raytcho Lazarov, Buyang Li, Yikan Liu, Joseph Pasciak, William Rundell, Yubin Yan and Zhi Zhou. Further several researchers have kindly provided constructive comments on the book at different stages. In particular, Yubin Yan and Lele Yuan kindly proofread several chapters of the book and gave many useful comments, which have helped eliminate many typos.

Last but not least, I would like to take this opportunity to thank my family for their constant support over the writing of the book, which has taken much of the time that I should have spent with them.

London, UK   *Bangti Jin*
March 2021

Contents

Part II   Fractional Ordinary Differential Equations

Part III   Time-Fractional Diffusion

Acronyms

$\Gamma(z)$	Gamma function
$\Gamma_{\epsilon,\varphi}$	contour in complex plane
$\theta_\alpha, \overline{\theta}_\alpha$	fractional θ functions
$AC^n(\overline{D})$	space of absolutely continuous functions
$B(\alpha, \beta)$	Beta function
$\mathbb{C}_+$	the set $\{z \in \mathbb{C} : \Re z > 0\}$
$\mathbb{C}_-$	the set $\{z \in \mathbb{C} : \Re(z) < 0\}$
$C(\alpha, \beta)$	binomial coefficient
$C^{k,\gamma}(\overline{\Omega})$	Hölder spaces
${}^C_a D^\alpha_x$	left-sided Djrbashian-Caputo fractional derivative of order α
${}^C_x D^\alpha_b$	right-sided Djrbashian-Caputo fractional derivative of order α
${}^C_a D^{\alpha*}_x$	regularized left-sided Djrbashian-Caputo fractional derivative of order α
${}^R_a D^\alpha_x$	left-sided Riemann-Liouville fractional derivative of order α
${}^R_x D^\alpha_b$	right-sided Riemann-Liouville fractional derivative of order α
${}^{GL}_a D^\alpha_x$	left-sided Grunwald-Letnikov fractional derivative
$E_{\alpha,\beta}(z)$	two-parameter Mittag-Leffler function
$\mathbb{E}$	expectation
${}_2F_1(a,b;c;x)$	hypergeometric function
$\mathcal{F}$	Fourier transform
$G_\alpha, \overline{G}_\alpha$	fundamental solutions for subdiffusion
$\dot{H}^s(\Omega)$	(fractional order) Sobolev space
$H^s(\Omega)$	Sobolev spaces
${}_a I^\alpha_x$	left-sided Riemann-Liouville fractional integral of order α
${}_x I^\alpha_b$	right-sided Riemann-Liouville fractional integral of order α
$\mathcal{L}$	Laplace transform
M_μ	Mainardi function
$\mathbb{N}$	the set of natural numbers
$\mathbb{N}_0$	the set of nonnegative integers, i.e., $\mathbb{N} = \mathbb{N} \cup \{0\}$

$\mathbb{R}_+$	the set $(0, \infty)$
$\mathbb{R}_-$	the set $(-\infty, 0)$
$W_{\rho,\mu}(z)$	Wright function
$W^{k,p}(\Omega)$	Sobolev spaces
∂_t^α	left-sided Djrbashian-Caputo fractional derivative of order α (in time)
$^R\partial_t^\alpha$	left-sided Riemann-Liouville fractional derivative of order α (in time)
ae	almost everywhere
erf	error function
erfc	complementary error function
BDF	backward differentiation formula
BVP	boundary value problem
CQ	convolution quadrature
CTRW	continuous time random walk
FDE	fractional differential equation
ODE	ordinary differential equation
PDE	partial differential equation
PDF	probability density function

Part I
Preliminaries

Chapter 1
Continuous Time Random Walk

The classical diffusion equation

$$\partial_t u(x,t) - \kappa \Delta u(x,t) = 0 \tag{1.1}$$

is widely used to describe transport phenomena observed in nature, where u denotes the concentration of a substance, κ is the diffusion coefficient, and (1.1) describes how the concentration evolves over time. The model (1.1) can be derived from a purely macroscopic argument, based on conservation of mass, i.e., $\partial_t u + \nabla \cdot J = 0$ (J is the flux), and Fick's first law of diffusion, i.e., $J = -\kappa \nabla u$. Alternatively, following the work of Albert Einstein in 1905 [Ein05], one can also derive it from the underlying stochastic process, under a Brownian motion assumption on the particle movement. Then the probability density function (PDF) $p(x,t)$ of the particle satisfies (1.1). In this chapter, with continuous time random walk, we show that the PDF $p(x,t)$ of the particle satisfies fractional differential equations (FDES) of the form

$$\begin{aligned}
&{}^C_0 D^\alpha_t p(x,t) = \kappa_\alpha \partial^2_{xx} p(x,t), \\
&\partial_t p(x,t) = \tfrac{\kappa_\mu}{2}\big((1-\beta)_{-\infty}^R D^\mu_x p(x,t) + (1+\beta)^R_x D^\mu_\infty p(x,t)\big),
\end{aligned}$$

where $0 < \alpha < 1$, $1 < \mu < 2$, $-1 \leq \beta \leq 1$, and the notation ${}^C_0 D^\alpha_t$ and $_{-\infty}^R D^\mu_x$, etc. denote Djrbashian-Caputo / Riemann-Liouville fractional derivatives, etc. in time / space; see Definitions 2.2 and 2.3 in Chapter 2 for the precise definitions. In physics, these models are used to describe so-called fractional kinetics [MK00, MK04, ZDK15], and will represent the main objects of this book.

1.1 Random Walk on a Lattice

Perhaps the simplest stochastic approach to derive equation (1.1) is to consider the random walk framework on a lattice, which is also known as a Brownian walk. At each time step (with a time step size Δt), the walker randomly jumps to one of its

© The Author(s), under exclusive license to Springer Nature Switzerland AG 2021 3
B. Jin, *Fractional Differential Equations*, Applied Mathematical Sciences 206,
https://doi.org/10.1007/978-3-030-76043-4_1

four nearest neighboring sites on a square lattice (with a space step Δx); see Fig. 1.1 for a schematic illustration.

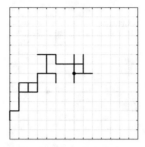

Fig. 1.1 Random walk on a square lattice, starting from the circle.

In the one-dimensional case, such a process can be modeled by

$$p_j(t + \Delta t) = \tfrac{1}{2}p_{j-1}(t) + \tfrac{1}{2}p_{j+1}(t),$$

where the index j denotes the position on the lattice (grid), and $p_j(t)$ the probability that the walker is at grid j at time t. It relates the probability of being at position j at time $t + \Delta t$ to that of the two adjacent sites $j \pm 1$ at time t. The factor $\tfrac{1}{2}$ expresses the directional isotropy of the jumps: the jumps to the left and right are equally likely. A rearrangement gives

$$\frac{p_j(t + \Delta t) - p_j(t)}{(\Delta x)^2} = \frac{1}{2}\frac{p_{j-1}(t) - 2p_j(t) + p_{j+1}(t)}{(\Delta x)^2}.$$

The right-hand side is the central finite difference approximation of the second-order derivative $\partial_{xx}^2 p(x, t)$ at grid j. In the limit $\Delta t, \Delta x \to 0^+$, if the PDF $p(x, t)$ is smooth in x and t, discarding the higher order terms in Δt and Δx in the following Taylor expansions

$$p(x, t + \Delta t) = p(x, t) + \Delta t \partial_t p(x, t) + O((\Delta t)^2),$$

$$p(x \pm \Delta x, t) = p(x, t) \pm \Delta x \partial_x p(x, t) + \tfrac{(\Delta x)^2}{2}\partial_{xx}^2 p(x, t) + O((\Delta x)^3)$$

leads to

$$\partial_t p(x, t) = \kappa \partial_{xx}^2 p(x, t). \tag{1.2}$$

The limit is taken such that the quotient

$$\kappa = \lim_{\Delta x \to 0^+, \Delta t \to 0^+} (2\Delta t)^{-1}(\Delta x)^2 \tag{1.3}$$

is a positive constant, and κ is called the diffusion coefficient, which connects the spatial and time scales.

Equation (1.2) can also be regarded as a consequence of the central limit theorem. Suppose that the jump length Δx has a PDF given by $\lambda(x)$ so that for $a < b$, $P(a < \Delta x < b) = \int_a^b \lambda(x)\mathrm{d}x$. Then Fourier transform gives

$$\widetilde{\lambda}(\xi) = \int_{-\infty}^{\infty} e^{-i\xi x} \lambda(x)\mathrm{d}x = \int_{-\infty}^{\infty} \left(1 - i\xi x - \tfrac{1}{2}\xi^2 x^2 + \ldots\right) \lambda(x)\mathrm{d}x$$

$$= 1 - i\xi\mu_1 - \tfrac{1}{2}\xi^2\mu_2 + \ldots,$$

where μ_j is the jth moment $\mu_j = \int_{-\infty}^{\infty} x^j \lambda(x)\mathrm{d}x$, provided that these moments do exist, which holds if $\lambda(x)$ decays sufficiently fast to zero as $x \to \pm\infty$. Further, assume that the PDF $\lambda(x)$ is normalized and even, i.e., $\mu_1 = 0$, $\mu_2 = 1$ and $\mu_3 = 0$. Then

$$\widetilde{\lambda}(\xi) = 1 - \tfrac{1}{2}\xi^2 + O(\xi^4).$$

Denote by Δx_i the jump length taken at the ith step. In the random walk model, the steps $\Delta x_1, \Delta x_2, \ldots$ are independent. The sum of the independent and identically distributed (i.i.d.) random variables Δx_n (according to the PDF with a rescaled $\lambda(x)$)

$$x_n = \Delta x_1 + \Delta x_2 + \ldots + \Delta x_n$$

gives the position of the walker after n steps. It is also a random variable. Now we recall a standard result on the sum of two independent random variables, where $*$ denotes convolution of f and g.

Theorem 1.1 *If X and Y are independent random variables with a PDF given by f and g, respectively, then the sum $Z = X + Y$ has a PDF $f * g$.*

By Theorem 1.1, the random variable X_n has a Fourier transform $\widetilde{p}_n(\xi) = (\widetilde{\lambda}(\xi))^n$, and the normalized sum $n^{-\frac{1}{2}} x_n$ has the Fourier transform

$$\widetilde{p}_n\left(n^{-\frac{1}{2}}\xi\right) = \left(1 - \tfrac{1}{2n}\xi^2 + O(n^{-2})\right)^n.$$

Taking the limit as $n \to \infty$ gives $\widetilde{p}(\xi) = e^{-\frac{\xi^2}{2}}$ and inverting the Fourier transform gives the standard Gaussian distribution $p(x) = (2\pi)^{-\frac{1}{2}} e^{-\frac{x^2}{2}}$, cf. Example A.4 in the appendix. This is precisely the central limit theorem, asserting that the long-term average behavior of i.i.d. random variables is Gaussian. One requirement for the whole procedure to work is the finiteness of the second moment μ_2 of $\lambda(x)$.

Now we interpret x_n as the particle position after n time steps, by correlating the time step size Δt with the variance of Δx according to the ansatz (1.3). This can be easily achieved by rescaling the variance of $\lambda(x)$ to $2\kappa t$. Then by the scaling rule for the Fourier transform, $\widetilde{p}_n(\xi)$ is given by

$$\widetilde{p}_n(n^{-\frac{1}{2}}\xi) = (1 - n^{-1}\kappa t\xi^2 + O(n^{-2}))^n$$

and taking limit as $n \to \infty$ gives $\widetilde{p}(\xi) = e^{-\xi^2\kappa t}$. By inverse Fourier transform, the PDF being at certain position x and time t is governed by the standard diffusion model

(1.2), and is given by the Gaussian

$$p(x, t) = (4\pi\kappa t)^{-\frac{1}{2}} e^{-\frac{x^2}{4\kappa t}}.$$

It is the fundamental solution, i.e., the solution $p(x, t)$ of (1.2) with the initial condition $p(x, 0) = \delta(x)$, the Dirac function concentrated at $x = 0$. Note that at any fixed time $t > 0$, $p(x, t)$ is a Gaussian distribution in x, with mean zero and variance $2\kappa t$, i.e., $\int_{-\infty}^{\infty} x^2 p(x, t) dx = 2\kappa t$, which scales linearly with the time t. This represents one distinct feature of normal diffusion processes.

1.2 Continuous Time Random Walk

There are several stochastic models for deriving differential equations involving a fractional-order derivative. We describe the continuous time random walk framework (CTRW) due to Montroll and Weiss [MW65]. CTRW generalizes the random walk model in Section 1.1, in which the length of each jump and the time elapsed between two successive jumps follow a given PDF. We assume that these two random variables are independent, though in theory one can allow correlation in order to achieve more flexible modeling. In one spatial dimension, the picture is as follows: a walker moves along the x-axis, starting at a position x_0 at time $t_0 = 0$. At time t_1, the walker jumps to x_1, then at time t_2 jumps to x_2, and so on. We assume that the temporal and spatial increments $\Delta t_n = t_n - t_{n-1}$ and $\Delta x_n = x_n - x_{n-1}$ are i.i.d. random variables, following PDFs $\psi(t)$ and $\lambda(x)$, respectively, known as the waiting time distribution and jump length distribution, respectively. Namely, the probability of Δt_n lying in any interval $[a, b] \subset \mathbb{R}_+$ and Δx_n lying in any interval $[c, d] \subset \mathbb{R}$ are given by

$$P(a < \Delta t_n < b) = \int_a^b \psi(t) dt \quad \text{and} \quad P(c < \Delta x_n < d) = \int_c^d \lambda(x) dx,$$

respectively. Now the goal is to determine the probability that the walker lies in a given spatial interval at time t. For given PDFs ψ and λ, the position x of the walker can be regarded as a step function of t.

Example 1.1 Suppose that the distribution $\psi(t)$ is exponential with a parameter $\tau > 0$, i.e., $\psi(t) = \tau^{-1} e^{-\frac{t}{\tau}}$ for $t \in \mathbb{R}_+$, and the jump length distribution $\lambda(x)$ is Gaussian with mean zero and variance σ^2, i.e., $\lambda(x) = (2\pi\sigma^2)^{-\frac{1}{2}} e^{-\frac{x^2}{2\sigma^2}}$ for $x \in \mathbb{R}$. Then the waiting time Δt_n and jump length Δx_n satisfy

$$\mathbb{E}[\Delta t_n] = \tau, \quad \mathbb{E}[\Delta t_n^2] = \tau^2, \quad \mathbb{E}[\Delta x_n] = 0, \quad \mathbb{E}[\Delta x_n^2] = \sigma^2,$$

where $\mathbb{E}$ denotes taking expectation with respect to the underlying distribution. The position $x(t)$ of the walker is a step function (in time t). Two sample trajectories of CTRW with $\tau = 1$ and $\sigma = 1$ are given in Fig. 1.2.

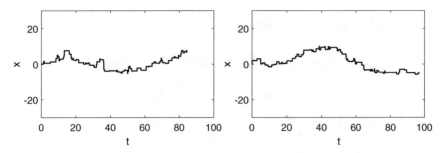

Fig. 1.2 Two realizations of 1D CTRW, with exponential waiting time distribution (with $\tau = 1$) and Gaussian jump length distribution (with $\sigma = 1$), starting from $x_0 = 0$.

Now we derive the PDF of the total waiting time t_n after n steps and the total jump length $x_n - x_0$ after n steps. The main tools in the derivation are Laplace and Fourier transforms in Appendices A.3.1 and A.3.2, respectively. We denote by $\psi_n(t)$ the PDF of $t_n = \Delta t_1 + \Delta t_2 + \ldots + \Delta t_n$, and $\psi_1 = \psi$. By Theorem 1.1, we have

$$\psi_n(t) = (\psi_{n-1} * \psi)(t) = \int_0^t \psi_{n-1}(s)\psi(t-s)ds.$$

Then the characteristic function $\widehat{\psi}_n(z)$ of $\psi_n(t)$ (i.e., its Laplace transform) $\widehat{\psi}_n(z) = \mathcal{L}[\psi_n](z) = \int_0^\infty e^{-zt}\psi_n(t)dt$, by the convolution rule for Laplace transform in (A.8), is given by $\widehat{\psi}_n(z) = (\widehat{\psi}(z))^n$. Let $\Psi(t)$ denote the survival probability, i.e., the probability of the walker not jumping within a time t (or equivalently, the probability of remaining stationary for at least a duration t). Then

$$\Psi(t) = \int_t^\infty \psi(s)ds = 1 - \int_0^t \psi(s)ds, \quad 0 < t < \infty.$$

The characteristic function for the survival probability Ψ is

$$\widehat{\Psi}(z) = z^{-1} - z^{-1}\widehat{\psi}(z).$$

Then the probability of taking exactly n steps up to time t is given by

$$\chi_n(t) = \int_0^t \psi_n(s)\Psi(t-s)ds = (\psi_n * \Psi)(t).$$

The convolution rule for Laplace transform yields

$$\widehat{\chi}_n(z) = \widehat{\psi}_n(z)\widehat{\Psi}(z) = \widehat{\psi}(z)^n z^{-1}(1 - \widehat{\psi}(z)).$$

Next we derive the PDF of the jump length $x_n - x_0$ after n steps. We denote by $\lambda_n(x)$ the PDF of the random variable $x_n - x_0 = \Delta x_1 + \Delta x_2 + \ldots + \Delta x_n$. Appealing to Theorem 1.1 again yields

$$\lambda_n(x) = (\lambda_{n-1} * \lambda)(x) = \int_{-\infty}^{\infty} \lambda_{n-1}(y)\lambda(x-y)dy,$$

and by the convolution rule for Fourier transform, $\widetilde{\lambda}_n(\xi)$ is given by

$$\widetilde{\lambda}_n(\xi) = (\widetilde{\lambda}(\xi))^n, \quad n \geq 0.$$

We denote by $p(x,t)$ the PDF of the walker at the position x at time t. Since $\chi_n(t)$ is the probability of taking n steps up to time t,

$$p(x,t) = \sum_{n=0}^{\infty} \lambda_n(x)\chi_n(t).$$

We denote the Fourier-Laplace transform of $p(x,t)$ by

$$\widehat{\widetilde{p}}(\xi,z) = \mathcal{LF}[p](\xi,z) = \int_0^{\infty} e^{-zt} \int_{-\infty}^{\infty} e^{-i\xi x} p(x,t)dxdt.$$

Then $\widehat{\widetilde{p}}(\xi,z)$ is given by

$$\widehat{\widetilde{p}}(\xi,z) = \sum_{n=0}^{\infty} \widetilde{\lambda}_n(\xi)\widehat{\chi}_n(z) = \frac{1-\widehat{\psi}(z)}{z} \sum_{n=0}^{\infty} [\widetilde{\lambda}(\xi)\widehat{\psi}(z)]^n. \tag{1.4}$$

Since λ and ψ are PDFS,

$$\widetilde{\lambda}(0) = \int_{-\infty}^{\infty} \lambda(x)dx = 1 \quad \text{and} \quad \widehat{\psi}(0) = \int_0^{\infty} \psi(t)dt = 1. \tag{1.5}$$

If $\xi \neq 0$ or $z > 0$, then $|\widetilde{\lambda}(\xi)\widehat{\psi}(z)| < 1$ so the series in (1.4) is absolutely convergent:

$$\sum_{n=0}^{\infty} [\widetilde{\lambda}(\xi)\widehat{\psi}(z)]^n = (1 - \widetilde{\lambda}(\xi)\widehat{\psi}(z))^{-1},$$

and hence we obtain the following fundamental relation for the PDF $p(x,t)$ in the Laplace-Fourier domain

$$\widehat{\widetilde{p}}(\xi,z) = \frac{1-\widehat{\psi}(z)}{z} \frac{1}{1-\widetilde{\lambda}(\xi)\widehat{\psi}(z)}. \tag{1.6}$$

Example 1.2 For the waiting time distribution $\psi(t) = \tau^{-1}e^{-\frac{t}{\tau}}$ and the jump length distribution $\lambda(x) = (2\pi\sigma^2)^{-\frac{1}{2}}e^{-\frac{x^2}{2\sigma^2}}$ in Example 1.1, we have $\widehat{\psi}(z) = (1 + \tau z)^{-1}$ and $\widetilde{\lambda}(\xi) = e^{-\frac{\sigma^2\xi^2}{2}}$. Thus, the Fourier-Laplace transform $\widehat{\widetilde{p}}(\xi,z)$ of the PDF $p(x,t)$ of the walker at position x (relative to x_0) at time t is given by

$$\widehat{\widetilde{p}}(\xi,z) = \tau(1 + \tau z - e^{-\frac{\sigma^2\xi^2}{2}})^{-1}.$$

Different types of CTRW processes can be classified according to the characteristic waiting time ϱ and jump length variance ς^2 defined by

$$\varrho =: \mathbb{E}[\Delta t_n] = \int_0^\infty t\psi(t)\mathrm{d}t \quad \text{and} \quad \varsigma^2 =: \mathbb{E}[(\Delta x_n)^2] = \int_{-\infty}^\infty x^2\lambda(x)\mathrm{d}x,$$

being finite or diverging. Below we shall discuss the following three scenarios: (i) Finite ϱ and ς^2, (ii) Diverging ϱ and finite ς^2 and (iii) Diverging ς^2 and finite ϱ.

(i) Finite characteristic waiting time and jump length variance. When both characteristic waiting time ϱ and jump length variance ς are finite, CTRW does not lead to anything new. Specifically, assume that the PDFs $\psi(t)$ and $\lambda(x)$ are normalized:

$$\int_0^\infty t\psi(t)\mathrm{d}t = 1, \quad \int_{-\infty}^\infty x\lambda(x)\mathrm{d}x = 0, \quad \int_{-\infty}^\infty x^2\lambda(x)\mathrm{d}x = 1. \quad (1.7)$$

These conditions are satisfied after suitable rescaling if the waiting time PDF $\psi(t)$ has a finite mean, and the jump length PDF $\lambda(x)$ has a finite variance. Recall the relation (1.5): $\widehat{\psi}(0) = 1 = \widetilde{\lambda}(0)$. Since

$$\frac{\mathrm{d}^k \widehat{\psi}}{\mathrm{d}z^k} = \int_0^\infty e^{-zt}(-t)^k\psi(t)\mathrm{d}t,$$

using the normalization condition (1.7), we have

$$\widehat{\psi}'(0) = -\int_0^\infty t\psi(t)\mathrm{d}t = -1.$$

Similarly, we deduce

$$\widetilde{\lambda}'(0) = -\mathrm{i}\int_{-\infty}^\infty x\lambda(x)\mathrm{d}x = 0, \quad \widetilde{\lambda}''(0) = -\int_{-\infty}^\infty x^2\lambda(x)\mathrm{d}x = -1.$$

Next, for $\tau > 0$ and $\sigma > 0$, let the random variables Δt_n and Δx_n follow the rescaled PDFs

$$\psi_\tau(t) = \tau^{-1}\psi(\tau^{-1}t) \quad \text{and} \quad \lambda_\sigma(x) = \sigma^{-1}\lambda(\sigma^{-1}x). \quad (1.8)$$

Now consider the limit of $\widehat{\widetilde{p}}(\xi, z; \sigma, \tau)$ as $\tau, \sigma \to 0^+$. Then simple computation shows $\varrho =: \mathbb{E}[\Delta t_n] = \tau$, $\mathbb{E}[\Delta x_n] = 0$, and $\varsigma^2 =: \mathbb{E}[\Delta x_n^2] = \sigma^2$. By the scaling rules for Fourier and Laplace transforms,

$$\widehat{\psi_\tau}(z) = \widehat{\psi}(\tau z) \quad \text{and} \quad \widetilde{\lambda_\sigma}(\xi) = \widetilde{\lambda}(\sigma\xi).$$

Consequently,

$$\widehat{\widetilde{p}}(\xi, z; \sigma, \tau) = \frac{1 - \widehat{\psi}(\tau z)}{z}\frac{1}{1 - \widehat{\psi}(\tau z)\widetilde{\lambda}(\sigma\xi)}.$$

The Taylor expansion of $\widehat{\psi}(z)$ around $z = 0$ yields

$$\widehat{\psi}(z) = \widehat{\psi}(0) + \widehat{\psi}'(0)z + \ldots = 1 - z + O(z^2) \quad \text{as } z \to 0.$$

Further we assume that the PDF $\lambda(x)$ is even, i.e., $\lambda(-x) = \lambda(x)$. Then $\widetilde{\lambda}'''(0) = 0$ and the Taylor expansion of $\widetilde{\lambda}(\xi)$ around $\xi = 0$ is given by

$$\widetilde{\lambda}(\xi) = \widetilde{\lambda}(0) + \widetilde{\lambda}'(0)\xi + \tfrac{1}{2}\widetilde{\lambda}''(0)\xi^2 + \ldots = 1 - \tfrac{1}{2}\xi^2 + O(\xi^4).$$

Using the last two relations, simple algebraic manipulation gives

$$\widehat{\widetilde{p}}(\xi, z; \sigma, \tau) = \frac{\tau}{\tau z + \tfrac{1}{2}\sigma^2\xi^2} \frac{1 + O(\tau z)}{1 + O(\tau z + \sigma^2\xi^2)}. \tag{1.9}$$

Now in the formula (1.9), we let $\sigma, \tau \to 0^+$ while keeping the relation $\lim_{\sigma \to 0^+, \tau \to 0^+}(2\tau)^{-1}\sigma^2 = \kappa$, for some $\kappa > 0$, and obtain

$$\widehat{\widetilde{p}}(\xi, z) = \lim \frac{\tau}{\tau z + \tfrac{1}{2}\sigma^2\xi^2} = \frac{1}{z + \kappa\xi^2}. \tag{1.10}$$

The scaling relation here for κ is identical to that in the random walk framework, cf. (1.3). Upon inverting the Laplace transform, we find

$$\widetilde{p}(\xi, t) = \frac{1}{2\pi i} \int_C \frac{e^{zt}}{z + \kappa\xi^2} dz = e^{-\kappa\xi^2 t},$$

and then inverting the Fourier transform gives the familiar Gaussian PDF $p(x, t) = (4\pi\kappa t)^{-\frac{1}{2}} e^{-\frac{x^2}{4\kappa t}}$. Last we verify that the PDF $p(x, t)$ satisfies (1.2):

$$\mathcal{LF}[\partial_t p - \kappa\partial_{xx}^2 p](\xi, z) = z\widehat{\widetilde{p}}(\xi, z) - \widetilde{p}(\xi, 0) + \kappa\xi^2\widehat{\widetilde{p}}(\xi, z)$$
$$= (z + \kappa\xi^2)\widehat{\widetilde{p}}(\xi, z) - \widetilde{p}(\xi, 0) = 1 - \widetilde{p}(\xi, 0) = 0,$$

where the last step follows by $\widetilde{p}(\xi, 0) = \widetilde{\delta}(\xi) = 1$. Hence, the PDF $p(x, t)$ for $x(t) - x_0$ at the position x (relative to x_0) of the walker at time t indeed satisfies (1.2). The CTRW framework recovers the classical diffusion model, as long as the waiting time PDF $\psi(t)$ has a finite mean and the jump length PDF $\lambda(x)$ has finite first and second moments. Further, the displacement variance is given by

$$\mu_2(t) = \mathbb{E}_{p(t)}[(x(t) - x_0)^2] = \int_{-\infty}^{\infty} x^2 p(x, t)dx = 2\kappa t,$$

which grows linearly with the time t. One striking thing is that any pair of PDFs with finite characteristic waiting time ϱ and jump length variance ς^2 lead to the same result, to the lowest order.

(ii) Divergent mean waiting time. Now we consider the situation where the characteristic waiting time ϱ diverges, but the jump length variance ς^2 is finite. This occurs for example when the particle might be trapped in a potential well, and it takes a long time for the particle to leave the well. To model such phenomena, we employ a heavy-tailed (i.e., the tail is not exponentially bounded) waiting time PDF with the asymptotic behavior

$$\psi(t) \sim at^{-1-\alpha}, \quad \text{as } t \to \infty \tag{1.11}$$

for some $\alpha \in (0, 1)$, where $a > 0$ is a constant. One example of such PDF is $\psi(t) = \alpha(1 + t)^{-1-\alpha}$. The (asymptotic) power law decay in (1.11) is heavy tailed and allows occasional very large waiting time between consecutive jumps. Again, the specific form of $\psi(t)$ is irrelevant, and only the power law decay at large time matters. The parameter α determines the asymptotic decay of the PDF $\psi(t)$: the closer is α to zero, the slower the decay is and more likely the long waiting time is.

Now the question is whether such occasional long waiting time will change completely the dynamics of the stochastic process in the large time. For the power law decay, the mean waiting time is divergent, i.e., $\int_0^\infty t\psi(t)\mathrm{d}t = +\infty$. So one cannot assume the normalizing condition on the PDF ψ in (1.7), and the above analysis breaks down. Nonetheless, in this part, the assumption on $\lambda(x)$ remains unchanged, i.e., $\int_{-\infty}^\infty x\lambda(x)\mathrm{d}x = 0$ and $\int_{-\infty}^\infty x^2\lambda(x)\mathrm{d}x = 1$.

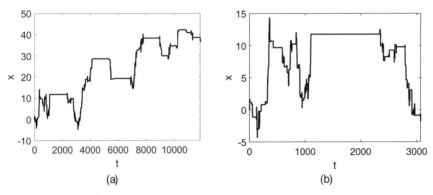

Fig. 1.3 CTRW with a power law waiting time PDF. Panel (a) gives one sample trajectory with waiting time PDF $\psi(t) = \alpha(1 + t)^{-1-\alpha}$, $\alpha = \frac{3}{4}$, and standard Gaussian jump length PDF; Panel (b) is a zoom-in of the first cluster of Panel (a).

In Fig. 1.3, we show one sample trajectory of CTRW with a power law waiting time PDF $\psi(t) = \alpha(1 + t)^{-1-\alpha}$, $\alpha = \frac{3}{4}$, and the standard Gaussian jump length PDF, and an enlarged version of the initial time interval. One observes clearly the occasional but large waiting time appearing at different time scales, cf. Fig. 1.3(b). This behavior is dramatically different from the case of finite mean waiting time in Fig. 1.2.

The following result bounds the Laplace transform $\widehat{\psi}(z)$ for $z \to 0$. The notation $\Gamma(z)$ denotes the Gamma function; see (2.1) in Section 2.1 for its definition.

Theorem 1.2 *Let $\alpha \in (0, 1)$, and $a > 0$. If $\psi(t) = at^{-1-\alpha} + O(t^{-2-\alpha})$ as $t \to \infty$, then*

$$\widehat{\psi}(z) = 1 - B_\alpha z^\alpha + O(z) \quad \text{as } z \to 0$$

with $B_\alpha = a\alpha^{-1}\Gamma(1 - \alpha)$.

Proof By the assumption on ψ, there exists $t_0 > 0$ such that

$$|t^{1+\alpha}\psi(t) - a| \le ct^{-1}, \quad \forall t_0 \le t < \infty \tag{1.12}$$

and we consider $0 < z < t_0^{-1}$ so that $zt_0 < 1$. Since $\widehat{\psi}(0) = 1$, we have

$$1 - \widehat{\psi}(z) = \int_0^\infty (1 - e^{-zt})\psi(t)\mathrm{d}t = \int_0^{t_0} (1 - e^{-zt})(\psi(t) - at^{-1-\alpha})\mathrm{d}t$$

$$+ \int_{t_0}^\infty (1 - e^{-zt})(\psi(t) - at^{-1-\alpha})\mathrm{d}t + \int_0^\infty (1 - e^{-zt})at^{-1-\alpha}\mathrm{d}t =: \sum_{i=1}^3 \mathrm{I}_i.$$

Since $0 \le 1 - e^{-s} \le s$ for all $s \in [0, 1]$, the term I_1 can be bounded by

$$|\mathrm{I}_1| \le \int_0^{t_0} zt\psi(t)\mathrm{d}t + \int_0^{t_0} zt at^{-1-\alpha}\mathrm{d}t$$

$$\le zt_0 \int_0^\infty \psi(t)\mathrm{d}t + az \int_0^{t_0} t^{-\alpha}\mathrm{d}t \le c_{\alpha,t_0} z.$$

For the term I_2, using (1.12) and changing variables $s = zt$, we deduce

$$|\mathrm{I}_2| \le \int_{t_0}^\infty (1 - e^{-zt})ct^{-2-\alpha}\mathrm{d}t$$

$$\le cz^{1+\alpha}\left(\int_{t_0z}^1 s^{-1-\alpha}\mathrm{d}s + \int_1^\infty s^{-2-\alpha}\mathrm{d}s \right)$$

$$\le cz^{1+\alpha}(\alpha^{-1}((t_0z)^{-\alpha} - 1) + (1 + \alpha)^{-1}) \le c_{\alpha,t_0}z.$$

Last, for the term I_3, by changing variables $s = zt$, we have

$$\mathrm{I}_3 = \int_0^\infty (1 - e^{-zt})at^{-1-\alpha}\mathrm{d}t = az^\alpha \int_0^\infty \frac{1 - e^{-s}}{s^{1+\alpha}}\mathrm{d}s.$$

Integration by parts and the identity (2.2) below give

$$\int_0^\infty \frac{1 - e^{-s}}{s^{1+\alpha}}\mathrm{d}s = \alpha^{-1} \int_0^\infty e^{-s}s^{-\alpha}\mathrm{d}s = \alpha^{-1}\Gamma(1 - \alpha).$$

Combining the preceding estimates yields the desired inequality. $\square$

Now we introduce the following rescaled PDFs for the incremental waiting time Δt_n and jump length Δx_n: $\psi_\tau(t) = \tau^{-1}\psi(\tau^{-1}t)$ and $\lambda_\sigma(x) = \sigma^{-1}\lambda(\sigma^{-1}x)$. By (1.6), $\widetilde{p}(\xi, z; \sigma, \tau)$ is given by

$$\widetilde{p}(\xi, z; \sigma, \tau) = \frac{1 - \widehat{\psi}(\tau z)}{z} \frac{1}{1 - \widehat{\psi}(\tau z)\widehat{\lambda}(\sigma\xi)}.$$

Under the power law waiting time PDF, from Theorem 1.2 we deduce $\widehat{\psi}(z) = 1 - B_\alpha z^\alpha + O(z)$ as $z \to 0$ and $\widetilde{\lambda}(\xi) = 1 - \frac{1}{2}\xi^2 + O(\xi^4)$ as $\xi \to 0$. Hence,

$$\widehat{\widetilde{p}}(\xi, z; \sigma, \tau) = \frac{B_\alpha \tau^\alpha z^{\alpha-1}}{B_\alpha \tau^\alpha z^\alpha + \frac{1}{2}\sigma^2 \xi^2} \frac{1 + O(\tau^{1-\alpha}z^{1-\alpha})}{1 + O(\tau^{1-\alpha}z^{1-\alpha} + \tau^\alpha z^\alpha + \sigma\xi)}. \tag{1.13}$$

Once again, in (1.13), we find the Fourier-Laplace transform $\widehat{\widetilde{p}}(\xi, z)$ by letting $\sigma, \tau \to 0^+$, while keeping $\lim_{\sigma\to0^+, \tau\to0^+}(2B_\alpha \tau^\alpha)^{-1}\sigma^2 = \kappa_\alpha$ for some $\kappa_\alpha > 0$. Thus we obtain

$$\widehat{\widetilde{p}}(\xi, z) = \lim \widehat{\widetilde{p}}(\xi, z; \sigma, \tau) = \lim \frac{B_\alpha \tau^\alpha z^{\alpha-1}}{B_\alpha \tau^\alpha z^\alpha + \frac{1}{2}\sigma^2 \xi^2} = \frac{z^{\alpha-1}}{z^\alpha + \kappa_\alpha \xi^2}. \tag{1.14}$$

Formally, we recover the formula (1.10) by setting $\alpha = 1$.

Last, we invert the Fourier-Laplace transform $\widehat{\widetilde{p}}(\xi, z)$ using the Mittag-Leffler function $E_{\alpha,1}(z)$ (cf. (3.1) in Chapter 3) and Wright function $W_{\rho,\mu}(z)$ (cf. (3.24) in Chapter 3). Using the Laplace transform formula of the Mittag-Leffler function $E_{\alpha,1}(z)$ in Lemma 3.2 below, we deduce

$$\widetilde{p}(\xi, t) = E_{\alpha,1}(-\kappa_\alpha t^\alpha \xi^2),$$

and then by the Fourier transform of the Wright function $W_{\rho,\mu}(z)$ in Proposition 3.4, we obtain an explicit expression of the PDF $p(x, t)$

$$p(x, t) = (4\kappa_\alpha t^\alpha)^{-\frac{1}{2}} W_{-\frac{\alpha}{2}, 1-\frac{\alpha}{2}}(-(\kappa_\alpha t^\alpha)^{-\frac{1}{2}}|x|).$$

With $\alpha = 1$, this formula recovers the familiar Gaussian density. Then Lemma 2.9 implies that the PDF $p(x, t)$ satisfies

$$\mathcal{LF}\left[{}_0^C D_t^\alpha p(x, t) - \kappa_\alpha \partial_{xx}^2 p(x, t) \right] = z^\alpha \widehat{\widetilde{p}}(\xi, z) - z^{\alpha-1}\widetilde{p}(\xi, 0) + \kappa_\alpha \xi^2 \widehat{\widetilde{p}}(\xi, z)$$
$$= (z^\alpha + \kappa_\alpha \xi^2)\widehat{\widetilde{p}}(\xi, z) - z^{\alpha-1}\widetilde{p}(\xi, 0) \equiv 0,$$

since $\widetilde{p}(\xi, 0) = \widetilde{\delta}(\xi) = 1$. Thus the PDF $p(x, t)$ satisfies the following time-fractional diffusion equation with a Djrbashian-Caputo fractional derivative ${}_0^C D_t^\alpha p$ in time t (see Definition 2.3 in Chapter 2 for the precise definition)

$$_0^C D_t^\alpha p(x, t) - \kappa_\alpha \partial_{xx}^2 p(x, t) = 0, \quad 0 < t < \infty, -\infty < x < \infty.$$

It can be rewritten using a Riemann-Liouville fractional derivative ${}_0^R D_t^{1-\alpha} p$ as

$$\partial_t p(x, t) = \kappa_\alpha \, {}_0^R D_t^{1-\alpha} \partial_{xx}^2 p(x, t).$$

In this book, we focus on the approach based on the Djrbashian-Caputo one.

Now we compute the mean square displacement $\mu_2(t) = \int_{-\infty}^\infty x^2 p(x, t)\,dx$ for $t > 0$. To derive an explicit formula, we resort to the Laplace transform

$$\widehat{\mu}_2(z) = \int_{-\infty}^{\infty} x^2 \widehat{p}(x, z) \mathrm{d}x = -\frac{\mathrm{d}^2}{\mathrm{d}\xi^2} \widehat{\widetilde{p}}(\xi, z)|_{\xi=0}$$

$$= -\frac{\mathrm{d}^2}{\mathrm{d}\xi^2}(z + \kappa_\alpha z^{1-\alpha}\xi^2)^{-1}|_{\xi=0} = 2\kappa_\alpha z^{-1-\alpha},$$

which upon inverse Laplace transform yields $\mu_2(t) = 2\Gamma(1 + \alpha)^{-1}\kappa_\alpha t^\alpha$. Thus the mean square displacement grows only sublinearly with the time t, which, at large time t, is slower than that in normal diffusion, and as $\alpha \to 1^-$, it recovers the formula for normal diffusion. Such a diffusion process is called subdiffusion or anomalously slow diffusion in the literature. Subdiffusion has been observed in a large number of physical applications, e.g., column experiments [HH98], thermal diffusion in fractal domains [Nig86], non-Fickian transport in geological formations [BCDS06], and protein transport within membranes [Kou08]. The mathematical theory for the subdiffusion model will be discussed in detail in Chapters 6 and 7.

(iii) Diverging jump length variance. Last we turn to the case of diverging jump length variance ς^2, but finite characteristic waiting time ϱ. To be specific, we assume an exponential waiting time $\psi(t)$, and that the jump length follows a (possibly asymmetric) Lévy distribution. See the monograph [Nol20] for an extensive treatment of univariate stable distributions and [GM98] for the use of stable distributions within fractional calculus.

The most convenient way to define Lévy random variables is via their characteristic function (Fourier transform)

$$\ln \widetilde{\lambda}(\xi; \mu, \beta) = \begin{cases} -|\xi|^\mu(1 - i\beta \operatorname{sign}(\xi) \tan \frac{\mu\pi}{2}), & \mu \neq 1, \\ -|\xi|(1 + i\beta \operatorname{sign}(\xi)\frac{2}{\pi} \ln |\xi|), & \mu = 1, \end{cases}$$

where $\mu \in (0, 2]$ and $\beta \in [-1, 1]$. The parameter μ determines the rate at which the tails of the PDF taper off. When $\mu = 2$, it recovers a Gaussian distribution, irrespective of the value of β, whereas with $\mu = 1$ and $\beta = 0$, the stable density is identical to the Cauchy distribution, i.e., $\lambda(x) = (\pi(1 + x^2))^{-1}$. The parameter β determines the degree of asymmetry of the distribution. When β is negative (respectively positive), the distribution is skewed to the left (respectively right). When $\beta = 0$, the expression simplifies to $\widetilde{\lambda}(\xi; \mu) = e^{-|\xi|^\mu}$.

Generally, the PDF of a Lévy stable distribution is not available in closed form. However, it is possible to compute the density numerically [Nol97] or to simulate random variables from such stable distributions, cf. Section 1.3. Nonetheless, following the proof of Theorem 1.2, when $1 < \mu < 2$, one can obtain an inverse power law asymptotic [Nol20, Theorem 1.2, p. 13]

$$\lambda(x) \sim a_{\mu,\beta}|x|^{-1-\mu} \quad \text{as } |x| \to \infty \tag{1.15}$$

for some $a_{\mu,\beta} > 0$, from which it follows directly that the jump length variance diverges, i.e.,

$$\int_{-\infty}^{\infty} x^2 \lambda(x) \, \mathrm{d}x = \infty. \tag{1.16}$$

In general, the pth moment of a stable random variable is finite if and only if $p < \mu$.

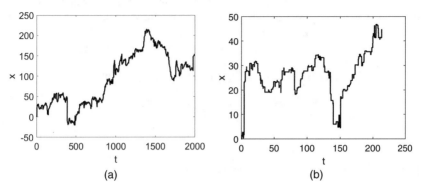

Fig. 1.4 CTRW with a Lévy jump length PDF. Panel (a) is one sample trajectory with an exponential waiting time PDF $\psi(t) = e^{-t}$, and a Lévy jump length PDF with $\mu = \frac{3}{2}$ and $\beta = 0$. Panel (b) is a zoom-in of Panel (a).

For CTRW, one especially relevant case is $\lambda(x; \mu, \beta)$ with $\mu \in (1, 2]$ for the jump length distribution. In Fig. 1.4, we show one sample trajectory of CTRW with a Lévy jump length distribution. Due to the asymptotic property (1.15) of the jump length PDF, very long jumps may occur with a much higher probability than for an exponentially decaying PDF like a Gaussian. The scaling nature of the Lévy jump length PDF leads to clusters along the trajectory, i.e., local motion is occasionally interrupted by long sojourns, at all length scales.

Below we derive the diffusion limit using the scaling technique, by appealing to the scaled PDFs $\psi_\tau(t)$ and $\lambda_\sigma(x)$, for some $\tau, \sigma > 0$. The PDF ψ is assumed to be normalized, i.e., $\int_0^\infty t\psi(t)\mathrm{d}t = 1$. Using Taylor expansion, for small ξ, we have ($\mu \neq 1$)

$$e^{-\tilde{\lambda}(\xi;\mu,\beta)} = 1 - |\xi|^\mu \left(1 - i\beta \operatorname{sign}(\xi) \tan \frac{\mu\pi}{2}\right) + O(|\xi|^{2\mu}).$$

Hence, for $\mu \neq 1$, we have

$$\widehat{\psi}_\tau(z) = 1 - \tau z + O(\tau^2 z^2) \quad \text{as } \tau \to 0,$$

$$\widetilde{\lambda}_\sigma(\xi) = 1 - \sigma^\mu |\xi|^\mu \left(1 - i\beta \operatorname{sign}(\xi) \tan \frac{\mu\pi}{2}\right) + O(\sigma^{2\mu}|\xi|^{2\mu}) \quad \text{as } \sigma \to 0.$$

These two identities and Fourier-Laplace convolution formulas yield

$$\widehat{\widetilde{p}}(\xi, z) = \lim \widehat{\widetilde{p}}(\xi, z; \sigma, \tau) = \frac{1}{z + \kappa_\mu |\xi|^\mu \left(1 - i\beta \operatorname{sign}(\xi) \tan \frac{\mu\pi}{2}\right)}, \qquad (1.17)$$

where the limit is taken under the condition $\kappa_\mu = \lim_{\sigma \to 0^+, \tau \to 0^+} \tau^{-1} \sigma^\mu$ for some $\kappa_\mu > 0$, which represents the diffusion coefficient.

Next we invert the Fourier transform of $\widehat{\widetilde{p}}(\xi, z)$ and obtain

$$\widetilde{p}(\xi, t) = e^{-\kappa_\mu |\xi|^\mu (1 - i\beta \,\text{sign}(\xi) \tan \frac{\mu\pi}{2})t},$$

which is the characteristic function of a Lévy stable distribution, for which an explicit expression is generally unavailable. This can be regarded as a generalized central limit theorem for stable distributions.

One can verify by Fourier and Laplace inversion that the PDF $p(x, t)$ satisfies

$$\partial_t p(x, t) = \frac{\kappa_\mu}{2} \left((1 - \beta)_{-\infty}^{R} D_x^\mu p(x, t) + (1 + \beta)_x^R D_\infty^\mu p(x, t) \right),$$

in view of Theorem 2.8 below. In the symmetric case $\beta = 0$, it simplifies to

$$\partial_t p(x, t) = -\kappa_\mu (-\Delta)_R^{\frac{\mu}{2}} p(x, t),$$

where $(-\Delta)_R^{\frac{\mu}{2}}$ denotes the Riesz fractional operator

$$-(-\Delta)_R^{\frac{\mu}{2}} f = \frac{1}{2} \left({}_{-\infty}^{R} D_x^\mu f + {}_x^R D_\infty^\mu f \right),$$

where the notation ${}_{-\infty}^{R} D_x^\mu$ and ${}_x^R D_\infty^\mu$ denote the left-sided and right-sided Riemann-Liouville fractional derivatives, cf. Definition 2.2 in Chapter 2 for the precise definitions. This operator can be viewed as the one-dimensional version of fractional Laplacian (see the survey [Gar19] for a nice introduction to fractional Laplacian). It recovers the Gaussian density as $\mu \to 2$. Using (1.15), we deduce the following power law asymptotic as $|x| \to \infty$

$$p(x, t) \sim \kappa_\mu t |x|^{-1-\mu}, \quad \mu < 2.$$

Due to this property, the mean squared displacement diverges $\int_{-\infty}^{\infty} |x|^2 p(x, t)\, dx = \infty$. That is, the particle following such a random walk model spreads faster than normal diffusion, and it is commonly known as superdiffusion in the literature. It has been proposed for several applications, including solute transport [BWM00] and searching patterns of animals (e.g., albatross, bacteria, plankton, jackals, spider monkey) [RFMM+04]. The divergent mean squares displacement has caused some controversy in practice, and various modifications have been proposed. Boundary value problems (BVPs) for superdiffusion are involved, since the long jumps make the very definition of a proper boundary condition actually quite intricate. In Chapter 5, we present mathematical theory for 1D stationary superdiffusion models.

1.3 Simulating Continuous Time Random Walk

To simulate CTRW, one needs to generate random numbers for a given PDF, e.g., power laws and Lévy stable distributions. The task of generating random numbers for an arbitrary PDF is often highly nontrivial, especially in high-dimensional spaces. There are a number of possible methods, and we describe only the transformation method, which is simple and easy to implement. For more advanced methods, we refer to the monograph [Liu08].

The transformation method requires that we have access to random numbers uniformly distributed in the unit interval $0 \leq r \leq 1$, which can be generated by any standard pseudo-random number generators. Now suppose that $p(x)$ is a continuous PDF from which we wish to draw samples. The PDFS $p(x)$ and $p(r)$ (the uniform distribution on $[0, 1]$) are related by

$$p(x) = p(r)\frac{\mathrm{d}r}{\mathrm{d}x} = \frac{\mathrm{d}r}{\mathrm{d}x},$$

where the second equality follows because $p(r) = 1$ over the interval $[0, 1]$. Integrating both sides with respect to x, we obtain the complementary cumulative distribution function $F(x)$ in terms of r:

$$F(x) = \int_x^\infty p(x')\,\mathrm{d}x' = \int_r^1 \mathrm{d}r' = 1 - r,$$

or equivalent $x = F^{-1}(1 - r)$, where F^{-1} denotes the inverse function of the function $F(x)$. For many PDFS, this function can be evaluated in closed form.

Example 1.3 In this example we illustrate the method on the power law density $\psi(x) = \alpha(1 + x)^{-1-\alpha}$. One can compute directly

$$F(x) = \int_x^\infty \psi(x')\,\mathrm{d}x' = \int_x^\infty \frac{\alpha}{(1 + x')^{1+\alpha}}\,\mathrm{d}x' = \frac{1}{(1 + x)^\alpha},$$

and the inverse function $F^{-1}(x)$ is given by $F^{-1}(x) = x^{-\frac{1}{\alpha}} - 1$. This last formula provides an easy way to generate random variables from the power law density $\psi(t)$, needed for simulating the subdiffusion model.

Generating samples from the Lévy stable distribution $\lambda(x; \mu, \beta)$ is more delicate. The main challenge lies in the fact that there are no analytic expressions for the inverse F^{-1}, and thus the transformation method is not easy to apply. Two well-known exceptions are the Gaussian ($\mu = 2$) and Cauchy ($\mu = 1, \beta = 0$). One popular algorithm is from [CMS76, WW95]. For $\mu \in (0, 2]$ and $\beta \in [-1, 1]$, it generates a random variable $X \sim \lambda(x; \mu, \beta)$ as in Algorithm 1.

Exercises

Exercise 1.1 Show formula (1.13).

Exercise 1.2 Prove the asymptotic (1.15). (Hint: [Nol20, Section 3.6])

Exercise 1.3 Show the jump length variance formula (1.16).

Algorithm 1 Generating stable random variables.

1: Generate a random variable V uniformly distributed on $(-\frac{\pi}{2}, \frac{\pi}{2})$ and an independent exponential random variable W with mean 1.

2: For $\mu \neq 1$, compute

$$X = C_{\mu,\beta} \frac{\sin(\mu(V + B_{\mu,\beta}))}{(\cos V)^{\frac{1}{\mu}}} \left(\frac{\cos(V - \mu(V + B_{\mu,\beta}))}{W} \right)^{\frac{1-\mu}{\mu}},$$

where

$$B_{\mu,\beta} = \frac{\arctan(\beta \tan \frac{\mu\pi}{2})}{\mu} \quad \text{and} \quad C_{\mu,\beta} = \left[1 + \beta^2 \tan^2 \frac{\mu\pi}{2} \right]^{\frac{1}{2\mu}}.$$

3: For $\mu = 1$, compute

$$X = \frac{2}{\pi} \left[\left(\frac{\pi}{2} + \beta V \right) \tan V - \beta \ln \frac{W \cos V}{\frac{\pi}{2} + \beta V} \right].$$

Exercise 1.4 Prove that the pth moment of a stable random variable is finite if and only if $p < \mu$.

Exercise 1.5 Derive the representation (1.17).

Exercise 1.6 Develop an algorithm for generating random variables from the Cauchy distribution $\lambda(x) = (\pi(1 + x^2))^{-1}$ and implement it.

Exercise 1.7 Implement the algorithm for generating random variables from Lévy stable distribution, with $\mu = 1.5$, and $\beta = 0$.

Exercise 1.8 Derive the diffusion limit for the CTRW with a power law waiting time and Lévy jump length distribution, and the corresponding governing equation.

Chapter 2
Fractional Calculus

In this chapter, we describe basic properties of fractional-order integrals and derivatives, including Riemann-Liouville fractional integral and derivative, Djrbashian-Caputo and Grünwald-Letnikov fractional derivatives. They will serve as the main modeling tools in FDES.

2.1 Gamma Function

This function will appear in nearly every formula in this book! The usual way to define the Gamma function $\Gamma(z)$ is

$$\Gamma(z) = \int_0^\infty t^{z-1} e^{-t} \mathrm{d}t, \quad \mathfrak{R}(z) > 0. \tag{2.1}$$

It is analytic in the right half complex plane $\mathbb{C}_+ := \{z \in \mathbb{C} : \mathfrak{R}(z) > 0\}$ and integration by parts shows the following recursion formula

$$\Gamma(z+1) = z\Gamma(z). \tag{2.2}$$

Since $\Gamma(1) = 1$, (2.2) implies that for any positive integer $n \in \mathbb{N}$, $\Gamma(n+1) = n!$.

The following reflection formula [AS65, 6.1.17, p. 256]

$$\Gamma(z)\Gamma(1-z) = \frac{\pi}{\sin(\pi z)}, \quad 0 < \mathfrak{R}(z) < 1, \tag{2.3}$$

and Legendre's duplication formula [AS65, 6.1.18, p. 256]

$$\Gamma(z)\Gamma(z + \tfrac{1}{2}) = \sqrt{\pi} 2^{1-2z} \Gamma(2z) \tag{2.4}$$

are very useful in practice. For example, one only needs to approximate $\Gamma(z)$ for $\mathfrak{R}(z) \geq \frac{1}{2}$ and uses (2.3) for $\mathfrak{R}(z) < \frac{1}{2}$. It also gives directly $\Gamma(\frac{1}{2}) = \sqrt{\pi}$.

© The Author(s), under exclusive license to Springer Nature Switzerland AG 2021
B. Jin, *Fractional Differential Equations*, Applied Mathematical Sciences 206,
https://doi.org/10.1007/978-3-030-76043-4_2

It is possible to continue $\Gamma(z)$ analytically into the left half complex plane $\mathbb{C}_- :=$ $\{z \in \mathbb{C} : \Re(z) < 0\}$. This is often done by representing the reciprocal Gamma function $\frac{1}{\Gamma(z)}$ as an infinite product (cf. Exercise 2.2)

$$\frac{1}{\Gamma(z)} = \lim_{n \to \infty} \frac{n^{-z}}{n!} z(z+1) \ldots (z+n), \quad \forall z \in \mathbb{C}. \tag{2.5}$$

Hence, $\frac{1}{\Gamma(z)}$ is an entire function of z with zeros at $z = 0, -1, -2, \ldots$, and accordingly, $\Gamma(z)$ has poles at $z = 0, -1, \ldots$. The integral representation for $\frac{1}{\Gamma(z)}$ due to Hermann Hankel [Han64] is very useful. It is often derived using Laplace transform (cf. Appendix A.3). Substituting $t = su$ in (2.1) yields

$$\frac{\Gamma(z)}{s^z} = \int_0^\infty u^{z-1} e^{-su} du,$$

which can be regarded as the Laplace transform of u^{z-1} for a fixed $z \in \mathbb{C}$. Hence u^{z-1} can be interpreted as an inverse Laplace transform

$$u^{z-1} = \mathcal{L}^{-1}\left[\frac{\Gamma(z)}{s^z}\right] = \frac{1}{2\pi i} \int_C e^{su} \frac{\Gamma(z)}{s^z} ds,$$

where C is any deformed Bromwich contour that winds around the negative real axis $\mathbb{R}_- = (-\infty, 0)$ in the anticlockwise sense. Now substituting $\zeta = su$ yields

$$\frac{1}{\Gamma(z)} = \frac{1}{2\pi i} \int_C \zeta^{-z} e^\zeta d\zeta. \tag{2.6}$$

The integrand ζ^{-z} has a branch cut on $\mathbb{R}_-$ but is analytic elsewhere so that (2.6) is independent of the contour C. The formula (2.6) is very useful for deriving integral representations of various special functions, e.g., Mittag-Leffler and Wright functions (cf. (3.1) and (3.24) in Chapter 3). In practice, C can be deformed to facilitate the numerical evaluation of the integral, which makes $\frac{1}{\Gamma(z)}$ actually easier to compute than $\Gamma(z)$ itself [ST07].

For large argument, the following Stirling's formula holds (for any $\epsilon > 0$)

$$\Gamma(z) = (2\pi)^{\frac{1}{2}} e^{-z} z^{z-\frac{1}{2}} (1 + O(z^{-1})) \quad \text{as } |z| \to \infty, |\arg(z)| < \pi - \epsilon. \tag{2.7}$$

Further, for any two real numbers x and s, the following identity holds

$$\lim_{x \to \infty} (x^s \Gamma(x))^{-1} \Gamma(x+s) = 1, \tag{2.8}$$

often known as Wendel's limit [Wen48], due to his elegant proof using only Hölder's inequality and the recursion identity (2.2); see Exercise 2.4 for the detail.

A closely related function, the Beta function $B(\alpha, \beta)$ for $\alpha, \beta > 0$ is defined by

$$B(\alpha, \beta) = \int_0^1 (1-s)^{\alpha-1} s^{\beta-1} ds. \tag{2.9}$$

It can be expressed by the Gamma function $\Gamma(z)$ as

$$B(\alpha, \beta) = \frac{\Gamma(\alpha)\Gamma(\beta)}{\Gamma(\alpha + \beta)}. \tag{2.10}$$

The (generalized) binomial coefficient $C(\alpha, \beta)$ is defined by

$$C(\alpha, \beta) = \frac{\Gamma(\alpha + 1)}{\Gamma(\beta + 1)\Gamma(\alpha - \beta + 1)},$$

provided $\beta \neq -1, -2, \ldots$ and $\alpha - \beta \neq -1, -2, \ldots$. For $\alpha, \beta \in \mathbb{N}_0$, it recovers the usual binomial coefficient. The following identities hold for the binomial coefficients.

Lemma 2.1 *Let* $j, k \in \mathbb{N}_0$, *and* $\alpha, \beta \neq 0, -1, \ldots$. *Then the following identities hold.*

$$(-1)^j C(\alpha, j) = C(j - \alpha - 1, j), \tag{2.11}$$

$$C(\alpha, k + j)C(k + j, k) = C(\alpha, k)C(\alpha - k, j), \tag{2.12}$$

$$\sum_{j=0}^{k} C(\alpha, j)C(\beta, k - j) = C(\alpha + \beta, k). \tag{2.13}$$

Proof The identities (2.11) and (2.12) are direct from the definition. Indeed,

$$\begin{aligned} C(\alpha, k + j)C(k + j, k) &= \frac{\Gamma(\alpha + 1)}{\Gamma(k + j + 1)\Gamma(\alpha - k - j + 1)} \frac{\Gamma(k + j + 1)}{\Gamma(k + 1)\Gamma(j + 1)} \\ &= \frac{\Gamma(\alpha + 1)}{\Gamma(k + 1)\Gamma(\alpha - k + 1)} \frac{\Gamma(\alpha - k + 1)}{\Gamma(j + 1)\Gamma(\alpha - k - j + 1)} \\ &= C(\alpha, k)C(\alpha - k, j), \end{aligned}$$

showing (2.12). The formula (2.13) can be found at [PBM86, p. 616, 4.2.15.13]. □

2.2 Riemann-Liouville Fractional Integral

The Riemann-Liouville fractional integral generalizes the well-known Cauchy's iterated integral formula. Let finite $a, b \in \mathbb{R}$, with $a < b$, and we denote by $D = (a, b)$. Then for any integer $k \in \mathbb{N}$, the k-fold integral ${}_aI_x^k$, based at the left end point $x = a$, is defined recursively by ${}_aI_x^0 f(x) = f(x)$ and

$${}_aI_x^k f(x) = \int_a^x {}_aI_s^{k-1} f(s)\, ds, \quad k = 1, 2, \ldots.$$

We claim the following Cauchy's iterated integral formula for $k \geq 1$

$${}_aI_x^k f(x) = \frac{1}{(k - 1)!} \int_a^x (x - s)^{k-1} f(s)\, ds \tag{2.14}$$

and verify it by mathematical induction. Clearly, it holds for $k = 1$. Now assume that the formula (2.14) holds for some $k \geq 1$, i.e.,

$$_aI_x^k f(x) = \frac{1}{(k-1)!} \int_a^x (x-s)^{k-1} f(s)\, ds.$$

Then by definition and changing integration order, we deduce

$$_aI_x^{k+1} f(x) = \int_a^x {_aI_s^k} f(s)\, ds = \frac{1}{(k-1)!} \int_a^x \int_a^s (s-t)^{k-1} f(t)\, dt ds$$

$$= \frac{1}{(k-1)!} \int_a^x \int_t^x (s-t)^{k-1}\, ds f(t)\, dt = \frac{1}{k!} \int_a^x (x-s)^k f(s)\, ds.$$

This shows the formula (2.14). Likewise, there holds

$$_xI_b^k f(x) = \frac{1}{(k-1)!} \int_x^b (s-x)^{k-1} f(s)\, ds. \tag{2.15}$$

One way to generalize these formulas to any $\alpha > 0$ is to use the Gamma function $\Gamma(z)$: the factorial $(k-1)!$ in the formula can be replaced by $\Gamma(k)$. Then for any $\alpha > 0$, by replacing k with α, and $(k-1)!$ with $\Gamma(\alpha)$, we arrive at the following definition of Riemann-Liouville fractional integrals. It is named after Bernhard Riemann's work in 1847 [Rie76] and Joseph Liouville's work in 1832 [Lio32], the latter of whom was the first to consider the possibility of fractional calculus [Lio32].

Definition 2.1 For any $f \in L^1(D)$, the left-sided Riemann-Liouville fractional integral of order $\alpha > 0$ based at $x = a$, denoted by $_aI_x^\alpha f$, is defined by

$$(_aI_x^\alpha f)(x) = \frac{1}{\Gamma(\alpha)} \int_a^x (x-s)^{\alpha-1} f(s)\, ds, \tag{2.16}$$

and the right-sided Riemann-Liouville fractional integral of order $\alpha > 0$ based at $x = b$, denoted by $_xI_b^\alpha f$, is defined by

$$(_xI_b^\alpha f)(x) = \frac{1}{\Gamma(\alpha)} \int_x^b (s-x)^{\alpha-1} f(s)\, ds. \tag{2.17}$$

When $\alpha = k \in \mathbb{N}$, they recover the k-fold integrals $_aI_x^k f$ and $_xI_b^k f$. Further, we define for $x \to a^+$ by

$$(_aI_x^\alpha f)(a^+) := \lim_{x \to a^+} (_aI_x^\alpha f)(x),$$

if the right-hand side exists, and likewise $(_xI_b^\alpha f)(b^-)$. In case $\alpha = 0$, we adopt the convention $_aI_x^0 f(x) = f(x)$, in view of the following result [DN68, (1.3)].

Lemma 2.2 *Let $f \in L^1(D)$. Then for almost all $x \in D$, there hold*

$$\lim_{\alpha \to 0^+} {_aI_x^\alpha} f(x) = \lim_{\alpha \to 0^+} {_xI_b^\alpha} f(x) = f(x).$$

Proof Let $x \in D$ be a Lebesgue point of f, i.e., $\lim_{h \to 0} \frac{1}{h} \int_0^h |f(x+s) - f(s)| ds = 0$. Then let $F_x(t) = \int_{x-t}^x f(s) ds$, for any $t \in [0, x - a]$. Then it can be represented as

$$F_x(t) = t[f(x) + \omega_x(t)], \quad 0 \le t \le x - a,$$

where $\omega_x(t) \to 0$ as $t \to 0$. Thus for each $\epsilon > 0$, there exists $\delta = \delta(\epsilon) > 0$ such that

$$|\omega_x(t)| < \epsilon, \quad \forall t \in (0, \delta). \tag{2.18}$$

Then by the definition of F and integration by parts, we deduce

$$
\begin{aligned}
_aI_x^\alpha f(x) &:= \frac{1}{\Gamma(\alpha)} \int_0^{x-a} s^{\alpha-1} f(x - s) ds \\
&= \frac{F_x(x - a)}{\Gamma(\alpha)} (x - a)^{\alpha-1} - \frac{\alpha - 1}{\Gamma(\alpha)} \int_0^{x-a} F_x(s) s^{\alpha-2} ds \\
&= \frac{F_x(x - a)}{\Gamma(\alpha)} (x - a)^{\alpha-1} - \frac{\alpha - 1}{\Gamma(\alpha + 1)} (x - a)^\alpha f(x) \\
&\quad - \frac{\alpha - 1}{\Gamma(\alpha)} \left[\int_0^\delta \omega_x(s) s^{\alpha-1} ds + \int_\delta^{x-a} \omega_x(s) s^{\alpha-1} ds \right].
\end{aligned}
$$

Then it follows from (2.18) and the fact 0 is a pole of $\Gamma(z)$ that $\lim_{\alpha \to 0^+} |_aI_x^\alpha f(x) - f(x)| \le \epsilon$. Since $\epsilon > 0$ is arbitrary, it proves the first identity for x. This completes the proof since the set of Lebesgue points of $f \in L^1(D)$ is dense in D. $\qquad \square$

The integral operators $_aI_x^\alpha$ and $_xI_b^\alpha$ inherit certain properties from integer-order integrals. Consider power functions $(x - a)^\gamma$ with $\gamma > -1$ (so that $(x - a)^\gamma \in L^1(D)$). It can be verified directly that for $\alpha > 0$ and $x > a$,

$$_aI_x^\alpha (x - a)^\gamma = \frac{\Gamma(\gamma + 1)}{\Gamma(\gamma + \alpha + 1)} (x - a)^{\gamma+\alpha}, \tag{2.19}$$

and similarly for $x < b$

$$_xI_b^\alpha (b - x)^\gamma = \frac{\Gamma(\gamma + 1)}{\Gamma(\gamma + \alpha + 1)} (b - x)^{\gamma+\alpha}.$$

Indeed, (2.19) follows from changing variables $s = a + t(x - a)$ and (2.10):

$$
\begin{aligned}
_aI_x^\alpha (x - a)^\gamma &= \frac{1}{\Gamma(\alpha)} \int_a^x (x - s)^{\alpha-1} (s - a)^\gamma ds = \frac{(x - a)^{\gamma+\alpha}}{\Gamma(\alpha)} \int_0^1 (1 - t)^{\alpha-1} t^\gamma dt \\
&= \frac{B(\alpha, \gamma + 1)}{\Gamma(\alpha)} (x - a)^{\gamma+\alpha} = \frac{\Gamma(\gamma + 1)}{\Gamma(\gamma + \alpha + 1)} (x - a)^{\gamma+\alpha},
\end{aligned}
$$

Clearly, for any integer $\alpha \in \mathbb{N}$, these formulas recover the integral counterparts.

Example 2.1 We compute the Riemann-Liouville fractional integral $_0I_x^\alpha f(x)$, $\alpha > 0$, of the exponential function $f(x) = e^{\lambda x}$, $\lambda \in \mathbb{R}$. By (2.19), we have

$$\begin{aligned}
{}_0I_x^\alpha f(x) &= {}_0I_x^\alpha \sum_{k=0}^\infty \frac{(\lambda x)^k}{\Gamma(k+1)} = \sum_{k=0}^\infty \lambda^k \frac{{}_0I_x^\alpha x^k}{\Gamma(k+1)} \\
&= \sum_{k=0}^\infty \lambda^k \frac{x^{k+\alpha}}{\Gamma(k+\alpha+1)} = x^\alpha \sum_{k=0}^\infty \frac{(\lambda x)^k}{\Gamma(k+\alpha+1)} = x^\alpha E_{1,\alpha+1}(\lambda x),
\end{aligned}$$

$$(2.20)$$

where the Mittag-Leffler function $E_{\alpha,\beta}(z)$ is defined by $E_{\alpha,\beta}(z) = \sum_{k=0}^\infty \frac{z^k}{\Gamma(k\alpha+\beta)}$, for $z \in \mathbb{C}$, cf. (3.1) in Chapter 3. The interchange of summation and integral is legitimate since $E_{\alpha,\beta}(z)$ is entire, cf. Proposition 3.1.

The following semigroup property holds for k-fold integrals: ${}_aI_x^k {}_aI_x^\ell f = {}_aI_x^{k+\ell} f$, $k, \ell \in \mathbb{N}_0$, which holds also for ${}_aI_x^\alpha$ and ${}_xI_b^\alpha$.

Theorem 2.1 *For $f \in L^1(D)$, $\alpha, \beta \geq 0$, there holds*

$${}_aI_x^\alpha {}_aI_x^\beta f = {}_aI_x^\beta {}_aI_x^\alpha f = {}_aI_x^{\alpha+\beta} f, \quad {}_xI_b^\alpha {}_xI_b^\beta f = {}_xI_b^\beta {}_xI_b^\alpha f = {}_xI_b^{\alpha+\beta} f.$$

Proof The case $\alpha = 0$ or $\beta = 0$ is trivial, since ${}_aI_x^0$ is the identity operator. It suffices to consider $\alpha, \beta > 0$. By changing integration order, we have

$$\begin{aligned}
({}_aI_x^\alpha {}_aI_x^\beta f)(x) &= \frac{1}{\Gamma(\alpha)\Gamma(\beta)} \int_a^x (x-s)^{\alpha-1} \int_a^s (s-t)^{\beta-1} f(t) \, dt \, ds \\
&= \frac{1}{\Gamma(\alpha)\Gamma(\beta)} \int_a^x f(t) \left(\int_t^x (x-s)^{\alpha-1}(s-t)^{\beta-1} \, ds \right) dt.
\end{aligned}$$

Now by changing variables, the integral in the bracket can be simplified to

$$\begin{aligned}
\int_t^x (x-s)^{\alpha-1}(s-t)^{\beta-1} ds &= (x-t)^{\alpha+\beta-1} \int_0^1 (1-s)^{\alpha-1} s^{\beta-1} ds \\
&= B(\alpha,\beta)(x-t)^{\alpha+\beta-1}.
\end{aligned}$$

By (2.10), we have

$$({}_aI_x^\alpha {}_aI_x^\beta f)(x) = \frac{B(\alpha,\beta)}{\Gamma(\alpha)\Gamma(\beta)} \int_a^x (x-s)^{\alpha+\beta-1} f(s) \, ds = ({}_aI_x^{\alpha+\beta} f)(x).$$

The other identity follows similarly. $\qquad\square$

The next result collects several mapping properties for ${}_aI_x^\alpha$, and similar results hold for ${}_xI_b^\alpha$. (i) indicates that ${}_aI_x^\alpha$ and ${}_xI_b^\alpha$ are bounded in $L^p(D)$ spaces, $1 \leq p \leq \infty$. Thus for any $f \in L^1(D)$, ${}_aI_x^\alpha f$ and ${}_xI_b^\alpha f$ of order $\alpha > 0$ exists almost everywhere (a.e.). The second part of (ii) is commonly known as the Hardy-Littlewood inequality. The spaces $AC(\overline{D})$ and $C^{k,\gamma}(\overline{D})$ are defined in Section A.1.1 in the appendix.

Theorem 2.2 *The following mapping properties hold for ${}_aI_x^\alpha$, and $\alpha > 0$.*

(i) *${}_aI_x^\alpha$ is bounded on $L^p(D)$ for any $1 \leq p \leq \infty$.*

(ii) *For* $1 \leq p < \alpha^{-1}$, $_aI_x^\alpha$ *is a bounded operator from* $L^p(D)$ *into* $L^r(D)$ *for* $1 \leq r < (1 - \alpha p)^{-1}$. *If* $1 < p < \alpha^{-1}$, *then* $_aI_x^\alpha$ *is bounded from* $L^p(D)$ *to* $L^{\frac{p}{1-\alpha p}}(D)$.

(iii) *For* $p > 1$ *and* $p^{-1} < \alpha < 1 + p^{-1}$ *or* $p = 1$ *and* $1 \leq \alpha < 2$, $_aI_x^\alpha$ *is bounded from* $L^p(D)$ *into* $C^{0, \alpha - \frac{1}{p}}(\overline{D})$, *and, moreover*, $_aI_x^\alpha f(0) = 0$, *for* $f \in L^p(D)$.

(iv) $_aI_x^\alpha$ *maps* $AC(\overline{D})$ *into* $AC(\overline{D})$.

(v) *If* $f \in L^1(D)$ *is nonnegative and nondecreasing then* $_aI_x^\alpha f$ *is nondecreasing.*

Proof By Young's inequality in (A.1), we deduce

$$\|_aI_x^\alpha f\|_{L^p(D)} \leq \frac{1}{\Gamma(\alpha)} \|(x - a)^{\alpha - 1}\|_{L^1(D)} \|f\|_{L^p(D)} = \frac{(b - a)^\alpha}{\Gamma(\alpha + 1)} \|f\|_{L^p(D)},$$

showing the assertion for $_aI_x^\alpha$. The first part of (ii) follows similarly, and the proof of the second part is technical and lengthy, and can be found in [HL28, Theorem 4]. To see (iii), consider first the case $p > 1$. Suppose first $a + h \leq x < x + h \leq b$. Then

$$\Gamma(\alpha)(_aI_x^\alpha f(x + h) - {}_aI_x^\alpha f(x))$$

$$= \int_a^{x+h} (x + h - s)^{\alpha - 1} f(s) ds - \int_a^x (x - s)^{\alpha - 1} f(s) ds$$

$$= \int_x^{x+h} (x + h - s)^{\alpha - 1} f(s) ds + \int_{x-h}^x ((x + h - s)^{\alpha - 1} - (x - s)^{\alpha - 1}) f(s) ds$$

$$+ \int_a^{x-h} ((x + h - s)^{\alpha - 1} - (x - s)^{\alpha - 1}) f(s) ds := \sum_{i=1}^3 I_i.$$

By Hölder's inequality, since $(\alpha - 1)p > -(p - 1)$ and $((p-1)^{-1}(\alpha - 1)p + 1)p^{-1}(p - 1) = \alpha - p^{-1}$,

$$|I_1| \leq \left(\int_x^{x+h} |f(s)|^p ds \right)^{\frac{1}{p}} \left(\int_x^{x+h} (x + h - s)^{\frac{p(\alpha-1)}{p-1}} ds \right)^{\frac{p-1}{p}} \leq c \|f\|_{L^p(D)} h^{\alpha - \frac{1}{p}}.$$

This argument also yields $|I_2| \leq c \|f\|_{L^p(D)} h^{\alpha - \frac{1}{p}}$. Now for the term I_3, since

$$|(x + h - s)^{\alpha - 1} - (x - s)^{\alpha - 1}| \leq h|1 - \alpha|(x - s)^{\alpha - 2},$$

we have

$$|I_3| \leq h|1 - \alpha| \int_a^{x-h} |f(s)|(x - s)^{\alpha - 2} ds$$

$$\leq h|1 - \alpha| \left(\int_a^{x-h} |f(s)|^p ds \right)^{\frac{1}{p}} \left(\int_a^{x-h} (x - s)^{\frac{p(\alpha-2)}{p-1}} ds \right)^{\frac{p-1}{p}}$$

$$\leq |1 - \alpha| \left(\frac{p - 1}{p + 1 - \alpha p} \right)^{\frac{p-1}{p}} h^{\alpha - \frac{1}{p}} \|f\|_{L^p(D)}.$$

Thus, the assertion follows. When $x \leq a + h$, the estimate is direct. If $p = 1$, we have $\alpha - 1 \geq 0$, and so

$$|I_1| \leq h^{\alpha-1} \int_x^{x+h} |f(s)| \mathrm{d}s,$$

and similarly

$$|I_3| \leq h(\alpha - 1)h^{\alpha-2} \int_a^{x-h} |f(s)| \mathrm{d}s.$$

Thus the estimate also holds. This shows (iii). Next, if $f \in AC(\overline{D})$, then $f' \in L^1(D)$ exists a.e. and $f(x) - f(a) = (_aI_x^1 f')(x)$ for all $x \in D$. Then

$$_aI_x^\alpha f(x) = {}_aI_x^\alpha {}_aI_x^1 f'(x) + (_aI_x^\alpha f(a))(x) = {}_aI_x^1(_aI_x^\alpha f)(x) + f(a)\frac{(x-a)^\alpha}{\Gamma(\alpha+1)}.$$

Both terms on the right-hand side belong to $AC(\overline{D})$, showing (iv). Last,

$$_aI_x^\alpha f(x) = \frac{(x-a)^\alpha}{\Gamma(\alpha)} \int_0^1 (1-s)^{\alpha-1} f((x-a)s + a) \mathrm{d}s,$$

which is nondecreasing in x when f is nonnegative and nondecreasing. □

Remark 2.1 The mapping properties of the Riemann-Liouville fractional integral were systematically studied by Hardy and Littlewood [HL28, HL32]. (ii) can be found in [HL28, Theorem 4]. They also showed that the result does not hold if $p = 1$ for any $\alpha \in (0, 1)$, and that if $0 < \alpha < 1$ and $p = \alpha^{-1}$, $_aI_x^\alpha f$ is not necessarily bounded. (iii) was proved in [HL28, Theorem 12], where it was also pointed out that the result is false in the cases $p > 1$, $\alpha = p^{-1}$ and $\alpha = 1 + p^{-1}$.

The next result asserts that $_aI_x^\alpha$ and $_xI_b^\alpha$ are adjoint to each other in $L^2(D)$. It is commonly referred to as the "fractional integration by parts" formula [SKM93, Corollary, p. 67].

Lemma 2.3 *For any $f \in L^p(D)$, $g \in L^q(D)$, $p, q \geq 1$ with $p^{-1} + q^{-1} \leq 1 + \alpha$ ($p, q \neq 1$, when $p^{-1} + q^{-1} = 1 + \alpha$), the following identity holds*

$$\int_a^b g(x)(_aI_x^\alpha f)(x) \, \mathrm{d}x = \int_a^b f(x)(_xI_b^\alpha g)(x) \, \mathrm{d}x.$$

Proof For $f, g \in L^2(D)$, the proof is direct: by Theorem 2.2(i), for $f \in L^2(D)$, $_aI_x^\alpha f \in L^2(D)$, and hence $\int_a^b |g(x)||(_aI_x^\alpha f)(x)| \, \mathrm{d}x \leq c\|f\|_{L^2(D)}\|g\|_{L^2(D)}$. Now the desired formula follows from Fubini's Theorem. The general case follows from Theorem 2.2(ii) instead. □

The next result discusses the mapping to $AC(\overline{D})$ [Web19a, Proposition 3.6]. It is useful for characterizing fractional derivatives and solution concepts for FDES.

Lemma 2.4 *Let $f \in L^1(D)$ and $\alpha \in (0, 1)$. Then $_aI_x^{1-\alpha} f \in AC(\overline{D})$ and $_aI_x^\alpha f(a) = 0$ if and only if there exists $g \in L^1(D)$ such that $f = {}_aI_x^\alpha g$.*

Proof If there exists $g \in L^1(D)$ such that $f = {}_aI_x^\alpha g$, then by Theorem 2.1,

$$_aI_x^{1-\alpha}f = {}_aI_x^{1-\alpha}{}_aI_x^\alpha g = {}_aI_x^1 g \in AC(\overline{D}),$$

and since $g \in L^1(D)$, by the absolute continuity of Lebesgue integral, $_aI_x^{1-\alpha}f(a) = \lim_{x \to a^+} \int_a^x g(s)\mathrm{d}s = 0$. Conversely, suppose $_aI_x^{1-\alpha}f \in AC(\overline{D})$ and $_aI_x^{1-\alpha}f(a) = 0$. Let $G(x) = {}_aI_x^{1-\alpha}f$ so that $G \in AC(\overline{D})$ and $G(a) = 0$. Then $g := G'$ exists for a.e. $x \in D$ with $g \in L^1(D)$, and $G(x) = {}_aI_x^1 g$. Furthermore, $_aI_x^\alpha G = {}_aI_x^1 f = {}_aI_x^\alpha{}_aI_x^1 g = {}_aI_x^1{}_aI_x^\alpha g$. Since $f, {}_aI_x^\alpha g \in L^1(D)$, $_aI_x^1 f, {}_aI_x^1{}_aI_x^\alpha g \in AC(\overline{D})$ and their derivatives exist a.e. as $L^1(D)$ functions and are equal, that is, $f = {}_aI_x^\alpha g$. $\qquad\square$

Last, we generalize the mean value theorem in calculus: if $f \in C(\overline{D})$, g is Lebesgue integrable on D and g does not change sign in D, then

$$\int_a^b f(x)g(x)\mathrm{d}x = f(\xi)\int_a^b g(x)\mathrm{d}x, \qquad (2.21)$$

for some $\xi \in (a, b)$. One fractional version is as follows [Die12, Theorem 2.1] (see also [Die17]). The classical case is recovered by setting $\alpha = 1$ and $x = b$.

Theorem 2.3 *Let $\alpha > 0$ and $f \in C(\overline{D})$, and let g be Lebesgue integrable on D and do not change its sign in D. Then for almost every $x \in (a, b]$, there exists some $\xi \in (a, x) \subset D$ such that $_aI_x^\alpha(fg)(x) = f(\xi)_aI_x^\alpha g(x)$. If additionally, $\alpha \geq 1$ or $g \in C(\overline{D})$, then this result holds for every $x \in (a, b]$.*

Proof Note that $_aI_x^\alpha(fg)(x) = \frac{1}{\Gamma(\alpha)}\int_a^x (x-s)^{\alpha-1}f(s)g(s)\mathrm{d}s$. If $\alpha \geq 1$ and $x \in (a, b]$, then $(x - s)^{\alpha-1}$ is continuous in s. Hence, $\tilde{g}(s) = \frac{(x-s)^{\alpha-1}}{\Gamma(\alpha)}g(s)$ is integrable on $[a, x]$ and does not change sign on $[a, x]$. Thus by (2.21),

$$_aI_x^\alpha(fg)(x) = \int_a^x f(s)\tilde{g}(s)\mathrm{d}s = f(\xi)\int_a^x \tilde{g}(s)\mathrm{d}s = f(\xi)_aI_x^\alpha g(x).$$

If $0 < \alpha < 1$ and g is continuous, the same line of proof works. Finally, if $0 < \alpha < 1$ and g is only integrable, then one can still argue in a similar way, but the integrability of $\tilde{g}$ holds only for almost all $x \in D$, cf. Theorem 2.2(i). $\qquad\square$

2.3 Fractional Derivatives

Now we discuss fractional derivatives, for which there are several different definitions. We only discuss Riemann-Liouville and Djrbashian-Caputo fractional derivatives, which represent two most popular choices in practice, and briefly mention Grünwald-Letnikov fractional derivative. There are several popular textbooks with extensive treatment on fractional derivatives [OS74, MR93, Pod99, KST06, Die10] and also monographs [Dzh66, SKM93]. An encyclopedic treatment of Riemann-Liouville fractional integral and derivatives is given in the monograph [SKM93], and the book [Die10] contains detailed discussions on the Djrbashian-Caputo fractional derivative. Like before, for any fixed $a, b \in \mathbb{R}$, $a < b$, which are assumed to be finite unless otherwise stated, we denote by $D = (a, b)$, and by $\overline{D} = [a, b]$. Further, for $k \in \mathbb{N}$, $f^{(k)}$ denotes the kth order derivative of f.

2.3.1 Riemann-Liouville fractional derivative

We begin with the definition of the Riemann-Liouville fractional derivative.

Definition 2.2 For $f \in L^1(D)$ and $n - 1 < \alpha \leq n$, $n \in \mathbb{N}$, its left-sided Riemann-Liouville fractional derivative of order α (based at $x = a$), denoted by $^{R}_{a}D^{\alpha}_{x}f$, is defined by

$$^{R}_{a}D^{\alpha}_{x}f(x) := \frac{d^n}{dx^n}(_aI^{n-\alpha}_x f)(x) = \frac{1}{\Gamma(n-\alpha)}\frac{d^n}{dx^n}\int_a^x (x-s)^{n-\alpha-1}f(s)\,ds,$$

if the integral on the right-hand side exists. Its right-sided Riemann-Liouville fractional derivative of order α (based at $x = b$), denoted by $^{R}_{x}D^{\alpha}_{b}f$, is defined by

$$^{R}_{x}D^{\alpha}_{b}f(x) := (-1)^n\frac{d^n}{dx^n}(_xI^{n-\alpha}_b f)(x) = \frac{(-1)^n}{\Gamma(n-\alpha)}\frac{d^n}{dx^n}\int_x^b (s-x)^{n-\alpha-1}f(s)\,ds,$$

if the integral on the right-hand side exists.

Definition 2.2 does not give the condition for the existence of a Riemann-Liouville fractional derivative. By Lemma 2.4, the condition $_aI^{n-\alpha}_x f \in AC^n(\overline{D})$ is needed to ensure $^{R}_{a}D^{\alpha}_{x}f \in L^1(D)$, and similarly $_xI^{n-\alpha}_b f \in AC^n(\overline{D})$ for $^{R}_{x}D^{\alpha}_{b}f \in L^1(D)$. We denote by

$$^{R}_{a}D^{\alpha}_{x}f(a^+) = \lim_{x \to a^+} {}^{R}_{a}D^{\alpha}_{x}f(x),$$

if the limit on the right-hand side exists. The quantity $^{R}_{x}D^{\alpha}_{b}f(b^-)$ is defined similarly.

Due to the presence of $_aI^{n-\alpha}_x$, $^{R}_{a}D^{\alpha}_{x}f$ is inherently nonlocal: the value of $^{R}_{a}D^{\alpha}_{x}f$ at $x > a$ depends on the values of f from a to x. The nonlocality dramatically influences its analytical properties. We compute the derivatives $^{R}_{a}D^{\alpha}_{x}(x-a)^{\gamma}$, $\alpha > 0$, of the function $(x - a)^{\gamma}$ with $\gamma > -1$. The identities (2.19) and (2.2) yield

$$^{R}_{a}D^{\alpha}_{x}(x-a)^{\gamma} = \frac{d^n}{dx^n}\frac{\Gamma(\gamma+1)(x-a)^{\gamma+n-\alpha}}{\Gamma(\gamma+1+n-\alpha)} = \frac{\Gamma(\gamma+1)(x-a)^{\gamma-\alpha}}{\Gamma(\gamma+1-\alpha)}. \tag{2.22}$$

One can draw a number of interesting observations. First, for $\alpha \notin \mathbb{N}$, $^{R}_{a}D^{\alpha}_{x}f$ of the constant function $f(x) \equiv 1$ (i.e., $\gamma = 0$) is not identically zero, since for $x > a$:

$$^{R}_{a}D^{\alpha}_{x}1 = \frac{(x-a)^{-\alpha}}{\Gamma(1-\alpha)}. \tag{2.23}$$

This fact is inconvenient for practical applications involving initial/boundary conditions where the physical meaning of (2.23) then becomes unclear. Nonetheless, when $\alpha \in \mathbb{N}$, since $0, -1, -2, \ldots$ are the poles of $\Gamma(z)$, the right-hand side of (2.23) vanishes, and thus it recovers the familiar identity $\frac{d^n}{dx^n} 1 = 0$. Second, given $\alpha > 0$, for any $\gamma = \alpha - 1, \alpha - 2, \ldots, \alpha - n$, $\gamma + 1 - \alpha$ is a negative integer or zero, we have

$$\,_{a}^{R}D_x^{\alpha}(x-a)^{\alpha-j} \equiv 0, \quad j = 1, \ldots, n. \tag{2.24}$$

In particular, for $\alpha \in (0, 1)$, $(x-a)^{\alpha-1}$ belongs to the kernel of the operator $\,_{a}^{R}D_x^{\alpha}$ and plays the same role as a constant function for the first-order derivative. Generally, it implies that if $f, g \in L^1(D)$ with $\,_{a}I_x^{n-\alpha}f, \,_{a}I_x^{n-\alpha}g \in AC(\overline{D})$ and $n - 1 < \alpha \le n$, then

$$\,_{a}^{R}D_x^{\alpha}f(x) = \,_{a}^{R}D_x^{\alpha}g(x) \quad \Leftrightarrow \quad f(x) = g(x) + \sum_{j=1}^{n} c_j(x-a)^{\alpha-j},$$

where c_j, $j = 1, 2, \ldots, n$, are arbitrary constants. Hence, we have the following useful result, where the notation $(\,_{a}^{R}D_x^{\alpha-n}f)(a^+)$ is identified with $(\,_{a}I_x^{n-\alpha}f)(a^+)$. When $\alpha \in \mathbb{N}$, the result recovers the familiar Taylor polynomial.

Theorem 2.4 If $n - 1 < \alpha \le n$, $n \in \mathbb{N}$, $f \in L^1(D)$, and $\,_{a}I_x^{n-\alpha}f \in AC^n(\overline{D})$, and satisfies $\,_{a}^{R}D_x^{\alpha}f(x) = 0$ in D, then it can be represented by

$$f(x) = \sum_{j=1}^{n} \frac{(\,_{a}^{R}D_x^{\alpha-j}f)(a^+)}{\Gamma(\alpha - j + 1)}(x-a)^{\alpha-j}.$$

Proof By the preceding discussion, we have $f(x) = \sum_{j=1}^{n} c_j(x-a)^{\alpha-j}$. Applying $\,_{a}^{R}D_x^{\alpha-k}$, $k = 1, \ldots, n$, to both sides gives

$$(\,_{a}^{R}D_x^{\alpha-k}f)(a^+) = \lim_{x \to a^+}(\,_{a}^{R}D_x^{\alpha-k}f)(x)$$

$$= \sum_{j=1}^{n} c_j \lim_{x \to a^+} \,_{a}^{R}D_x^{\alpha-k}(x-a)^{\alpha-j} = c_k \Gamma(\alpha - k + 1),$$

in view of the identities (2.22) and (2.24), which gives directly the assertion. $\qquad \square$

Example 2.2 The integer-order derivative of $e^{\lambda x}$, $\lambda \in \mathbb{R}$, is still a multiple of $e^{\lambda x}$. Now we compute $\,_{0}^{R}D_x^{\alpha}f$ of $f(x) = e^{\lambda x}$. From (2.22), we deduce

$$\,_{0}^{R}D_x^{\alpha}e^{\lambda x} = \,_{0}^{R}D_x^{\alpha}\sum_{k=0}^{\infty}\frac{(\lambda x)^k}{\Gamma(k+1)} = \sum_{k=0}^{\infty}\lambda^k\frac{\,_{0}^{R}D_x^{\alpha}x^k}{\Gamma(k+1)} \tag{2.25}$$

$$= \sum_{k=0}^{\infty}\frac{\lambda^k x^{k-\alpha}}{\Gamma(k+1-\alpha)} = x^{-\alpha}\sum_{k=0}^{\infty}\frac{(\lambda x)^k}{\Gamma(k+1-\alpha)} = x^{-\alpha}E_{1,1-\alpha}(\lambda x).$$

Generally the right-hand side is no longer an exponential function if $\alpha \notin \mathbb{N}$.

Example 2.3 For $\alpha > 0$ and $\lambda \in \mathbb{C}$, let $f(x) = x^{\alpha-1} E_{\alpha,\alpha}(\lambda x^{\alpha})$. Then the Riemann-Liouville fractional derivative ${}^{R}_{0}D^{\alpha}_{x} f$ is given by

$$
{}^{R}_{0}D^{\alpha}_{x} f(x) = {}^{R}_{0}D^{\alpha}_{x} x^{\alpha-1} \Big(\sum_{k=0}^{\infty} \frac{(\lambda x^{\alpha})^k}{\Gamma(k\alpha + \alpha)} \Big) = \sum_{k=0}^{\infty} \lambda^k \frac{{}^{R}_{0}D^{\alpha}_{x} x^{(k+1)\alpha-1}}{\Gamma(k\alpha + \alpha)}
$$

$$
= \sum_{k=0}^{\infty} \frac{\lambda^k x^{k\alpha-1}}{\Gamma(k\alpha)} = \lambda x^{\alpha-1} E_{\alpha,\alpha}(\lambda x^{\alpha}).
$$

The interchange of the fractional derivative and summation is legitimate since $E_{\alpha,\beta}(z)$ is an entire function, cf. Proposition 3.1 in Chapter 3. That is, $x^{\alpha-1} E_{\alpha,\alpha}(\lambda x^{\alpha})$ is invariant under ${}^{R}_{0}D^{\alpha}_{x}$, and is a candidate for an eigenfunction of the operator ${}^{R}_{0}D^{\alpha}_{x}$ (when equipped with suitable boundary conditions).

Remark 2.2 Note that in Examples 2.2 and 2.3, the obtained results depend on the starting value a. For example, if we take $a = -\infty$, then direct computation shows

$$
{}^{R}_{-\infty}D^{\alpha}_{x} e^{\lambda x} = \lambda^{\alpha} e^{\lambda x}, \quad \lambda > 0.
$$

Next we examine analytical properties of ${}^{R}_{a}D^{\alpha}_{x}$ more closely. In calculus, both integral and differential operators satisfy the commutativity and semigroup property. This is also true for ${}_{a}I^{\alpha}_{x}$. However, the operator ${}^{R}_{a}D^{\alpha}_{x}$ satisfies neither property. The composition of two Riemann-Liouville fractional derivatives may lead to totally unexpected results. Further, for $f \in L^1(D)$ and $\alpha, \beta > 0$ with $\alpha + \beta = 1$, the following statements hold:

(i) If ${}^{R}_{a}D^{\alpha}_{x}({}^{R}_{a}D^{\beta}_{x} u) = f$, then $u(x) = {}_{a}I^{1}_{x} f(x) + c_0 + c_1(x - a)^{\beta-1}$;
(ii) If ${}^{R}_{a}D^{\beta}_{x}({}^{R}_{a}D^{\alpha}_{x} v) = f$, then $v(x) = {}_{a}I^{1}_{x} f(x) + c_2 + c_3(x - a)^{\alpha-1}$;
(iii) If ${}^{R}_{a}D^{1}_{x} w = f$, then $w(x) = {}_{a}I^{1}_{x} f(x) + c_4$,

where $c_i, i = 0, \ldots, 4$ are arbitrary constants. Indeed, for (i), one first applies ${}_{a}I^{\alpha}_{x}$ to both sides, and then ${}_{a}I^{\beta}_{x}$. Thus two extra terms have to be included: (i) and (ii) include two constants, whereas (iii) includes only one. Note that if we restrict $f \in C(\overline{D})$, then the singular term $(x - a)^{\alpha-1}$ (or $(x - a)^{\beta-1}$) must disappear, indicating that a composition rule might still be possible on suitable subspaces. Indeed, it does hold on the space ${}_{a}I^{\gamma}_{x}(L^p(D))$, $\gamma > 0$ and $p \in [1, \infty]$, defined by

$$
{}_{a}I^{\gamma}_{x}(L^p(D)) = \{ f \in L^p(D) : f = {}_{a}I^{\gamma}_{x} \varphi \text{ for some } \varphi \in L^p(D) \},
$$

and similarly the space ${}_{x}I^{\gamma}_{b}(L^p(D))$. A function in ${}_{a}I^{\gamma}_{x}(L^p(D))$ has the property that its function value and a sufficient number of derivatives vanishing at $x = a$.

Lemma 2.5 *For any $\alpha, \beta \geq 0$, there holds*

$$
{}^{R}_{a}D^{\alpha}_{x}{}^{R}_{a}D^{\beta}_{x} f = {}^{R}_{a}D^{\alpha+\beta}_{x} f, \quad \forall f \in {}_{a}I^{\alpha+\beta}_{x}(L^1(D)).
$$

Proof Since $f \in {}_{a}I^{\alpha+\beta}_{x}(L^1(D))$, $f = {}_{a}I^{\alpha+\beta}_{x} \varphi$ for some $\varphi \in L^1(D)$. Now the desired assertion follows directly from Theorem 2.6(i) below and Theorem 2.1. □

There is no simple formula for the Riemann-Liouville fractional derivative of the product of two functions. In the integer-order case, the Leibniz rule asserts

$$(fg)^{(k)} = \sum_{i=0}^{k} C(k,i) f^{(i)} g^{(k-i)}.$$

This formula is fundamental to many powerful tools in PDE analysis. In contrast, generally, for any $\alpha \in (0,1)$, ${}_{d}^{R}D_x^{\alpha}(fg) \neq ({}_{d}^{R}D_x^{\alpha}f)g + f{}_{d}^{R}D_x^{\alpha}g$, although there is a substantially modified version. Hence, many useful tools derived from this, e.g., integration by parts formula, are either invalid or require substantial modification. This result directly implies

$$_{d}^{R}D_x^{\alpha}((x-a)f) = (x-a)({}_{d}^{R}D_x^{\alpha}f) + C(\alpha,1)({}_{d}^{R}D_x^{\alpha-1}f).$$

Theorem 2.5 *Let f and g be analytic on $\overline{D}$. Then for any $\alpha > 0$ and $\beta \in \mathbb{R}$,*

$$_{d}^{R}D_x^{\alpha}(fg) = \sum_{k=0}^{\infty} C(\alpha,k)({}_{d}^{R}D_x^{\alpha-k}f)g^{(k)},$$

$$_{d}^{R}D_x^{\alpha}(fg) = \sum_{k=-\infty}^{\infty} C(\alpha,k+\beta)({}_{d}^{R}D_x^{\alpha-\beta-k}f)({}_{d}^{R}D_x^{\beta+k}g).$$

Proof Since f,g are analytic, the product $h = fg$ is also analytic. Hence, there exists a power series representation of h based at x,

$$h(s) = \sum_{k=0}^{\infty} (-1)^k \frac{h^{(k)}(x)}{k!}(x-s)^k,$$

and then applying the operator $_{d}^{R}D_x^{\alpha}$ termwise (which is justified by the uniform convergence of the corresponding series), we obtain for $n-1 < \alpha \leq n, n \in \mathbb{N}$,

$$\begin{aligned}
_{d}^{R}D_x^{\alpha}h(x) &= \frac{1}{\Gamma(n-\alpha)} \frac{d^n}{dx^n} \int_a^x (x-s)^{n-\alpha-1} \sum_{k=0}^{\infty} \frac{(-1)^k (x-s)^k}{\Gamma(k+1)} h^{(k)}(x) ds \\
&= \sum_{k=0}^{\infty} \frac{d^n}{dx^n} \frac{(-1)^k h^{(k)}(x)}{\Gamma(n-\alpha)\Gamma(k+1)} \int_a^x (x-s)^{k+n-\alpha-1} ds \\
&= \sum_{k=0}^{\infty} \frac{d^n}{dx^n} \frac{(-1)^k (x-a)^{k+n-\alpha} h^{(k)}(x)}{\Gamma(n-\alpha)\Gamma(k+1)(k+n-\alpha)}.
\end{aligned}$$

Next we simplify the constant on the right-hand side. By the identity (2.2),

$$\frac{1}{\Gamma(n-\alpha)\Gamma(k+1)(k+n-\alpha)} = \frac{C(k+n-\alpha-1,k)}{\Gamma(k+n-\alpha+1)}.$$

Now by the combinatorial identity (2.11), we further deduce

$$\frac{(-1)^k}{\Gamma(n-\alpha)\Gamma(k+1)(k+n-\alpha)} = \frac{C(\alpha-n,k)}{\Gamma(k+n-\alpha+1)}.$$

Consequently,

$$^R_a D^\alpha_x h(x) = \sum_{k=0}^\infty C(\alpha-n,k) \frac{d^n}{dx^n} \frac{(x-a)^{k+n-\alpha} h^{(k)}(x)}{\Gamma(n+k-\alpha+1)}.$$

Then by the standard Leibniz rule, we obtain

$$^R_a D^\alpha_x h(x) = \sum_{k=0}^\infty C(\alpha-n,k) \sum_{j=0}^n C(n,j) \frac{(x-a)^{j+k-\alpha} h^{(k+j)}(x)}{\Gamma(j+k-\alpha+1)}$$

$$= \sum_{k=0}^\infty \sum_{j=0}^\infty C(\alpha-n,k) C(n,j) \frac{(x-a)^{j+k-\alpha} h^{(k+j)}(x)}{\Gamma(j+k-\alpha+1)}.$$

Now introducing a new variable $\ell = j + k$ of summation and changing the order of summation lead to

$$^R_a D^\alpha_x h(x) = \sum_{\ell=0}^\infty \left(\sum_{k=0}^\ell C(\alpha-n,k) C(n,\ell-k) \right) \frac{(x-a)^{\ell-\alpha} h^{(\ell)}(x)}{\Gamma(\ell-\alpha+1)}$$

$$= \sum_{\ell=0}^\infty C(\alpha,\ell) \frac{(x-a)^{\ell-\alpha} h^{(\ell)}(x)}{\Gamma(\ell-\alpha+1)}, \tag{2.26}$$

where we have used the identity (2.13). Applying the classical Leibniz rule, interchanging of summation order and the identity (2.12) give

$$^R_a D^\alpha_x (fg) = \sum_{k=0}^\infty g^{(k)}(x) \sum_{j=0}^\infty C(\alpha, k+j) C(k+j,k) \frac{(x-a)^{k+j-\alpha}}{\Gamma(k+j+1-\alpha)} f^{(j)}(x)$$

$$= \sum_{k=0}^\infty C(\alpha,k) g^{(k)}(x) \left(\sum_{j=0}^\infty C(\alpha-k,j) \frac{(x-a)^{k+j-\alpha}}{\Gamma(k+j+1-\alpha)} f^{(j)}(x) \right).$$

Now using the identity (2.26) completes the proof of the first assertion. The second assertion can be proved in a similar but more tedious manner, and thus we refer to [SKM93, Section 15] or [Osl70] for a complete proof. □

The next result gives the fundamental theorem of calculus for $^R_a D^\alpha_x$: it is the left inverse of $_a I^\alpha_x$ on $L^1(D)$, but generally not a right inverse.

Theorem 2.6 *Let $\alpha > 0$, $n - 1 < \alpha \le n$, $n \in \mathbb{N}$. Then*

(i) *For any $f \in L^1(D)$, $^R_a D^\alpha_x (_a I^\alpha_x f) = f$.*
(ii) *If $_a I^{n-\alpha}_x f \in AC^n(\overline{D})$, then*

$$_aI_x^{\alpha}{}_d^R D_x^{\alpha} f = f - \sum_{j=0}^{n-1} {}_d^R D_x^{\alpha-j-1} f(a^+) \frac{(x-a)^{\alpha-j-1}}{\Gamma(\alpha-j)}.$$

Proof (i) If $f \in L^1(D)$, by Theorem 2.2(i), $_aI_x^{\alpha} f \in L^1(D)$, and by Theorem 2.1,

$$_aI_x^{n-\alpha}{}_aI_x^{\alpha} f = {}_aI_x^n f \in AC^n(\overline{D}).$$

Upon differentiating, we obtain the desired assertion.

(ii) If $_aI_x^{n-\alpha} f \in AC^n(\overline{D})$, by the characterization of the space $AC^n(\overline{D})$ (cf. Theorem A.1 in the appendix), we deduce

$$_aI_x^{n-\alpha} f = \sum_{j=0}^{n-1} c_j \frac{(x-a)^j}{\Gamma(j+1)} + {}_aI_x^n \varphi_f, \tag{2.27}$$

for some $\varphi_f \in L^1(D)$, with $c_j = {}_d^R D_x^{j-n+\alpha} f(a^+)$. Consequently,

$$_aI_x^{\alpha}{}_d^R D_x^{\alpha} f = {}_aI_x^{\alpha} \varphi_f. \tag{2.28}$$

Now applying $_aI_x^{\alpha}$ to both sides of (2.27) and using Theorem 2.1, we obtain

$$_aI_x^n f = \sum_{j=0}^{n-1} c_j \frac{(x-a)^{\alpha+j}}{\Gamma(\alpha+j+1)} + {}_aI_x^{\alpha+n} \varphi_f.$$

Differentiating both sides n times gives

$$f = \sum_{j=0}^{n-1} c_j \frac{(x-a)^{\alpha+j-n}}{\Gamma(\alpha+j-n+1)} + {}_aI_x^{\alpha} \varphi_f.$$

This together with (2.28) yields assertion (ii). Alternatively, by definition, we have

$$_aI_x^{\alpha R}{}_d D_x^{\alpha} f(x) = \frac{1}{\Gamma(\alpha)} \int_a^x (x-s)^{\alpha-1} \frac{d^n}{ds^n}({}_aI_s^{n-\alpha} f)(s)ds$$

$$= \frac{d}{dx}\left\{ \frac{1}{\Gamma(\alpha+1)} \int_a^x (x-s)^{\alpha} \frac{d^n}{ds^n}({}_aI_s^{n-\alpha} f)(s)ds \right\}.$$

Then applying integration by parts n times to the term in bracket gives

$$\frac{1}{\Gamma(\alpha-n)} \int_0^x (x-s)^{\alpha-n-1} {}_aI_s^{n-\alpha} f(s)ds - \sum_{j=1}^n \frac{d^{n-j}}{ds^{n-j}} {}_aI_s^{n-\alpha} f(s)|_{s=a} \frac{(x-a)^{\alpha-j}}{\Gamma(\alpha+1-j)}.$$

Note that the expression makes sense due to the conditions on $f(x)$.   □

The next result gives a generalization of Theorem 2.6 [DN68, (1.13)–(1.15)].

Proposition 2.1 Let $f \in L^1(D)$.

(i) For $\alpha, \beta \geq 0$, if $^{R}_{a}D^{\alpha-\beta}_{x} f \in L^{1}(D)$, then

$$^{R}_{a}D^{\alpha}_{x}\,_{a}I^{\beta}_{x} f(x) = ^{R}_{a}D^{\alpha-\beta}_{x} f(x).$$

(ii) If $^{R}_{a}D^{\beta}_{x} f \in L^{1}(D)$, with $0 \leq n - 1 \leq \beta \leq n$, then for any $\alpha \geq 0$, there holds

$$_{a}I^{\alpha R}_{x\,a}D^{\beta}_{x} f(x) = ^{R}_{a}D^{\beta-\alpha}_{x} f(x) - \sum_{j=1}^{n} {}^{R}_{a}D^{\beta-j}_{x} f(a)\frac{(x-a)^{\alpha-j}}{\Gamma(\alpha+1-j)}.$$

(iii) If $\alpha, \beta \in (0, 1]$, then

$$^{R}_{a}D^{\alpha R}_{x\,a}D^{\beta}_{x} f(x) = \frac{d}{dx}\left[{}^{R}_{a}D^{\alpha+\beta-1}_{x} f - {}_{a}I^{1-\beta}_{x} f(a)\frac{(x-a)^{-\alpha}}{\Gamma(1-\alpha)} \right],$$

provided that all the terms make sense in $L^{1}(D)$.

Proof Not that for $\beta \geq \alpha$, (i) follows directly from Theorems 2.1 and 2.6(i) by

$$^{R}_{a}D^{\alpha}_{x\,a}I^{\beta}_{x} f(x) = ^{R}_{a}D^{\alpha}_{x\,a}I^{\alpha}_{x\,a}I^{\beta-\alpha}_{x} f(x).$$

Meanwhile, for $\beta < \alpha$, there exist integers $m, n > 0$ such that $\alpha - \beta = m + \theta$, $0 \leq \theta < 1$, $\alpha - m = n - r$, $r \geq 0$. Then the preceding discussion and the definition of the Riemann-Liouville fractional derivative give

$$^{R}_{a}D^{\alpha-\beta}_{x} f(x) = \frac{d^{m}}{dx^{m}}\,_{a}I^{\theta}_{x} f(x) = \frac{d^{m}}{dx^{m}} {}^{R}_{a}D^{\alpha-m}_{x}\,_{a}I^{\beta}_{x} f(x)$$

$$= \frac{d^{m}}{dx^{m}}\frac{d^{n}}{dx^{n}}\,_{a}I^{r}_{x\,a}I^{\beta}_{x} f(x) = ^{R}_{a}D^{\alpha}_{x\,a}I^{\beta}_{x} f(x).$$

The identity in (ii) follows from Theorem 2.6(ii), since $^{R}_{a}D^{\beta}_{x} f(x) = ^{R}_{a}D^{\alpha R}_{x\,a}D^{\beta-\alpha}_{x} f(x)$. Last, (iii) follows from (ii) since by definition

$$^{R}_{a}D^{\alpha R}_{x\,a}D^{\beta}_{x} f(x) = \frac{d}{dx}(_{a}I^{1-\alpha R}_{x\,a}D^{\beta}_{x} f)(x).$$

This completes the proof of the proposition. $\square$

The next result due to Tamarkin [Tam30] is about the inversion of Abel's integral equation. Abel [Abe81, Abe26] solved the integral equation now named after him in connection with the tautochrone problem (corresponding to $\alpha = \frac{1}{2}$), and gave the solution for any $\alpha \in (0, 1)$. The proper history of fractional calculus began with the papers by Abel and Liouville.

Theorem 2.7 Let $g \in L^{1}(D)$. The integral equation with $n - 1 < \alpha < n$, $n \in \mathbb{N}$,

$$g(x) = \frac{1}{\Gamma(\alpha)} \int_{a}^{x} (x-s)^{\alpha-1} f(s)ds, \tag{2.29}$$

has a solution $f \in L^{1}(D)$ if and only if

(i) *The function $\omega(x) = {}_aI_x^{n-\alpha}g \in AC^n(\overline{D})$.*
(ii) $\omega(a) = \omega'(a) = \ldots = \omega^{(n-1)}(a) = 0$.

If conditions (i) and (ii) are satisfied, then the solution to (2.29) is unique and can be represented a.e. on $\overline{D}$ by

$$f(x) = \frac{d^n}{dx^n}\,{}_aI_x^{n-\alpha}g(x). \tag{2.30}$$

Proof If (2.29) has a solution $f \in L^1(D)$, then applying to both sides the integral operator ${}_aI_x^{n-\alpha}$ and then appealing to Theorem 2.1 give $\omega(x) = {}_aI_x^{n-\alpha}g(x) = {}_aI_x^{n-\alpha}{}_aI_x^{\alpha}f(x) = {}_aI_x^n f(x)$. Thus the formula (2.30) and conditions (i) and (ii) hold. Next we prove the sufficiency of the conditions. It follows from (i) that

$$\frac{d^n}{dx^n}\omega(x) = \frac{d^n}{dx^n}\,{}_aI_x^{n-\alpha}g(x) \in L^1(D).$$

Meanwhile, the semigroup property in Theorem 2.1 implies

$${}_aI_x^1 g(x) = {}_aI_x^{\alpha-n+1}\omega(x) = \frac{1}{\Gamma(\alpha-n+1)}\int_a^x (x-s)^{\alpha-n}\omega(s)\mathrm{d}s.$$

By integration by parts and condition (ii), the last integral can be rewritten as

$$\int_a^x g(s)\mathrm{d}s = \frac{1}{\Gamma(\alpha+1)}\int_a^x (x-s)^\alpha \omega^{(n)}\mathrm{d}s.$$

Thus the function $f(x)$ of the form (2.30) satisfies equation

$$\int_a^x g(s)\mathrm{d}s = \frac{1}{\Gamma(\alpha+1)}\int_a^x (x-s)^\alpha f(s)\mathrm{d}s.$$

Thus, the sufficiency part follows. □

We have the following fractional version of integration by parts formula.

Lemma 2.6 *Let $\alpha > 0$, $p, q \geq 1$, and $p^{-1} + q^{-1} \leq 1 + \alpha$ ($p, q \neq 1$ when $p^{-1} + q^{-1} = 1 + \alpha$). Then if $f \in {}_aI_x^{\alpha}(L^p(D))$ and $g \in {}_xI_b^{\alpha}(L^q(D))$, there holds*

$$\int_a^b g(x)\,{}_a^R D_x^\alpha f(x)\,\mathrm{d}x = \int_a^b f(x)\,{}_x^R D_b^\alpha g(x)\,\mathrm{d}x.$$

Proof Since $f \in {}_aI_x^{\alpha}(L^p(D))$, $f = {}_aI_x^{\alpha}\varphi_f$ for some $\varphi_f \in L^p(D)$, and similarly $g = {}_xI_b^{\alpha}\varphi_g$ for some $\varphi_g \in L^q(D)$. By Theorem 2.6(i), it suffices to show

$$\int_a^b \varphi_f(x)\,{}_xI_b^{\alpha}\varphi_g(x)\,\mathrm{d}x = \int_a^b {}_aI_x^{\alpha}\varphi_f\varphi_g(x)\,\mathrm{d}x,$$

which holds in view of Lemma 2.3. □

The next result gives the Laplace transform of the Riemann-Liouville fractional integral and derivatives. See Section A.3.1 of the appendix for Laplace transform. The Laplace transform $\mathcal{L}[{}_0^R D_x^\alpha f]$ involves ${}_0^R D_x^{\alpha+k-n} f(0^+)$. Thus applying Laplace transform to solve relevant FDES requires such initial conditions.

Lemma 2.7 *Let* $\alpha > 0$, $f \in L^1(0, b)$ *for any* $b > 0$, *and* $|f(x)| \le ce^{p_0 x}$, *for all* $x > b > 0$, *for some constant* c *and* $p_0 > 0$.

(i) *The relation* $\mathcal{L}[{}_0 I_x^\alpha f](z) = z^{-\alpha} \mathcal{L}[f](z)$ *holds for any* $\mathfrak{R}(z) > p_0$

(ii) *If* $n-1 < \alpha \le n, n \in \mathbb{N}, f \in AC^n[0, b]$ *for any* $b > 0$, *and* $\lim_{x \to \infty} {}_0^R D_x^{\alpha+j-n} f(x) = 0$, $j = 0, 1, \ldots, n-1$. *Then for any* $\mathfrak{R}(z) > p_0$, *there holds*

$$\mathcal{L}[{}_0^R D_x^\alpha f](z) = z^\alpha \mathcal{L}[f](z) - \sum_{j=0}^{n-1} z^{n-j-1} ({}_0^R D_x^{\alpha+j-n} f)(0^+).$$

Proof The condition on f ensures that the Laplace transform is well defined. We write ${}_0 I_x^\alpha$ in a convolution form as ${}_0 I_x^\alpha f(x) = (\omega_\alpha * f)(x)$, with $\omega_\alpha(x) = \Gamma(\alpha)^{-1} \max(0, x)^{\alpha-1}$. Since $\mathcal{L}[\omega_\alpha] = z^{-\alpha}$ (cf. Example A.2), part (i) follows from the convolution rule (A.8) for Laplace transform. Likewise, ${}_0^R D_x^\alpha f(x) = g^{(n)}(x)$ with $g(x) = {}_0 I_x^{n-\alpha} f(x)$, Then $\mathcal{L}[g](z) = z^{\alpha-n} \mathcal{L}[f](z)$. By the convolution rule (A.8),

$$\mathcal{L}[{}_0^R D_x^\alpha f(x)](z) = \mathcal{L}[g^{(n)}(x)] = z^n \mathcal{L}[g](z) - \sum_{j=0}^{n-1} z^{n-j-1} g^{(j)}(0^+)$$

$$= z^\alpha \mathcal{L}[f](z) - \sum_{j=0}^{n-1} z^{n-j-1} {}_0^R D_x^{\alpha+j-n} f(0^+),$$

which shows the desired formula. $\square$

A similar relation holds for the Fourier transform (see Appendix A.3.2) of the Liouville fractional integrals ${}_{-\infty} I_x^\alpha f$ and ${}_x I_\infty^\alpha f$ and derivatives ${}_{-\infty}^R D_x^\alpha f$ and ${}_x^R D_\infty^\alpha f$.

Theorem 2.8 *For* $0 < \alpha < 1$ *and* f *sufficiently regular, the following relations hold*

$$\mathcal{F}[{}_{-\infty} I_x^\alpha f](\xi) = (i\xi)^{-\alpha} \mathcal{F}[f](\xi) \quad and \quad \mathcal{F}[{}_x I_\infty^\alpha f](\xi) = (-i\xi)^{-\alpha} \mathcal{F}[f](\xi), \quad (2.31)$$

where the notation $(\pm i\xi)^\alpha$, $\xi \in \mathbb{R}$, *denotes* $|\xi|^\alpha e^{\pm\alpha\pi i \frac{\text{sign}(\xi)}{2}}$. *Similarly, for any* $\alpha > 0$, *the following relations hold for the Liouville fractional derivatives:*

$$\mathcal{F}[{}_{-\infty}^R D_x^\alpha f](\xi) = (i\xi)^\alpha \mathcal{F}[f](\xi) \quad and \quad \mathcal{F}[{}_x^R D_\infty^\alpha f](\xi) = (-i\xi)^\alpha \mathcal{F}[f](\xi). \quad (2.32)$$

Proof Note that

$$\mathcal{F}[_{-\infty}I_x^\alpha f](\xi) = \frac{1}{\Gamma(\alpha)} \int_{-\infty}^{\infty} e^{-i\xi x} \int_{-\infty}^{x} (x-t)^{\alpha-1} f(t) dt dx$$

$$= \frac{1}{\Gamma(\alpha)} \frac{d}{d\xi} \int_{-\infty}^{\infty} \frac{1-e^{-i\xi x}}{ix} \int_{-\infty}^{x} (x-t)^{\alpha-1} f(t) dt dx$$

$$= \frac{1}{\Gamma(\alpha)} \frac{d}{d\xi} \int_{-\infty}^{\infty} f(t) \int_{t}^{\infty} \frac{1-e^{-i\xi x}}{ix} (x-t)^{\alpha-1} dx dt.$$

The interchange of order of integration is made possible by Fubini's theorem. Further, after differentiation, we obtain

$$\mathcal{F}[_{-\infty}I_x^\alpha f](\xi) = \frac{1}{\Gamma(\alpha)} \int_{-\infty}^{\infty} f(t) \int_{t}^{\infty} e^{-ix\xi} (x-t)^{\alpha-1} dx dt$$

$$= \frac{1}{\Gamma(\alpha)} \int_{-\infty}^{\infty} f(t) e^{-i\xi t} dt \int_{0}^{\infty} e^{-i\xi x} x^{\alpha-1} dx = \frac{\mathcal{F}[f](\xi)}{\Gamma(\alpha)} \int_{0}^{\infty} x^{\alpha-1} e^{-ix\xi} dx.$$

Hence, we obtain the desired assertion for $_{-\infty}I_x^\alpha$ from the following identity

$$\frac{1}{\Gamma(\alpha)} \int_{0}^{\infty} x^{\alpha-1} e^{-ix\xi} dx = (i\xi)^{-\alpha}, \quad \forall \alpha \in (0,1).$$

Indeed, the condition $\alpha \in (0,1)$ provides the convergence of the left-hand side integral at infinity. Then with the substitution $ix\xi = s$,

$$\int_{0}^{\infty} x^{\alpha-1} e^{-ix\xi} dx = (i\xi)^{-\alpha} \int_{\mathcal{L}} s^{\alpha-1} e^{-s} ds,$$

where $\mathcal{L}$ is the imaginary half-axis $(0, i\infty)$ for $\xi > 0$ and the half-axis $(-i\infty, 0)$ for $\xi < 0$. Since e^{-s} exponentially vanishes in the right half plane as $|\xi| \to \infty$, we have $\int_{\mathcal{L}} s^{\alpha-1} e^{-s} ds = \int_{0}^{\infty} s^{\alpha-1} e^{-s} ds = \Gamma(\alpha)$ by the Cauchy integral theorem. The other assertions follow in a similar manner, and thus the proof is omitted. $\qquad\square$

Remark 2.3 For $\alpha \in (0,1)$, $f \in L^1(\mathbb{R})$ is sufficient for the identities in (2.31). For $\alpha \geq 1$, the left-hand side in may not exist even for very smooth functions, e.g., $f \in C_0^\infty(\mathbb{R})$. Indeed, if $\alpha = 1$, then $_{-\infty}I_x^1 f(x) = \int_{-\infty}^{x} f(t) dt$, so it tends to a constant as $x \to +\infty$, and thus the Fourier transform $\mathcal{F}[_{-\infty}I_x f]$ does not exist in the usual sense. In contrast, (2.32) is valid for all sufficiently smooth functions, e.g., those which are differentiable up to order n, and vanish sufficiently rapidly at infinity together with their derivatives. One may verify this by writing $_{-\infty}^R D_x^\alpha f$ as $_{-\infty}^R D_x^\alpha f = {}_{-\infty}I_x^{n-\alpha} f^{(n)}$ and then applying (2.31).

The next nonnegativity result, direct from Theorem 2.8, is useful.

Proposition 2.2 *For any $f \in L^p(D)$, with $p \geq 2(1+\alpha)^{-1}$, and $\alpha \in (0,1)$, there holds*

$$\int_{a}^{b} (_aI_x^\alpha f)(x) f(x) dx \geq \cos\frac{\alpha\pi}{2} \|_aI_x^{\frac{\alpha}{2}} f\|_{L^2(D)}^2.$$

Proof Let $\bar{f}$ be the zero extension of f to $\mathbb{R}$. Then by Theorem 2.1 and the regularity assumption on f (cf. Lemma 2.3)

$$\int_a^b (_aI_x^\alpha f)f\,\mathrm{d}x = \int_{-\infty}^\infty (_{-\infty}I_x^\alpha \bar{f})\bar{f}\,\mathrm{d}x = \int_{-\infty}^\infty (_{-\infty}I_x^{\frac{\alpha}{2}} \bar{f})(_xI_\infty^{\frac{\alpha}{2}} \bar{f})\,\mathrm{d}x.$$

Then by Theorem 2.2(ii), Theorem 2.8 and Parseval's theorem

$$\int_a^b (_aI_x^\alpha f)f\,\mathrm{d}x = \frac{1}{2\pi} \int_{-\infty}^\infty (i\xi)^{-\alpha}|\widetilde{f}(\xi)|^2\,\mathrm{d}\xi = \frac{\cos\frac{\alpha\pi}{2}}{2\pi} \int_{-\infty}^\infty |\xi|^{-\alpha}|\widetilde{f}(\xi)|^2\,\mathrm{d}\xi.$$

Then the desired result follows from Theorem 2.8.   □

Now we discuss mapping properties in Sobolev spaces [JLPR15, Theorems 2.1 and 3.1]. See also [BLNT17] for relevant results in the space of bounded variations. These properties are important for the study of related BVPs in Chapter 5. We use the spaces $H_0^s(D)$ and $H_{0,L}^s(D)$, etc. defined in Section A.2.3 in the appendix. The following smoothing property holds for the operators $_aI_x^\alpha$ and $_xI_b^\alpha$.

Theorem 2.9 *For any $s \geq 0$ and $0 < \alpha < 1$, the operators $_aI_x^\alpha$ and $_xI_b^\alpha$ are bounded from $H_0^s(D)$ into $H_{0,L}^{s+\alpha}(D)$ and $H_{0,R}^{s+\alpha}(D)$, respectively.*

Proof It suffices to prove the result for $D = (0,1)$. The key idea of the proof is to extend $f \in H_0^s(D)$ to a function $\bar{f} \in H_0^s(0,2)$ whose moments up to $(k-1)$th order vanish with $k > \alpha - \frac{1}{2}$. To this end, we employ orthogonal polynomials $\{p_0, p_1, \ldots, p_{k-1}\}$ with respect to the inner product $\langle \cdot, \cdot \rangle$ defined by

$$\langle f, g \rangle = \int_1^2 ((x-1)(2-x))^\ell f(x)g(x)\,\mathrm{d}x,$$

where the integer ℓ satisfies $\ell > s - \frac{1}{2}$ so that $((x-1)(2-x))^\ell p_i \in H_0^s(1,2)$, $i = 0, \ldots, k-1$. Then we set $w_j = \gamma_j((x-1)(2-x))^\ell p_j$ with γ_j chosen so that

$$\int_1^2 w_j p_j\,\mathrm{d}x = 1 \quad \text{so that} \quad \int_1^2 w_j p_\ell\,\mathrm{d}x = \delta_{j,\ell}, \quad j,\ell = 0, \ldots, k-1,$$

where $\delta_{j,\ell}$ is the Kronecker symbol. Next we extend both f and w_j, $j = 0, \ldots, k-1$ by zero to $(0,2)$ by setting

$$f_e = f - \sum_{j=0}^{k-1} \left(\int_0^1 f p_j\,\mathrm{d}x \right) w_j.$$

The resulting function f_e has vanishing moments for $j = 0, \ldots, k-1$ and by construction it is in the space $H_0^s(0,2)$. Further, obviously the inequality $\|f_e\|_{L^2(0,2)} \leq c\|f\|_{L^2(D)}$ holds, i.e., the extension is bounded in $L^2(D)$. We denote by $\bar{f}_e$ the extension of f_e to $\mathbb{R}$ by zero. Now for $x \in (0,1)$, we have $(_0I_x^\alpha f)(x) = (_{-\infty}I_x^\alpha \bar{f}_e)(x)$,

By Theorem 2.8, we have $\mathcal{F}[_{-\infty}I_x^\alpha \bar{f}_e](\xi) = (i\xi)^{-\alpha}\mathcal{F}(\bar{f}_e)(\xi)$ and hence by Plancherel's identity (A.9),

$$\|_{-\infty}I_x^\alpha \bar{f}_e\|_{L^2(\mathbb{R})}^2 = \frac{1}{2\pi}\int_{\mathbb{R}}|\xi|^{-2\alpha}|\mathcal{F}(\bar{f}_e)(\xi)|^2\,d\xi.$$

By Taylor expansion centered at 0, there holds

$$e^{-i\xi x} - 1 - (-i\xi)x - \cdots - \frac{(-i\xi)^{k-1}x^{k-1}}{(k-1)!}$$

$$= \frac{(-i\xi x)^k}{k!} + \frac{(-i\xi x)^{k+1}}{(k+1)!} + \frac{(-i\xi x)^{k+2}}{(k+2)!} + \cdots = (-i\xi)^k {}_0I_x^k(e^{-i\xi x}).$$

Clearly, there holds

$$|{}_0I_x^k(e^{-i\xi x})| \le {}_0I_x^k(1) = (k!)^{-1}x^k, \quad \forall x \in (0,1).$$

Since the first k moments of $\bar{f}_e$ vanish, multiplying the identity by $\bar{f}_e$ and integrating over $\mathbb{R}$ gives

$$\mathcal{F}(\bar{f}_e)(\xi) = (-i\xi)^k \int_{\mathbb{R}} {}_0I_x^k(e^{-i\xi x})\bar{f}_e(x)\,dx,$$

and upon noting $\text{supp}(\bar{f}_e) \subset (0,2)$, we deduce

$$|\mathcal{F}(\bar{f}_e)(\xi)| \le 2^k(k!)^{-1}|\xi|^k\|f_e\|_{L^2(0,2)} \le c|\xi|^k\|f\|_{L^2(D)}.$$

We then have

$$\|_{-\infty}I_x^\alpha \bar{f}_e\|_{H^{\alpha+s}(\mathbb{R})}^2 = \int_{\mathbb{R}}(1+|\xi|^2)^{\alpha+s}|\xi|^{-2\alpha}|\mathcal{F}(\bar{f}_e)|^2 d\xi$$

$$\le c_1\|f\|_{L^2(D)}\int_{|\xi|<1}|\xi|^{-2\alpha+2k}d\xi + c_2\int_{|\xi|>1}|\xi|^{2s}|\mathcal{F}(\bar{f}_e)|^2 d\xi \le c\|f\|_{\tilde{H}^s(D)}.$$

The desired assertion follows from this and the inequality

$$\|{}_0I_x^\alpha f\|_{H^{\alpha+s}(D)} \le \|_{-\infty}I_x^\alpha \bar{f}_e\|_{H^{\alpha+s}(\mathbb{R})}^2.$$

This completes the proof of the theorem.   □

Remark 2.4 The restriction $\alpha \in (0,1)$ in Theorem 2.9 stems from Theorem 2.8: for $\alpha \ge 1$, $_{-\infty}I_x^\alpha f$ has to be understood as a distribution, due to the slow decay at infinity; see [SKM93, Section 2.8]. Nonetheless, $_0I_x^\alpha$ is bounded from $H_{0,L}^s(D)$ to $H_{0,L}^{\alpha+s}(D)$, and $_xI_1^\alpha$ is bounded from $H_{0,R}^s(D)$ to $H_{0,R}^{\alpha+s}(D)$, and repeatedly applying Theorem 2.9 shows that the statement actually holds for any $\alpha \ge 0$.

We have the following immediate corollary.

Corollary 2.1 *For $\gamma \ge 0$, the functions $(x-a)^\gamma$ and $(b-x)^\gamma$ belong to $H_{0,L}^\alpha(D)$ and $H_{0,R}^\alpha(D)$, respectively, for any $0 \le \alpha < \gamma + \frac{1}{2}$.*

Proof Since $(x - a)^\gamma = c_\gamma \, _aI_x^\gamma(1)$ and $(b - x)^\gamma = c_\gamma \, _xI_b^\gamma(1)$, the assertion follows from Theorem 2.9 and the fact that $1 \in H_{0,L}^\delta(D)$ and $1 \in H_{0,R}^\delta(D)$ for any $\delta \in [0, \frac{1}{2})$. $\qquad\square$

The next result gives the mapping property of the operators $_0^RD_x^\alpha$ and $_x^RD_1^\alpha$.

Theorem 2.10 *For any $\alpha \in (n - 1, n)$, the operators $_a^RD_x^\alpha f$ and $_x^RD_b^\alpha f$ defined for $f \in C_0^\infty(D)$ extend continuously from $H_0^\alpha(\Omega)$ to $L^2(D)$.*

Proof We consider the left-sided case, since the right side case follows similarly. For $f \in C_0^\infty(\mathbb{R})$, Theorem 2.8 and Plancherel's identity (A.9) imply

$$\| _{-\infty}^R D_x^\alpha f\|_{L^2(\mathbb{R})} = (2\pi)^{-\frac{1}{2}} \|\mathcal{F}\big(_{-\infty}^R D_x^\alpha f\big)\|_{L^2(\mathbb{R})} \le c\|f\|_{H^\alpha(\mathbb{R})}.$$

Thus, we can continuously extend $_{-\infty}^R D_x^\alpha$ to an operator from $H^\alpha(\mathbb{R})$ into $L^2(\mathbb{R})$. Note that for $f \in C_0^\infty(D)$, there holds (with $\bar{f}$ the zero extension of f to $\mathbb{R}$)

$$_a^R D_x^\alpha f = \, _{-\infty}^R D_x^\alpha \bar{f}|_D. \tag{2.33}$$

By definition, $f \in H_0^\alpha(D)$ implies $\bar{f} \in H^\alpha(\mathbb{R})$, and hence $\| _{-\infty}^R D_x^\alpha \bar{f}\|_{L^2(\mathbb{R})} \le c\|f\|_{H_0^\alpha(D)}$. Thus, formula (2.33) provides an extension of the operator $_a^R D_x^\alpha$ defined on $C_0^\infty(D)$ to a bounded operator from the space $H_0^\alpha(D)$ into $L^2(D)$. $\qquad\square$

The next result slightly relaxes the condition in Theorem 2.10.

Corollary 2.2 *The operator $_a^R D_x^\alpha f$ defined for $f \in \widetilde{C}_0^\alpha(\overline{D}) = \{v \in C^n(\overline{D}) : v^{(j)}(0) = 0, j = 0, \ldots, [\alpha - \frac{1}{2}]\}$ extends continuously from $H_{0,L}^\alpha(D)$ to $L^2(D)$.*

Proof It suffices to consider $D = (0, 1)$. For any $f \in H_{0,L}^\alpha(D)$, let f_e be a bounded extension of f to $H_0^\alpha(0, 2)$ and $\bar{f}_e$ be its extension by zero to $\mathbb{R}$, and then set $_a^R D_x^\alpha f = \, _a^R D_x^\alpha \bar{f}_e|_D$. Note that $_a^R D_x^\alpha f$ is independent of the extension f_e and coincides with the formal definition of $_a^R D_x^\alpha f$ when $f \in \widetilde{C}_0^\alpha(\overline{D})$. Obviously we have

$$\| _a^R D_x^\alpha f\|_{L^2(D)} \le \| _{-\infty}^R D_x^\alpha \bar{f}_e\|_{L^2(\mathbb{R})} \le c\|f\|_{H_{0,L}^\alpha(D)}.$$

This completes the proof. $\qquad\square$

Last, we give an alternative representation of $_a^R D_x^\alpha f$. It is often called the left-sided Marchaud fractional derivative.

Lemma 2.8 *If $f \in C^{0,\gamma}(\overline{D})$ for some $\gamma \in (0, 1]$, then for any $\alpha \in (0, \gamma)$,*

$$_a^R D_x^\alpha f(x) = \frac{(x - a)^{-\alpha}}{\Gamma(1 - \alpha)} f(x) + \frac{\alpha}{\Gamma(1 - \alpha)} \int_a^x (x - s)^{-\alpha-1}(f(x) - f(s)) \, ds.$$

Proof Note the identity

$$_aI_x^{1-\alpha} f(x) = \frac{(x - a)^{1-\alpha}}{\Gamma(2 - \alpha)} f(x) + \frac{1}{\Gamma(1 - \alpha)} \int_a^x (x - s)^{-\alpha}(f(s) - f(x)) \, ds.$$

The derivative of the integral is given by

$$\lim_{s \to x^-} \frac{f(s) - f(x)}{(x-s)^\alpha} - \alpha \int_a^x (x-s)^{-\alpha-1}(f(s) - f(x))\,ds - \frac{(x-a)^{1-\alpha}}{1-\alpha} f'(x).$$

Since $f \in C^{0,\gamma}(\overline{D})$, and $\alpha < \gamma$, we deduce $\lim_{s \to x^-}(x-s)^{-\alpha}(f(s) - f(x)) = 0$. Then collecting the terms gives desired identity. $\qquad\square$

This result can be used to derive an extremal principle for the Riemann-Liouville fractional derivative at an extremum; see Exercise 2.16.

2.3.2 Djrbashian-Caputo fractional derivative

Now we turn to the Djrbashian-Caputo fractional derivative, often known as the Caputo fractional derivative in the engineering and physics literature. It is one of the most popular fractional derivatives for time-fractional problems due to its ability to handle initial conditions in a physically transparent way. The possibility of this derivative was known from the nineteenth century. Such a form can even be found in a paper by Joseph Liouville himself [Lio32, p. 10, formula (B)], but Liouville, not recognizing its role, disregarded this notion. It was extensively studied by the Armenian mathematician Mkhitar M. Djrbashian (also spelt as Dzhrbashyan, Dzsrbasjan, Džarbašjan, and Džrbašjan, etc., translated from Russian) in the late 1950s and his students (see the monograph [Dzh66] (in Russian) for a summary; see also [Djr93]). The Italian geophysicist Michele Caputo rediscovered this version in 1967 [Cap67] as a tool for understanding seismological phenomena, and later together with Francesco Mainardi in viscoelasticity [CM71a, CM71b].

Definition 2.3 For $f \in L^1(D)$ and $n - 1 < \alpha \le n$, $n \in \mathbb{N}$, its left-sided Djrbashian-Caputo fractional derivative ${}^C_a D^\alpha_x f$ of order α based at $x = a$ is defined by

$$\prescript{C}{a}{D}^\alpha_x f(x) := ({}_a I^{n-\alpha}_x f^{(n)})(x) = \frac{1}{\Gamma(n-\alpha)} \int_a^x (x-s)^{n-\alpha-1} f^{(n)}(s)\,ds,$$

if the integral on the right-hand side exists. Likewise, its right-sided Djrbashian-Caputo fractional derivative ${}^C_x D^\alpha_b f$ of order α based at $x = b$ is defined by

$$\prescript{C}{x}{D}^\alpha_b f(x) := (-1)^n ({}_x I^{n-\alpha}_b f^{(n)})(x) = \frac{(-1)^n}{\Gamma(n-\alpha)} \int_x^b (s-x)^{n-\alpha-1} f^{(n)}(s)\,ds,$$

if the integral on the right-hand side exists.

The Djrbashian-Caputo fractional derivative is more restrictive than the Riemann-Liouville one, since the definition requires $f \in AC^n(\overline{D})$, which is always implicitly assumed when using this definition of the fractional derivative. Note that the limits of ${}^C_a D^\alpha_x f$ as α approaches $(n-1)^+$ and n^- are not quite as expected. Actually, by integration by parts, for sufficiently regular f, we deduce

$$
\begin{aligned}
{}^{C}_{a}D^{\alpha}_{x} f(x) &\to ({}_{a}I^{1}_{x} f^{(n)})(x) = f^{(n-1)}(x) - f^{(n-1)}(a^{+}) \quad \text{as } \alpha \to (n-1)^{+}, \\
{}^{C}_{a}D^{\alpha}_{x} f(x) &\to f^{(n)}(x) \quad \text{as } \alpha \to n^{-}.
\end{aligned}
\tag{2.34}
$$

The first limit is direct, and similarly, for $f \in C^{n+1}(\overline{D})$, integration by parts gives

$$
\begin{aligned}
{}^{C}_{a}D^{\alpha}_{x} f(x) &= \frac{1}{\Gamma(n-\alpha)} \int_{a}^{x} (x-s)^{n-\alpha-1} f^{(n)}(s)\,ds \\
&= \frac{f^{(n)}(a)(x-a)^{n-\alpha}}{\Gamma(n+1-\alpha)} + {}_{a}I^{n+1-\alpha}_{x} f^{(n+1)}(x) \to f^{(n)}(x) \quad \text{as } \alpha \to n^{-}.
\end{aligned}
$$

For $n-1 < \alpha \le n$, we compute ${}^{C}_{a}D^{\alpha}_{x}(x-a)^{\gamma}$, $\gamma > n-1$, to gain some insight. The condition $\gamma > n-1$ ensures that the nth derivative inside the integral is integrable, so that the operator ${}_{0}I^{n-\alpha}_{x}$ can be applied. Then direct computation shows

$$
{}^{C}_{a}D^{\alpha}_{x}(x-a)^{\gamma} = \frac{\Gamma(\gamma+1)}{\Gamma(\gamma+1-\alpha)}(x-a)^{\gamma-\alpha}.
$$

For the case $\gamma \le n-1$, ${}^{C}_{a}D^{\alpha}_{x}(x-a)^{\gamma}$ is generally undefined, except for $\gamma = 0, 1, \ldots, n-1$, for which it vanishes identically, i.e.,

$$
{}^{C}_{a}D^{\alpha}_{x}(x-a)^{j} = 0, \quad j = 0, 1, \ldots, n-1.
$$

Thus, for $\alpha \in (0,1)$, $f(x) \equiv 1$ lies in the kernel of ${}^{C}_{a}D^{\alpha}_{x}$, exactly as the first-order derivative, which contrasts sharply with the Riemann-Liouville case. The following result holds. The proof is identical with Theorem 2.4 and hence omitted.

Theorem 2.11 *If $n-1 < \alpha \le n$, $n \in \mathbb{N}$, $f \in AC^{n}(\overline{D})$, and satisfies ${}^{C}_{a}D^{\alpha}_{x} f(x) = 0$ in D, then it can be represented by $f(x) = \sum_{k=0}^{n-1} \frac{f^{(k)}(a^{+})}{k!}(x-a)^{k}$.*

Example 2.4 Let us repeat Example 2.2 for the Djrbashian-Caputo fractional derivative. With $f(x) = e^{\lambda x}$, $\lambda \in \mathbb{R}$, for $n-1 < \alpha \le n$, $n \in \mathbb{N}$,

$$
\begin{aligned}
{}^{C}_{0}D^{\alpha}_{x} f(x) &= {}^{C}_{0}D^{\alpha}_{x} \sum_{k=0}^{\infty} \frac{\lambda^{k} x^{k}}{k!} = \sum_{k=0}^{\infty} \lambda^{k} \frac{{}^{C}_{0}D^{\alpha}_{x} x^{k}}{k!} = \sum_{k=n}^{\infty} \lambda^{k} \frac{x^{k-\alpha}}{\Gamma(k-\alpha+1)} \\
&= \lambda^{n} x^{n-\alpha} \sum_{k=0}^{\infty} \frac{(\lambda x)^{k}}{\Gamma(k+n-\alpha+1)} = \lambda^{n} x^{n-\alpha} E_{1,n-\alpha+1}(\lambda x).
\end{aligned}
$$

This result is completely different from that in Example 2.2, except the case $\alpha = 1$.

For the operator ${}^{C}_{a}D^{\alpha}_{x}$, neither the composition rule nor the product rule holds, and there is also no known analogue of Theorem 2.5. We leave the details to the exercises. Generally, we have $({}^{R}_{a}D^{\alpha}_{x} f)(x) \ne ({}^{C}_{a}D^{\alpha}_{x} f)(x)$, even when both fractional derivatives are defined: for $f(x) \equiv 1$, ${}^{C}_{a}D^{\alpha}_{x} f(x) = 0$, but ${}^{R}_{a}D^{\alpha}_{x} f(x)$ is nonzero by (2.23). Nonetheless, they are closely related to each other.

Theorem 2.12 *Let $n - 1 < \alpha \le n$, $n \in \mathbb{N}$, and $f \in AC^n(\overline{D})$. Then for the left-sided fractional derivatives, there holds*

$$(^R_aD^\alpha_x f)(x) = (^C_aD^\alpha_x f)(x) + \sum_{j=0}^{n-1} \frac{(x-a)^{j-\alpha}}{\Gamma(j-\alpha+1)} f^{(j)}(a^+), \qquad (2.35)$$

and likewise for the right-sided fractional derivatives, there holds

$$(^R_xD^\alpha_b f)(x) = (^C_xD^\alpha_b f)(x) + \sum_{j=0}^{n-1} \frac{(-1)^j(b-x)^{j-\alpha}}{\Gamma(j-\alpha+1)} f^{(j)}(b^-).$$

Proof Let D denote taking the usual first-order derivative. Then we claim that for any $\beta > 0$, $f \in AC(\overline{D})$, there holds $(D_aI^\beta_x - {_aI^\beta_x}D)f(x) = \Gamma(\beta)^{-1}f(a)(x-a)^{\beta-1}$. Indeed, by the fundamental theorem of calculus, $_aI^1_x Df(x) = f(x) - f(a)$, i.e., $f(x) = {_aI^1_x}Df(x) + f(a)$. Thus, by applying the operator $_aI^\beta_x$, $\beta > 0$, to both sides and using Theorem 2.1, we deduce $_aI^\beta_x f(x) = {_aI^{\beta+1}_x}Df(x) + f(a)(_aI^\beta_x 1)(x)$, which upon differentiation yields the desired claim. The case $\alpha \in (0, 1)$ is trivial by setting $\beta = 1 - \alpha$ in the claim. Now for $\alpha \in (1, 2)$, the claim with $\beta = 2 - \alpha$ yields

$$^R_aD^\alpha_x f(x) = D^2 {_aI^{2-\alpha}_x} f(x) = D\left(_aI^{2-\alpha}_x Df(x) + f(a)\frac{(x-a)^{1-\alpha}}{\Gamma(2-\alpha)}\right)$$

$$= {_aI^{2-\alpha}_x}D^2 f(x) + Df(a)\frac{(x-a)^{1-\alpha}}{\Gamma(2-\alpha)} + f(a)\frac{x^{-\alpha}}{\Gamma(1-\alpha)},$$

thereby showing (2.37) for $n = 2$. The general case follows in a similar manner. $\qquad\square$

Together with the formula (2.22), we have $n - 1 < \alpha \le n$,

$$^C_aD^\alpha_x f(x) = {^R_aD^\alpha_x}(f - T_{n-1}f)(x),$$

with

$$T_{n-1}f(x) = \sum_{k=0}^{n-1} \frac{(x-a)^k}{k!} f^{(k)}(a^+).$$

Hence $^C_aD^\alpha_x f$ can be regarded as $^R_aD^\alpha_x f$ with an initial correction of its Taylor expansion $T_{n-1}f(x)$ of order $n - 1$ (at the base point $x = a$). This is a form of regularization to remove the singular behavior at the base point $x = a$, cf. (2.23). Very often, it is also employed as the defining identity for the Djrbashian-Caputo fractional derivative $^C_aD^\alpha_x f$ (see, e.g., [EK04, Die10]), and below it is termed as the regularized Djrbashian-Caputo fractional derivative.

Definition 2.4 For $n - 1 < \alpha \ge n$, $n \in \mathbb{N}$, and $f \in C^{n-1}(\overline{D})$ with $_aI^{n-\alpha}_x(f - T_{n-1}f)(x) \in AC^n(\overline{I})$, the regularized Djrbashian-Caputo fractional derivative $^C_aD^{\alpha*}_x f$ is defined by

$$^C_aD^{\alpha*}_x f(x) = {^R_aD^\alpha_x}(f - T_{n-1}f)(x), \qquad (2.36)$$

and similarly one can define the right-sided version by

$$\,^{C}_{x}D^{\alpha*}_{b} f(x) = \,^{R}_{x}D^{\alpha}_{b}\left(f(x) - \sum_{k=0}^{n-1} \frac{(-1)^k (b-x)^k}{k!} f^{(k)}(b^-)\right).$$

If $n-1 < \alpha < n$, $n \in \mathbb{N}$, then $\,^{C}_{a}D^{\alpha*}_{x} f = \,^{C}_{a}D^{\alpha}_{x} f$ for $f \in AC^n(\overline{D})$. Generally, $\,^{C}_{a}D^{\alpha}_{x} f$ and $\,^{C}_{a}D^{\alpha*}_{x} f$ can be different, and the latter requires less regularity assumption on f. We shall use the two definitions interchangeably. The relations also imply

$$\,^{C}_{a}D^{\alpha}_{x} f(x) = \,^{R}_{a}D^{\alpha}_{x} f(x), \quad \text{if } f^{(j)}(a) = 0, \ j = 0, 1, \ldots, n-1. \tag{2.37}$$

Note that there are several different proposals to relax the regularity assumption, e.g., operator interpolation [GLY15, KRY20] and convolution groups [LL18a].

The next result summarizes several properties of $\,^{C}_{a}D^{\alpha*}_{x}$ [Web19a, Section 4]. (i) shows that $\,^{C}_{a}D^{\alpha*}_{x}$ is the left inverse of $\,_{a}I^{\alpha}_{x}$ for continuous functions. $\,^{C}_{a}D^{\alpha*}_{x}$ generally do not commutes, but there is a positive result due to [Die10, Lemma 3.13], given in (ii), for which the existence of ℓ is important, without which the assertion can fail.

Proposition 2.3 *The following statements hold.*

(i) *For $n-1 < \alpha < n$, $n \in \mathbb{N}$, and $f \in C(\overline{D})$, $\,^{C}_{a}D^{\alpha*}_{x} \,_{a}I^{\alpha}_{x} f(x) = f(x)$ for all $x \in \overline{D}$. Similarly, if $f \in L^{\infty}(D)$, then $\,^{C}_{a}D^{\alpha*}_{x} \,_{a}I^{\alpha}_{x} f = f$ a.e. D.*

(ii) *Let $f \in C^k(\overline{D})$ for some $k \in \mathbb{N}$. Moreover let $\alpha, \beta > 0$ be such that there exists some $\ell \in \mathbb{N}$ with $\ell \le k$ and $\beta, \alpha + \beta \in [\ell - 1, \ell]$. Then*

$$\,^{C}_{a}D^{\alpha*}_{x} \,^{C}_{a}D^{\beta*}_{x} f = \,^{C}_{a}D^{\alpha+\beta*}_{x} f.$$

(iii) *For $n-1 < \alpha < n$, $n \ge 2$, if $f \in AC(\overline{D})$, then $\,^{C}_{a}D^{\alpha*}_{x} f(x) = \,^{C}_{a}D^{\alpha-1*}_{x} f'(x)$, whenever both fractional derivatives exist.*

Proof (i) Since $f \in C(\overline{D})$, $\,_{a}I^{\alpha}_{x} f$ is continuous, and $(\,_{a}I^{\alpha}_{x} f)^{(j)}(a) = 0$, $j = 0, 1, \ldots, n-1$. Thus,

$$\,^{C}_{a}D^{\alpha*}_{x} \,_{a}I^{\alpha}_{x} f(x) = \,^{R}_{a}D^{\alpha}_{x}(\,_{a}I^{\alpha}_{x} f - T_{n-1}(\,_{a}I^{\alpha}_{x} f))(x)$$
$$= (\,_{a}I^{n-\alpha}_{x} \,_{a}I^{\alpha}_{x} f)^{(n)}(x) = (\,_{a}I^{n}_{x} f)^{(n)}(x) = f(x).$$

This identity holds for any $x \in \overline{D}$, since $f \in C(\overline{D})$. The argument also works for $f \in L^{\infty}(D)$ with a.e. $x \in D$. Note that requirement $f \in L^{\infty}(D)$ can be relaxed to $f \in L^p(D)$, with $p > \frac{1}{\alpha - n + 1}$.

(ii) The statement is trivial in the case $\beta = \ell - 1$ and $\alpha + \beta = \ell$. So it suffices to treat the remaining cases. Note that the assumption implies $\alpha \in (0, 1)$. There are three possible cases (a) $\ell := \alpha + \beta \in \mathbb{N}$, (b) $\beta \in \mathbb{N}$ and (c) $\ell - 1 < \beta < \beta + \alpha < \ell$. For (a), we have $\alpha = n - \beta \in [0, 1)$. Since $f \in C^k(\overline{D})$, $k \ge \ell$, $\,^{C}_{a}D^{k*}_{x} f(a) = 0$, and thus

$$\,^{C}_{a}D^{\alpha*}_{x} \,^{C}_{a}D^{\beta*}_{x} f = \,^{R}_{a}D^{\alpha}_{x} \,^{C}_{a}D^{\beta*}_{x} f = \,^{R}_{a}D^{\alpha}_{x} \,_{a}I^{\alpha}_{x} f^{(\ell)} = f^{(\ell)} = \,^{C}_{a}D^{\alpha+\beta*}_{x} f.$$

For (b), $\,^{C}_{a}D^{\alpha*}_{x} \,^{C}_{a}D^{\beta*}_{x} f = \,^{C}_{a}D^{\alpha*}_{x} f^{(\beta)} = \,_{a}I^{1-\alpha}_{x} f^{(\beta+1)} = \,^{C}_{a}D^{\alpha+\beta*}_{x} f$. For (c), there holds

$$\substack{C\\a}D_x^{\alpha*}\,\substack{C\\a}D_x^{\beta*} f = \substack{R\\a}D_x^{\alpha}\,\substack{C\\a}D_x^{\beta*} f = \substack{R\\a}D_x^{\alpha}\,{_aI_x^{\ell-\beta}} f^{(\ell)} = ({_aI_x^{1-\alpha}}\,{_aI_x^{\ell-\beta}} f^{(\ell)})'$$
$$= ({_aI_{xa}^{1}I_x^{\ell-(\beta+\alpha)}} f^{(\ell)})' = \substack{C\\a}D_x^{\alpha+\beta*} f.$$

(iii) By the definition of $\substack{C\\a}D_x^{\alpha*} f$, the assertion follows by

$$\substack{C\\a}D_x^{\alpha*} f(x) = ({_aI_x^{n-\alpha}}(f - T_{n-1}f))^{(n)}(x)$$
$$= ({_aI_x^{n-\alpha}}\,{_aI_x^{1}}(f' - T_{n-2}f))^{(n)}(x) = ({_aI_{xa}^{1}I_x^{n-\alpha}}(f' - T_{n-2}f))^{(n)}(x)$$
$$= ({_aI_x^{n-\alpha}}(f' - T_{n-2}f))^{(n-1)}(x) = \substack{C\\a}D_x^{\alpha-1*} f'(x).$$

This completes the proof of the proposition. □

The next result gives the fundamental theorem of calculus for $\substack{C\\a}D_x^{\alpha}$. (i) shows that when $\alpha \notin \mathbb{N}$, $\substack{C\\a}D_x^{\alpha}$ provides a left inverse to $_aI_x^{\alpha}$, if $_aI_x^{\alpha-n+1} f \in AC^n(\overline{D})$. However, it is not a left inverse when $\alpha \in \mathbb{N}$. The conditions in Theorem 2.13 are more stringent than that in Theorem 2.6; this is to be expected from its more restrictive definition.

Theorem 2.13 *Let* $\alpha > 0$, $n - 1 < \alpha < n$, $n \in \mathbb{N}$. *Then the following statements hold.*

(i) *If* $f \in L^1(D)$ *with* $_aI_x^{\alpha-n+1} f \in AC(\overline{D})$ *and* $_aI_x^{\alpha-n+1} f(a) = 0$, *then*

$$\substack{C\\a}D_x^{\alpha}\,{_aI_x^{\alpha}} f = f, \quad a.e.\ D.$$

(ii) *If* $f \in AC^n(\overline{D})$, *then*

$$_aI_x^{\alpha}\,\substack{C\\a}D_x^{\alpha} f = f - T_{n-1}f, \quad a.e.\ D.$$

If $f \in C^n(\overline{D})$, *then the identity holds for all* $x \in \overline{D}$.

Proof (i) Since $f \in L^1(D)$ with $_aI_x^{\alpha-n+1} f \in AC(\overline{D})$, by Lemma 2.4, there exists a $\varphi_f \in L^1(D)$ such that $f = {_aI_x^{n-\alpha}} \varphi_f$. Then by Theorem 2.1,

$$_aI_x^{\alpha} f = {_aI_x^{\alpha}}\,{_aI_x^{n-\alpha}} \varphi_f = {_aI_x^n} \varphi_f \in AC^n(\overline{D}).$$

Differentiating both sides n times gives assertion (i).
(ii) If $f \in AC^n(\overline{D})$, then by Theorem 2.1, we deduce

$$({_aI_x^{\alpha}}\,\substack{C\\a}D_x^{\alpha}) f(x) = ({_aI_x^{\alpha}}\,{_aI_x^{n-\alpha}} f^{(n)})(x) = ({_aI_x^n} f^{(n)})(x)$$

from which assertion (ii) follows. The proof for $f \in C^n(\overline{D})$ follows similarly. □

The following lemma gives the Laplace transform of $\substack{C\\0}D_x^{\alpha} f$.

Lemma 2.9 *Let* $\alpha > 0$, $n - 1 < \alpha \le n$, $n \in \mathbb{N}$, *such that* $f \in C^n(\mathbb{R}_+)$,

$$|f(x)| \le c e^{p_0 x} \quad \text{for large } x,$$

for some $p_0 \in \mathbb{R}$, *and* $f \in AC^n[0, b]$ *for any* b, *the Laplace transforms* $\widehat{f}(z)$ *and* $\widehat{f^{(n)}}(z)$ *exist, and* $\lim_{x \to \infty} f^{(k)}(x) = 0$, $k = 0, 1, \dots, n-1$. *Then the following relation holds for* $\mathcal{R}(z) > p_0$:

$$\mathcal{L}[{}_0^C D_x^\alpha f](z) = z^\alpha \widehat{f}(z) - \sum_{k=0}^{n-1} z^{\alpha-k-1} f^{(k)}(0).$$

Proof For $n - 1 < \alpha < n$, then ${}_0^C D_x^\alpha f(x) = {}_0 I_x^{n-\alpha} g(x)$ with $g(x) = f^{(n)}(x)$. By the convolution rule (A.8) for Laplace transform, we deduce

$$\mathcal{L}[{}_0^C D_x^\alpha f](z) = z^{-(n-\alpha)} \mathcal{L}[g](z) = z^{\alpha-n} \Big(z^n \mathcal{L}[f](z) - \sum_{k=0}^{n-1} z^{n-1-k} f^{(k)}(0) \Big)$$

$$= z^\alpha \mathcal{L}[f](z) - \sum_{k=0}^{n-1} z^{\alpha-1-k} f^{(k)}(0),$$

which shows the desired formula.   □

The classical Djrbashian-Caputo fractional derivative ${}_a^C D_x^\alpha f$ is defined in terms of $f^{(n)}(x)$, requiring $f \in AC^n(\overline{D})$. One can give an alternative representation using $f^{(n-1)}(x)$, analogous to Lemma 2.8. The right-hand side is well defined for any $f \in C^{n-1,\gamma}(\overline{D})$. This representation appeared several times; see, e.g., [ACV16].

Lemma 2.10 *Let $n - 1 < \alpha < n$, $n \in \mathbb{N}$. If $f \in C^{n-1,\gamma}(\overline{D}) \cap AC^n(\overline{D})$, $\gamma \in (0, 1]$, then for any $\alpha - n + 1 \in (0, \gamma)$*

$$
{}_a^C D_x^\alpha f(x) = \frac{f^{(n-1)}(x) - f^{(n-1)}(a)}{\Gamma(n-\alpha)(x-a)^{\alpha-n+1}} + \frac{\alpha-n+1}{\Gamma(n-\alpha)} \int_a^x \frac{f^{(n-1)}(x) - f^{(n-1)}(s)}{(x-s)^{\alpha-n+2}} \, ds.
$$

Proof Applying integration by parts to the definition of ${}_a^C D_x^\alpha f$ gives

$$
\Gamma(n-\alpha) {}_a^C D_x^\alpha f(x) = \int_a^x (x-s)^{n-\alpha-1} \frac{d}{ds} (f^{(n-1)}(s) - f^{(n-1)}(x)) \, ds
$$

$$
= \frac{f^{(n-1)}(s) - f^{(n-1)}(x)}{(x-s)^{\alpha-n+1}} \Big|_{s=a}^{s=x} + (n-\alpha-1) \int_a^x \frac{f^{(n-1)}(s) - f^{(n-1)}(x)}{(x-s)^{n-\alpha-2}} \, ds.
$$

Since $f \in C^{n-1,\gamma}(\overline{D})$, with $\gamma > \alpha - n + 1$, $\lim_{s \to x^-} (x-s)^{n-\alpha-1} (f^{(n-1)}(s) - f^{(n-1)}(x)) = 0$. This shows the assertion.   □

The next lemma due to Alikhanov [Ali10, Lemma 1] is very useful. In the case $\alpha = 1$, the identity holds trivially: $f f' = \frac{1}{2}(f^2)'$.

Lemma 2.11 *For $\alpha \in (0, 1)$, and $f \in AC(\overline{D})$, there holds*

$$f(x) {}_a^C D_x^\alpha f(x) \geq \frac{1}{2} {}_a^C D_x^\alpha f^2(x).$$

Proof We rewrite the inequality as

$$\Gamma(1-\alpha)\left[f(x)\,{}^C_aD^\alpha_x f(x) - \tfrac{1}{2}\,{}^C_aD^\alpha_x f^2(x)\right]$$

$$= f(x)\int_a^x (x-s)^{-\alpha} f'(s)\,ds - \int_a^x (x-s)^{-\alpha} f(s)f'(s)\,ds$$

$$= \int_a^x (x-s)^{-\alpha} f'(s)(f(x)-f(s))\,ds = \int_a^x (x-s)^{-\alpha} f'(s)\int_s^x f'(\eta)\,d\eta\,ds$$

$$= \int_a^x f'(\eta)\left(\int_a^\eta (x-s)^{-\alpha} f'(s)\,ds\right)d\eta =: \mathrm{I}.$$

Then by changing the integration order and integration by parts, we have

$$\mathrm{I} = \int_a^x (x-\eta)^\alpha \frac{f'(\eta)}{(x-\eta)^\alpha}\int_a^\eta (x-s)^{-\alpha} f'(s)\,ds\,d\eta$$

$$= \frac{1}{2}\int_a^x (x-\eta)^\alpha \frac{d}{d\eta}\left(\int_a^\eta (x-s)^{-\alpha} f'(s)\,ds\right)^2 d\eta$$

$$= \frac{\alpha}{2}\int_a^x (x-\eta)^{\alpha-1}\left(\int_a^\eta (x-s)^{-\alpha} f'(s)\,ds\right)^2 d\eta \geq 0,$$

which completes the proof of the lemma. $\qquad\square$

The next result extends Lemma 2.11 to convex energies [LL18b, Lemma 2.4], where $\langle\cdot,\cdot\rangle$ denotes the duality pairing between a Banach space X and its dual X'.

Lemma 2.12 *If $f \in C^{0,\gamma}(\overline{D};X)$, $0 < \gamma < 1$ and $f \mapsto E(f)$ is a C^1 convex function on X. Then for $\alpha \in (0,\gamma)$,*

$$\,{}^C_aD^\alpha_x E(f(x)) \leq \langle \nabla_f E(f(x)), {}^C_aD^\alpha_x f(x)\rangle.$$

Proof By the convexity of $E(f)$, we have

$$E(f(x)) - E(f_0) \leq \langle \nabla_u E(f(x)), f(x) - f_0\rangle$$

for any $f_0 \in X$. This, Lemma 2.10 and the assumption $f \in C^{0,\gamma}(\overline{D})$ imply

$$\,{}^C_aD^\alpha_x E(f(x)) = \frac{1}{\Gamma(1-\alpha)}\left(\frac{E(f(x))-E(f(a))}{(x-a)^\alpha} + \alpha\int_a^x \frac{E(f(x))-E(f(s))}{(x-s)^{\alpha+1}}\,ds\right)$$

$$\leq \langle \nabla_f E(f(x)), {}^C_aD^\alpha_x f(x)\rangle.$$

This completes the proof of the lemma. $\qquad\square$

Below we discuss a few fractional versions of properties from calculus. First, for a smooth f, the starting value ${}^C_aD^\alpha_x f(a^+)$ is implied by the smoothness. This property has far-reaching impact on the study of FDEs: if one assumes the solution is smooth, the initial- or boundary-condition can only be trivial. This result is well known, greatly popularized by Stynes [Sty16, Lemma 2.1] in the context of subdiffusion.

Lemma 2.13 *Let $\gamma \in (0,1]$ and $\alpha \in (n-1, n-1+\gamma)$, $n \in \mathbb{N}$. Then for $f \in C^{n-1,\gamma}(\overline{D}) \cap AC^n(\overline{D})$, $\lim_{x\to a^+} {}^C_aD^\alpha_x f(x) = 0$.*

Proof We prove only for $\alpha \in (0, 1)$. Since $f \in C^{0,\gamma}(\overline{D}) \cap AC(\overline{D})$, $|f(x) - f(s)| \leq L|x - s|^\gamma$ for any $x, s \in \overline{D}$, and thus by Lemma 2.10, for any $x \in (a, b]$,

$$|{}^C_aD^\alpha_x f(x)| \leq \frac{1}{\Gamma(1-\alpha)} \left(\frac{|f(x) - f(a)|}{(x-a)^\alpha} + \alpha \int_a^x \frac{|f(x) - f(s)|}{(x-s)^{\alpha+1}} ds \right)$$

$$\leq \frac{L}{\Gamma(1-\alpha)} \left((x-a)^{\gamma-\alpha} + \alpha \int_a^x (x-s)^{\gamma-\alpha-1} ds \right)$$

$$\leq \frac{L}{\Gamma(1-\alpha)} \frac{\gamma}{\gamma-\alpha}(x-a)^{\gamma-\alpha} \to 0 \quad \text{as } x \to a^+.$$

This completes the proof of the lemma.   □

The next result characterizes the monotonicity of f [Web19a, Proposition 7.2].

Proposition 2.4 *Let* $0 < \alpha < 1$. *Suppose that* $f \in C(\overline{I})$ *with* ${}_aI^{1-\alpha}_x u \in AC(\overline{I})$. *Then the following statements hold.*

(i) *If* f *is nondecreasing, then* ${}^C_aD^{\alpha*}_x f(x) \geq 0$ *for a.e.* $x \in D$.
(ii) *If* ${}^C_aD^{\alpha*}_x f \in C(\overline{D})$ *and* ${}^C_aD^{\alpha*}_x f(x) \geq 0$ *for every* $x \in \overline{D}$, *then* $f(x) \geq f(a)$ *for every* $x \in \overline{D}$. *If* ${}^C_aD^{\alpha*}_x f(x) > 0$ *for* $x \in D$, *then* $f(x) > f(a)$ *for every* $x > a$.

Proof (i) Let $f_a \equiv f(a)$ in D. Then ${}^C_aD^{\alpha*}_x f(x) = ({}_aI^{1-\alpha}_x(f - f_a))'(x)$ and ${}_aI^{1-\alpha}_x(f - f_a)(x)$ is nondecreasing in x, by Theorem 2.2(v), since $f - f_a$ is nonnegative and nondecreasing. Thus, ${}^C_aD^{\alpha*}_x f \geq 0$ a.e.
(ii) Let $g(x) = {}^C_aD^{\alpha*}_x f$. By assumption, $f(x) = f(a) + ({}_aI^\alpha_x g)(x)$, and thus $f(x) \geq f(a)$, since $g(x) \geq 0$ for a.e. $x \in D$. The strict inequality follows similarly.   □

The next result gives an extremal principle for the Djrbashian-Caputo derivative; see [Die16, Theorem 2.2] for (iii). It partially generalizes Fermat's theorem in calculus: at every local extremum x_0 of f, $f'(x_0) = 0$. However, one cannot determine the local behavior of f near $x = x_0$ by ${}^C_aD^\alpha_x f(x_0)$, and (ii) gives a partial converse to (i). If $f \in AC(\overline{D})$, then ${}^C_aD^\alpha_x f = {}^C_aD^{\alpha*}_x f$, and hence the statements in Proposition 2.4 hold also for ${}^C_aD^\alpha_x f$, and (iii) gives a partial converse to this statement. A slightly weaker version of the extremal principle was first given by Luchko [Luc09, Theorem 1]; see also [AL14, Theorem 2.1] for an analogue in the Riemann-Liouville case.

Proposition 2.5 *Let* $f \in C^{0,\gamma}(\overline{D}) \cap AC(\overline{D})$, $0 < \alpha < \gamma \leq 1$.

(i) *If* $f(x) \leq f(x_0)$ *for all* $x \in [a, x_0]$ *for some* $x_0 \in (a, b)$. *Then*

$$\,{}^C_aD^\alpha_x f(x_0) \geq \frac{(x_0 - a)^{-\alpha}}{\Gamma(1-\alpha)}(f(x_0) - f(a)) \geq 0.$$

(ii) *If* $f(x) \leq f(x_0)$ *for all* $a \leq x \leq x_0 \leq b$, *and* ${}^C_aD^\alpha_x f(x_0) = 0$. *Then* $f(x) = f(x_0)$ *for all* $x \in [a, x_0]$.
(iii) *If* $f \in C^1(\overline{D})$ *satisfy* ${}^C_aD^\alpha_x f(x) \geq 0$ *for all* $x \in \overline{D}$ *and all* $\alpha \in (\alpha_0, 1)$ *for some* $\alpha_0 \in (0, 1)$. *Then* f *is monotonically increasing.*

Proof (i). By Lemma 2.10, for $f \in C^{0,\gamma}(\overline{D})$, there holds for any $x \in (a, b]$,

$$
{}^C_a D^\alpha_x f(x) = \frac{1}{\Gamma(1-\alpha)} \left(\frac{f(x) - f(a)}{(x-a)^\alpha} + \alpha \int_a^x \frac{f(x) - f(s)}{(x-s)^{\alpha+1}} \, ds \right).
$$

Since $f(x_0) - f(s) \geq 0$ for all $a \leq s \leq x_0$, the integral is nonnegative.
(ii). Under the given conditions, we obtain

$$
\frac{f(x_0) - f(a)}{(x_0 - a)^\alpha} = 0 \quad \text{and} \quad \int_a^{x_0} (x_0 - s)^{-\alpha-1}(f(x_0) - f(s)) \, ds = 0.
$$

Since f is continuous, it follows that $f(x_0) - f(s) = 0$ for all $s \in [a, x_0]$.
(iii) By assumption $0 \leq {}^C_a D^\alpha_x f(x) = {}_a I^{1-\alpha}_x f'(x)$, since f' is continuous, we apply
(2.34) to find the limit of the right-hand side as $\alpha \to 1^-$, which yields

$$
0 \leq \lim_{\alpha \to 1^-} {}^C_a D^\alpha_x f(x) = \lim_{\alpha \to 1^-} {}_a I^{1-\alpha}_x f'(x) = f'(x).
$$

Thus, f' is nonnegative on $(a, b]$. The continuity of f' on the closed interval $\overline{D}$ implies
$f'(a) \geq 0$. Hence f' is nonnegative on $\overline{D}$, and f is monotonically increasing. $\qquad\square$

The extremal property can be generalized to the case $\alpha \in (1, 2)$ [Al-12, Theorem
2.1]. It generalizes the criterion that at every local minimum x_0 of a smooth f, the
second derivative is nonnegative, i.e., $f''(x_0) \geq 0$.

Lemma 2.14 *Let $\alpha \in (1, 2)$, and $f \in C^2(\overline{D})$ attain its minimum at $x_0 \in \overline{D}$. Then*

$$
{}^C_a D^\alpha_x f(x_0) \geq \frac{(x_0 - a)^{-\alpha}}{\Gamma(2-\alpha)} \left[(\alpha - 1)(f(a) - f(x_0)) - (x_0 - a)f'(a) \right].
$$

Proof Let $g(x) = f(x) - f(x_0)$. Then $g(x) \geq 0$ on $[a, x_0]$, $g(x_0) = g'(x_0) = 0$,
$g''(x_0) \geq 0$ and ${}^C_a D^\alpha_x g(x) = {}^C_a D^\alpha_x f(x)$. Then integration by parts yields

$$
\Gamma(2-\alpha) {}^C_a D^\alpha_x g(x_0) = (x_0 - s)^{1-\alpha} g'(s)|_a^{x_0} - (\alpha - 1) \int_0^{x_0} (x_0 - s)^{-\alpha} g'(s) ds.
$$

Since $g(x_0) = g'(x_0) = 0$ and $g''(x_0)$ is bounded, we have

$$
\lim_{x \to x_0^-} (x_0 - x)^{1-\alpha} g'(x) = \lim_{x \to x_0^-} (x_0 - x)^{-\alpha} g(x) = 0,
$$

and thus

$$
\Gamma(2-\alpha) {}^C_a D^\alpha_x g(x_0) = -(x_0 - a)^{1-\alpha} g'(a) - (\alpha - 1) \int_a^{x_0} (x_0 - s)^{-\alpha} g'(s) ds.
$$

Integrating by parts again, since $g(x) \geq 0$ on $[a, x_0]$, yields

$$\int_a^{x_0} (x_0 - s)^{-\alpha} g'(s)ds = -(x_0 - a)^{-\alpha} g(a) - \alpha \int_a^{x_0} (x_0 - s)^{-\alpha-1} g(s)ds$$

$$\leq -(x_0 - a)^{-\alpha} g(a).$$

Last combining the preceding relations yields

$$\Gamma(2 - \alpha) {}_a^C D_x^\alpha g(x_0) \geq -(x_0 - a)^{1-\alpha} f'(a) + (\alpha - 1)(x_0 - a)^{-\alpha} (f(a) - f(x_0))$$

from which the desired result directly follows.   □

The result is weaker than ${}_a^C D_x^\alpha f(x_0) \geq 0$. In the classical case: at a local minimum x_0, $f''(x_0) \geq 0$. Unfortunately, this is generally not true for ${}_a^C D_x^\alpha f(x_0)$, $\alpha \in (1, 2)$.

Example 2.5 Consider the function $f(x) = x(x - \frac{1}{2})(x - 1)$ on $[0, 1]$. Clearly, $f(x)$ has an absolute minimum value at $x_0 = \frac{3+\sqrt{3}}{6} < 1$. For any $\alpha \in (1, 2)$, we have

$$_0^C D_x^\alpha f(x) = \frac{\Gamma(4)}{\Gamma(4 - \alpha)} x^{3-\alpha} - \frac{3}{2} \frac{\Gamma(3)}{\Gamma(3 - \alpha)} x^{2-\alpha}.$$

Thus, $_0^C D_x^{1.1} f(x_0) = -0.427 < 0$. Thus, at the global minimum x_0, $_a^C D_x^\alpha f(x_0) \geq 0$ does not hold for $\alpha \in (1, 2)$. See Fig. 2.1 for an illustration of f and $_0^C D_x^\alpha f$.

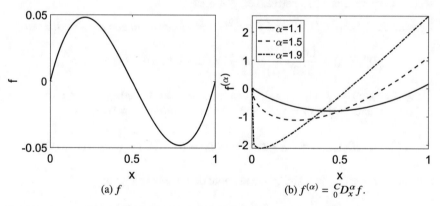

Fig. 2.1 The plot of f and $_0^C D_x^\alpha f$ at $\alpha = 1.1, 1.5$ and 1.9.

Next we generalize the mean value theorem: if $f \in C^1(D) \cap C(\overline{D})$, $\frac{f(b)-f(a)}{b-a} = f'(\xi)$ for some $\xi \in (a, b)$. One fractional version reads [Die12, Theorem 2.3]:

Theorem 2.14 *Let $n - 1 < \alpha < n$, $n \in \mathbb{N}$, $f \in C^n(\overline{D})$ with $_a^C D_x^\alpha f \in C(\overline{D})$. Then there exists some $\xi \in (a, b)$ such that*

$$\frac{f(b) - \sum_{k=0}^{n-1} \frac{(b-a)^k}{k!} f^{(k)}(a)}{(b - a)^\alpha} = \frac{_a^C D_x^\alpha f(\xi)}{\Gamma(\alpha + 1)}.$$

Proof By Theorem 2.13(ii), there holds $f(b) - (T_{n-1}f)(b) = {}_aI_x^\alpha\,{}_a^CD_x^\alpha f(b)$, and by assumption, ${}_a^CD_x^\alpha f$ is continuous. Thus applying Theorem 2.3 to the right-hand side, with ${}_a^CD_x^\alpha f$ in place of f, the claim follows. □

The existence of a Djrbashian-Caputo fractional derivative implies the existence of a limit at $x = a$ in the sense of the Lebesgue point [LL18b, Corollary 2.16]. The limit is sometimes called the assignment of the initial value for f.

Theorem 2.15 *For $\alpha \in (0, 1)$, if ${}_a^CD_x^{\alpha*} f \in L_{loc}^{\frac{1}{\alpha}}(D)$, then a is a Lebesgue point, i.e., $f(a^+) = c$ for some $c \in \mathbb{R}$ in the sense of*

$$\lim_{x \to a^+} (x - a)^{-1} \int_a^x |f(s) - c|\mathrm{d}s = 0.$$

Proof With the relation $f(x) - c = {}_aI_x^\alpha\,{}_a^CD_x^{\alpha*} f$, we deduce

$$\int_a^x |f(s) - c|\mathrm{d}s \le \frac{1}{\Gamma(\alpha)} \int_a^x \int_a^s (s - \xi)^{\alpha-1} |{}_a^CD_x^{\alpha*} f(\xi)|\mathrm{d}\xi\mathrm{d}s$$

$$= \frac{1}{\Gamma(\alpha)} \int_a^x |{}_a^CD_x^{\alpha*} f(\xi)| \int_\xi^x (s - \xi)^{\alpha-1}\mathrm{d}s\mathrm{d}\xi.$$

Since $\int_\xi^x (s - \xi)^{\alpha-1}\mathrm{d}s = \alpha^{-1}(x - \xi)^\alpha$,

$$(x - a)^{-1} \int_a^x |f(s) - c|\mathrm{d}s \le \frac{1}{(x - a)\Gamma(1 + \alpha)}\|(x - \xi)^\alpha\|_{L^{\frac{1}{1-\alpha}}(a,x)} \|{}_a^CD_x^{\alpha*} f\|_{L^{\frac{1}{\alpha}}(a,x)}$$

$$= \Gamma(1 + \alpha)^{-1}(1 - \alpha)^{-\frac{1}{1-\alpha}} \|{}_a^CD_x^{\alpha*} f\|_{L^{\frac{1}{\alpha}}(a,x)} \to 0,$$

as $x \to a^+$, due to the absolute continuity of Lebesgue integral. □

2.3.3 Grünwald-Letnikov fractional derivative

The derivative $f'(x)$ of a function $f : D \to \mathbb{R}$ at a point $x \in D$ is defined by

$$f'(x) = \lim_{h \to 0} h^{-1}\Delta_h f(x), \quad \text{with } \Delta_h f(x) = f(x) - f(x - h).$$

Similarly one can construct higher order derivatives directly by higher order backward differences. By induction on k, one can verify that

$$\Delta_h^k f(x) = \sum_{j=0}^k C(k, j)(-1)^j f(x - jh) \quad \text{for } k \in \mathbb{N}_0.$$

An induction on k shows the following identity

$$\Delta_h^k f(x) = \int_0^h \cdots \int_0^h f^{(k)}(x - s_1 - \cdots - s_k) \, ds_1 \cdots ds_k, \tag{2.38}$$

and thus by the mean value theorem, we obtain that for $f \in C^k(\overline{D})$:

$$f^{(k)}(x) = \lim_{h \to 0} h^{-k} \Delta_h^k f(x). \tag{2.39}$$

It is natural to ask whether one can define a fractional derivative (or integral) in a similar manner, without resorting to the integer-order derivatives and integrals. This line of pursuit leads to the Grünwald-Letnikov definition of a fractional derivative, introduced by Anton Karl Grünwald from Prague in 1867 [Grü67], and independently by Aleksey Vasilievich Letnikov in Moscow in 1868 [Let68]. It proceeds as the case of the Riemann-Liouville fractional integral by replacing the integer k by a real number α. Specifically, one may define the fractional backward difference formula of order $\alpha \in \mathbb{R}$, denoted by $\Delta_{h,k}^\alpha f$, by

$$\Delta_{h,k}^\alpha f(x) = \sum_{j=0}^k C(\alpha, j)(-1)^j f(x - jh).$$

In view of the identity (2.39), given x and a, we therefore define

$${}_a^{GL}D_x^\alpha f(x) = \lim h^{-\alpha} \Delta_{h,k}^\alpha f(x),$$

where the limit is obtained by sending $k \to \infty$ and $h \to 0^+$ while keeping $h = \frac{x-a}{k}$, so that $kh = x - a$ is constant. The next result shows that the Grünwald-Letnikov fractional derivative coincides with the Riemann-Liouville one for $\alpha < 0$.

Theorem 2.16 *For $\alpha > 0$, $f \in C(\overline{D})$, there holds*

$${}_a^{GL}D_x^{-\alpha} f(x) = {}_a I_x^\alpha f(x).$$

Proof By (2.11), $C(-\alpha, j) = (-1)^j C(j + \alpha - 1, j)$, and we may rewrite $\Delta_{h,k}^{-\alpha} f(x)$ as

$$h^\alpha \Delta_{h,k}^{-\alpha} f(x) = h^\alpha \sum_{j=0}^k C(j + \alpha - 1, j) f(x - jh) = \frac{1}{\Gamma(\alpha)} \sum_{j=0}^k B_j A_{k,j}$$

with $B_j = \Gamma(\alpha) j^{1-\alpha} C(j + \alpha - 1, j)$ and $A_{k,j} = h(jh)^{\alpha-1} f(x - jh)$. By the definition of $C(j + \alpha - 1, j)$ and Wendel's limit (2.8),

$$\lim_{j \to \infty} B_j = \lim_{j \to \infty} \frac{\Gamma(\alpha)}{j^{\alpha-1}} \frac{\Gamma(j + \alpha)}{\Gamma(j + 1)\Gamma(\alpha)} = \lim_{j \to \infty} \frac{\Gamma(j + \alpha)}{\Gamma(j + 1)j^{\alpha-1}} = 1.$$

Meanwhile, for $f \in C(\overline{D})$,

$$\lim_{k\to\infty}\sum_{j=0}^{k}A_{k,j}=\lim_{h\to 0}\sum_{j=0}^{k}h(jh)^{\alpha-1}f(x-jh)=\int_{a}^{x}(x-s)^{\alpha-1}f(s)ds.$$

Now the desired assertion follows from the property of limit. □

The next result shows that the Grünwald-Letnikov fractional derivative of order $\alpha > 0$ coincides with the Riemann-Liouville one for $\alpha > 0$.

Theorem 2.17 *If $n - 1 < \alpha < n$, $n \in \mathbb{N}$, and $f \in C^n(\overline{D})$, then*

$$^{GL}_aD_x^\alpha f(x)=\sum_{j=0}^{n-1}f^{(j)}(a)\frac{(x-a)^{j-\alpha}}{\Gamma(j+1-\alpha)}+\int_{a}^{x}\frac{(x-s)^{n-\alpha-1}}{\Gamma(n-\alpha)}f^{(n)}(s)\,ds.$$

Proof It follows from the identity $C(\alpha, j) = C(\alpha - 1, j) + C(\alpha - 1, j - 1)$ that

$$\Delta_{h,k}^\alpha f(x)=\sum_{j=0}^{k}(-1)^jC(\alpha-1,j)f(x-jh)+\sum_{j=1}^{k}(-1)^jC(\alpha-1,j-1)f(x-jh)$$

$$=(-1)^kC(\alpha-1,k)f(a)+\sum_{j=0}^{k-1}(-1)^jC(\alpha-1,j)\Delta_h f(x-jh).$$

Repeating the procedure n times yields

$$\Delta_{h,k}^\alpha f(x)=\sum_{j=0}^{n-1}(-1)^{k-j}C(\alpha-j-1,k-j)\Delta_h^j f(a+jh)$$

$$+\sum_{j=0}^{k-n}(-1)^jC(\alpha-n,j)\Delta_h^n f(x-jh).$$

It remains to find the limits of the terms. First, by taking limit as $h \to 0^+$, with $kh = x - a$, and from the identity (2.11), we deduce

$$\lim h^{-\alpha}(-1)^{k-j}C(\alpha-j-1,k-j)\Delta_h^j(a+jh)$$

$$=\lim C(k-\alpha,k-j)(k-j)^{\alpha-j}\left(\frac{k}{k-j}\right)^{\alpha-j}(kh)^{-\alpha+j}\frac{\Delta_h^j f(a+jh)}{h^j}$$

$$=\frac{(x-a)^{j-\alpha}}{\Gamma(j+1-\alpha)}f^{(j)}(a^+),$$

where the last line follows from the identity

$$\lim_{k\to\infty}\frac{C(k-\alpha,k-j)}{(k-j)^{j-\alpha}}=\lim_{k\to\infty}\frac{\Gamma(k-\alpha+1)(k-j)^{\alpha-j}}{\Gamma(k-j+1)\Gamma(j-\alpha+1)}=\frac{1}{\Gamma(j+1-\alpha)},$$

cf. Wendel's limit (2.8). Using the identity (2.11), we have

$$h^{-\alpha} \sum_{j=0}^{k-n} (-1)^j C(\alpha - n, j) \Delta_h^n f(t - jh) = \sum_{j=0}^{k-n} B_j A_{k,j}$$

with $B_j = C(j + n - \alpha - 1, j) j^{-n+\alpha+1}$ and $A_{k,j} = h(jh)^{n-\alpha-1} \frac{\Delta^n f(t-jh)}{h^n}$. Then direct computation with Wendel's limit (2.8) gives

$$\lim_{j \to \infty} B_j = \lim_{j \to \infty} \frac{\Gamma(j + n - \alpha)}{\Gamma(j + 1)\Gamma(n - \alpha)} j^{-n+\alpha+1} = \frac{1}{\Gamma(n - \alpha)}.$$

Meanwhile, we have

$$\lim_{k \to \infty} \sum_{j=0}^{k-n} A_{k,j} = \int_a^x (x - a)^{n-\alpha-1} f^{(n)}(x) dx.$$

Combining these identities with the property of the limit yields the assertion. □

By Theorems 2.16 and 2.17, the Grünwald-Letnikov fractional integral and derivative coincide with the Riemann-Liouville counterparts, under suitable regularity condition on f, which is slightly stronger than that for the Riemann-Liouville counterparts. Nonetheless, the Grünwald-Letnikov definition, when fixed at a finite $h > 0$, serves as one simple way to construct numerical approximations of the Riemann-Liouville fractional integrals and derivatives. See Fig. 2.2 for an illustration on $f(x) = x^3$. Generally, the accuracy of the approximation is at best first-order $O(h)$.

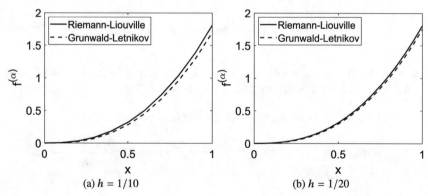

Fig. 2.2 The Grünwald-Letnikov approximation of $_0^R D_x^\alpha f$ of $f(x) = x^3$ over $[0, 1]$, with $\alpha = 0.5$.

Exercises

Exercise 2.1 Show the identity (2.2) by integration by parts.

Exercise 2.2 Using the identity $\lim_{n\to\infty}(1 - \frac{t}{n})^n = e^{-t}$, prove the identity

$$\Gamma(z) = \lim_{n\to\infty} \frac{n!n^z}{z(z+1)\dots(z+n)}.$$

Euler originally defines the Gamma function by this product formula in a letter to Goldbach, dated October 13th 1729 [Eul29].

Exercise 2.3 Prove that $\Gamma(x)$ is log-convex on $\mathbb{R}_+$, i.e., for any $x_1, x_2 \in \mathbb{R}_+$ and $\theta \in [0, 1]$, there holds

$$\Gamma(\theta x_1 + (1 - \theta)x_2) \le \Gamma(x_1)^\theta \Gamma(x_2)^{1-\theta}.$$

Exercise 2.4 James Wendel [Wen48] presented a strengthen version of the identity (2.8), by proving the following double inequality

$$\left(\frac{x}{x+s}\right)^{1-s} \le \frac{\Gamma(x+s)}{x^s\Gamma(x)} \le 1,$$

for $0 < s < 1$ and $x > 0$. The proof employs only Hölder's inequality and the recursion formula (2.2). This exercise is to outline the proof.

(i) Use Hölder's inequality to prove

$$\Gamma(x+s) \le \Gamma(x+1)^s\Gamma(x)^{1-s}.$$

This and the recursion formula yield

$$\Gamma(x+s) \le x^s\Gamma(x). \tag{2.40}$$

(ii) Substitute s by $1 - s$ in (2.40) and then $x - s$ by x to get

$$\Gamma(x+1) \le (x+s)^{1-s}\Gamma(x+s).$$

This proves the double inequality, and also the identity (2.8).

(iii) Extend the proof to $s \in \mathbb{R}$ by the recursion formula (2.2).

Exercise 2.5 Use the log-convexity of $\Gamma(x)$ to prove Gautschi's inequality [Gau60] for the Gamma function $\Gamma(x)$, i.e., for any positive $x \in \mathbb{R}_+$ and any $s \in (0, 1)$, there holds

$$x^{1-s} < \frac{\Gamma(x+1)}{\Gamma(x+s)} < (x+1)^{1-s}.$$

This inequality is closely connected with Wendel's identity; see the survey [Qi10] for a comprehensive survey on inequalities related to the ratio of two Gamma functions.

Exercise 2.6 Prove the identity (2.10), using the convolution rule for the Laplace transform on the function $\int_0^t (t-s)^{\alpha-1} s^{\beta-1} ds$.

Exercise 2.7 Prove the identity for $f \in C(\overline{D})$,

$$f(x) = \lim_{\alpha \to 0^+} {}_a I_x^\alpha f(x), \quad \forall x \in \overline{D}.$$

Exercise 2.8 Let $f \in L^1(D)$. Prove the following identity

$$\lim_{\alpha \to 0^+} \|{}_a I_x^\alpha f - f\|_{L^1(D)} = 0.$$

Exercise 2.9 Repeat Examples 2.1 and 2.2 for any fixed $a \in \mathbb{R}$.

Exercise 2.10 Find (i) ${}_0 I_x^\alpha \cos(\lambda x)$, $\lambda \in \mathbb{R}$ and (ii) ${}_0 I_x^\alpha \Gamma(x)$.

Exercise 2.11 Prove Theorem 2.2 (iii).

Exercise 2.12 Find ${}_0^R D_x^\alpha \sin(\lambda x)$ and ${}_0^R D_x^\alpha \cos(\lambda x)$, $\lambda \in \mathbb{R}$.

Exercise 2.13 Prove that for $\alpha \in (0, 1)$ and $p \in [1, \alpha^{-1}]$, if $f \in AC(\overline{D})$, then ${}_a^R D_x^\alpha f \in L^p(D)$.

Exercise 2.14 This exercise generalizes Lemma 2.6. Let $f, g \in L^1(D)$ and $\alpha \in (0, 1)$ such that (i) ${}_a^R D_x^\alpha f$ and ${}_x^R D_b^\alpha g$ exist and $g {}_a^R D_x^\alpha f \in L^1(D)$; and (ii) the functions $g, {}_a I_x^{1-\alpha} f$ are continuous at $x = a$ and the functions $f, {}_x I_b^{1-\alpha} g$ are continuous at $x = b$. Then prove the following identity

$$\int_D g {}_a^R D_x^\alpha f dx = \int_D f {}_x^R D_b^\alpha g dx + (f {}_x I_b^{1-\alpha} g)|_{x=b} - (g {}_a I_x^{1-\alpha} f)|_{x=a}.$$

Exercise 2.15 Generalize Lemma 2.8 to the case $\alpha \in (1, 2)$.

Exercise 2.16 [Al-12] If $f \in C^1(\overline{D})$ attains its maximum at $x_0 \in (a, b]$, then for any $\alpha \in (0, 1)$, there holds

$$({}_a^R D_x^\alpha f)(x_0) \geq \frac{(x_0 - a)^{-\alpha}}{\Gamma(1 - \alpha)} f(x_0).$$

Exercise 2.17 This exercise is about the continuity of the Riemann-Liouville fractional derivative in the order α. Let $n - 1 < \alpha < n, n \in \mathbb{N}$.

(i) For $f \in C^n(\overline{D})$

$$\lim_{\alpha \to (n-1)^+} {}_a^R D_x^\alpha f(x) = f^{(n-1)}(x), \quad \forall x \in \overline{D}.$$

(ii) For $f \in AC^n(\overline{D})$,

$$\lim_{\alpha \to (n-1)^+} {}_a^R D_x^\alpha f(x) = f^{(n-1)}(x), \quad \text{a.e. } D.$$

Discuss how to further relax the regularity on f.

Exercise 2.18 [RSL95] Consider the Weierstrass function defined by (with $s > 0$)

$$W_\lambda(x) = \sum_{k=0}^{\infty} \lambda^{(s-2)k} \sin \lambda^k x, \quad \lambda > 1.$$

It is known that for $s \in (1, 2)$, W_λ is continuous but nowhere differentiable. Prove the following two statements.

(i) ${}_0^R D_x^\alpha W_\lambda(x)$ of $W_\lambda(x)$ exists for every order $\alpha < 2 - s$.
(ii) ${}_0^R D_x^\alpha W_\lambda(x)$ of $W_\lambda(x)$ does not exist for any order $\alpha > 2 - s$.

Exercise 2.19 The way Djrbashian introduced his version of the fractional derivative (see, e.g., [DN68]) is actually more general than that given in Definition 2.3. It is defined for a set of indices $\{\gamma_k\}_{k=0}^n \subset (0, 1]$. Denote

$$\sigma_k := \sum_{j=0}^{k} \gamma_j - 1,$$

and suppose $\sigma_n = \sum_{j=0}^n \gamma_j - 1 > 0$. Then for a function f defined on D, he introduced the following differential operations associated with $\{\gamma_k\}_{k=0}^n$ using the Riemann-Liouville fractional derivative

$$
{}_a D_x^{\sigma_0} f(x) = {}_a I_x^{1-\gamma_0} f(x),
$$

$$
{}_a D_x^{\sigma_k} f(x) = {}_a I_x^{1-\gamma_k} {}_a^R D_x^{\gamma_{k-1}} \cdots {}_a^R D_x^{\gamma_1} {}_a^R D_x^{\gamma_0} f, \quad k = 1, \ldots, n,
$$

supposing that these operations have a sense at least in a.e. on D. Clearly, with the choice $\gamma_0 = \gamma_1 = \ldots = \gamma_{n-1} = 1$, it recovers the classical Djrbashian-Caputo fractional derivative ${}_a^C D_x^\alpha f$ with $\alpha = \sigma_n$ in Definition 2.3 (associated with the set $\{\gamma_k\}_{k=0}^n$). With this definition, he and his collaborators studied the Cauchy problem [DN68] and boundary value problems [Džr70] for the associated ODEs (see the book [Djr93, Chapter 10] for some relevant results).

(i) Prove the following identity

$$
{}_a D_x^{\sigma_s} \frac{(x-a)^{\sigma_s}}{\Gamma(\sigma_s + 1)} =
\begin{cases}
0, & 0 \le k \le s - 1, 0 \le s \le n, \\[2mm]
\dfrac{(x-a)^{\sigma_k - \sigma_s}}{\Gamma(\sigma_k - \sigma_s + 1)}, & s \le k \le n.
\end{cases}
$$

(ii) Derive the Laplace transform relation for ${}_0 D_x^{\sigma_n} f$.
(iii) Derive the fundamental theorem for ${}_a D_x^{\sigma_n}$, in a manner similar to Theorem 2.6.

Exercise 2.20 Show the following identity for $\alpha > 0$ and $\lambda \in \mathbb{R}$:

$$
{}_0^C D_x^\alpha E_{\alpha,1}(\lambda x^\alpha) = \lambda E_{\alpha,1}(\lambda x^\alpha).
$$

Exercise 2.21 Find one function $f(x)$ such that for $0 < \alpha, \beta < 1$

$$_a^C D_x^\alpha \, _a^C D_x^\beta f(x) = {}_a^C D_x^\beta \, _a^C D_x^\alpha f(x) \neq {}_a^C D_x^{\alpha+\beta} f(x).$$

Exercise 2.22 [NSY10] Prove that for $f \in C^2(\overline{D})$ and $0 < \alpha, \beta < 1, \alpha + \beta < 1$

$$_a^C D_x^\beta ({}_a^C D_x^\alpha f) = {}_a^C D_x^{\alpha+\beta} f.$$

Exercise 2.23 Prove that for $f \in C^2(\overline{D})$ and $\alpha \in (1, 2)$

$$({}_a^C D_x^{\frac{\alpha}{2}})^2 f(x) = {}_a^C D_x^\alpha f(x) - \frac{f'(a)}{\Gamma(2 - \alpha)}(x - a)^{1-\alpha}.$$

Exercise 2.24 Let $\alpha \in (0, 1)$, $f \in AC(\overline{D})$, and $f_+ = \max(f, 0)$. Prove the following inequality

$$f_+(x) {}_a^C D_x^\alpha f(x) \geq \tfrac{1}{2} {}_a^C D_x^\alpha f_+^2(x).$$

This inequality generalizes Alikhanov's inequality in Lemma 2.11.

Exercise 2.25 In Proposition 2.5(iii), can the condition $\alpha \in (\alpha_0, 1)$ be relaxed to $\alpha \in (\alpha_0, \alpha_1)$ for some $0 < \alpha_0 < \alpha_1 < 1$?

Exercise 2.26 Show the formula (2.38).

Exercise 2.27 Give the full argument for "the basic property of limit" in the proof of Theorem 2.16.

Exercise 2.28 One generalization of the fractional integral operator $_a I_x^\alpha$ is the Erdélyi-Köber fractional integral operator, for $\alpha > 0$ and $\eta > -\frac{1}{2}$, defined as

$$(I^{\eta,\alpha} f)(x) = \frac{2x^{-2\eta-2\alpha}}{\Gamma(\alpha)} \int_0^x (x^2 - s^2)^{\alpha-1} s^{2\eta+1} f(s) \, ds.$$

 (i) Show that if f is integrable, the integral on the right-hand side exists.
 (ii) Show that the semigroup-type relation $I^{\eta,\alpha} I^{\eta+\alpha,\beta} = I^{\eta,\alpha+\beta}$ holds.

Chapter 3
Mittag-Leffler and Wright Functions

The exponential function e^z plays an extremely important role in the theory of integer-order differential equations. For FDEs, its role is subsumed by the Mittag-Leffler and Wright functions. In this chapter, we discuss their basic analytic properties and numerical computation.

3.1 Mittag-Leffler Function

Magnus Gustaf Mittag-Leffler treated the function that bears his name in a series of five papers, in connection with his method of summation of divergent series; the two most frequently referenced ones are [ML03, ML05]. These considered the one-parameter version $E_{\alpha,1}(z)$, initially for $\alpha \in \mathbb{R}$ and later for $\alpha \in \mathbb{C}$. The two-parameter version was introduced by Wiman [Wim05] in 1905, but majority of the analysis for this case appeared only nearly 50 years later in the (independent) works of Agarwal and Humbert [Aga53, Hum53, HA53] and independently also Djrbashian [Džr54a, Džr54b, Džr54c] (see the monograph by Djrbashian [Dzh66] for an overview of many results). This function was initially treated under "Miscellaneous Functions" in Volume III of the classical Bateman project [EMOT55, pp. 206–212], but its central role in FDEs is now widely recognized and the function becomes so prominent that sometimes it is called the "queen function of fractional calculus" [MG07]. The survey [HMS11] and the monograph [GKMR14] provide a comprehensive treatment of Mittag-Leffler-type functions. For $\alpha > 0$, and $\beta \in \mathbb{R}$, the two-parameter Mittag-Leffler function $E_{\alpha,\beta}(z)$ is defined by

$$E_{\alpha,\beta}(z) = \sum_{k=0}^{\infty} \frac{z^k}{\Gamma(\alpha k + \beta)}, \quad z \in \mathbb{C}. \tag{3.1}$$

Often we write $E_\alpha(z) = E_{\alpha,1}(z)$, when $\beta = 1$.

We begin with several elementary facts of the function $E_{\alpha,\beta}(z)$. It generalizes the familiar exponential function e^z:

© The Author(s), under exclusive license to Springer Nature Switzerland AG 2021   59
B. Jin, *Fractional Differential Equations*, Applied Mathematical Sciences 206,
https://doi.org/10.1007/978-3-030-76043-4_3

$$E_{1,1}(z) = \sum_{k=0}^{\infty} \frac{z^k}{\Gamma(k+1)} = \sum_{k=0}^{\infty} \frac{z^k}{k!} = e^z.$$

There are a few other easy-to-verify special cases

$$E_{1,2}(z) = z^{-1}(e^z - 1), \quad E_{2,1}(z) = \cosh(z^{\frac{1}{2}}), \quad E_{2,2}(z) = z^{-\frac{1}{2}} \sinh z^{\frac{1}{2}}, \tag{3.2}$$

where $\cosh z = \frac{e^z + e^{-z}}{2}$ and $\sinh z = \frac{e^z - e^{-z}}{2}$ denote the hyperbolic cosine and sine, respectively. Here $z^{\frac{1}{2}}$ means the principal value of the square root of $z \in \mathbb{C}$ cut along $\mathbb{R}_- = (-\infty, 0)$. The relation to the complementary error function is also well known

$$E_{\frac{1}{2},1}(z) = \sum_{k=0}^{\infty} \frac{z^k}{\Gamma(\frac{k}{2}+1)} = e^{z^2}(1 + \mathrm{erf}(z)) = e^{z^2}\mathrm{erfc}(-z), \tag{3.3}$$

where erf and erfc denote the error function and the complementary error function, respectively, for any $z \in \mathbb{C}$, defined by

$$\mathrm{erfc}(z) := \frac{2}{\sqrt{\pi}} \int_0^z e^{-s^2} \mathrm{d}s \quad \text{and} \quad \mathrm{erfc}(z) := 1 - \mathrm{erf}(z) = \frac{2}{\sqrt{\pi}} \int_z^\infty e^{-s^2} \mathrm{d}s. \tag{3.4}$$

The Riemann-Liouville fractional integrals and derivatives of a Mittag-Leffler function are still Mittag-Leffler functions: for any $\gamma > 0$ and $\lambda \in \mathbb{R}$,

$$_0I_x^\gamma x^{\beta-1} E_{\alpha,\beta}(\lambda x^\alpha) = x^{\beta+\gamma-1} E_{\alpha,\beta+\gamma}(\lambda x^\alpha), \tag{3.5}$$

$$_0^R D_x^\gamma x^{\beta-1} E_{\alpha,\beta}(\lambda x^\alpha) = x^{\beta-\gamma-1} E_{\alpha,\beta-\gamma}(\lambda x^\alpha). \tag{3.6}$$

Indeed, since $E_{\alpha,\beta}(z)$ is entire (cf. Proposition 3.1), we can integrate termwise and use (2.19) to obtain

$$_0I_x^\gamma x^{\beta-1} E_{\alpha,\beta}(\lambda x^\alpha) = {}_0I_x^\gamma \sum_{k=0}^{\infty} \frac{\lambda^k x^{k\alpha+\beta-1}}{\Gamma(k\alpha+\beta)} = \sum_{k=0}^{\infty} \frac{\lambda^k {}_0I_x^\gamma x^{k\alpha+\beta-1}}{\Gamma(k\alpha+\beta)}$$

$$= \sum_{k=0}^{\infty} \frac{\lambda^k x^{k\alpha+\beta+\gamma-1}}{\Gamma(k\alpha+\beta+\gamma)} = x^{\beta+\gamma-1} E_{\alpha,\beta+\gamma}(\lambda x^\alpha).$$

The identity (3.6) follows similarly.

Recall that the order ρ and the corresponding type of an entire function $f(z)$ are defined, respectively, by

$$\rho := \limsup_{R \to \infty} \frac{\ln\left(\ln \max_{|z| \le R} |f(z)|\right)}{\ln R} \quad \text{and} \quad \limsup_{R \to \infty} \frac{\ln \max_{|z| \le R} |f(z)|}{R^\rho}.$$

Proposition 3.1 *For $\alpha > 0$ and $\beta \in \mathbb{R}$, $E_{\alpha,\beta}(z)$ is entire of order $\frac{1}{\alpha}$ and type 1.*

Proof It suffices to show that the series (3.1) converges for any $z \in \mathbb{C}$. Using Stirling's formula (2.7), we have

$$\frac{\Gamma((k+1)\alpha + \beta)}{\Gamma(k\alpha + \beta)} = (k\alpha)^\alpha (1 + O(k^{-1})) \quad \text{as } k \to \infty.$$

Thus, by Cauchy's criterion, the series (3.1) converges for any $z \in \mathbb{C}$. The order and type of $E_{\alpha,\beta}(z)$ follow similarly. $\square$

3.1.1 Basic analytic properties

Integral representations play a prominent role in the study of entire functions, e.g., analytical properties and practical computations. For $E_{\alpha,\beta}(z)$, such representations in the form of an improper integral along the Hankel contour C were treated in the case $\beta = 1$ and arbitrary β by [EMOT55] and Djrbashian [Dzh66], respectively. We shall focus on the case $\alpha \in (0, 2)$, since the case $\alpha \geq 2$ can be reduced to $\alpha \in (0, 2)$, cf. Lemma 3.1. First, we define the contour $\Gamma_{\epsilon,\varphi}$ by

$$\Gamma_{\epsilon,\varphi} = \{z \in \mathbb{C} : z = re^{\pm i\varphi}, r \geq \epsilon\} \cup \{z \in \mathbb{C} : z = \epsilon e^{i\theta} : |\theta| \leq \varphi\},$$

oriented with an increasing imaginary part. The contour $\Gamma_{\epsilon,\varphi}$ divides $\mathbb{C}$ into two parts $G^\pm(\epsilon, \varphi)$, where $G^-(\epsilon, \varphi)$ and $G^+(\epsilon, \varphi)$ are the regions to the left and right of the contour path, respectively, see Fig. 3.1 for an illustration.

The derivation is based on the representation (2.6) of the reciprocal Gamma function $\frac{1}{\Gamma(z)}$. First, we perform the substitution $\tilde{\zeta} = \zeta^{\frac{1}{\alpha}}$, $\alpha \in (0, 2)$ in (2.6) and, for $1 \leq \alpha < 2$, consider only the contours $\Gamma_{\epsilon,\varphi}$ for which $\frac{\pi}{2} < \varphi < \frac{\pi}{\alpha}$. Since ϵ is arbitrary, we obtain

$$\frac{1}{\Gamma(z)} = \frac{1}{2\alpha\pi i} \int_{\Gamma_{\epsilon,\varphi}} e^{\zeta^{\frac{1}{\alpha}}} \zeta^{\frac{1-z-\alpha}{\alpha}} d\zeta, \tag{3.7}$$

where $\alpha \in (0, 2)$ and $\frac{\alpha}{2}\pi < \varphi < \min(\pi, \alpha\pi)$.

For $\alpha \in (0, 2)$, $E_{\alpha,\beta}(z)$ admits the following integral representations.

Theorem 3.1 *Let $\alpha \in (0, 2)$ and $\beta \in \mathbb{C}$. Then for any $\epsilon > 0$ and φ such that $\frac{\pi\alpha}{2} < \varphi < \min(\pi, \alpha\pi)$, there hold*

$$E_{\alpha,\beta}(z) = \frac{1}{2\alpha\pi i} \int_{\Gamma_{\epsilon,\varphi}} \frac{e^{\zeta^{\frac{1}{\alpha}}} \zeta^{\frac{1-\beta}{\alpha}}}{\zeta - z} d\zeta, \quad \forall z \in G^-(\epsilon, \varphi), \tag{3.8}$$

$$E_{\alpha,\beta}(z) = \frac{1}{\alpha} z^{\frac{1-\beta}{\alpha}} e^{z^{\frac{1}{\alpha}}} + \frac{1}{2\alpha\pi i} \int_{\Gamma_{\epsilon,\varphi}} \frac{e^{\zeta^{\frac{1}{\alpha}}} \zeta^{\frac{1-\beta}{\alpha}}}{\zeta - z} d\zeta, \quad \forall z \in G^+(\epsilon, \varphi). \tag{3.9}$$

Proof The proof employs the integral representation of $\frac{1}{\Gamma(z)}$ in (3.7). If $|z| < \epsilon$, then $|\frac{z}{\zeta}| < 1$ for $\zeta \in \Gamma_{\epsilon,\varphi}$. It follows from (3.7) that for $0 < \alpha < 2$ and $|z| < \epsilon$, there holds

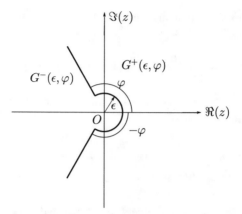

Fig. 3.1 The contour $\Gamma_{\epsilon,\varphi}$, which divides $\mathbb{C}$ into two regions $G^{\pm}(\epsilon, \varphi)$

$$E_{\alpha,\beta}(z) = \sum_{k=0}^{\infty} \frac{1}{2\alpha\pi i} \int_{\Gamma_{\epsilon,\varphi}} e^{\zeta^{\frac{1}{\alpha}}} \zeta^{\frac{1-\beta}{\alpha}-k-1} d\zeta z^k$$

$$= \frac{1}{2\alpha\pi i} \int_{\Gamma_{\epsilon,\varphi}} e^{\zeta^{\frac{1}{\alpha}}} \zeta^{\frac{1-\beta}{\alpha}-1} \sum_{k=0}^{\infty} \left(\frac{z}{\zeta}\right)^k d\zeta = \frac{1}{2\alpha\pi i} \int_{\Gamma_{\epsilon,\varphi}} \frac{e^{\zeta^{\frac{1}{\alpha}}} \zeta^{\frac{1-\beta}{\alpha}}}{\zeta - z} d\zeta.$$

Under the condition on φ, the last integral is absolutely convergent and defines a function of z, which is analytic in both $G^-(\epsilon, \varphi)$ and $G^+(\epsilon, \varphi)$. Since for every $\varphi \in (\frac{\alpha}{2}\pi, \min(\pi, \alpha\pi))$, the circle $\{z \in \mathbb{C} : |z| < \epsilon\}$ lies in the region $G^-(\epsilon, \varphi)$, by analytic continuation, the integral is equal to $E_{\alpha,\beta}(z)$ not only in the circle $\{z \in \mathbb{C} : |z| < \epsilon\}$, but also in the entire domain $G^-(\epsilon, \varphi)$, and thus (3.8) follows. Now we turn to $z \in G^+(\epsilon, \varphi)$. For any $\epsilon_1 > |z|$, we have $z \in G^-(\epsilon_1, \varphi)$, and thus (3.8) holds. However, if $\epsilon < |z| < \epsilon_1$ and $|\arg(z)| < \varphi$, then the Cauchy theorem gives

$$\frac{1}{2\alpha\pi i} \int_{\Gamma_{\epsilon_1,\varphi}-\Gamma_{\epsilon,\varphi}} \frac{e^{\zeta^{\frac{1}{\alpha}}} \zeta^{\frac{1-\beta}{\alpha}}}{\zeta - z} d\zeta = \frac{1}{\alpha} z^{\frac{1-\beta}{\alpha}} e^{z^{\frac{1}{\alpha}}}.$$

From the preceding two identities, the representation (3.9) follows directly. □

Remark 3.1 For the special case $\beta = 1$, the integral representation is given by

$$E_{\alpha,1}(z) = \frac{1}{2\alpha\pi i} \int_{\Gamma_{\epsilon,\varphi}} \frac{e^{\zeta^{\frac{1}{\alpha}}}}{\zeta - z} d\zeta, \quad \forall z \in G^-(\epsilon, \varphi). \tag{3.10}$$

This formula was known to Mittag-Leffler in 1905 [ML05, (56), p. 135]. There are alternative ways, by deforming the contour properly. The following is one variant

$$E_{\alpha,\beta}(z) = \frac{1}{2\pi i} \int_C \frac{\zeta^{\alpha-\beta} e^\zeta}{\zeta^\alpha - z} d\zeta, \tag{3.11}$$

where the contour C is a loop which starts and ends at $-\infty$, and encircles the circular disk $|\zeta| \leq |z|^{\frac{1}{\alpha}}$ in the positive sense; $|\arg(\zeta)| \leq \pi$ on C. The integrand has a branch point at $\zeta = 0$. The complex plane $\mathbb{C}$ is cut along $\mathbb{R}_-$, and in the cut plane, the integrand is single valued: the principal branch of ζ^α is taken in the cut plane.

Theorem 3.1 is fundamental to the study of $E_{\alpha,\beta}(z)$. One important property of $E_{\alpha,\beta}(z)$ is its asymptotic behavior as $z \to \infty$ in various sectors of $\mathbb{C}$. It was first derived by Djrbashian [Dzh66], and subsequently refined [WZ02, Par02]. The asymptotic for $E_{\alpha,\beta}(z)$ with $\alpha \in (0, 2)$ is given below.

Theorem 3.2 Let $\alpha \in (0, 2)$, $\beta \in \mathbb{R}$ and $\varphi \in (\frac{\alpha}{2}\pi, \min(\pi, \alpha\pi))$, and $N \in \mathbb{N}$. Then for $|\arg(z)| \leq \varphi$ with $|z| \to \infty$, we obtain

$$E_{\alpha,\beta}(z) = \frac{1}{\alpha} z^{\frac{1-\beta}{\alpha}} e^{z^{\frac{1}{\alpha}}} - \sum_{k=1}^{N} \frac{1}{\Gamma(\beta - \alpha k)} \frac{1}{z^k} + O\left(\frac{1}{z^{N+1}}\right), \qquad (3.12)$$

and for $\varphi \leq |\arg(z)| \leq \pi$ with $|z| \to \infty$

$$E_{\alpha,\beta}(z) = -\sum_{k=1}^{N} \frac{1}{\Gamma(\beta - \alpha k)} \frac{1}{z^k} + O\left(\frac{1}{z^{N+1}}\right). \qquad (3.13)$$

Proof To show (3.12), we take $\phi \in (\varphi, \min(\pi, \alpha\pi))$. By the identity

$$\frac{1}{\zeta - z} = -\sum_{k=1}^{N} \frac{\zeta^{k-1}}{z^k} + \frac{\zeta^N}{z^N(\zeta - z)}$$

and the representation (3.8) with $\epsilon = 1$, for any $z \in G^+(1, \phi)$, there holds

$$E_{\alpha,\beta}(z) = \frac{1}{\alpha} z^{\frac{1-\beta}{\alpha}} e^{z^{\frac{1}{\alpha}}} - \sum_{k=1}^{N} \frac{1}{2\alpha\pi i} \int_{\Gamma_{1,\phi}} e^{\zeta^{\frac{1}{\alpha}}} \zeta^{\frac{1-\beta}{\alpha}+k-1} d\zeta z^{-k} + I_N,$$

where the integral I_N is defined by

$$I_N = \frac{1}{2\alpha\pi i z^N} \int_{\Gamma_{1,\phi}} e^{\zeta^{\frac{1}{\alpha}}} \frac{\zeta^{\frac{1-\beta}{\alpha}+N}}{\zeta - z} d\zeta. \qquad (3.14)$$

By (3.7), the first integral can be evaluated by

$$\frac{1}{2\alpha\pi i} \int_{\Gamma_{1,\phi}} e^{\zeta^{\frac{1}{\alpha}}} \zeta^{\frac{1-\beta}{\alpha}+k-1} d\zeta = \frac{1}{\Gamma(\beta - \alpha k)}, \qquad k \geq 1.$$

It remains to bound I_N. For sufficiently large $|z|$ and $|\arg(z)| \leq \varphi$, we have $\min_{\zeta \in \Gamma_{1,\varphi}} |\zeta - z| = |z| \sin(\phi - \varphi)$, and hence

$$|I_N(z)| \leq \frac{|z|^{-N-1}}{2\alpha\pi \sin(\phi - \varphi)} \int_{\Gamma_{1,\phi}} |e^{\zeta^{\frac{1}{\alpha}}}| |\zeta^{\frac{1-\beta}{\alpha}+N}| d\zeta.$$

For $\zeta \in \Gamma_{1,\phi}$, $\arg(\zeta) = \pm\phi$ and $|\zeta| \geq 1$, we have $|e^{\zeta^{\frac{1}{\alpha}}}| = e^{|\zeta|^{\frac{1}{\alpha}} \cos\frac{\phi}{\alpha}}$ and due to the choice of ϕ, $\cos(\frac{\phi}{\alpha}) < 0$. Hence, the last integral converges. These estimates together yield (3.12). To show (3.13), we take $\phi \in (\frac{\alpha}{2}\pi, \varphi)$, and the representation (3.8) with $\epsilon = 1$. Then, we obtain

$$E_{\alpha,\beta}(z) = -\sum_{k=1}^{N} \frac{z^{-k}}{\Gamma(\beta - \alpha k)} + I_N(z), \quad z \in G^-(1, \phi),$$

where I_N is defined in (3.14). For large $|z|$ with $\varphi \leq |\arg(z)| \leq \pi$, there holds $\min_{\zeta \in \Gamma_{1,\phi}} |\zeta - z| = |z| \sin(\varphi - \phi)$, and consequently

$$|I_N(z)| \leq \frac{|z|^{-1-N}}{2\pi\alpha \sin(\varphi - \phi)} \int_{\Gamma_{1,\phi}} |e^{\zeta^{\frac{1}{\alpha}}}||\zeta^{\frac{1-\beta}{\alpha}+N}|d\zeta,$$

where the integral converges like before, thereby showing (3.13). $\qquad\square$

Remark 3.2 In the area around the Stokes line $|\arg(z)| \leq \alpha\pi \pm \delta$, with $\delta < \frac{\alpha}{2}\pi$, $E_{\alpha,\beta}(z)$ exhibits the so-called Stokes phenomenon and there exists Berry-type smoothing [WZ02, Section 2]. With the function erfc(z), cf. (3.4), it is given by

$$E_{\alpha,\beta}(z) \sim \frac{1}{2\alpha} z^{\frac{1-\beta}{\alpha}} e^{z^{\frac{1}{\alpha}}} \mathrm{erfc}\left(-c(\theta)\sqrt{\tfrac{1}{2}|z|^{\frac{1}{\alpha}}}\right) - \sum_{k=1}^{\infty} \frac{z^{-k}}{\Gamma(\beta - \alpha k)}, \tag{3.15}$$

around the lower Stokes line for $-\frac{3\alpha}{2}\pi < \arg(z) < -\frac{\alpha}{2}\pi$, where the parameter $c(\theta)$ is given by the relation $\frac{c^2}{2} = 1 + i\theta - e^{i\theta}$, with $\theta = \arg(z^{\frac{1}{\alpha}}) + \pi$ and the principal branch of c is chosen such that $c \approx \theta + i\frac{\theta^2}{6} - \frac{\theta^3}{36}$ for small θ. Around the upper Stokes line $\frac{\alpha}{2}\pi < \arg(z) < \frac{3\alpha}{2}\pi$, one finds

$$E_{\alpha,\beta}(z) \sim \frac{1}{2\alpha} z^{\frac{1-\beta}{\alpha}} e^{z^{\frac{1}{\alpha}}} \mathrm{erfc}\left(c(\theta)\sqrt{\tfrac{1}{2}|z|^{\frac{1}{\alpha}}}\right) - \sum_{k=1}^{\infty} \frac{z^{-k}}{\Gamma(\beta - \alpha k)}, \tag{3.16}$$

with $\frac{c^2}{2} = 1 + i\theta - e^{i\theta}$, $\theta = \arg(z^{\frac{1}{\alpha}}) - \pi$, and the same condition as before for small θ. This exponentially improved asymptotic series converges very rapidly for most value of $z \in \mathbb{C}$ with $|z| > 1$. See also the work [Par20] for discussions on the Mittag-Leffler function $E_{\alpha,1}(z)$ on $\mathbb{R}_-$ as $\alpha \to 1^-$.

Theorem 3.2 indicates that $E_{\alpha,\beta}(z)$, with $\alpha \in (0, 2)$ and $\beta - \alpha \notin \mathbb{Z}^- \cup \{0\}$, decays only linearly on $\mathbb{R}_-$, much slower than the exponential decay for e^z. However, on $\mathbb{R}_+$, it can grow super-exponentially, and the growth rate increases dramatically as $\alpha \to 0^+$. We show some plots of $E_{\alpha,1}(x)$, $x \in \mathbb{R}$, in Figs. 3.2 and 3.3. On $\mathbb{R}_+$, $E_{\alpha,1}(x)$ is monotonically increasing, and for fixed x, it is decreasing in α. The behavior on $\mathbb{R}_-$ is more complex. For $\alpha \in (0, 1]$, $E_{\alpha,1}(x)$ is monotonically decreasing, but the rate of decay differs substantially with α: for $\alpha = 1$, it decays exponentially, but for any other $\alpha \in (0, 1)$, it decays only linearly, concurring with Theorem 3.2, with the monotonicity to be proved in Theorem 3.5. For $\alpha \in (1, 2]$, $E_{\alpha,1}(x)$ is no longer positive any more on $\mathbb{R}_-$, and instead it oscillates wildly. For α away from 2, $E_{\alpha,1}(x)$ decays as $x \to -\infty$, and the closer α is to one, the faster the decay is. These behaviors underpin many important properties of the related FDES. In the limiting case $\alpha = 2$, there is no decay: by the identity (3.2),

$$E_2(x) = \cosh i\sqrt{-x} = \cos \sqrt{-x},$$

i.e., it recovers the cosine function, and thus Theorem 3.2 does not apply to $\alpha = 2$.

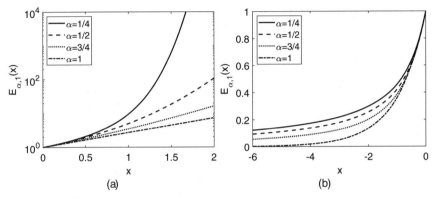

Fig. 3.2 Plots of the function $E_{\alpha,1}(x)$ for $\alpha \in (0, 1]$

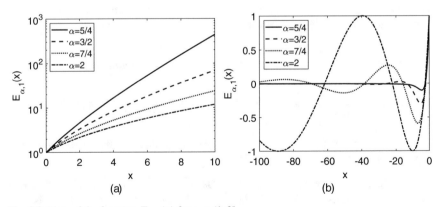

Fig. 3.3 Plots of the function $E_{\alpha,1}(x)$ for $\alpha \in (1, 2]$.

The next result is a direct corollary of Theorem 3.2.

Corollary 3.1 *For $0 < \alpha < 2$, $\beta \in \mathbb{R}$ and $\frac{\alpha}{2}\pi < \varphi < \min(\pi, \alpha\pi)$, there holds*

$$|E_{\alpha,\beta}(z)| \le c_1(1 + |z|)^{\frac{1-\beta}{\alpha}} e^{\Re(z^{\frac{1}{\alpha}})} + c_2(1 + |z|)^{-1}, \quad |\arg(z)| \le \varphi,$$

$$|E_{\alpha,\beta}(z)| \le c(1 + |z|)^{-1}, \quad \varphi \le |\arg(z)| \le \pi.$$

Last, the case $\alpha \ge 2$ can be reduced to the case $\alpha \in (0, 2)$ via the following duplication formula. The notation $[\cdot]$ denotes taking the integral part of a real number.

Lemma 3.1 *For all* $\alpha > 0$, $\beta \in \mathbb{R}$, $z \in \mathbb{C}$ *and* $m = [\frac{(\alpha-1)}{2}] + 1$, *there holds*

$$E_{\alpha,\beta}(z) = \frac{1}{2m+1} \sum_{j=-m}^{m} E_{\frac{\alpha}{2m+1},\beta}(z^{\frac{1}{2m+1}} e^{2\pi \frac{ij}{2m+1}}). \tag{3.17}$$

Proof Recall the following elementary identity:

$$\sum_{j=-m}^{m} e^{\frac{j}{2m+1} 2ki\pi} = \begin{cases} 2m + 1, & \text{if } k \equiv 0 \pmod{2m+1}, \\ 0, & \text{if } k \not\equiv 0 \pmod{2m+1}. \end{cases}$$

This and the definition of the function $E_{\alpha,\beta}(z)$ give

$$\sum_{j=-m}^{m} E_{\alpha,\beta}(ze^{2\pi i \frac{j}{2m+1}}) = (2m + 1)E_{(2m+1)\alpha,\beta}(z^{2m+1}).$$

Substituting $m\alpha$ by α and $z^{\frac{1}{m}}$ by z gives the desired identity. $\qquad\square$

Using Lemma 3.1, the following result holds. The statement is taken from [PS11, Theorem 1.2.2], which corrects one minor error in the summation range that has appeared in many popular textbooks (see [PS11, pp. 220–221] for a counterexample).

Theorem 3.3 *Let* $\alpha \ge 2$, $\beta \in \mathbb{R}$, $N \in \mathbb{N}$, *the following asymptotic behavior holds:*

$$E_{\alpha,\beta}(z) = \frac{1}{\alpha} \sum (z^{\frac{1}{\alpha}} e^{\frac{2m\pi i}{\alpha}})^{1-\beta} e^{z^{\frac{1}{\alpha}} e^{\frac{2m\pi i}{\alpha}}} - \sum_{k=1}^{N} \frac{z^{-k}}{\Gamma(\beta - k\alpha)} + O(|z|^{-N-1}),$$

as $z \to \infty$, *where the first sum is taken for integer* m *satisfying the condition* $|\arg(z) + 2m\pi| \le \frac{3\alpha\pi}{4}$.

Last, we discuss the distribution of zeros of $E_{\alpha,\beta}(z)$, which is of independent interest because of its crucial role in the study of Fourier-Laplace-type integrals with Mittag-Leffler kernels [Dzh66] and spectral theory of FDES [Nak77], to name a few. Thus the issue has been extensively discussed. Actually, shortly after the appearance of the work of Mittag-Leffler [ML03] in 1903, Wiman [Wim05] showed in 1905 that for $\alpha \ge 2$, all zeros of the function $E_{\alpha,1}(z)$ are real, negative and simple, and later Pólya [Pól18] reproved this result for $2 \le \alpha \in \mathbb{N}$ by a different method. It was revisited by Djrbashian [Dzh66], and many deep results were derived. There are many further refinements [Sed94, OP97, Sed00, Sed04, Psk05a], see [PS11] for an overview. A general result for the case $\alpha \in (0, 2)$ and $\beta \in \mathbb{R}$ is given below, see [PS11, Theorem 2.1.1] for the technical proof.

Theorem 3.4 *Let* $\alpha \in (0, 2)$, *and* $\beta \in \mathbb{R}$, *where* $\beta \neq 1, 0, -1, -2, \ldots$ *for* $\alpha = 1$. *Then all sufficiently large zeros* z_n *(in modulus) of the function* $E_{\alpha,\beta}(z)$ *are simple and the following asymptotic formula holds:*

$$z_n^{\frac{1}{\alpha}} = 2\pi i n + \alpha \tau_\beta \ln 2\pi i n + \ln c_\beta + \frac{d_\beta}{c_\beta(2\pi i n)^\alpha} + (\alpha \tau_\beta)^2 \frac{\ln 2\pi i n}{2\pi i n} - \alpha \tau_\beta \frac{c_\beta}{2\pi i n} + r_n,$$

as $n \to \pm\infty$, *where the constants* c_β, d_β *and* τ_β *are defined by*

$$c_\beta = \frac{\alpha}{\Gamma(\beta - \alpha)}, \ d_\beta = \frac{\alpha}{\Gamma(\beta - 2\alpha)}, \ \tau_\beta = 1 + \frac{1 - \beta}{\alpha}, \ \text{if } \beta \neq \alpha - \ell, \ell \in \mathbb{N},$$

$$c_\beta = \frac{\alpha}{\Gamma(\beta - 2\alpha)}, \ d_\beta = \frac{\alpha}{\Gamma(\beta - 3\alpha)}, \ \tau_\beta = 2 + \frac{1 - \beta}{\alpha}, \ \text{if } \beta = \alpha - \ell, \ell \in \mathbb{N}, \alpha \neq \mathbb{N},$$

and the remainder r_n *is given by*

$$r_n = O\left(\frac{\ln |n|}{|n|^{1+\alpha}}\right) + O\left(\frac{1}{|n|^{2\alpha}}\right) + O\left(\frac{\ln^2 |n|}{n^2}\right).$$

The following result specializes Theorem 3.4. We sketch a short derivation to give a flavor of the proof of Theorem 3.4, see [JR12].

Proposition 3.2 *Let* $\alpha \in (0, 2)$ *and* $\alpha \neq 1$. *Then all sufficiently large zeros* $\{z_n\}$ *(in modulus) of the function* $E_{\alpha,2}(z)$ *are simple, and as* $n \to \pm\infty$, *there holds*

$$z_n^{\frac{1}{\alpha}} = 2n\pi i - (\alpha - 1)\left(\ln 2\pi |n| + \frac{\pi}{2}\text{sign}(n)\, i\right) + \ln \frac{\alpha}{\Gamma(2 - \alpha)} + r_n,$$

where the remainder r_n *is* $O\left(\frac{\ln |n|}{|n|}\right)$.

Proof Taking $N = 1$ in (3.12) gives

$$E_{\alpha,2}(z) = \frac{1}{\alpha} z^{-\frac{1}{\alpha}} e^{z^{\frac{1}{\alpha}}} - \frac{1}{\Gamma(2 - \alpha)}\frac{1}{z} + O\left(z^{-2}\right), \quad \text{as } |z| \to \infty.$$

Hence we have $z^{1-\frac{1}{\alpha}} e^{z^{\frac{1}{\alpha}}} = \frac{\alpha}{\Gamma(2-\alpha)} + O(z^{-1})$. Next with $\zeta = z^{\frac{1}{\alpha}}$ and $w = \zeta + (\alpha-1)\ln\zeta$, it can be rewritten as $e^w = \frac{\alpha}{\Gamma(2-\alpha)} + O(w^{-\alpha})$. The roots w_n for all sufficiently large n satisfy

$$w_n = 2\pi n\, i + \ln \frac{\alpha}{\Gamma(2 - \alpha)} + O(|n|^{-\alpha}),$$

or equivalently

$$\zeta_n + (\alpha - 1)\ln\zeta_n = 2\pi n\, i + \ln \frac{\alpha}{\Gamma(2 - \alpha)} + O(|n|^{-\alpha}).$$

Then the desired assertion follows from this identity.   □

In Fig. 3.4, we show the contour plot of $|E_{\alpha,2}(z)|$ and $|E_{\alpha,\alpha}(z)|$, with $\alpha = \frac{4}{3}$, over $\mathbb{C}_-$ in a logarithmic scale. They are eigenfunctions of the fractional Sturm-Liouville problem, and thus the behavior of the roots is important, see Section 5.3 in Chapter 5. The deep wells in the plots correspond to the zeros. In both cases, the zeros always appear in complex conjugate pairs, and there is only a finite number of real zeros. This latter is predicted by Theorem 3.4: for any $\alpha \in (0, 2)$, asymptotically the zeros with large magnitude must lie on two rays in the complex plane $\mathbb{C}$. However, the existence of a real zero for general $E_{\alpha,\beta}(z)$ remains poorly understood. The wells run much deeper for $E_{\alpha,\alpha}(z)$ than $E_{\alpha,2}(z)$, which agrees with the faster asymptotic decay predicted by Theorem 3.2.

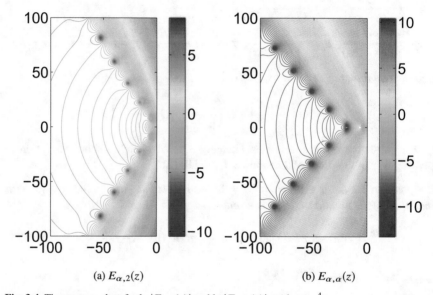

(a) $E_{\alpha,2}(z)$ (b) $E_{\alpha,\alpha}(z)$

Fig. 3.4 The contour plots for $\ln |E_{\alpha,2}(z)|$ and $\ln |E_{\alpha,\alpha}(z)|$, with $\alpha = \frac{4}{3}$.

Besides the distribution of roots, the basis property or completeness of Mittag-Leffler functions has also been extensively studied in the Russian literature [Džr74, DM75, Hač75, DM77, DM83], and a necessary and sufficient condition generalizing the classical Müntz theorem to Mittag-Leffler functions was established [Džr74].

3.1.2 Mittag-Leffler function $E_{\alpha,1}(-x)$

The function $E_{\alpha,1}(-x)$ with $x \in \mathbb{R}_+$ is especially important in practice. We describe its complete monotonicity and asymptotics. A function $f(x) : \overline{\mathbb{R}_+} \to \mathbb{R}$ is called completely monotone if $(-1)^n f^{(n)}(x) \geq 0$, $n = 0, 1, \ldots$. For example, e^{-x} and

$f(x) = (x + c_0)^\beta$, with $c_0 > 0$ and $\beta \in [-1, 0)$, are completely monotone. The next result due to Pollard [Pol48] gives the complete monotonicity of $E_{\alpha,1}(-x)$.

Theorem 3.5 *For $\alpha \in [0, 1]$, $E_{\alpha,1}(-x)$ is completely monotone, and*

$$E_{\alpha,1}(-x) = \int_0^\infty e^{-sx} F_\alpha'(s) ds,$$

with

$$F_\alpha'(s) = \frac{1}{\alpha\pi} \sum_{k=1}^\infty \frac{(-1)^k}{k!} \sin \pi\alpha k \Gamma(\alpha k + 1) s^{k-1}.$$

Proof Since $E_{0,1}(-x) = \frac{1}{1+x}$ and $E_{1,1}(-x) = e^{-x}$, these two cases hold trivially. It suffices to show the case $0 < \alpha < 1$. By the integral representation (3.10), we have

$$E_{\alpha,1}(-x) = \frac{1}{2\alpha\pi i} \int_{\Gamma_{\epsilon,\mu}} \frac{e^{\zeta^{\frac{1}{\alpha}}}}{\zeta + x} d\zeta, \tag{3.18}$$

with $\mu \in (\frac{\alpha}{2}\pi, \alpha\pi)$ and arbitrary but fixed $\epsilon > 0$. Using the identity $(x + \zeta)^{-1} = \int_0^\infty e^{-(\zeta+x)s} ds$ in (3.18) and the absolute convergence of the double integral, it follows from Fubini's theorem that

$$E_{\alpha,1}(-x) = \int_0^\infty e^{-xs} F_\alpha'(s) ds$$

with

$$F_\alpha'(s) = \frac{1}{2\alpha\pi i} \int_{\Gamma_{\epsilon,\mu}} e^{\zeta^{\frac{1}{\alpha}}} e^{-\zeta s} d\zeta.$$

By Bernstein's theorem in Theorem A.8, it remains to show $F_\alpha'(s) \geq 0$ for all $s \geq 0$. Integration by parts yields

$$F_\alpha'(s) = \frac{1}{2\alpha\pi i s} \int_{\Gamma_{\epsilon,\mu}} e^{-\zeta s} \frac{\zeta^{\frac{1}{\alpha}-1}}{\alpha} e^{\zeta^{\frac{1}{\alpha}}} d\zeta.$$

Substituting $\zeta s = z^\alpha$ gives

$$F_\alpha'(s) = \frac{s^{-1-\frac{1}{\alpha}}}{\alpha} \frac{1}{2\pi i} \int_{\Gamma_{\epsilon,\mu}'} e^{-z^\alpha} e^{zs^{-\frac{1}{\alpha}}} dz, \tag{3.19}$$

where $\Gamma_{\epsilon,\mu}'$ is the image of $\Gamma_{\epsilon,\mu}$ under the mapping. Let $\phi_\alpha(\zeta) = \frac{1}{2\pi i} \int_{\Gamma_{\epsilon,\mu}'} e^{-z^\alpha} e^{z\zeta} dz$, which is the inverse Laplace transform of $e^{-z^\alpha} = \int_0^\infty e^{-z\zeta} \phi_\alpha(\zeta) d\zeta$, which is completely monotone for any $\alpha \in (0, 1)$ [Pol46], see also Exercise 3.13 for related discussions. Hence

$$F_\alpha'(s) = \alpha^{-1} s^{-1-\frac{1}{\alpha}} \phi_\alpha(s^{-\frac{1}{\alpha}}) \geq 0,$$

and $E_{\alpha,1}(-x)$ is completely monotone. Finally, we derive the expression of $F_\alpha'(s)$. Indeed, one can deform the contour $\Gamma_{\epsilon,\mu}'$ to $\Gamma_{\epsilon,\varphi}$, with $\epsilon > 0$ and $\varphi \in (\frac{\pi}{2}, \pi)$, without traversing any singularity of the integrand,

$$\phi_\alpha(s) = \frac{1}{2\pi i} \int_{\Gamma_{\epsilon,\varphi}} e^{-z^\alpha} e^{zs} dz.$$

Now substituting $z = \epsilon e^{i\theta}$ on the circular arc (with $\theta \in [-\varphi, \varphi]$) and $z = re^{\pm i\varphi}$ (with $r \geq \epsilon$) on the two rays gives

$$\phi_\alpha(s) = \frac{1}{2\pi i} \int_{-\varphi}^{\varphi} e^{-\epsilon^\alpha e^{\alpha\theta i}} e^{\epsilon e^{i\theta} s} \epsilon i e^{i\theta} d\theta + \frac{1}{2\pi i} \int_{\epsilon}^{\infty} e^{-r^\alpha e^{i\alpha\varphi}} e^{re^{i\varphi} s} e^{i\varphi} dr$$
$$+ \frac{1}{2\pi i} \int_{\epsilon}^{\infty} e^{-r^\alpha e^{-i\alpha\varphi}} e^{re^{-i\varphi} s} e^{-i\varphi} dr.$$

Letting $\epsilon \to 0^+$ and $\varphi \to \pi^-$ yields

$$\phi_\alpha(s) = \frac{1}{\pi} \Im\left(\int_0^\infty e^{-sr} e^{-(e^{-i\pi} r)^\alpha} dr \right) = \frac{1}{\pi} \Im\left(\sum_{k=0}^\infty \frac{(-1)^k}{k!} e^{-i\pi\alpha k} \frac{\Gamma(k\alpha + 1)}{s^{k\alpha+1}} \right)$$
$$= -\frac{1}{\pi} \sum_{k=0}^\infty \frac{(-1)^k}{k!} \sin(k\alpha\pi) \frac{\Gamma(k\alpha + 1)}{s^{k\alpha+1}}.$$

Substituting this expression, we find the formula of $F'_\alpha(s)$. $\square$

Schneider [Sch96] improved Theorem 3.5 that $E_{\alpha,\beta}(-x)$ is completely monotone on $\mathbb{R}_+$ for $\alpha > 0, \beta > 0$ if and only if $\alpha \in (0, 1]$ and $\beta \geq \alpha$. His proof employs the corresponding probability measures and the Hankel contour integration. Miller and Samko [MS98] presented an alternative (and shorter) proof using Abelian transformation given below, see also [Djr93, Theorem 1.3-8]. These results were extended by [Sim15] to several functions related to Mittag-Leffler functions.

Corollary 3.2 *For $\alpha \in [0, 1]$ and $\beta \geq \alpha$, $E_{\alpha,\beta}(-x)$ is completely monotone on $\mathbb{R}_+$.*

Proof Observe that $E_{0,\beta}(-x) = (\Gamma(\beta)(1 + x)^{-1}$ and $E_{0,0}(-x) = 0$. In either case, $E_{0,\beta}(-x)$ is completely monotone. More generally, for $\beta > \alpha > 0$, there holds

$$E_{\alpha,\beta}(-x) = \frac{1}{\alpha\Gamma(\beta - \alpha)} \int_0^1 (1 - t^{\frac{1}{\alpha}})^{\beta-\alpha-1} E_{\alpha,\alpha}(-tx) dt. \tag{3.20}$$

Indeed, with the series representation of $E_{\alpha,\alpha}(-tx)$, changing variables $t = s^\alpha$ and the identity (2.10), the right-hand side can be evaluated to be

$$\frac{1}{\alpha\Gamma(\beta-\alpha)}\sum_{k=0}^{\infty}\frac{(-x)^k}{\Gamma(k\alpha+\alpha)}\int_0^1(1-t^{\frac{1}{\alpha}})^{\beta-\alpha-1}t^k\,dt$$

$$=\frac{1}{\Gamma(\beta-\alpha)}\sum_{k=0}^{\infty}\frac{(-x)^k}{\Gamma(k\alpha+\alpha)}\int_0^1(1-s)^{\beta-\alpha-1}s^{k\alpha+\alpha-1}\,ds$$

$$=\frac{1}{\Gamma(\beta-\alpha)}\sum_{k=0}^{\infty}\frac{B(\beta-\alpha,(k+1)\alpha)(-x)^k}{\Gamma(k\alpha+\alpha)}=\sum_{k=0}^{\infty}\frac{(-x)^k}{\Gamma(k\alpha+\beta)}=E_{\alpha,\beta}(-x).$$

This shows (3.20). Next since $E_{\alpha,1}(-x)$ is entire, using the recursion formula (2.2) for Gamma function $\Gamma(z)$, we deduce

$$\frac{d}{dx}E_{\alpha,1}(-x)=\sum_{k=0}^{\infty}\frac{d}{dx}\frac{(-x)^k}{\Gamma(k\alpha+1)}=\sum_{k=1}^{\infty}\frac{(-1)^k kx^{k-1}}{k\alpha\Gamma(k\alpha)}=-\alpha^{-1}E_{\alpha,\alpha}(-x). \quad (3.21)$$

The identities (3.20) and (3.21) and Theorem 3.5 complete the proof.   □

The positivity of $E_{\alpha,\alpha}(-x)$ is a simple consequence of Theorem 3.5.

Corollary 3.3 *For $0<\alpha<1$, we have $E_{\alpha,\alpha}(-x)>0$ on $\mathbb{R}_+$.*

Proof Assume the contrary that it vanishes at some $x=x_0>0$. By (3.21) and Theorem 3.5, $E_{\alpha,\alpha}(-x)$ is completely monotone, and thus it vanishes for all $x\geq x_0$. The analyticity of $E_{\alpha,\alpha}(-x)$ in x implies then it vanishes identically over $\mathbb{R}_+$, which contradicts the fact that $E_{\alpha,\alpha}(0)=\Gamma(\alpha)^{-1}$.   □

The following two-sided bounds on $E_{\alpha,1}(-x)$ are useful. Simon [Sim14, Theorem 4] derived the bounds using a probabilistic argument. The current proof is due to [VZ15, Lemma 6.1], which applies to more general completely positive kernels, using the corresponding resolvent.

Theorem 3.6 *For every $\alpha\in(0,1)$, the following bounds hold:*

$$(1+\Gamma(1-\alpha)x)^{-1}\leq E_{\alpha,1}(-x)\leq(1+\Gamma(1+\alpha)^{-1}x)^{-1},\quad\forall x\in\mathbb{R}_+.$$

Proof By changing variables $x=t^\alpha$ and letting $g(t)=E_{\alpha,1}(-t^\alpha)$, it suffices to prove

$$(1+\Gamma(1-\alpha)t^\alpha)^{-1}\leq g(t)\leq(1+\Gamma(1+\alpha)^{-1}t^\alpha)^{-1},\quad\forall t\geq0.$$

For the function $g(t)$, it follows directly from (3.5) that

$$g(t)+{}_0I_t^\alpha g(t)=1,\quad t\geq0. \quad (3.22)$$

By Theorem 3.5, $g(t)$ is monotonically decreasing, which implies

$${}_0I_t^\alpha g(t)=\frac{1}{\Gamma(\alpha)}\int_0^t(t-s)^{\alpha-1}g(s)\,ds\geq\frac{g(t)}{\Gamma(\alpha)}\int_0^t(t-s)^{\alpha-1}\,ds=\frac{t^\alpha}{\Gamma(\alpha+1)}g(t),$$

and therefore

$$1 = g(t) + {}_0I_t^\alpha g(t) \geq [1 + \Gamma(\alpha + 1)^{-1}t^\alpha]g(t).$$

This gives the desired upper bound. Next, let

$$h(t) = -g'(t) = t^{\alpha-1}E_{\alpha,\alpha}(-t^\alpha) > 0, \tag{3.23}$$

cf. Corollary 3.3. Consequently, using (3.5) again, we deduce

$${}_0I_t^{1-\alpha}h(t) = {}_0I_t^{1-\alpha}[t^{\alpha-1}E_{\alpha,\alpha}(-t^\alpha)] = g(t).$$

Then by Theorem 2.1 and the identity (3.22),

$$g(t) = {}_0I_t^{1-\alpha}h(t) = \frac{1}{\Gamma(1-\alpha)}\int_0^t (t-s)^{-\alpha}h(s)\mathrm{d}s \geq \frac{t^{-\alpha}}{\Gamma(1-\alpha)}\int_0^t h(t-s)\mathrm{d}s$$

$$= \frac{t^{-\alpha}}{\Gamma(1-\alpha)}{}_0I_t^\alpha {}_0I_t^{1-\alpha}h(t) = \frac{t^{-\alpha}{}_0I_t^\alpha g(t)}{\Gamma(1-\alpha)} = \frac{t^{-\alpha}(1-g(t))}{\Gamma(1-\alpha)},$$

where the inequality follows from the positivity of h over $\mathbb{R}_+$ in (3.23). Rearranging this inequality gives the lower bound. $\qquad\square$

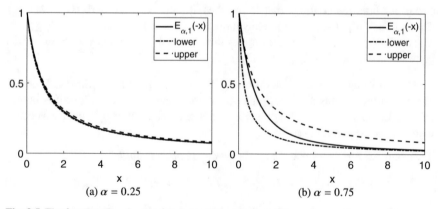

Fig. 3.5 The function $E_{\alpha,1}(-x)$ and its lower and upper bounds

The result can be viewed as rational approximations in t^α:

$$\frac{1}{1 + \Gamma(1+\alpha)^{-1}t^\alpha} \sim 1 - \frac{t^\alpha}{\Gamma(1+\alpha)} \sim E_{\alpha,1}(-t^\alpha), \quad \text{as } t \to 0^+,$$

$$\frac{1}{1 + \Gamma(1-\alpha)t^\alpha} \sim \frac{t^{-\alpha}}{\Gamma(1-\alpha)} \sim E_{\alpha,1}(-t^\alpha), \quad \text{as } t \to \infty.$$

The asymptotics as $t \to 0^+$ and $t \to \infty$ are also known as stretched exponential decay and negative power law, respectively. The lower and upper bounds were also empirically observed by Mainardi [Mai14]. In Fig. 3.5, we plot $E_{\alpha,1}(-x)$ and its lower and upper bounds given in Theorem 3.6. The approximations are quite sharp for α close to zero, but they are less sharp for α close to one.

In general, the Laplace transform of $E_{\alpha,\beta}(x)$ is a Wright function $W_{\rho,\mu}(z)$, but that of $x^{\beta-1}E_{\alpha,\beta}(-\lambda x^{\alpha})$ does take a simple form. The minus sign in the argument is to ensure the existence of Laplace transform, in view of the exponential asymptotics of $E_{\alpha,\beta}(z)$ in Theorem 3.2. The functions $E_{\alpha,1}(-\lambda x^{\alpha})$ and $x^{\alpha-1}E_{\alpha,\alpha}(-\lambda x^{\alpha})$, $\lambda \in \mathbb{R}$ often appear in the study of fractional ODEs. Their Laplace transforms are given by

$$\mathcal{L}[E_{\alpha,1}(-\lambda x^{\alpha})](z) = \frac{z^{\alpha-1}}{\lambda + z^{\alpha}} \quad \text{and} \quad \mathcal{L}[x^{\alpha-1}E_{\alpha,\alpha}(-\lambda x^{\alpha})](z) = \frac{1}{\lambda + z^{\alpha}}.$$

Lemma 3.2 *Let $\lambda \geq 0$, $\alpha > 0$ and $\beta > 0$. Then*

$$\mathcal{L}[x^{\beta-1}E_{\alpha,\beta}(-\lambda x^{\alpha})](z) = \frac{z^{\alpha-\beta}}{z^{\alpha} + \lambda}, \quad \mathcal{R}(z) > 0,$$

$$\mathcal{L}[E_{\alpha,\beta}(-\lambda x)](z) = z^{-1}W_{\alpha,\beta}(-\lambda z^{-1}).$$

Proof By the series expansion of $x^{\beta-1}E_{\alpha,\beta}(-\lambda x^{\alpha})$, we have

$$\mathcal{L}[x^{\beta-1}E_{\alpha,\beta}(-\lambda x^{\alpha})](z) = \int_0^{\infty} e^{-zx} x^{\beta-1} \sum_{k=0}^{\infty} \frac{(-\lambda x^{\alpha})^k}{\Gamma(k\alpha + \beta)}dx$$

$$= \sum_{k=0}^{\infty}(-\lambda)^k \int_0^{\infty} \frac{x^{k\alpha+\beta-1}}{\Gamma(k\alpha + \beta)}e^{-zx}dx = \sum_{k=0}^{\infty}(-\lambda)^k z^{-k\alpha-\beta} = \frac{z^{\alpha-\beta}}{z^{\alpha} + \lambda}.$$

Clearly, this identity holds for all $\mathcal{R}(z) > \lambda^{\frac{1}{\alpha}}$, for which changing the order of summation and integral is justified. Since $\int_0^{\infty} e^{-zx} x^{\beta-1} E_{\alpha,\beta}(-\lambda x^{\alpha})dx$ is analytic with respect to $z \in \mathbb{C}_+$, analytic continuation yields the identity for $z \in \mathbb{C}_+$. Similarly, by the series expansion of $E_{\alpha,\beta}(-\lambda x)$, we have

$$\mathcal{L}[E_{\alpha,\beta}(-\lambda x)](z) = \int_0^{\infty} e^{-zx} \sum_{k=0}^{\infty} \frac{(-\lambda x)^k}{\Gamma(k\alpha + \beta)}dx = \sum_{k=0}^{\infty}(-\lambda)^k \int_0^{\infty} \frac{x^k e^{-zx}}{\Gamma(k\alpha + \beta)}dx$$

$$= \sum_{k=0}^{\infty} \frac{(-\lambda)^k}{z^{k+1}\Gamma(k + 1)\Gamma(k\alpha + \beta)} = z^{-1}W_{\alpha,\beta}(-\lambda z^{-1}).$$

This completes the proof of the lemma. $\square$

The next example illustrates the occurrence of the function $E_{\alpha,1}(-\lambda x^{\alpha})$.

Example 3.1 Consider the following fractional ODE, with $0 < \alpha < 1$ and $\lambda > 0$: find u satisfying

$${}_0^C D_x^{\alpha} u(x) + \lambda u(x) = 0, \quad \text{for } x > 0, \quad \text{with } u(0) = 1.$$

It is commonly known as the fractional relaxation equation. We claim that the solution $u(x)$ is given by $u(x) = E_{\alpha,1}(-\lambda x^{\alpha})$. In fact, the initial condition $u(0) = 1$

holds trivially, and further, we have

$$
{}_0^C D_x^\alpha u(x) = {}_0^C D_x^\alpha \sum_{k=0}^\infty (-\lambda)^k \frac{x^{k\alpha}}{\Gamma(k\alpha + 1)} = \sum_{k=1}^\infty (-\lambda)^k \frac{x^{(k-1)\alpha}}{\Gamma((k-1)\alpha + 1)}
$$

$$
= -\lambda \sum_{k=0}^\infty (-\lambda)^k \frac{x^{k\alpha}}{\Gamma(k\alpha + 1)} = -\lambda u(x).
$$

Alternatively, by Lemma 2.9, the Laplace transform $\widehat{u}$ of u satisfies $z^\alpha \widehat{u}(z) - z^{\alpha-1} + \lambda \widehat{u}(z) = 0$, and thus $\widehat{u}(z) = \frac{z^{\alpha-1}}{z^\alpha + \lambda}$. Then Lemma 3.2 gives the desired expression. Note that in the limit of $\alpha = 1$, it recovers the familiar formula $u(x) = e^{-\lambda x}$. The function $E_{\alpha,1}(-\lambda x^\alpha)$ is continuous but not differentiable at $x = 0$, which contrasts sharply with $C^\infty[0, \infty)$ regularity of $e^{-\lambda x}$ for $\alpha = 1$.

3.2 Wright Function

For μ, $\rho \in \mathbb{R}$ with $\rho > -1$, the Wright function $W_{\rho,\mu}(z)$ is defined by

$$
W_{\rho,\mu}(z) = \sum_{k=0}^\infty \frac{z^k}{k!\Gamma(k\rho + \mu)}, \quad z \in \mathbb{C}. \tag{3.24}
$$

It was first introduced in connection with a problem in number theory (the asymptotic theory of partitions) by Edward M. Wright in a series of notes starting from 1933 [Wri33, Wri35b, Wri35a]. Originally, Wright assumed that $\rho \geq 0$ and, only in 1940 [Wri40], did he consider the case $-1 < \rho < 0$, for which the function is still entire but exhibits certain different features. An early overview of the properties of $W_{\rho,\mu}(z)$ is given by Stankovic [Sta70], including integral representations, various special cases, functional relation, sign and majorants, etc. The function $W_{\rho,\mu}(z)$ generalizes the exponential function $W_{0,1}(z) = e^z$, and the Bessel function of first kind J_μ and the modified Bessel function of first kind I_μ:

$$
W_{1,\mu+1}\left(-\frac{z^2}{4}\right) = \left(\frac{z}{2}\right)^{-\mu} J_\mu(z), \quad W_{1,\mu+1}\left(\frac{z^2}{4}\right) = \left(\frac{z}{2}\right)^{-\mu} I_\mu(z). \tag{3.25}
$$

Indeed, they follow from direct computation (and the series definitions of J_μ and I_μ)

$$
\left(\frac{z}{2}\right)^\mu W_{1,\mu+1}\left(\mp \frac{z^2}{4}\right) := \left(\frac{z}{2}\right)^\mu \sum_{k=0}^\infty (\mp 1)^k \frac{z^{2k}}{4^k k!\Gamma(k + \mu + 1)}.
$$

Hence in the literature it is often called the generalized Bessel function. Further, it follows from direct computation that the following differentiation relations hold:

$$\frac{d}{dz} W_{\rho,\mu}(z) = W_{\rho,\mu+\rho}(z), \tag{3.26}$$

$$\frac{d^n}{dt^n}(t^{\mu-1}W_{\rho,\mu}(ct^\rho)) = t^{\mu-1-n}W_{\rho,\mu-n}(ct^\rho). \tag{3.27}$$

Actually, since the Wright function is entire, termwise differentiation leads to

$$\frac{d}{dz} W_{\rho,\mu}(z) = \sum_{k=0}^{\infty} \frac{d}{dz} \frac{z^k}{k!\Gamma(k\rho + \mu)} = \sum_{k=0}^{\infty} \frac{z^k}{k!\Gamma(k\rho + \rho + \mu)} = W_{\rho,\rho+\mu}(z),$$

$$\frac{d^n}{dt^n}(t^{\mu-1}W_{\rho,\mu}(ct^\rho)) = \sum_{k=0}^{\infty} \frac{d^n}{dt^n} \frac{c^k t^{k\rho+\mu-1}}{k!\Gamma(k\rho + \mu)}$$

$$= \sum_{k=0}^{\infty} \frac{c^k t^{k\rho+\mu-1-n}}{k!\Gamma(k\rho + \mu - n)} = t^{\mu-n-1}W_{\rho,\mu-n}(ct^\rho).$$

More generally we have the following identity: for all $\alpha, \mu \in \mathbb{R}$, $\rho \in (-1, 0)$ and $c > 0$, (for $\alpha < 0$, $^R_0 D_x^\alpha$ is identified with $_0 I_x^{-\alpha}$)

$$^R_0 D_x^\alpha(x^{\mu-1}W_{\rho,\mu}(-cx^\rho)) = x^{\mu-\alpha-1}W_{\rho,\mu-\alpha}(-cx^\rho). \tag{3.28}$$

3.2.1 Basic analytic properties

The next result gives the order of $W_{\rho,\mu}(z)$. Note that for $\rho \in (-1, 0)$, the function is not of exponential order.

Lemma 3.3 *For any $\rho > -1$ and $\mu \in \mathbb{R}$, the function $W_{\rho,\mu}(z)$ is entire of order $\frac{1}{1+\rho}$ and type $(1 + \rho)|\rho|^{-\frac{1}{1+\rho}}$.*

Proof We sketch the proof only for $\rho \in (-1, 0)$, since the case $\rho \geq 0$ is simpler. Using the reflection formula (2.3), we rewrite $E_{\rho,\mu}(z)$ as

$$W_{\rho,\mu}(z) = \frac{1}{\pi} \sum_{k=0}^{\infty} \frac{\Gamma(1 - k\rho - \mu) \sin(\pi(k\rho + \mu))}{k!} z^k.$$

Next we introduce a majorizing sequence $\frac{1}{\pi} \sum_{k=0}^{\infty} c_k|z|^k$, with $c_k = \frac{|\Gamma(1-k\rho-\mu)|}{k!}$. By Stirling's formula (2.7), $\lim_{k\to\infty} \frac{c_k}{c_{k+1}} = \lim_{k\to\infty} \frac{k+1}{|\rho|^\rho k^{-\rho}} = \infty$. Hence the auxiliary sequence and also the series (3.24) converge over the whole complex plane $\mathbb{C}$. The order and type follow analogously. □

The asymptotic expansions are based on a suitable integral representation [Wri33]

$$W_{\rho,\mu}(z) = \frac{1}{2\pi i} \int_C e^{\zeta + z\zeta^{-\rho}} \zeta^{-\mu} d\zeta, \tag{3.29}$$

where C is any deformed Hankel contour. It follows from (2.6) by

$$W_{\rho,\mu}(z) = \sum_{k=0}^{\infty} \frac{z^k}{k!} \frac{1}{2\pi i} \int_C e^{\zeta} \zeta^{-\rho k - \mu} d\zeta = \frac{1}{2\pi i} \int_C e^{\zeta + z\zeta^{-\rho}} \zeta^{-\mu} d\zeta.$$

The interchange between series and integral is legitimate by the uniform convergence of the series, since $W_{\rho,\mu}(z)$ is an entire function.

We state only the asymptotic expansions without the detailed proof, taken from [Luc08, Theorems 3.1 and 3.2]. Such expansions were first studied by Wright, and then later refined [WZ99a, WZ99b] by including Stokes phenomena. These estimates can be used to deduce the distribution of its zeros [Luc00].

Theorem 3.7 *The following asymptotic formulae hold:*

(i) *Let* $\rho > 0$, $\arg(z) = \theta$, $|\theta| \leq \pi - \epsilon$, $\epsilon > 0$. *Then*

$$W_{\rho,\mu}(z) = Z^{\frac{1}{2}-\mu} e^{\frac{1+\rho}{\rho}Z} \left(\sum_{m=0}^{M} \frac{(-1)^m a_m}{Z^m} + O(|Z|^{-M-1}) \right), \quad Z \to \infty,$$

where $Z = (\rho|z|)^{\frac{1}{\rho+1}} e^{i\frac{\theta}{\rho+1}}$ *and the coefficients* a_m, $m = 0, 1, \ldots$, *are defined as the coefficients of* v^{2m} *in the expansion*

$$\frac{\Gamma(m+\frac{1}{2})}{2\pi} \left(\frac{2}{\rho+1} \right)^{m+\frac{1}{2}} (1-v)^{-\beta} \left(1 + \frac{\rho+2}{3}v + \frac{(\rho+2)(\rho+3)}{3\cdot4}v^2 + \ldots \right)^{-\frac{2m+1}{2}}.$$

(ii) *Let* $-1 < \rho < 0$, $y = -z$, $\arg(z) \leq \pi$, $-\pi < \arg(y) \leq \pi$, $|\arg(y)| \leq \min(\frac{3(1+\rho)}{2}\pi, \pi) - \epsilon$, $\epsilon > 0$. *Then*

$$W_{\rho,\mu}(z) = Y^{\frac{1}{2}-\mu} e^{-Y} \left(\sum_{m=0}^{M-1} A_m Y^{-m} + O(Y^{-M}) \right), \quad Y \to \infty,$$

with $Y = (1+\rho)((-\rho)^{-\rho}y)^{\frac{1}{1+\rho}}$ *and the coefficients* A_m, $m = 0, 1, \ldots$, *defined by*

$$\frac{\Gamma(1-\mu-\rho t)}{2\pi(-\rho)^{-\rho t}(1+\rho)^{(1+\rho)(t+1)}\Gamma(t+1)}$$

$$= \sum_{m=0}^{M-1} \frac{(-1)^m A_m}{\Gamma((1+\rho)t + \mu + \frac{1}{2} + m)} + O\left(\frac{1}{\Gamma((1+\rho)t + \beta + \frac{1}{2} + M)} \right),$$

valid for $\arg(t), \arg(-\rho t)$ *and* $\arg(1 - \mu - \rho t)$ *all lying between* $-\pi$ *and* π *and* t *tending to infinity.*

Remark 3.3 For $\rho > 0$, the case $z = -x$, $x > 0$ is not covered by Theorem 3.7. Then the following asymptotic formula holds:

$$W_{\rho,\mu}(-x) = x^{p(\frac{1}{2}-\mu)} e^{\sigma x^p \cos p\pi} \cos((\tfrac{1}{2} - \mu)p\pi + \sigma x^p \sin p\pi)(c_1 + O(x^{-p})),$$

where $p = (1+\rho)^{-1}$, $\sigma = (1+\rho)\rho^{-\frac{\rho}{1+\rho}}$ and the constant c_1 can be evaluated exactly. Similarly, for $-\frac{1}{3} < \rho < 0$, the asymptotic expansion as $z = x \to +\infty$ is not covered, and it is given by

$$W_{\rho,\mu}(-x) = x^{p(\frac{1}{2}-\mu)}e^{-\sigma x^p \cos p\pi} \cos((\tfrac{1}{2} - \mu)p\pi - \sigma x^p \sin p\pi)(c_2 + O(x^{-p})),$$

where $p = (1 + \rho)^{-1}$, $\sigma = (1 + \rho)(-\rho)^{-\frac{\rho}{1+\rho}}$ and the constant c_2 can be evaluated exactly. It follows from Theorem 3.7 that for $z \in \mathbb{C}$ and $|\arg(z)| \leq \pi - \epsilon$ (with $0 < \epsilon < \pi$),

$$W_{\rho,\mu}(z) = (2\pi(1 + \rho))^{-\frac{1}{2}}(\rho z)^{\frac{1-\mu}{1+\rho}} e^{\frac{1+\rho}{\rho}(\rho z)^{\frac{1}{1+\rho}}} (1 + O(z^{-\frac{1}{1+\rho}})) \quad \text{as } z \to \infty.$$

The Laplace transform of a Wright function is a Mittag-Leffler function (cf. also Lemma 3.2). The relation (3.31) can be found in [Sta70, (3), p. 114]. Note that for $\rho > 0$, the series representation of $W_{\rho,\mu}(z)$ can be transformed term by term. However, for $\rho \in (-1, 0)$, $W_{\rho,\mu}(z)$ is an entire function of order great than 1, so that this approach is no longer legitimate. Thus care is required in this case in establishing the existence of Laplace transform, which the minus sign in the argument is to ensure, by the asymptotic formula valid about $\mathbb{R}_-$ in Theorem 3.7.

Proposition 3.3 *The following Laplace transform relations hold:*

$$\mathcal{L}[W_{\rho,\mu}(-x)](z) = \begin{cases} z^{-1}E_{\rho,\mu}(-z^{-1}), & \rho \geq 0, \\ E_{-\rho,\mu-\rho}(-z), & -1 < \rho < 0, \end{cases} \tag{3.30}$$

$$\mathcal{L}[x^{\mu-1}W_{\rho,\mu}(-cx^\rho)](z) = z^{-\mu}e^{-cz^{-\rho}}, \quad \forall c > 0. \tag{3.31}$$

Proof The identity (3.30) is direct from the integral representation (3.29):

$$\mathcal{L}[W_{\rho,\mu}(-x)](z) = \int_0^\infty e^{-zx}\left(\frac{1}{2\pi i}\int_C e^{\zeta-x\zeta^{-\rho}}\zeta^{-\mu}d\zeta\right)dx$$

$$= \frac{1}{2\pi i}\int_C e^\zeta \zeta^{-\mu}\left(\int_0^\infty e^{-(z+\zeta^{-\rho})x}dx\right)d\zeta = \frac{1}{2\pi i}\int_C e^\zeta \frac{\zeta^{-\mu}}{z+\zeta^{-\rho}}d\zeta$$

$$= \frac{1}{2\pi i z}\int_C e^\zeta \frac{\zeta^{\rho-\mu}}{\zeta^\rho + z^{-1}}d\zeta = \begin{cases} z^{-1}E_{\rho,\mu}(-z^{-1}), & \rho > 0, \\ E_{-\rho,\mu-\rho}(-z), & 0 > \rho > -1, \end{cases}$$

where changing integration order is justified by the asymptotics of the Wright function in Theorem 3.7, and the last identity is due to the integral representation (3.11) of $E_{\alpha,\beta}(z)$. When $\rho = 0$, $W_{\rho,\mu}(-x) = \frac{e^{-x}}{\Gamma(\mu)}$, and thus

$$\mathcal{L}[W_{\rho,\mu}(-x)](z) = (\Gamma(\mu)(z+1))^{-1} = z^{-1}E_{\rho,\mu}(-z^{-1}),$$

i.e., the desired identity also holds. The identity (3.31) follows similarly by changing variables $\xi = x^{-1}\zeta$ and then applying the Cauchy integral formula

$$\mathcal{L}[x^{\mu-1}W_{\rho,\mu}(-cx^\rho)](z) = \int_0^\infty e^{-zx}x^{\mu-1}\left(\frac{1}{2\pi i}\int_C e^{\zeta-cx^\rho\zeta^{-\rho}}\zeta^{-\mu}d\zeta\right)dx$$

$$= \int_0^\infty e^{-zx}\left(\frac{1}{2\pi i}\int_C e^{x\xi-c\xi^{-\rho}}\xi^{-\mu}d\xi\right)dx = \frac{1}{2\pi i}\int_C\left(\int_0^\infty e^{(\xi-z)x}dx\right)e^{-c\xi^{-\rho}}\xi^{-\mu}d\xi$$

$$= \frac{1}{2\pi i}\int_C\frac{1}{z-\xi}e^{-c\xi^{-\rho}}\xi^{-\mu}d\xi = z^{-\mu}e^{-cz^{-\rho}}.$$

This completes the proof of the proposition.   □

The following convolution rule holds for $x^{\mu-1}W_{\rho,\mu}(-cx^\rho)$.

Corollary 3.4 *For any $c_1, c_2 > 0$, the following convolution identity holds:*

$$x^{\mu-1}W_{\rho,\mu}(-c_1x^\rho) * x^{\nu-1}W_{\rho,\nu}(-c_2x^\rho) = x^{\mu+\nu-1}W_{\rho,\mu+\nu}(-(c_1+c_2)x^\rho).$$

Proof By the Laplace transform relation (3.31), we have

$$\mathcal{L}[x^{\mu-1}W_{\rho,\mu}(-c_1x^\rho) * x^{\nu-1}W_{\rho,\nu}(-c_2x^\rho)](z)$$

$$= \mathcal{L}[x^{\mu-1}W_{\rho,\mu}(-c_1x^\rho)](z)\mathcal{L}[x^{\nu-1}W_{\rho,\nu}(-c_2x^\rho)](z) = z^{-\mu-\nu}e^{-(c_1+c_2)z^\rho}.$$

This directly shows the desired identity.   □

We also have a Fourier transform relation.

Proposition 3.4 *For any $\rho \in (-1,0)$ and $\mu \in \mathbb{R}$, the following identity holds:*

$$\mathcal{F}[W_{\rho,\mu}(-|x|)](\xi) = 2E_{-2\rho,\mu-\rho}(-\xi^2).$$

Proof It follows from the series expansion $\cos\xi x = \sum_{k=0}^\infty(-1)^k\frac{(\xi x)^{2k}}{(2k)!}$ that

$$\mathcal{F}[W_{\rho,\mu}(-|x|)](\xi) = \int_{-\infty}^\infty e^{-i\xi x}W_{\rho,\mu}(-|x|)dx = 2\int_0^\infty W_{\rho,\mu}(-x)\cos\xi xdx$$

$$= 2\sum_{k=0}^\infty(-1)^k\frac{\xi^{2k}}{(2k)!}\int_0^\infty x^{2k}W_{\rho,\mu}(-x)dx.$$

Now we claim the identity

$$\int_0^\infty x^{\nu-1}W_{\rho,\mu}(-x)dx = \frac{\Gamma(\nu)}{\Gamma(-\rho\nu+\mu)}. \tag{3.32}$$

Indeed, it follows from the integral representations of the Wright function $W_{\rho,\mu}(-x)$ and the reciprocal Gamma function $\frac{1}{\Gamma(z)}$, cf. (3.29) and (2.6), that

$$\int_0^\infty x^{\nu-1}W_{\rho,\mu}(-x)dx = \int_0^\infty x^{\nu-1}\frac{1}{2\pi i}\int_C e^{\zeta-x\zeta^{-\rho}}\zeta^{-\mu}d\zeta dx$$

$$= \frac{1}{2\pi i}\int_C e^\zeta\zeta^{-\mu}\left[\int_0^\infty e^{-x\zeta^{-\rho}}x^{\nu-1}dx\right]d\zeta = \frac{\Gamma(\nu)}{2\pi i}\int_C e^\zeta\zeta^{\nu\rho-\mu}d\zeta = \frac{\Gamma(\nu)}{\Gamma(-\rho\nu+\mu)}.$$

The change of integration order is legitimate due to the asymptotics of $W_{\rho,\mu}(-|x|)$ in Theorem 3.7. Thus,

$$\mathcal{F}[W_{\rho,\mu}(-|x|)](\xi) = 2\sum_{k=0}^{\infty}(-1)^k \frac{\xi^{2k}}{\Gamma(-(2k+1)\rho+\mu)} = 2E_{-2\rho,\mu-\rho}(-\xi^2).$$

This completes the proof of the proposition. $\qquad\qquad\qquad\qquad\qquad\qquad$ □

3.2.2 Wright function $W_{\rho,\mu}(-x)$

The function $W_{\rho,\mu}(-x)$, with $-1 < \rho < 0$ and $x \in \mathbb{R}_+$, is of independent interest in the study of the subdiffusion model in Chapter 7. This case has been extensively studied in Pskhu [Psk05b] (in Russian). Most of the estimates in Theorem 3.8(i) (and relevant lemmas) are taken from [Psk05b, Section 2.2.7]. We begin with a useful integral representation of $W_{\rho,\mu}(-x)$. It allows deriving various useful upper bounds.

Lemma 3.4 *For $x \geq 0$ and $\rho \in (-1, 0)$, there holds with $k(\varphi) = \frac{\sin(1+\rho)\varphi}{\sin\varphi}$,*

$$W_{\rho,\mu}(-x) = \frac{1}{\pi(1+\rho)}\int_0^\pi r^{1-\mu}\left(\frac{-\rho\sin(1+\rho-\mu)\varphi}{\sin(-\rho\varphi)} + \frac{\sin\mu\varphi}{\sin\varphi}\right)e^{-xr^{-\rho}k(\varphi)}d\varphi.$$

Proof We take a contour $\Gamma' = \{z : z = r(\varphi)e^{i\varphi}, -\pi < \varphi < \pi\}$, with $r(\varphi) = \left(\frac{x\sin(-\rho\varphi)}{\sin\varphi}\right)^{\frac{1}{1+\rho}}$. Then we deform the Hankel contour C to Γ' and obtain

$$\int_C e^z z^{-\mu}e^{-xz^{-\rho}}dz = \int_{\Gamma'} e^z z^{-\mu}e^{-xz^{-\rho}}dz.$$

The parametrization of the contour Γ' implies $dz = (r'(\varphi) + ir(\varphi))e^{i\varphi}d\varphi$, and by the choice of $r(\varphi)$, $r\sin\varphi - xr^{-\rho}\sin(-\rho\varphi) = 0$. Consequently,

$$z - xz^{-\rho} = r\cos\varphi - xr^{-\rho}\cos(-\rho\varphi) = r(\cos\varphi - \sin\varphi\cot(-\rho\varphi))$$

$$= -r\frac{\sin(1+\rho)\varphi}{\sin(-\rho\varphi)} = xr^{-\rho}\frac{\sin(1+\rho)\varphi}{\sin\varphi}.$$

Then by the definition of $k(\varphi)$,

$$W_{\rho,\mu}(-x) = \frac{1}{2\pi i}\int_{-\pi}^\pi r^{-\mu}e^{i\varphi(1-\mu)}(r' + ir)e^{-xr^{-\rho}k(\varphi)}d\varphi.$$

Since $r(\varphi)$ is even, and $r'(\varphi)$ is odd, we have

$$e^{i\varphi(1-\mu)}(r'(\varphi) + ir(\varphi)) + e^{-i\varphi(1-\mu)}(r'(-\varphi) + ir(-\varphi))$$

$$=2i\Im(e^{i\varphi(1-\mu)}(r'(\varphi) + ir(\varphi))) = 2i(r'(\varphi)\sin(1-\mu)\varphi + r(\varphi)\cos(1-\mu)\varphi).$$

Thus, the last integral can be reduced to

$$W_{\rho,\mu}(-x) = \frac{1}{\pi} \int_0^\pi r^{-\mu}(r' \sin(1-\mu)\varphi + r \cos(1-\mu)\varphi) e^{-xr^{-\rho}k(\varphi)} d\varphi.$$

Since $r'(\varphi) = \frac{r(\varphi)}{1+\rho}(-\rho \cot(-\rho\varphi) - \cot\varphi)$, we obtain

$$r' \sin(1-\mu)\varphi + r \cos(1-\mu)\varphi = \frac{r}{1+\rho}\left(\frac{-\rho \sin(1+\rho-\mu)\varphi}{\sin(-\rho\varphi)} + \frac{\sin\mu\varphi}{\sin\varphi}\right).$$

Upon substituting the expression, we obtain the desired integral representation.   □

The next result gives the positivity and monotonicity of $W_{\rho,\mu}(-x)$. (i) can also be found in [Sta70, Theorem 8].

Lemma 3.5 *For $\rho \in (-1,0)$, the following assertions hold:*

(i) *If $\mu \geq 0$, then $W_{\rho,\mu}(-x) > 0$ for all $x \in \mathbb{R}_+$.*

(ii) *If $\mu \geq -\rho$, $W_{\rho,\mu}(-x)$ is monotonically decreasing in x over $\mathbb{R}_+$.*

(iii) *Let $\rho \in (-1,0)$, and $\gamma \geq \mu \geq -\rho$. Then for any $x \geq 0$,*

$$W_{\rho,\gamma}(-x) \leq \frac{\Gamma(\mu)}{\Gamma(\gamma)} W_{\rho,\mu}(-x).$$

Proof It is direct to verify the following identity:

$$x^{\mu-1} W_{\rho,\mu}(-x^\rho) = {}_0I_x^\mu x^{-1} W_{\rho,0}(-x^\rho),$$

cf. (3.28). By the integral representation in Lemma 3.4, $W_{\rho,0}(-x^\rho) > 0$. This shows (i). Similarly, (ii) follows from the identity $\frac{d}{dx} W_{\rho,\mu}(-x) = -W_{\rho,\mu+\rho}(-x)$, cf. (3.26), and the assertion of (i). Last, by changing variables $y = x^{\frac{1}{\rho}}$,

$$W_{\rho,\gamma}(-x) = W_{\rho,\gamma}(-y^\rho) = y^{1-\gamma} {}_0I_y^{\gamma-\mu} y^{\mu-1} W_{\rho,\mu}(-y^\rho),$$

cf. (3.28). Since $\mu \geq -\rho$, by (ii), the function $W_{\rho,\mu}(-x)$ is monotonically decreasing, and hence for $\rho \in (-1,0)$,

$${}_0I_y^{\gamma-\mu} y^{\mu-1} W_{\rho,\mu}(-y^\rho) \leq W_{\rho,\mu}(-y^\rho) {}_0I_y^{\gamma-\mu} y^{\mu-1} = W_{\rho,\mu}(-y^\rho) y^{\gamma-1} \Gamma(\gamma)^{-1} \Gamma(\mu).$$

The last two relations together imply (iii).   □

Lemma 3.6 *Let $\rho \in (-1,0)$. Then for any $\mu \geq 0$, $x \geq 0$, there holds*

$$W_{\rho,\mu}(-x) \leq P_n(x) W_{\rho,1}(-x)$$

with P_n given by

$$P_n(x) = \sum_{k=0}^n \frac{(-\rho x)^k}{\Gamma(k(1+\rho)+\mu)},$$

and $n \in \mathbb{N}_0$ is such that the condition $\mu + n(1+\rho) \geq 1$ holds.

Proof The proof is based on mathematical induction. The case $n = 0$ follows directly from Lemma 3.5(iii). Next assume that the assertion holds for $n = N - 1$. Suppose $\mu + N(1+\rho) \geq 1$ for some integer N. Let $\mu' = \mu + 1 + \rho$ so that $\mu' + (N-1)(1+\rho) \geq 1$. First we claim the recursion identity

$$W_{\rho,\mu}(-x) = -\rho x W_{\rho,\mu+1+\rho}(-x) + \mu W_{\rho,\mu+1}(-x). \tag{3.33}$$

Actually, by the definition of the function $W_{\rho,\mu}(z)$,

$$- \rho x W_{\rho,\mu+1+\rho}(-x) + \mu W_{\rho,\mu+1}(-x)$$

$$= \sum_{k=0}^{\infty} \frac{\rho(-x)^{k+1}}{k!\,\Gamma(k\rho + \mu + 1 + \rho)} + \sum_{k=0}^{\infty} \frac{\mu(-x)^k}{k!\,\Gamma(k\rho + \mu + 1)}$$

$$= \frac{\mu(-x)^0}{\Gamma(\mu + 1)} + \sum_{k=1}^{\infty} \left[\frac{\rho}{(k-1)!\,\Gamma(k\rho + \mu + 1)} + \frac{\mu}{k!\,\Gamma(k\rho + \mu + 1)} \right](-x)^k$$

$$= \frac{(-x)^0}{\Gamma(\mu)} + \sum_{k=1}^{\infty} \frac{(-x)^k}{k!\,\Gamma(k\rho + \mu)} = W_{\rho,\mu}(-x),$$

since by the recursion formula (2.2),

$$\frac{\rho}{(k-1)!\,\Gamma(k\rho + \mu + 1)} + \frac{\mu}{k!\,\Gamma(k\rho + \mu + 1)} = \frac{1}{k!}\,\frac{k\rho + \mu}{\Gamma(k\rho + \mu + 1)} = \frac{1}{k!\,\Gamma(k\rho + \mu)}.$$

It follows from this identity, and the induction hypothesis that

$$W_{\rho,\mu}(-x) \leq \left[(-\rho x) \sum_{k=0}^{N-1} \frac{(-\rho x)^k}{\Gamma(k(1+\rho) + \mu + 1 + \rho)} + \frac{\mu}{\Gamma(1+\mu)} \right] W_{\rho,1}(-x)$$

$$= \left[\sum_{k=1}^{N} \frac{(-\rho x)^k}{\Gamma(k(1+\rho) + \mu)} + \frac{1}{\Gamma(\mu)} \right] W_{\rho,1}(-x).$$

This shows the assertion for $n = N$, completing the induction step.   □

Proposition 3.5 *If $\rho \in (-1, 0)$ and $\mu < 1$, then for all $x, y \geq 0$, the following statements hold for any $a \in (0, x)$, $\xi \in [-\rho, 1]$ and $\omega \in (\frac{1}{2}, \min(1, \frac{1}{-2\rho}))$:*

$$|y^{\mu-1} W_{\rho,\mu}(-xy^\rho)| \leq \frac{1}{\xi\pi} [c_\rho(\xi, \omega)]^{\frac{\mu-1}{\xi}} \Gamma\left(\frac{1-\mu}{\xi}\right) [x^{\frac{1-\xi}{1+\rho}} y^{\frac{\xi+\rho}{1+\rho}}]^{\frac{\mu-1}{\xi}},$$

$$|y^{\mu-1} W_{\rho,\mu}(-xy^\rho)| \leq \frac{1}{-\rho\pi} [c_\rho(\rho, \omega)]^{\frac{\mu-1}{-\rho}} \Gamma\left(\frac{1-\mu}{-\rho}\right) (x - a)^{\frac{\mu-1}{-\rho}} W_{\rho,1}(-ay^\rho),$$

where

$$c_\rho(\xi, \omega) = (1 + \rho) \left(\frac{-\cos(\omega\pi)}{\xi + \rho} \right)^{\frac{\xi+\rho}{1+\rho}} \left(\frac{\cos(-\rho\omega\pi)}{1 - \xi} \right)^{\frac{1-\xi}{1+\rho}}.$$

Proof By deforming the contour C in (3.29) to $\Gamma_{\epsilon,\omega\pi}$, $W_{\rho,\mu}(-x)$ is represented by

$$W_{\rho,\mu}(-x) = \frac{1}{2\pi i} \int_{-\omega\pi}^{\omega\pi} \epsilon^{1-\mu} e^{\epsilon e^{i\varphi} - xe^{-\rho}e^{-i\rho\varphi} + (1-\mu)\varphi} d\varphi$$

$$+ \frac{1}{2\pi i} \int_{\epsilon}^{\infty} (r^{-\mu} e^{re^{i\omega\pi} - xr^{-\rho}e^{-i\rho\omega\pi} + (1-\mu)\omega\pi} - r^{-\mu} e^{re^{-i\omega\pi} - xr^{-\rho}e^{i\rho\omega\pi} - (1-\mu)\omega\pi}) dr.$$

Since $\mu < 1$, upon sending $\epsilon \to 0^+$, the first term vanishes. Thus, we obtain

$$W_{\rho,\mu}(-x) = \frac{1}{\pi} \int_0^{\infty} r^{-\mu} e^{r\cos\omega\pi - xr^{-\rho}\cos(-\rho\omega\pi)}$$

$$\times \sin(r\sin\omega\pi - xr^{-\rho}\sin(-\rho\omega\pi) + (1-\mu)\omega\pi) dr.$$

This directly implies

$$|W_{\rho,\mu}(-x)| \le \frac{1}{\pi} \int_0^{\infty} r^{-\mu} e^{r\cos\omega\pi - xr^{-\rho}\cos(-\rho\omega\pi)} dr.$$

Now recall Young's inequality, $A + B \ge (\frac{A}{\gamma})^{\gamma}(\frac{B}{1-\gamma})^{1-\gamma}$, for all $A, B \ge 0$ and $\gamma \in [0, 1]$. Thus, with $\omega \in (\frac{1}{2}, \min(1, \frac{1}{-2\rho}))$, there holds

$$r\cos\omega\pi - xr^{-\rho}\cos(-\rho\omega\pi) \le -\left(-\frac{\cos\omega\pi}{\gamma}\right)^{\gamma}\left(\frac{x\cos(-\rho\omega\pi)}{1-\gamma}\right)^{1-\gamma} r^{\gamma+(1-\gamma)(-\rho)}.$$

Hence, with $\xi = \gamma + (1-\gamma)(-\rho)$ and $K = \left(-\frac{\cos\omega\pi}{\gamma}\right)^{\gamma}\left(\frac{x\cos(-\rho\omega\pi)}{1-\gamma}\right)^{1-\gamma} r^{\gamma+(1-\gamma)(-\rho)}$,

$$|W_{\rho,\mu}(-x)| \le \frac{1}{\pi} \int_0^{\infty} r^{-\mu} e^{-Kr^{\xi}} dr = \frac{1}{\xi\pi} \int_0^{\infty} s^{\frac{1-\mu}{\xi}-1} e^{-Ks} ds = \frac{K^{\frac{\mu-1}{\xi}}}{\xi\pi} \Gamma\left(\frac{1-\mu}{\xi}\right).$$

Now the first assertion follows by substituting x with xy^{ρ}. In particular, with the choices $\xi = -\rho$ and $\xi = 1$, the inequality implies

$$|y^{\mu-1} W_{\rho,\mu}(-xy^{\rho})| \le \frac{1}{-\rho\pi} [c_{\rho}(\rho, \omega)]^{\frac{\mu-1}{-\rho}} \Gamma\left(\frac{1-\mu}{-\rho}\right) x^{\frac{\mu-1}{-\rho}},$$

$$|y^{\mu-1} W_{\rho,\mu}(-xy^{\rho})| \le \frac{1}{\pi} [c_{\rho}(1, \omega)]^{\mu-1} \Gamma(1-\mu) y^{\mu-1},$$

with

$$c_{\rho}(\rho, \omega) = \frac{\cos(-\rho\omega\pi)}{-\rho\pi} \quad \text{and} \quad c_{\rho}(1, \omega) = -\frac{\cos(\omega\pi)}{\pi}.$$

Next, using Corollary 3.4, we rewrite $y^{\mu-1} W_{\rho,\mu}(-xy^{\rho})$ as

$$y^{\mu-1} W_{\rho,\mu}(-xy^{\rho}) = y^{\mu-1} W_{\rho,\mu}(-(x-a)y^{\rho}) * y^{-1} W_{\rho,0}(-ay^{\rho}).$$

Consequently,

$$|y^{\mu-1}W_{\rho,\mu}(-xy^\rho)| \le \frac{1}{-\rho\pi}[c_\rho(\rho,\omega)]^{\frac{\mu-1}{-\rho}}\Gamma\left(\frac{1-\mu}{-\rho}\right)(x-a)^{\frac{\mu-1}{-\rho}}\int_0^y t^{-1}W_{\rho,0}(-at^\rho)dt.$$

This directly gives the second assertion. $\square$

The next result gives useful bounds on $W_{\rho,\mu}(-x)$, see also [KV13, Lemma 3.1] for (ii) and [Kra16, Lemma 2.1] for (iv).

Theorem 3.8 *For $\rho \in (-1,0)$, the following assertions hold for $x \ge 0$:*

(i) *If $\mu \ge 1$, then*

$$0 < W_{\rho,\mu}(-x) \le \Gamma(\mu)^{-1}e^{-(-\rho)^{\frac{-\rho}{1+\rho}}(1+\rho)x^{\frac{1}{1+\rho}}}.$$

(ii) *If $\mu \in (-\rho, 1]$, then*

$$0 < W_{\rho,\mu}(-x) \le \left(1 + 2^{1+\rho}\Gamma(1+\mu+\rho)^{-1}(-\rho)^{1+\rho}e^{-(1+\rho)}\right)e^{-\frac{1+\rho}{2}(-\rho)^{\frac{-\rho}{1+\rho}}x^{\frac{1}{1+\rho}}}.$$

(iii) *For any $\mu \in (0, 1)$, there holds for all $x > 0$,*

$$|W_{\rho,\mu}(-x)| \le c(1 + x^{\frac{\mu-1}{\rho}})^{-1}.$$

(iv) *If $\mu \in \mathbb{R}$, there holds for any $x > 0$*

$$|W_{\rho,\mu}(-x)| \le ce^{-\sigma x^{\frac{1}{1+\rho}}}\begin{cases} 1, & \mu \notin \mathbb{Z}_- \cup \{0\}, \\ x, & \mu \in \mathbb{Z}_- \cup \{0\}. \end{cases}$$

Proof The positivity in (i) and (ii) is direct from Lemma 3.5, and thus we prove only upper bounds. We claim that for all $\gamma \in (0, 1)$, $\frac{d}{d\varphi}\frac{\sin(\gamma\varphi)}{\sin\varphi} \ge 0$ for $\varphi \in [0, \pi]$. Indeed,

$$\frac{d}{d\varphi}\frac{\sin(\gamma\varphi)}{\sin\varphi} = \frac{\gamma\cos(\gamma\varphi)\sin\varphi - \sin(\gamma\varphi)\cos\varphi}{\sin^2\varphi}.$$

Clearly,

$$\gamma\cos(\gamma\varphi)\sin\varphi - \sin(\gamma\varphi)\cos\varphi|_{\varphi=0} = 0,$$

and

$$\frac{d}{d\varphi}(\gamma\cos(\gamma\varphi)\sin\varphi - \sin(\gamma\varphi)\cos\varphi) = (1-\gamma^2)\sin(\gamma\varphi)\sin\varphi \ge 0.$$

Thus the desired claim follows. Now by Lemma 3.4, for $\mu = 1$, there holds

$$W_{\rho,1}(-x) = \frac{1}{\pi}\int_0^\pi e^{-xr^{-\rho}k(\varphi)}d\varphi \le e^{-x^{\frac{1}{1+\rho}}(-\rho)^{\frac{-\rho}{1+\rho}}(1+\rho)}.$$

The case $\mu > 1$ follows from this and Lemma 3.5(iii), showing (i).
Part (ii), i.e., $\mu \in [-\rho, 1]$. We use the recursion formula (3.33). Since $\mu + \rho \ge 0$ and $\mu > 0$, $1 + \mu > 1$ and $1 + \mu + \rho \ge 1$. Thus, we can apply (i) to the two terms on the right-hand side and obtain with $\kappa(\rho) = (-\rho)^{\frac{-\rho}{1+\rho}}(1+\rho)$,

$$W_{\rho,\mu}(-x) \leq \frac{\mu e^{-\kappa(\rho)x^{\frac{1}{1+\rho}}}}{\Gamma(\mu+1)} + \frac{-\rho x e^{-\kappa(\rho)x^{\frac{1}{1+\rho}}}}{\Gamma(1+\mu+\rho)} = \frac{e^{-\kappa(\rho)x^{\frac{1}{1+\rho}}}}{\Gamma(\mu)} + \frac{-\rho x e^{-\kappa(\rho)x^{\frac{1}{1+\rho}}}}{\Gamma(1+\mu+\rho)}.$$

For $\mu \in [-\rho, 1)$, there holds $\Gamma(\mu)^{-1} \leq 1$. Then direct computation leads to

$$\sup_{x \in \mathbb{R}_+} x e^{-\frac{\kappa(\rho)}{2}x^{\frac{1}{1+\rho}}} \leq 2^{1+\rho}(-\rho)^\rho e^{-(1+\rho)}.$$

The last inequalities show (ii). (iii) is direct from the first estimate in Proposition 3.5. Last, for (iv), the estimate

$$|W_{\rho,\mu}(-x)| \leq c e^{-\sigma x^{\frac{1}{1+\rho}}}, \quad \forall \mu \in \mathbb{R},$$

follows from (i)-(ii) and repeatedly applying the recursion (3.33). Indeed, for $-1 < \mu < 0$, with m being the smallest integer such that $\mu + \mu(1 + \rho) > 0$,

$$W_{\rho,\mu}(-x) = (-\rho x)^2 W_{\rho,\mu+2(1+\rho)}(-x) - \rho x \mu W_{\rho,\mu+1+\rho+1}(-x)) + \mu W_{\rho,\mu+1}(-x)$$

$$= (-\rho x)^m W_{\rho,\mu+m(1+\rho)}(-x) + \sum_{k=0}^{m-1} \mu(-\rho x)^k W_{\rho,\mu+k(1+\rho)}(-x).$$

The terms now can be bounded using triangle inequality and (i)-(ii). The case of general μ follows similarly. Thus the desired inequality for $\mu \notin \mathbb{Z}_- \cup \{0\}$ is proved. For $\mu \in \mathbb{Z}_- \cup \{0\}$, we use the recursion formula (cf. (3.33))

$$W_{\rho,-k}(-x) = -\rho x W_{\rho,-k+1+\rho}(-x) - k W_{\rho,-k+1}(-x).$$

Using this formula and mathematical induction, we deduce

$$W_{\rho,-k}(-x) = -\rho x \sum_{m=0}^{k} (-1)^{k-m} \frac{k!}{m!} W_{\rho,-m+1+\rho}(-x).$$

Then the desired assertion follows. $\square$

3.3 Numerical Algorithms

In this section, we describe algorithms for evaluating $E_{\alpha,\beta}(z)$ and $W_{\rho,\mu}(z)$.

3.3.1 Mittag-Leffler function $E_{\alpha,\beta}(z)$

The computation of $E_{\alpha,\beta}(z)$ over the whole complex plane $\mathbb{C}$ is delicate, and has received much attention [GLL02, SH09]. We describe an algorithm from [SH09] and the relevant error estimates. See also the work [Gar15, McL21] for the computation of Mittag-Leffler functions, by the inverse Laplace transform with deformed contours and suitable quadrature rules. Since in different regions of $\mathbb{C}$, $E_{\alpha,\beta}(z)$ exhibits very different behavior, different numerical schemes are needed. First, we introduce some notation. We denote by $\overline{D(r)} = \{z \in \mathbb{C} : |z| \le r\}$ the closed disk of radius r centered at the origin, and by $W(\phi_1, \phi_2)$ and $\overline{W}(\phi_1, \phi_2)$ the wedges

$$W(\phi_1, \phi_2) = \{z \in \mathbb{C} : \phi_1 < \arg(z) < \phi_2\},$$
$$\overline{W}(\phi_1, \phi_2) = \{z \in \mathbb{C} : \phi_1 \le \arg(z) \le \phi_2\},$$

where $\phi_2 - \phi_1$ is the opening angle measured in a positive sense, and $\phi_1, \phi_2 \in (-\pi, \pi)$. The complex plane $\mathbb{C}$ is divided into seven regions $\{G_i, i = 0, \ldots, 6\}$, cf. Fig. 3.6.

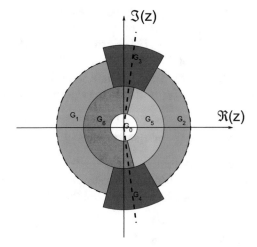

Fig. 3.6 The partition of $\mathbb{C}$, with the thick dashed lines being the Stokes line at $\arg(z) = \alpha\pi$.

On a disk of radius $r_0 < 1$, i.e., $G_0 = \overline{D(r_0)}$, for any $\alpha > 0$, the following Taylor series expansion

$$T_N(z) = \sum_{k=0}^{N-1} \frac{z^k}{\Gamma(\alpha k + \beta)} \tag{3.34}$$

gives a good approximation, provided that the truncation point N is determined suitably, e.g., by (3.40). In the algorithm, one chooses $r_0 = 0.95$.

Below we consider only the case $\alpha \in (0, 1)$, since the case $\alpha \ge 1$ can be reduced to the case $\alpha < 1$ using a recursion relation, cf. Lemma 3.1.

For large values of $z \in \mathbb{C}$, the exponential asymptotics and Berry-type smoothing can be used to compute $E_{\alpha,\beta}(z)$. Specifically, we choose the constant $r_1 > r_0$ defined in (3.41), and employ the exponential asymptotic for $|z| \ge r_1$. Away from the Stokes lines $|\arg(z)| = \alpha\pi$, the exponential asymptotics (3.12) and (3.13) are very accurate.

However, close to the Stokes lines, these approximations become less stable, and so the Berry-type smoothed asymptotic series (3.15) and (3.16) are employed instead. Hence, the region $\mathbb{C} \setminus D(r_1)$ is further divided into four subregions. Let δ and $\tilde{\delta}$ be positive numbers smaller than $\frac{\alpha}{2}\pi$. In the algorithm, δ and $\tilde{\delta}$ are chosen to be $\delta = \frac{\alpha}{8}\pi$ and $\tilde{\delta} = \min(\frac{\alpha}{8}\pi, \frac{\alpha+1}{2}\pi)$. Then in different regions of $\mathbb{C}$, the scheme employs different approximations as follows:

- The asymptotics (3.13) in the wedge $G_1 = [\mathbb{C} \setminus D(r_1)] \cap W(\alpha\pi + \tilde{\delta}, -\alpha\pi - \tilde{\delta})$.
- The asymptotics (3.12) in the wedge $G_2 = [\mathbb{C} \setminus D(r_1)] \cap W(-\alpha\pi + \delta, \alpha\pi - \delta)$.
- The Berry smoothing (3.16) in the area around the upper Stokes line $G_3 = [\mathbb{C} \setminus D(r_1)] \cap \overline{W}(\alpha\pi - \delta, \alpha\pi + \tilde{\delta})$.
- The Berry smoothing (3.15) in the area around the lower Stokes line $G_4 = [\mathbb{C} \setminus D(r_1)] \cap \overline{W}(-\alpha\pi - \tilde{\delta}, -\alpha\pi + \delta)$.

In the exponentially improved asymptotics, there are two parameters, i.e., truncation point N of the series and a lower limit r_1 of $|z|$, that have to be determined such that the error is smaller than ϵ for $|z| > r_1$. The error bound in Theorem 3.9 suggests the optimal parameters. The optimal truncation for the series will increase like $\sim \alpha^{-1}|z|^{\frac{1}{\alpha}}$. However, for $z \to \infty$, the asymptotic series becomes better and better, so that we apply an upper limit of $N < 100$. In the algorithm, the constant c is set to

$$c = \frac{1}{2\pi}\left(\frac{1}{\sin \alpha\pi} + \frac{1}{\min\{\sin \alpha\pi, \sin \xi\}}\right) \approx \frac{1}{\pi \sin \alpha\pi}, \tag{3.35}$$

where the angle ξ describes the distance of z to the Stokes line.

In the transition region $D(r_1) \setminus D(r_0)$, we employ the integral representations (3.8) and (3.9) with properly deformed contours. We use the following two regions:

- $G_5 = D(r_1) \cap \overline{W}(-\frac{5}{6}\alpha\pi, \frac{5}{6}\alpha\pi) \setminus G_0$
- $G_6 = D(r_1) \cap \overline{W}(\frac{5}{6}\alpha\pi, -\frac{5}{6}\alpha\pi) \setminus G_0$.

However, these formulas become numerically unstable when z is close to the contour path, due to the singularity of the integrand at $r = |z|e^{\pm i\theta}$. Hence, we take (3.9) with $\mu = \alpha\pi$ and (3.8) with $\mu = \frac{2}{3}\alpha\pi$ such that $G^+(\epsilon, \alpha\pi)$ and $G^-(\epsilon, \frac{2}{3}\alpha\pi)$ have nonempty overlap, which allows one to avoid their use close to the contour path. The value of ϵ is set to $\frac{1}{2}$, and thus lies in G_1, where the Taylor series is used for the calculation. Equation (3.9) is used with $\mu = \alpha\pi$ for $z \in G_5$, and (3.8) with $\mu = \frac{2}{3}\alpha\pi$ for $z \in G_6$.

When evaluating the integrals, several cases arise. We distinguish the cases $\beta \leq 1$ and $\beta > 1$. Next we give the explicit forms of the integrals involved in the computation. For $z \in G_5$, upon inserting the parametrization of the contour path for $\mu = \alpha\pi$ and $\epsilon = \frac{1}{2}$ yields for $z \in G^+(\epsilon, \mu)$ that for $\beta \leq 1$

$$E_{\alpha,\beta}(z) = A(z; \alpha, \beta, 0) + \int_0^\infty B(r; \alpha, \beta, z, \alpha\pi)dr, \tag{3.36}$$

and for $\beta > 1$

$$E_{\alpha,\beta}(z) = A(z; \alpha, \beta, 0) + \int_{\frac{1}{2}}^{\infty} B(r; \alpha, \beta, z, \alpha\pi)dr + \int_{-\pi\alpha}^{\pi\alpha} C(\varphi; \alpha, \beta, z, \tfrac{1}{2})d\varphi, \quad (3.37)$$

where

$$A(z; \alpha, \beta, x) = \alpha^{-1} z^{\frac{1-\beta}{\alpha}} e^{z^{\frac{1}{\alpha}}} \cos(\tfrac{x}{\alpha}),$$

$$B(r; \alpha, \beta, z, \phi) = \pi^{-1} A(r; \alpha, \beta, \phi) \frac{r \sin[\omega(r, \phi, \alpha, \beta) - \phi] - z \sin[\omega(r, \phi, \alpha, \beta)]}{r^2 - 2rz\cos\phi + z^2},$$

$$C(\varphi; \alpha, \beta, z, \varrho) = \frac{\varrho}{2\pi} A(\varrho; \alpha, \beta, \varphi) \frac{\cos[\omega(\varrho, \varphi, \alpha, \beta)] + i\sin[\omega(\varrho, \varphi, \alpha, \beta)]}{\varrho(\cos\varphi + i\sin\varphi) - z},$$

$$\omega(x, y, \alpha, \beta) = x^{\frac{1}{\alpha}} \sin(\alpha^{-1} y) + \alpha^{-1}(1 + \alpha - \beta)y.$$

In the case $\beta \leq 1$, we have applied the limit $\epsilon \to 0$. These equations are used to calculate $E_{\alpha,\beta}(z)$ for $z \in G_5$. For $z \in G_6$, with a contour with $\theta = \frac{2}{3}\alpha\pi$, the integral representations read

$$E_{\alpha,\beta}(z) = \int_0^{\infty} B(r; \alpha, \beta, z, \tfrac{2}{3}\alpha\pi)dr, \quad \beta \leq 1, \tag{3.38}$$

$$E_{\alpha,\beta}(z) = \int_{\frac{1}{2}}^{\infty} B(r; \alpha, \beta, z, \tfrac{2}{3}\alpha\pi)dr + \int_{-\frac{2}{3}\alpha\pi}^{\frac{2}{3}\alpha\pi} C(\varphi; \alpha, \beta z, \tfrac{1}{2})d\varphi, \quad \beta > 1. \tag{3.39}$$

The integrand $C(\varphi; \alpha, \beta, z, \varrho)$ is oscillatory but bounded over the integration interval. Thus the integrals over $C(\varphi; \alpha, \beta, z, \varrho)$ can be evaluated numerically using an appropriate quadrature formula. The integrals over $B(r; \alpha, \beta, z, \phi)$ involve unbounded intervals and have to be treated more carefully, especially suitable truncation of the integration interval is needed, which is specified in Theorem 3.9.

The complete procedure is listed in Algorithm 2, which is used throughout this book. The following result summarizes the recommended algorithmic parameters for the given accuracy and also the errors incurred in the approximation [SH09]. The notation $[\cdot]$ denotes taking integral part of a real number.

Theorem 3.9 *The following error estimates hold.*

(i) *Let $\epsilon > 0$. If $|z| < 1$ and*

$$N \geq \max\left(\left[\tfrac{2-\beta}{\alpha}\right] + 1, \left[\tfrac{\ln(\epsilon(1-|z|))}{\ln(|z|)}\right] + 1\right), \tag{3.40}$$

then the error $R_N := |E_{\alpha,\beta}(z) - T_N(z)|$ of the Taylor series approximation (3.34) is smaller than the given accuracy ϵ.

(ii) *Let $\alpha \in (0, 1)$. For $N \approx \alpha^{-1}|z|^{\frac{1}{\alpha}}$ and*

$$|z| \geq r_1 := \left(-2\ln\tfrac{\epsilon}{c}\right)^{\alpha} \tag{3.41}$$

the error term of the asymptotic series (3.12) and (3.13) fulfills the condition

$$R_N(z) = \left| E_{\alpha,\beta}(z) - \left(-\sum_{k=1}^{N-1} \frac{z^{-k}}{\Gamma(\beta - \alpha k)} \right) \right| \leq \epsilon,$$

where c is a constant dependent only on α, β, chosen according to (3.35).

(iii) *Suppose for $\phi = \alpha\pi$*

$$R_{\max} \geq \begin{cases} \max\left(1, 2|z|, \left(-\ln\frac{\epsilon\pi}{6}\right)^{\alpha}\right), & \beta \geq 0, \\ \max\left((|\beta|+1)^{\alpha}, 2|z|, \left(-2\ln\frac{\epsilon\pi}{6(|\beta|+2)(2|\beta|)^{|\beta|}}\right)^{\alpha}\right), & \beta < 0 \end{cases}$$

and for $\phi = \frac{2}{3}\alpha\pi$

$$R_{\max} \geq \begin{cases} \max\left(2^{\alpha}, 2|z|, (-\ln\frac{2^{\beta}\epsilon\pi}{12})^{\alpha}\right), & \beta \geq 0, \\ \max\left([2(|\beta|+1)]^{\alpha}, 2|z|, [-4\ln\frac{2^{\beta}\epsilon\pi}{12(|\beta|+2)(4|\beta|)^{|\beta|}}]^{\alpha}\right), & \beta < 0. \end{cases}$$

Then the following truncation error holds

$$R(R_{\max}, \alpha, \beta, z, \phi) = |\int_{R_{\max}}^{\infty} B(r, \alpha, \beta, z, \phi)dr| < \epsilon.$$

Algorithm 2 Computing the Mittag-Leffler function $E_{\alpha,\beta}(z)$

1: Input α, β, z and tolerance ϵ.
2: **if** $|z| < 0.95$ **then**
3:     Compute $E_{\alpha,\beta}(z)$ by Taylor series (3.34) of order N, cf. (3.40).
4: **else if** $\alpha > 1$ **then**
5:     Compute $E_{\alpha,\beta}(z)$ by the recursive formula (3.17).
6: **end if**
7: Compute r_1 by (3.41).
8: **if** $|z| < r_1$ **then**
9:     **if** $|\arg(z)| < 5\alpha\pi/6$ **then**
10:         Compute $E_{\alpha,\beta}(z)$ by (3.36) or (3.37).
11:     **else**
12:         Compute $E_{\alpha,\beta}(z)$ by (3.38) or (3.39).
13:     **end if**
14: **else**
15:     **if** $|\arg(z)| \geq \alpha\pi + \delta$ **then**
16:         Compute $E_{\alpha,\beta}(z)$ by (3.13).
17:     **else if** $|\arg(z)| \leq \alpha\pi - \delta$ **then**
18:         Compute $E_{\alpha,\beta}(z)$ by (3.12).
19:     **else if** $\arg(z) > \alpha\pi - \delta$ and $\arg(z) < \alpha\pi + \delta$ **then**
20:         Compute $E_{\alpha,\beta}(z)$ by (3.16).
21:     **else**
22:         Compute $E_{\alpha,\beta}(z)$ by (3.15).
23:     **end if**
24: **end if**

3.3.2 Wright function $W_{\rho,\mu}(x)$

The computation of $W_{\rho,\mu}(z)$ is fairly delicate. In principle, like $E_{\alpha,\beta}(z)$, it can be computed using power series for small values of the argument and a known asymptotic formula for large values, while for the intermediate case, using an integral representation. However, an algorithm that works for any $z \in \mathbb{C}$ with rigorous error bounds is still unavailable. For $z \in \mathbb{R}$, some preliminary analysis has been done [Luc08], which we describe below. We have the following representation formulas, derived from the representation (3.29), by suitably deforming the Hankel contour C.

Theorem 3.10 *The following integral representations hold.*

(i) *Let $z = -x$, $x > 0$. Then $W_{\rho,\mu}(-x)$ is given by*

$$W_{\rho,\mu}(-x) = \frac{1}{\pi} \int_0^\infty K(\rho, \mu, -x, r) \mathrm{d}r,$$

if $-1 < \rho < 0$ and $\mu < 1$, or $0 < \rho < \frac{1}{2}$, or $\rho = \frac{1}{2}$ and $\mu < 1 + \rho$,

$$W_{\rho,\mu}(-x) = e + \frac{1}{\pi} \int_0^\infty K(\rho, \mu, -x, r) \mathrm{d}r.$$

if $-1 < \rho < 0$ and $\mu = 1$, and

$$W_{\rho,\mu}(-x) = \frac{1}{\pi} \int_1^\infty K(\rho, \mu, -x, r) \mathrm{d}r + \frac{1}{\pi} \int_0^\pi \tilde{P}(\rho, \mu, -x, \varphi) \mathrm{d}\varphi,$$

in all other cases, with

$$K(\rho, \mu, x, r) = e^{-r + xr^{-\rho} \cos \rho\pi} r^{-\mu} \sin(xr^{-\rho} \sin \rho\pi + \mu\pi),$$
$$\tilde{P}(\rho, \mu, x, \varphi) = e^{\cos \varphi + x \cos \rho\varphi + \cos(\mu-1)\varphi} \cos(\sin \varphi - x \sin \rho\varphi - \sin(\mu-1)\varphi).$$

(ii) *Let $z = x > 0$. Then $W_{\rho,\mu}(x)$ has the following integral representation:*

$$W_{\rho,\mu}(x) = \begin{cases} \frac{1}{\pi} \int_0^\infty K(\rho, \mu, x, r) \mathrm{d}r, & -1 < \rho < 0, \mu < 1, \\ e + \frac{1}{\pi} \int_0^\infty K(\rho, \mu, x, r) \mathrm{d}r, & -1 < \rho < 0, \mu = 1, \\ \frac{1}{\pi} \int_1^\infty K(\rho, \mu, x, r) \mathrm{d}r + \frac{1}{\pi} \int_0^\pi \tilde{P}(\rho, \mu, x, \varphi) \mathrm{d}\varphi, & \text{otherwise,} \end{cases}$$

where the functions K and $\tilde{P}$ are the same as before.

The integral representations in Theorem 3.10 give one way to compute $W_{\rho,\mu}(x)$, $x \in \mathbb{R}$. In principle, these integrals can be evaluated by any quadrature rule. However, the kernel $K(\rho, \mu, x, r)$ is singular with a leading order $r^{-\mu}$, with successive singularities. Hence, a direct treatment via numerical quadrature can be inefficient. Dependent on the parameter ρ, a suitable transformation might be useful. For example, for $\rho > 0$, a more efficient approach is to use the change of variable $s = r^{-\rho}$, i.e., $r = s^{-\frac{1}{\rho}}$, and the transformed kernel is given by

$$\widetilde{K}(\rho, \mu, x, s) = (-\rho)^{-1} s^{(-\rho)^{-1}(-\mu+1)-1} e^{-s^{-\frac{1}{\rho}} + x \cos(\pi\rho)s} \sin(x \sin(\pi\rho)s + \pi\mu).$$

Example 3.2 The case directly relevant to subdiffusion is $\rho = -\frac{\alpha}{2} < 0$, $0 < \mu = 1 + \rho < 1$ (and $z = -x$, $x > 0$). Then the kernel is given by

$$\widetilde{K}(x, \rho, 1 + \rho, s) = (-\rho)^{-1} e^{-s^{-\frac{1}{\rho}} + x \cos(\pi\rho)s} \sin(x \sin(\pi\rho)s + (1 + \rho)\pi).$$

This kernel is free from grave singularities. The integral can be computed efficiently using Gauss-Jacobi quadrature with the weight function $s^{(-\rho)^{-1}(-\mu+1)-1}$. In Fig. 3.7, we plot the function $W_{-\mu, 1-\mu}(-|x|)$ over the interval $[-5, 5]$, which is the basic building block of the fundamental solution for subdiffusion, see Proposition 7.1 in Chapter 7 for detailed discussions. Except for the case $\mu = \frac{1}{2}$, for which it is infinitely differentiable, generally it shows a clear "kink" at the origin, indicating the limited smoothing property of related solution operators.

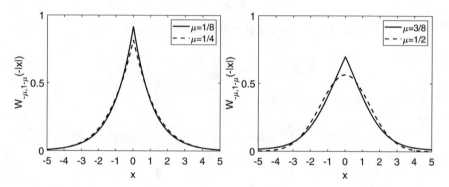

Fig. 3.7 The Wright function $W_{-\mu, 1-\mu}(-|x|)$.

Exercises

Exercise 3.1 Verify the identities in (3.2).

Exercise 3.2 Prove the identity (3.3), and that it can be rewritten as

$$E_{\frac{1}{2}, 1}(z) = e^{z^2} \left(1 + \frac{2}{\sqrt{\pi}} \int_0^z e^{-s^2} ds\right).$$

Exercise 3.3 Prove the following duplication formula for $\alpha > 0$ and $\beta \in \mathbb{R}$:

$$E_{2\alpha, \beta}(z^2) = \frac{1}{2}(E_{\alpha, \beta}(z) + E_{\alpha, \beta}(-z)), \quad \forall z \in \mathbb{C}.$$

Exercise 3.4 Prove the order and type in Proposition 3.1.

Exercise 3.5 Similar to Lemma 3.1, prove the following version of reduction formula:

$$E_{\alpha,\beta}(z) = \frac{1}{m}\sum_{j=0}^{m-1} E_{\frac{\alpha}{m},\beta}(z^{\frac{1}{m}}e^{2\pi i\frac{j}{m}}), \quad m \geq 1,$$

from the elementary formula

$$\sum_{j=0}^{m-1} e^{2\pi i\frac{jk}{m}} = \begin{cases} m, & \text{if } k \equiv 0 \ (\mathrm{mod}\ m), \\ 0, & \text{if } k \not\equiv 0 \ (\mathrm{mod}\ m). \end{cases}$$

Exercise 3.6 Show that the condition $0 \leq \alpha \leq 1$ in Theorem 3.5 cannot be dropped.

Exercise 3.7 Prove that for $\lambda > 0$, $\alpha > 0$ and $m \in \mathbb{N}$,

$$\frac{d^m}{dx^m}E_{\alpha,1}(-\lambda x^\alpha) = -\lambda x^{\alpha-m}E_{\alpha,\alpha-m+1}(-\lambda x^\alpha).$$

Is there an analogous identity for ${}^R_0 D^\beta_x E_{\alpha,1}(-\lambda x^\alpha)$ of order $\beta \geq 0$?

Exercise 3.8 By Laplace transform, there holds for $x > 0$

$$E_{\alpha,1}(-\lambda x^\alpha) = \frac{1}{2\pi i}\int_C \frac{e^{zx}}{z + \lambda z^{1-\alpha}}dz.$$

(i) By collapsing the Hankel contour C, show

$$E_{\alpha,1}(-\lambda x^\alpha) = \frac{\sin\alpha\pi}{\pi}\int_0^\infty \frac{e^{-rx}\lambda r^{\alpha-1}}{(r^\alpha + \lambda\cos\alpha\pi)^2 + (\lambda\sin\alpha\pi)^2}dr.$$

(ii) Using part (i) to show

$$E_{\alpha,1}(-\lambda x^\alpha) > 0 \quad \text{and} \quad \frac{d}{dx}E_{\alpha,1}(-\lambda x^\alpha) < 0, \quad \forall x > 0.$$

(iii) Use Bernstein theorem in Theorem A.8 to prove that $E_{\alpha,1}(-x^\alpha)$ is completely monotone.

Exercise 3.9 The exponential function e^t satisfies the identity $e^{\lambda(s+t)} = e^{\lambda s}e^{\lambda t}$. Is there a similar relation for $E_{\alpha,1}(\lambda t)$, i.e., $E_{\alpha,1}(\lambda(s + t)) = E_{\alpha,1}(\lambda s)E_{\alpha,1}(\lambda t)$?

Exercise 3.10 [BS05] Prove that for any $\alpha \in [0, 1]$, there holds

$$E_{\alpha,1}(-x) = \frac{2x}{\pi}\int_0^\infty \frac{E_{2\alpha,1}(-t^2)}{x^2 + t^2}dt.$$

Exercise 3.11 Show the following identity for $\beta > 0$ and $t > 0$:

$$\int_0^\infty e^{-\frac{x^2}{4t}} x^{\beta-1} E_{\alpha,\beta}(x^\alpha) dx = \sqrt{\pi} t^{\frac{\beta}{2}} E_{\frac{\alpha}{2}, \frac{\beta+1}{2}}(t^{\frac{\alpha}{2}}).$$

Exercise 3.12 Show the following Cristoffel-Darboux-type formula:

$$\int_0^t s^{\gamma-1} E_{\alpha,\gamma}(ys^\alpha)(t-s)^{\beta-1} E_{\alpha,\beta}(z(t-s)^\alpha) ds$$
$$= \frac{y E_{\alpha,\gamma+\beta}(yt^\alpha) - z E_{\alpha,\gamma+\beta}(zt^\alpha)}{y-z} t^{\gamma+\beta-1},$$

where $y \neq z$ are any complex numbers.

Exercise 3.13 This exercise is concerned with completely monotone functions.

(i) Let f be completely monotone, and h be nonnegative with a complete monotone derivative (which is sometimes called a Bernstein function). Prove that $f \circ h$ is completely monotone.

(ii) Using the fact that x^α is a Bernstein function, prove that the function $f(x) = e^{-|x|^\alpha}$, $\alpha \in (0,1)$, is completely monotone.

(iii) Prove that for $\alpha \in (0,1)$, $E_{\alpha,1}(-x^\alpha)$ is completely monotone using the complete monotonicity of $E_{\alpha,1}(-x)$ in Theorem 3.5.

Exercise 3.14 [MS01, Theorem 11] Prove that for any $\alpha, \beta > 0$, the function $E_{\alpha,\beta}(x^{-1})$ is completely monotone on $\mathbb{R}_+$.

Exercise 3.15 [MS18a] This exercise is concerned with Turán-type inequality for Mittag-Leffler functions. Suppose $\alpha, \beta > 0$ and $x \in \mathbb{R}_+$. Let

$$\mathbf{E}_{\alpha,\beta}(x) = \Gamma(\beta) E_{\alpha,\beta}(x).$$

(i) Show that the digamma function $\psi(x) = (\ln \Gamma(x))' = \frac{\Gamma'(x)}{\Gamma(x)}$ is concave.
(ii) Show that the function $\beta \to \mathbf{E}_{\alpha,\beta}(x)$ is logconvex on $\mathbb{R}_+$.
(iii) Prove the following Turán-type inequality

$$\mathbf{E}_{\alpha,\beta}(z) \mathbf{E}_{\alpha,\beta+2}(z) \geq \mathbf{E}_{\alpha,\beta+1}(z)^2.$$

Exercise 3.16 [MS18b] This exercise is concerned with the log-convexity of Mittag-Leffler function, which arises in the study of the maximum principle for two-point boundary value problems with a Djrbashian-Caputo fractional derivative.

(i) For $0 < \alpha \leq 1$, $z \in \mathbb{C}$, $z \neq 0$ and $|\arg z| > \alpha \pi$, there holds

$$E_{\alpha,\alpha+1}(z) = -\frac{\sin \alpha \pi}{\alpha \pi} \int_0^{+\infty} \frac{e^{-r^{\frac{1}{\alpha}}}}{r^2 - 2rz \cos \alpha \pi + z^2} dr - \frac{1}{z}.$$

(ii) Prove the following integral identities for $0 < \alpha \leq 1$ and $x > 0$:

$$x^\alpha E_{\alpha,\alpha+1}(-x^\alpha) = -\frac{\sin \alpha\pi}{\pi} \int_0^\infty \frac{r^{\alpha-1}e^{-xr}}{r^{2\alpha} + 2r^\alpha \cos(\alpha\pi) + 1}\,dr + 1,$$

$$x^{\alpha-1} E_{\alpha,\alpha}(-x^\alpha) = \frac{\sin \alpha\pi}{\pi} \int_0^\infty \frac{r^\alpha e^{-xr}}{r^{2\alpha} + 2r^\alpha \cos(\alpha\pi) + 1}\,dr,$$

$$E_{\alpha,1}(-x^\alpha) = \frac{\sin \alpha\pi}{\pi} \int_0^\infty \frac{r^{\alpha-1}e^{-xr}}{r^{2\alpha} + 2r^\alpha \cos(\alpha\pi) + 1}\,dr.$$

(iii) Using (ii) of this exercise to prove that for $\alpha \in (0, 1]$, the function $x^\alpha E_{\alpha,\alpha+1}(-x^\alpha)$ is log concave, and $E_{\alpha,1}(-x^\alpha)$ and $x^{\alpha-1} E_{\alpha,\alpha}(-x^\alpha)$ are log convex.

Exercise 3.17 Prove the following identity:

$$\frac{d}{dz}W_{\rho,\mu}(z) = \frac{1}{\rho z}(W_{\rho,\mu-1}(z) + (1 - \mu)W_{\rho,\mu}(z)).$$

Exercise 3.18 Determine the values for

$$\lim_{|z|\to\infty} W_{\rho,\mu}(z) \quad \text{and} \quad \lim_{|z|\to\infty} zW_{\rho,\mu}(z),$$

where $\arg(z) > \pi - \epsilon$, for small ϵ.

Exercise 3.19 The following version of the Wright function:

$$M_\mu(z) = W_{-\mu,1-\mu}(-z) = \sum_{k=0}^\infty \frac{(-1)^k z^k}{k!\Gamma(1 - \mu(k + 1))},$$

with $\mu \in (0, 1)$, was first introduced by Francesco Mainardi [Mai96] in his study of time-fractional diffusion, unaware of the prior work of Wright [Wri40]. It is often called the Mainardi M-function or M-Wright function in the literature, and serves the family of similarity solutions for one-dimensional subdiffusion [GLM99, GLM00, MMP10]. Using the reflection identity (2.3) of $\Gamma(z)$, it is equivalent to

$$M_\mu(z) = \frac{1}{\pi} \sum_{k=0}^\infty \frac{(-z)^{k-1}}{(k - 1)!}\Gamma(k\mu) \sin(k\mu\pi).$$

Prove that this function has the following properties:

(i) The case $\mu = \frac{1}{2}$ recovers a Gaussian function $M_{\frac{1}{2}}(z) = \frac{1}{\sqrt\pi}e^{-\frac{z^2}{4}}$.

(ii) The following asymptotic behavior holds [MT95]

$$M_\mu(x) \sim Ax^a e^{-bx^c}, \quad x \to \infty$$

with $A = (2\pi(1 - \mu)\mu^{\frac{1-2\mu}{1-\mu}})^{-\frac{1}{2}}$, $a = (2\mu - 1)(2 - 2\mu)^{-1}$, $b = (1 - \mu)\mu^{\frac{\mu}{1-\mu}}$ and $c = (1 - \mu)^{-1}$.

(iii) For $\mu \in (0, 1)$, there holds

$$\mathcal{L}[M_\mu(x)](z) = E_{\mu,1}(-z).$$

(iv) For $\mu \in (0, 1)$, the Fourier transform of $M_\mu(|x|)$ is given by

$$\mathcal{F}[M_\mu(|x|)](\xi) = 2E_{2\mu,1}(-\xi^2).$$

Exercise 3.20 Prove Theorem 3.9(i).

Part II
Fractional Ordinary Differential Equations

Chapter 4
Cauchy Problem for Fractional ODEs

In this chapter, we discuss the existence, uniqueness and regularity of solutions to fractional ordinary differential equations (ODEs). The existence and uniqueness can be analyzed using the method of successive approximations and fixed point argument. The former, pioneered by Cauchy, Lipschitz, Peano and Picard for classical ODEs in various settings, shows the existence by constructing suitable approximations, and lends itself to a constructive algorithm, although not necessarily efficient. The latter is an abstraction of the former and converts the existence issue into the existence of a fixed point for a certain (nonlinear) map, often derived via suitable integral transform. It is often easier to analyze but less constructive. The equivalence between an ODE and a suitable Volterra integral equation is key to both approaches. While these arguments largely parallel that of the classical ODE theory, extra care is needed in the fractional case due to the often limited solution regularity. Generally, the regularity issue in the fractional case is more delicate and not yet fully understood.

Notation: Throughout, the notation $^{R}\partial_{t}^{\alpha}$ and ∂_{t}^{α} denote the (left sided) Riemann-Liouville and Djrbashian-Caputo fractional derivative, respectively, bases at $t = 0$, $I = (0, T]$ for some $T > 0$, which will be evident from the context, and $\overline{I}$ its closure, $L^{p}(I) = L^{p}(0, T)$, and $C(I) = C(0, T]$.

4.1 Gronwall's Inequalities

Integral inequalities play an important role in the qualitative analysis of the solutions to (nonlinear) differential equations. The classical Gronwall inequality provides explicit bounds on solutions of a class of linear integral / differential inequalities. It is often used to establish *a priori* bounds which are then used in proving global existence, uniqueness and stability results. There are two different forms, the differential form of Thomas Gronwall [Gro19], and the integral form of Richard Bellman [Bel43]. Below we state only the integral form. First we recall the standard version: if the functions $u, a, k \in C(\overline{I})$ for some $0 < T \leq \infty$ and are nonnegative, and satisfy

© The Author(s), under exclusive license to Springer Nature Switzerland AG 2021
B. Jin, *Fractional Differential Equations*, Applied Mathematical Sciences 206,
https://doi.org/10.1007/978-3-030-76043-4_4

$$u(t) \le a(t) + \int_0^t k(s)u(s)\mathrm{d}s, \quad t \in \bar{I},$$

then $u(t)$ satisfies

$$u(t) \le a(t) + \int_0^t a(s)k(s)e^{\int_s^t k(\tau)\mathrm{d}\tau}\mathrm{d}s, \quad t \in \bar{I}. \tag{4.1}$$

If, in addition, $a(t)$ is nondecreasing, then

$$u(t) \le a(t)e^{\int_0^t k(s)\mathrm{d}s}, \quad t \in \bar{I}.$$

By an approximation argument, the condition $k \in C(\bar{I})$ can be relaxed to $k \in L^1(I)$, and we have the following result. Below for any $1 \le p \le \infty$, we denote by

$$L_+^p(I) = \{u \in L^p(I) : u(t) \ge 0 \quad \text{a.e. } I\},$$

and the inequality should be understood in almost everywhere (a.e.) sense. There is also a uniform Gronwall's inequality; see Exercise 4.2.

Theorem 4.1 *Let $a(t) \in L_+^\infty(I)$ be nondecreasing, and $k(t) \in L_+^1(I)$. If $u \in L_+^\infty(I)$ satisfies*

$$u(t) \le a(t) + \int_0^t k(s)u(s)\mathrm{d}s, \quad \text{a.e. } t \in \bar{I}.$$

Then there holds

$$u(t) \le a(t)e^{\int_0^t k(s)\mathrm{d}s} \quad \text{a.e. } t \in \bar{I}.$$

Now we develop several variants of the inequality involving weak singularities, e.g., $a(t) = t^{-\alpha}$, with $\alpha \in (0, 1)$, which appears often in the study of fractional ODEs. Such inequalities were first systematically developed by Daniel Henry [Hen81, Chapter 7], using an iterative process and Mittag-Leffler function $E_{\alpha,\beta}(z)$, cf. (3.1) in Chapter 3. The next result [YGD07, Theorem 1] is one useful version. Case (ii) can be found in Henry [Hen81, Lemma 7.1.1] and is well known. There is also an extended version involving an additional L^p integrable factor inside the integral [LY20, Lemma 2.4]; see Exercise 4.4 for details.

Theorem 4.2 *Let $\beta > 0$, $a(t) \in L_+^1(I)$, and $0 \le b(t) \in C(\bar{I})$ be nondecreasing. Let $u(t) \in L_+^1(I)$ satisfy*

$$u(t) \le a(t) + b(t){}_0I_t^\beta u(t), \quad t \in I.$$

Then there holds

$$u(t) \le a(t) + \int_0^t \sum_{n=1}^\infty \frac{(b(t))^n}{\Gamma(n\beta)}(t - s)^{n\beta - 1}a(s)\mathrm{d}s, \quad t \in I.$$

In particular, the following assertions hold.

(i) If $a(t)$ is nondecreasing on $\bar{I}$ and $b(t) \equiv b$, then

$$u(t) \leq a(t)E_{\beta,1}(bt^{\beta}) \quad \text{on } I.$$

(ii) *If $a(t) \equiv at^{-\alpha}$, with $\alpha \in (0,1)$ and $a > 0$, and $b(t) \equiv b > 0$, then*

$$u(t) \leq a\Gamma(1 - \alpha)E_{\beta,1-\alpha}(bt^{\beta})t^{-\alpha} \quad \text{on } I.$$

Proof For any $v \in L^1_+(I)$, let $Av(t) = b(t)_0I^{\beta}_t v(t)$. Then the condition $u(t) \leq a(t) + Au(t)$ and the nonnegativity of a and b imply

$$u(t) \leq \sum_{k=0}^{n-1} A^k a(t) + A^n u(t).$$

The assertion follows from the claims that $A^n u(t) \leq b(t)^n {}_0I^{n\beta}_t u(t)$ and $A^n u(t) \to 0$ as $n \to \infty$ for each $t \in I$. The first claim holds for $n = 1$. Assume that it is true for some $n \geq 1$. Then for $n + 1$, the induction hypothesis implies $A^{n+1}u(t) \leq b(t)_0I^{\beta}_t (b(\cdot)^n {}_0I^{n\beta}_t u)$. Since $b(t)$ is nondecreasing, by Theorem 2.1,

$$A^{n+1}u(t) \leq b(t)^{n+1} {}_0I^{\beta}_t {}_0I^{n\beta}_t u = b(t)^{n+1} {}_0I^{(n+1)\beta}_t u(t).$$

This shows the first claim. Further, since $A^n u(t) \leq \|b\|^n_{L^{\infty}(I)} {}_0I^{n\beta}_t u(t) \to 0$ as $n \to \infty$ for any $t \in \bar{I}$, this shows the second claim. (i) follows similarly as

$$u(t) \leq a(t) \sum_{n=0}^{\infty} b^n ({}_0I^{n\beta}_t 1)(t) = a(t) \sum_{n=0}^{\infty} \frac{(bt^{\beta})^n}{\Gamma(n\beta + 1)} = a(t)E_{\beta,1}(bt^{\beta}),$$

using the definition of $E_{\beta,1}(z)$. Similarly, (ii) follows from the elementary identity

$$\frac{1}{\Gamma(k\beta)} \int_0^t (t - s)^{k\beta - 1} s^{-\alpha} ds = \frac{\Gamma(1 - \alpha)}{\Gamma(k\beta - \alpha + 1)} t^{k\beta - \alpha}$$

and the definition of the Mittag-Leffler function $E_{\beta,\alpha}(z)$, cf. (3.1) in Chapter 3. $\square$

Next we give several Gronwall type inequalities involving double singularities [Web19b, Theorem 3.2]. We begin with case of constant coefficients. For a function $u \in L^{\infty}(I)$, the nondecreasing function u^* is defined by $u^*(t) = \text{esssup}_{s \in [0,t]} u(s)$.

Theorem 4.3 *Let $a \geq 0$ and $b > 0$, and suppose that $\beta > 0, \gamma \geq 0$ and $\beta + \gamma < 1$, and the function $u \in L^{\infty}_+(I)$ satisfies*

$$u(t) \leq a + b \int_0^t (t - s)^{-\beta} s^{-\gamma} u(s) ds, \quad t \in \bar{I}.$$

For $r > 0$, let $t_r = [bB(1 - \beta, 1 - \gamma)]^{-\frac{1}{1-\beta-\gamma}} r^{\frac{1}{1-\beta-\gamma}}$. Then, if $r \leq bB(1 - \beta, 1 - \gamma)T^{1-\beta-\gamma}$ and also $r < 1$, there holds

$$u(t) \leq a(1 - r)^{-1} e^{\frac{bt_r^{-\beta}}{(1-r)(1-\gamma)} t^{1-\gamma}}, \quad t \in \bar{I}. \tag{4.2}$$

Further, for $r_* = (1 - \gamma)^{-1}\beta$, there holds

$$u(t) \leq a(1 - \gamma)(1 - \beta - \gamma)^{-1}e^{\frac{bt_{r_*}^{-\beta}}{1-\beta-\gamma}t^{1-\gamma}}, \quad t \in \bar{I}. \tag{4.3}$$

Moreover, the exponent $(1 - \beta - \gamma)^{-1}bt_{r_*}^{-\beta}$ is optimal in the sense that it is the smallest possible for admissible choices of r.

Proof Let $c_{\beta,\gamma} = B(1 - \beta, 1 - \gamma)$ and $v(t) = a + b\int_0^t (t - s)^{-\beta}s^{-\gamma}u(s)ds$. Since $u \in L^\infty(I)$ and $\beta + \gamma < 1$, v is continuous, $u(t) \leq v(t)$ for a.e. $t \in \bar{I}$ and

$$v(t) \leq a + b\int_0^t (t - s)^{-\beta}s^{-\gamma}v(s)ds, \quad t \in \bar{I}.$$

Clearly, it suffices to prove (4.2) and (4.3) for v, and we may assume that u is continuous. Choose $r \leq bc_{\beta,\gamma}T^{1-\beta-\gamma}$ with $r < 1$, hence $t_r \leq T$. Let $t \in I$ and fix any $\xi \in (0, t]$. If $\xi \leq t_r$, then from the identity

$$\int_0^\xi (\xi - s)^{-\beta}s^{-\gamma}ds = c_{\beta,\gamma}\xi^{1-\beta-\gamma} \qquad \cdot \tag{4.4}$$

and the definitions of u^* and t_r, we deduce

$$u(\xi) \leq a + b\int_0^\xi (\xi - s)^{-\beta}s^{-\gamma}u(s)ds \leq a + b\int_0^\xi (\xi - s)^{-\beta}s^{-\gamma}u^*(t)ds$$

$$= a + bc_{\beta,\gamma}\xi^{1-\beta-\gamma}u^*(t) \leq a + bc_{\beta,\gamma}t_r^{1-\beta-\gamma}u^*(t) = a + ru^*(t). \tag{4.5}$$

Next, if $t_r \leq \xi \leq t$, then

$$u(\xi) \leq a + b\int_0^{\xi-t_r} (\xi - s)^{-\beta}s^{-\gamma}u(s)ds + b\int_{\xi-t_r}^\xi (\xi - s)^{-\beta}s^{-\gamma}u(s)ds.$$

Using the inequalities $(\xi - s)^{-\beta} \leq t_r^{-\beta}$ and $s^{-\gamma} \leq (s - \xi + t_r)^{-\gamma}$ in the first and second integrals, respectively, and applying the identity (4.4), we obtain

$$u(\xi) \leq a + bt_r^{-\beta}\int_0^{\xi-t_r} s^{-\gamma}u(s)ds + b\int_{\xi-t_r}^\xi (\xi - s)^{-\beta}(s - \xi + t_r)^{-\gamma}u^*(t)ds$$

$$\leq a + bt_r^{-\beta}\int_0^{\xi-t_r} s^{-\gamma}u^*(s)ds + bc_{\beta,\gamma}t_r^{1-\beta-\gamma}u^*(t),$$

and consequently,

$$u(\xi) \leq a + bt_r^{-\beta}\int_0^t s^{-\gamma}u^*(s)ds + ru^*(t). \tag{4.6}$$

It follows from (4.5) that (4.6) holds for $t \in I$ and for all $\xi \in \bar{I}$. Hence by taking the supremum for $\xi \in [0, t]$, since $r < 1$, we obtain

$$(1 - r)u^*(t) \le a + bt_r^{-\beta} \int_0^t s^{-\gamma} u^*(s) ds,$$

i.e.,

$$u^*(t) \le a(1 - r)^{-1} + b(1 - r)^{-1} t_r^{-\beta} \int_0^t s^{-\gamma} u^*(s) ds.$$

Then Theorem 4.1 implies

$$u(t) \le u^*(t) \le a(1 - r)^{-1} e^{\frac{bt_r^{-\beta}}{(1-r)(1-\gamma)} t^{1-\gamma}}, \quad t \in \bar{I}.$$

Now we show that the optimality of the choice $r_* := (1 - \gamma)^{-1}\beta$ in the sense that the exponent $[(1 - r)(1 - \gamma)]^{-1} bt_r^{-\beta}$ is smallest possible. The set of admissible r is the interval $(0, bc_{\beta,\gamma} T^{1-\beta-\gamma}]$. Define a function f by $f(r) := (1 - r) t_r^{\beta} = (1 - r) r^{\frac{\beta}{1-\beta-\gamma}} (bc_{\beta,\gamma})^{-\frac{\beta}{1-\beta-\gamma}}$. By differentiating f, we deduce $f'(r) = 0$ at $r_* = (1 - \gamma)^{-1}\beta$ and it is a maximum of f, and we have $(1 - r_*)(1 - \gamma) = 1 - \beta - \gamma$. Note that the choice $r = r_*$ is only valid if $r_* \le bc_{\beta,\gamma} T^{1-\beta-\gamma}$, i.e., $bc_{\beta,\gamma} T^{1-\beta-\gamma} \ge (1 - \gamma)^{-1}\beta$. But, if $bc_{\beta,\gamma} T^{1-\beta-\gamma} < (1 - \gamma)^{-1}\beta$, we have for any $\xi \le t \le T$,

$$u(\xi) \le a + b \int_0^\xi (\xi - s)^{-\beta} s^{-\gamma} u(s) ds = a + bc_{\beta,\gamma} \xi^{1-\beta-\gamma} u^*(t)$$

$$\le a + bc_{\beta,\gamma} T^{1-\beta-\gamma} u^*(t) < a + (1 - \gamma)^{-1}\beta u^*(t).$$

Taking the supremum for $\xi \in [0, t]$ gives $u(t) \le u^*(t) \le a(1 - \beta - \gamma)^{-1}(1 - \gamma)$ for $t \in \bar{I}$. Thus the conclusion holds also in this case. $\qquad\square$

Next we relax a, b to be bounded functions instead of constants using a^* and b^*.

Corollary 4.1 *Let $\beta > 0$, $\gamma \ge 0$ and $\beta + \gamma < 1$, and $a, b \in L_+^\infty(I)$. If $u \in L_+^\infty(I)$ satisfies*

$$u(t) \le a(t) + b(t) \int_0^t (t - s)^{-\beta} s^{-\gamma} u(s) ds, \quad t \in \bar{I},$$

then with $r_* = (1 - \beta - \gamma)^{-1}(1 - \gamma)$,

$$u(t) \le r_* a^*(t) e^{\frac{b^*(t) t_{r_*}^{-\beta}}{1-\beta-\gamma} t^{1-\gamma}}, \quad t \in \bar{I}.$$

Proof For any $t_1 \in I$, the following inequality holds:

$$u(t) \le a^*(t_1) + b^*(t_1) \int_0^t (t - s)^{-\beta} s^{-\gamma} u(s) ds, \quad t \in [0, t_1].$$

Then Theorem 4.3 gives $u(t) \le r_* a^*(t_1) e^{\frac{b^*(t_1) t_{r_*}^{-\beta}}{1-\beta-\gamma} t^{1-\gamma}}$, for $t \in [0, t_1]$. This estimate holds with $u^*(t_1)$ on the left and t replaced by t_1 on the right. Since t_1 is arbitrary in I, the desired assertion follows. $\qquad\square$

Example 4.1 If a function $u \in L_+^\infty(I)$ satisfies

$$u(t) \leq a + b \int_0^t (t-s)^{-\frac{1}{2}} u(s) ds \quad \text{on } I,$$

then by Theorem 4.3, we have $u(t) \leq 2ae^{8b^2 t}$ for a.e. $t \in \bar{I}$. In fact, with $\beta = \frac{1}{2}$ and $\gamma = 0$, $t_{r_*} = (16b^2)^{-1}$ and $(1-\beta)^{-1} bt_{r_*}^{-\beta} = 8b^2$. Theorem 4.2 gives $u(t) \leq aE_{\frac{1}{2},1}(bt^{\frac{1}{2}} \Gamma(\frac{1}{2}))$. Using a property of $E_{\frac{1}{2},1}(z)$ in (3.3), we have $u(t) \leq aE_{\frac{1}{2},1}(bt^{\frac{1}{2}} \Gamma(\frac{1}{2})) \leq 2ae^{\pi b^2 t}$. This bound is sharper than $u(t) \leq 2ae^{8b^2 t}$, but of the same type.

Next we discuss the case where the function u may be singular. The next result is immediate from Theorem 4.3.

Corollary 4.2 *Let $a, b \geq 0$, and $c > 0$. Let $0 < \alpha, \beta < 1$, $\gamma \geq 0$ with $\alpha + \beta + \gamma < 1$. Suppose that a function $u(t)t^\alpha \in L_+^\infty(I)$ and u satisfies*

$$u(t) \leq at^{-\alpha} + b + c \int_0^t (t-s)^{-\beta} s^{-\gamma} u(s) ds, \quad t \in \bar{I}.$$

Then with $r_* = (1 - \alpha - \gamma)^{-1}\beta$ and $c_{\alpha,\beta,\gamma} = (1 - \gamma - \alpha)(1 - \alpha - \beta - \gamma)^{-1}$,

$$u(t) \leq c_{\alpha,\beta,\gamma}(at^{-\alpha} + b)e^{\frac{ct_{r_*}^{-\beta}}{1-\alpha-\beta-\gamma}t^{1-\gamma}}, \quad t \in \bar{I}.$$

Proof Let $v(t) = t^\alpha u(t)$ so that $v \in L^\infty(I)$ and v satisfies

$$v(t) \leq a + bt^\alpha + ct^\alpha \int_0^t (t-s)^{-\beta} s^{-(\gamma+\alpha)} v(s) ds.$$

Then applying Corollary 4.1 with $\alpha + \gamma$ in place of γ, we obtain

$$v(t) \leq c_{\alpha,\beta,\gamma}(a + bt^\alpha)e^{ct^\alpha \frac{t_{r_*}^{-\beta}}{1-\alpha-\beta-\gamma}t^{1-\gamma-\alpha}}.$$

Then the desired assertion follows by dividing both sides by t^α. $\qquad\square$

The next result refines Corollary 4.2 with a few leading singular terms, under weaker conditions on the exponents α, β and γ [Web19b, Theorem 3.9].

Theorem 4.4 *Let $a, b \geq 0$ and $c > 0$ be constants. Let $0 < \alpha, \beta, \gamma < 1$ with $\alpha + \gamma < 1$ and $\beta + \gamma < 1$. Suppose that a function $u(t)t^\alpha \in L_+^\infty(I)$ satisfies*

$$u(t) \leq at^{-\alpha} + b + c \int_0^t (t-s)^{-\beta} s^{-\gamma} u(s) ds, \quad t \in \bar{I}.$$

Then, for $t \in \bar{I}$,

$$u(t) \leq at^{-\alpha} + acb_1 t^{-\alpha+1-\beta-\gamma} + ac^2 b_1 b_2 t^{-\alpha+2(1-\beta-\gamma)} + \dots$$

$$+ (1-\gamma)(1-\beta-\gamma)^{-1}(b + ac^m b_1 b_2 \dots b_m t^{-\alpha+m(1-\beta-\gamma)})e^{-\frac{ct_{r_*}^{-\beta}}{1-\beta-\gamma}t^{1-\gamma}}$$

with $m = \lceil (1 - \beta - \gamma)^{-1} \alpha \rceil$, $r_* = (1 - \gamma)^{-1} \beta$, *and for* $n \in \mathbb{N}$, $b_n := b(1 - \beta, 1 - \alpha - \gamma + (n - 1)(1 - \beta - \gamma))$.

Proof The proof is based on mathematical induction. Let $w_0(t) = u(t)$, $c_0 = \|w_0(t) t^\alpha\|_{L^\infty(I)}$ and $w_1(t) := b + c \int_0^t (t - s)^{-\beta} s^{-\gamma} w_0(s) ds$. Then

$$w_0(t) \leq at^{-\alpha} + w_1(t), \quad t \in \bar{I}. \tag{4.7}$$

First, direct computation gives

$$w_1(t) \leq b + cc_0 \int_0^t (t - s)^{-\beta} s^{-(\alpha+\gamma)} ds = b + cc_0 t^{1-\beta-\alpha-\gamma} b_1.$$

Thus, $w_1 \in L^\infty(I)$ if $\alpha + \beta + \gamma \leq 1$, and $t^{\alpha+\beta+\gamma-1} w_1(t) \in L^\infty(I)$ when $\alpha + \beta + \gamma > 1$. If $\alpha + \beta + \gamma \leq 1$, then w_1 satisfies

$$w_1(t) \leq b + c \int_0^t (t - s)^{-\beta} s^{-\gamma} (as^{-\alpha} + w_1(s)) ds$$

$$= b + act^{1-\alpha-\beta-\gamma} b_1 + c \int_0^t (t - s)^{-\beta} s^{-\gamma} w_1(s) ds. \tag{4.8}$$

By Corollary 4.1 (since $1 - \alpha - \beta - \gamma \geq 0$), we obtain

$$w_1(t) \leq (1 - \gamma)(1 - \beta - \gamma)^{-1} (b + act^{1-\alpha-\beta-\gamma} b_1) e^{\frac{ct_{r_*}^{-\beta}}{1-\beta-\gamma} t^{1-\gamma}}, \quad t \in \bar{I}.$$

Thus, when $\alpha + \beta + \gamma \leq 1$, we obtain the desired assertion. Meanwhile, when $\alpha + \beta + \gamma > 1$, we define w_2 by $w_2(t) := b + c \int_0^t (t - s)^{-\beta} s^{-\gamma} w_1(s) ds$, and we have $t^{\alpha+\beta+\gamma-1} w_1(t) \leq c_1$ for some $c_1 > 0$, i.e., $w_1(t) \leq c_1 t^{1-\alpha-\beta-\gamma}$, and so $w_2(t) \leq b + cc_1 b_2 t^{-\alpha+2(1-\beta-\gamma)}$. Thus, if $2(\beta+\gamma)+\alpha \leq 2$, then $w_2 \in L^\infty(I)$, and if $2\beta+2\gamma+\alpha > 2$, then $t^{2\beta+2\gamma+\alpha-2} w_2 \in L^\infty(I)$. When $2\beta + 2\gamma + \alpha \leq 2$, i.e., $-\alpha + 2(1 - \beta - \gamma) \geq 0$, we deduce

$$w_1(t) \leq acb_1 t^{-\alpha+1-\beta-\gamma} + w_2(t). \tag{4.9}$$

Therefore, w_2 satisfies

$$w_2(t) \leq b + c \int_0^t (t - s)^{-\beta} s^{-\gamma} w_1(s) ds$$

$$\leq b + ac^2 b_1 b_2 t^{-\alpha+2(1-\beta-\gamma)} + c \int_0^t (t - s)^{-\beta} s^{-\gamma} w_2(s) ds.$$

Since $-\alpha + 2(1 - \beta - \gamma) \geq 0$ and $w_2 \in L^\infty(I)$, by Corollary 4.1, we deduce

$$w_2(t) \leq (1 - \gamma)(1 - \beta - \gamma)^{-1} (b + ac^2 b_1 b_2 t^{-\alpha+2(1-\beta-\gamma)}) e^{\frac{ct_{r_*}^{-\beta}}{1-\beta-\gamma} t^{1-\gamma}}, \quad t \in \bar{I}.$$

From (4.7) and (4.9), this gives, for $-\alpha+2(1-\beta-\gamma) \geq 0$, the desired assertion. Note that the procedure can be repeated. At each step, we gain $1 - \beta - \gamma$ in the exponent and the process can be continued for a finite number of steps until the power of t becomes nonnegative, i.e., for $m = \lceil \frac{\alpha}{1-\beta-\gamma} \rceil$. Thus we obtain the desired assertion. $\square$

4.2 ODEs with a Riemann-Liouville Fractional Derivative

In this section, we describe the solution theory for the following Cauchy problem with a Riemann-Liouville fractional derivative (with $n - 1 < \alpha < n$, $n \in \mathbb{N}$)

$$
\begin{cases}
{}^R\partial_t^\alpha u = f(t, u), & t \in I, \\
({}^R\partial_t^{\alpha-j} u)(0) = c_j, & j = 1, \ldots, n
\end{cases}
\tag{4.10}
$$

with $\{c_j\}_{j=1}^n \subset \mathbb{R}$. The notation ${}^R\partial_t^{\alpha-n} u(0)$ should be identified with ${}_0 I_t^{n-\alpha} u(0)$, and $({}^R\partial_t^{\alpha-j} u)(0) = \lim_{t \to 0^+}({}^R\partial_t^{\alpha-j} u)(t)$. The initial conditions are inherently nonlocal, which are needed for the well-posedness of the problem, but their physical interpretation is generally unclear. The Djrbashian-Caputo fractional derivative $\partial_t^\alpha u$ does not have this drawback. First we characterize the nonlocal initial condition ${}_0 I_t^{1-\alpha} u(0) = c$ [KST06, Lemmas 3.2 and 3.5] and [BBP15, Theorems 6.1 and 6.2].

Proposition 4.1 *Let $0 < \alpha < 1$, and $u \in L^1(I) \cap C(I)$.*

(i) *If there exists a limit $\lim_{t \to 0^+} t^{1-\alpha} u(t) = c\Gamma(\alpha)^{-1}$, then there also exists a limit $\lim_{t \to 0^+} {}_0 I_t^{1-\alpha}(t) = c$.*

(ii) *If there exists a limit $\lim_{t \to 0^+} {}_0 I_t^{1-\alpha}(t) = c$ and if there exists the limit $\lim_{t \to 0^+} t^{1-\alpha} u(t)$, then $\lim_{t \to 0^+} t^{1-\alpha} u(t) = c\Gamma(\alpha)^{-1}$.*

Proof (i) If $\lim_{t \to 0^+} t^{1-\alpha} u(t) = c\Gamma(\alpha)^{-1}$, then for any $\epsilon > 0$, there exists $\delta > 0$ such that $t \in (0, \delta)$ implies $|t^{1-\alpha} u(t) - c\Gamma(\alpha)^{-1}| < \epsilon\Gamma(\alpha)^{-1}$, or $(c - \epsilon)\Gamma(\alpha)^{-1} t^{\alpha-1} < u(t) < (c + \epsilon)\Gamma(\alpha)^{-1} t^{\alpha-1}$, for $t \in (0, \delta)$. This implies $|u(t)| < \Gamma(\alpha)^{-1}(|c| + \epsilon)t^{\alpha-1}$ for any $t \in (0, \delta)$. Hence, for $0 < s < t < \delta$, $(t - s)^{-\alpha}|u(s)| < (|c| + \epsilon)\Gamma(\alpha)^{-1} s^{\alpha-1}(t - s)^{-\alpha}$. Since the right-hand side is integrable, the integral $\int_0^t (t - s)^{-\alpha} u(s) ds$ converges absolutely for any $t \in (0, \delta)$. Hence,

$$
\frac{c - \epsilon}{\Gamma(\alpha)} \int_0^t (t - s)^{-\alpha} s^{\alpha-1} ds \leq \int_0^t (t - s)^{-\alpha} u(s) ds \leq \frac{c + \epsilon}{\Gamma(\alpha)} \int_0^t (t - s)^{-\alpha} s^{\alpha-1} ds.
$$

Direct computation gives that for any $t \in (0, \delta)$, $c - \epsilon \leq {}_0 I_t^{1-\alpha} u(t) \leq c + \epsilon$, i.e., $|{}_0 I_t^{1-\alpha} u(t) - c| < \epsilon$.

(ii) Since $u \in C(I) \cap L^1(I)$ and $\alpha \in (0, 1)$, we deduce that ${}_0 I_t^{1-\alpha} u(t)$ exists at each $t \in [0, T]$. Since $\lim_{t \to 0^+} {}_0 I_t^{1-\alpha} u(t) = c$, for each $\epsilon > 0$, there exists $\delta > 0$ such that $|{}_0 I_t^{1-\alpha} u(t) - c| < \epsilon$ or $c - \epsilon < {}_0 I_t^{1-\alpha} u(t) < c + \epsilon$, for $t \in (0, \delta)$. Applying the operator ${}_0 I_t^\alpha$ to both sides, by Theorem 2.1, gives that for any $t \in (0, \delta)$, $(c - \epsilon)\Gamma(\alpha + 1)^{-1} t^\alpha \leq {}_0 I_t^1 u(t) \leq (c + \epsilon)\Gamma(\alpha + 1)^{-1} t^\alpha$, i.e.,

$$
\lim_{t \to 0^+} \Gamma(\alpha + 1) t^{-\alpha} \int_0^t u(s) ds = c.
$$

Since $u \in L^1(I)$, $\int_0^t u(s)ds \to 0$ as $t \to 0^+$, and for $\alpha \in (0,1)$, $t^\alpha \to 0$ as $t \to 0^+$. Since $u \in C(I)$, and $t^{\alpha-1} \neq 0$ in a neighborhood of $t = 0$, L'Hôpital's rule gives

$$\lim_{t \to 0^+} \Gamma(\alpha + 1)t^{-\alpha} \int_0^t u(s)ds = \Gamma(\alpha) \lim_{t \to 0^+} t^{1-\alpha}u(t).$$

This completes the proof of the proposition. □

Remark 4.1 The sufficiency part requires only $u \in L^1(I)$. The necessity employs the L'Hôpital's rule, which assumes the limit exists and thus stronger conditions. In view of Proposition 4.1, for $\alpha \in (0,1)$, problem (4.10) may be rewritten as

$$^R\partial_t^\alpha u = f(t, u), \quad t \in I, \quad \text{with} \lim_{t \to 0^+} t^{1-\alpha}u(t) = c.$$

The theory developed below is also applicable to this problem.

Now we prove existence and uniqueness for problem (4.10), for which, parallel to classical ODEs, there are several different ways: reduction to Volterra integral equations, Laplace transform and operational calculus. We employ the reduction to Volterra integral equations and then apply a version of fixed point theorem (e.g., Banach contraction mapping theorem) to obtain a unique solution. This key step is to establish the equivalence between the FDE and the Volterra integral equation. This idea, if $f(t, u)$ is Lipschitz in u and bounded, goes back to Pitcher and Sewell [PS38] in 1938 for $^R\partial_t^\alpha u = f(t, u)$ (with $0 < \alpha < 1$). However, the integral operator used in [PS38] to invert the fractional derivative omitted an essential term from the initial value. In 1965, Al-Bassam [AB65] established the existence of a global continuous solution, using the correct Volterra integral equation and the method of successive approximations. Nonetheless, the boundedness assumption on f is restrictive, excluding even $f(t, u) = u$. In 1996, Delbosco and Rodino [DR96] considered a problem with $u(0) = c_0$ instead of $_0I_t^{1-\alpha}u(0^+) = c_0$, and proved local existence under a continuity assumption and also uniqueness under a global Lipschitz continuity assumption. The work [HTR+99] obtained the same result for the initial condition $_0I_t^{1-\alpha}u(0^+) = c_0$, by the same argument. Kilbas et al [KBT00] established existence and uniqueness results in spaces of integrable functions.

One main step is to prove the equivalence between the solutions of the formulations, which is closely connected with mapping properties of fractional integral and derivative in the spaces of interest. Below we state two equivalence results under slightly different assumptions on $f(t, u(t))$, under which problem (4.10) is equivalent to the following Volterra integral equation

$$u(t) = \sum_{j=1}^n \frac{c_j t^{\alpha-j}}{\Gamma(\alpha - j + 1)} + \frac{1}{\Gamma(\alpha)} \int_0^t (t - s)^{\alpha-1} f(s, u(s)) \, ds. \qquad (4.11)$$

The first result works in the space $L^1(I)$: if $u \in L^1(I)$ with $_0I_t^{n-\alpha}u \in AC^n(\bar{I})$ and f satisfies $t \mapsto f(t, u(t)) \in L^1(I)$. The result imposes only integrability of f, needed for defining $_0I_t^{n-\alpha} f(\cdot, u(\cdot))$, but no boundedness assumption. The latter is problematic for problem (4.10), due to the presence of the factor $O(t^{\alpha-n})$.

Proposition 4.2 *Let $n - 1 < \alpha < n$, $n \in \mathbb{N}$, $u \in L^1(I)$ with $_0I_t^{n-\alpha}u \in AC^n(\bar{I})$ and $t \mapsto f(t, u(t)) \in L^1(I)$. Then u solves problem (4.10) if and only if it solves (4.11).*

Proof Given a function $u \in L^1(I)$ with $_0I_t^{n-\alpha}u \in AC^n(\bar{I})$, by assumption, $g(t) :=$ $f(t, u(t)) \in L^1(I)$. Thus, $_0I_t^{n-\alpha}u(t)$ is n times differentiable a.e. on I, and by (4.10), $g(t) = {}^R\partial_t^\alpha u(t) = (_0I_t^{n-\alpha}u(t))^{(n)} \in L^1(I)$. Integrating the identity yields

$$\int_\eta^t g(s)ds = (_0I_t^{n-\alpha}u)^{(n-1)}(t) - (_0I_t^{n-\alpha}u)^{(n-1)}(\eta), \quad 0 < \eta < t \leq T.$$

Taking the limit $\eta \to 0^+$ and using the initial condition lead to $_0I_t^1 g(t) = (_0I_t^{n-\alpha}u)^{(n-1)}(t) - c_1$. Performing the integration n times leads to

$$_0I_t^{n-\alpha}u(t) = \sum_{j=1}^n \frac{c_j t^{n-j}}{\Gamma(n-j+1)} + {}_0I_t^n g(t).$$

Since $g(t) \in L^1(I)$, applying $_0I_t^\alpha$ to both sides and using Theorem 2.1 give

$$_0I_t^n u(t) = \sum_{j=1}^n \frac{c_j t^{\alpha+n-j}}{\Gamma(\alpha+n-j+1)} + {}_0I_t^{n+\alpha} g(t).$$

Differentiating both sides n times gives (4.11). Next we show the converse. By hypothesis, $u(t)$ solves (4.11). Applying $_0I_t^{1-\alpha}$ to both sides of (4.11) gives

$$_0I_t^{n-\alpha}u(t) = {}_0I_t^{n-\alpha}\left(\sum_{j=1}^n \frac{c_j t^{\alpha-j}}{\Gamma(\alpha-j+1)}\right) + {}_0I_t^{n-\alpha}{}_0I_t^\alpha f(t, u(t))$$

$$= \sum_{j=1}^n \frac{c_j t^{n-j}}{\Gamma(n-j+1)} + {}_0I_t^n f(t, u(t)).$$

Since $f(t, u(t)) \in L^1(I)$, $_0I_t^{n-\alpha}u(t) \in AC^n(\bar{I})$. By the absolute continuity of Lebesgue integral and repeated differentiation, we deduce $\lim_{t\to 0^+}(_0I_t^{n-\alpha})^{n-j}u(t) = c_j$, $j = 1, \ldots, n$. Further, differentiating both sides n times yields $^R\partial_t^\alpha u(t) = f(t, u(t))$ a.e. I, and thus it solves (4.10). This completes the proof of the proposition. $\qquad\square$

In Proposition 4.2, the conditions $f \mapsto f(t, u(t)) \in L^1(I)$ for $u \in L^1(I)$ and $_0I_t^{n-\alpha}u \in AC^n(\bar{I})$ are central. The former holds for every $u \in L^1(I)$ if and only if $|f(t, u)| \leq a(t) + b|u|$ for all (t, u) for some $a \in L^1(I)$ and $b > 0$.

Now we can give a unique existence result for problem (4.10).

Theorem 4.5 *Let $n - 1 < \alpha < n$, $n \in \mathbb{N}$, $J \subset \mathbb{R}$ be an open subset, $f : I \times J \to \mathbb{R}$ be a function such that $f(t, u) \in L^1(I)$ for any $u \in J$ and Lipschitz in u:*

$$|f(t, u) - f(t, v)| \leq L|u - v|, \quad \forall u, v \in J, t \in \bar{I}, \tag{4.12}$$

where $L > 0$. Then there exists a unique solution $u \in L^1(I)$ to problem (4.10).

Proof We give two slightly different proofs, both based on Banach fixed point theorem, cf. Theorem A.10 in the appendix.

(i) By Proposition 4.2, it suffices to show the unique existence of a solution $u \in L^1(I)$ to problem (4.11). Clearly, (4.11) makes sense for any $[0, t_1]$, $t_1 < T$. Choose $t_1 > 0$ such that $Lt_1^\alpha < \Gamma(\alpha + 1)$. We define an operator $A : L^1(I) \to L^1(I)$ by

$$Au = u_0(t) + {}_0I_t^\alpha f(t, u(t)), \quad \text{with } u_0(t) = \sum_{j=1}^n \frac{c_j}{\Gamma(\alpha - j + 1)} t^{\alpha - j}. \tag{4.13}$$

Clearly, $u_0 \in L^1(0, t_1)$. By assumption, $f(t, u(t)) \in L^1(I)$, ${}_0I_t^\alpha f(t, u(t)) \in L^1(I)$, cf. Theorem 2.2(i). Hence $Au \in L^1(I)$. Next, by the Lipschitz continuity of $f(t, u)$ in u in (4.12) and the estimate in Theorem 2.2(i), we deduce

$$\|Au_1 - Au_2\|_{L^1(0,t_1)} \leq \|{}_0I_t^\alpha (f(t, u_1(t)) - f(t, u_2(t)))\|_{L^1(0,t_1)}$$
$$\leq L\|{}_0I_t^\alpha |u_1(t) - u_2(t)|\|_{L^1(0,t_1)} \leq \Gamma(\alpha + 1)^{-1} Lt_1^\alpha \|u_1 - u_2\|_{L^1(0,t_1)}.$$

By the choice of t_1, A is contractive on $L^1(0, t_1)$. Hence, by Banach fixed point theorem, there exists a unique solution $u \in L^1(0, t_1)$ to (4.11) on $[0, t_1]$. (Further, the solution u can be obtained as the limit of the convergence sequence $A^m u_0$.) Next consider the interval $[t_1, 2t_1]$ (assuming $2t_1 \leq T$). Then we rewrite (4.11) as

$$u(t) = u_1(t) + \frac{1}{\Gamma(\alpha)} \int_{t_1}^t (t - s)^{\alpha - 1} f(s, u(s)) ds$$

with $u_1(t) = u_0(t) + \frac{1}{\Gamma(\alpha)} \int_0^{t_1} (t - s)^{\alpha - 1} f(s, u(s)) ds$. Note that the function u_1 belongs to $L^1(\Omega)$ and is known, since $u(t)$ is already uniquely defined on the interval $[0, t_1]$. Then the preceding argument shows that there exists a unique solution $u \in L^1(t_1, 2t_1)$ to problem (4.11) on the interval $[t_1, 2t_1]$. We complete the proof by repeating the process for a finite number of steps. By the very construction, it can be verified that the constructed solution u does belong to $L^1(I)$ with ${}_0I_t^{n-\alpha} u \in AC^n(\bar{I})$.

(ii). We employ the approach of Bielecki [Bie56]. We equip the space $L^1(I)$ with a family of weighted norms $\| \cdot \|_\lambda$, defined by $\|u\|_\lambda = \int_0^T e^{-\lambda t} |u(t)| dt$, with $\lambda > 0$. For any fixed $\lambda > 0$, $\| \cdot \|_\lambda$ is equivalent to $\| \cdot \|_{L^1(I)}$. It has already been verified that the operator A defined in (4.13) maps $L^1(I)$ into itself. Further, by the Lipschitz continuity of $f(t, u)$ in u in (4.12), there holds

$$\|Au_1 - Au_2\|_\lambda \leq \frac{1}{\Gamma(\alpha)} \int_0^T e^{-\lambda t} \int_0^t (t - s)^{\alpha - 1} |f(s, u_1(s)) - f(s, u_2(s))| ds dt$$
$$\leq \frac{L}{\Gamma(\alpha)} \int_0^T e^{-\lambda t} \int_0^t (t - s)^{\alpha - 1} |u_1(s) - u_2(s)| ds dt$$
$$= \frac{L}{\Gamma(\alpha)} \int_0^T e^{-\lambda s} |u_1(s) - u_2(s)| \left[\int_s^T e^{-\lambda(t-s)} (t - s)^{\alpha - 1} dt \right] ds.$$

Direct computation shows $\int_s^T e^{-\lambda(t-s)}(t-s)^{\alpha-1}dt \leq \lambda^{-\alpha}\int_0^\infty e^{-t}t^{\alpha-1}dt = \lambda^{-\alpha}\Gamma(\alpha)$.
Combining the preceding two estimates yields

$$\|Au_1 - Au_2\|_\lambda \leq \lambda^{-\alpha}L\|u_1 - u_2\|_\lambda.$$

For λ sufficiently large, A is contractive on $L^1(I)$ in the norm $\|\cdot\|_\lambda$, and by Banach
fixed point theorem, there exists a unique solution $u \in L^1(I)$ to problem (4.11). $\square$

In view of the Volterra reformulation (4.11), the solution $u(t)$ to problem (4.10)
can be singular at $t = 0$, i.e., $u(t) \sim c_n t^{\alpha-n}$ as $t \to 0^+$. The singularity precludes
working in the space $C(\overline{I})$ (unless $c_n = 0$). Nonetheless, one may employ a weighted
space $C_\gamma(\overline{D})$ (of continuous functions), $\gamma \geq 0$, defined by

$$C_\gamma(\overline{I}) = \{u \in C(I) : t^\gamma u(t) \in C(\overline{I})\},$$

equipped with the weighted norm $\|u\|_{C_\gamma(\overline{I})} = \|u(t)t^\gamma\|_{C(\overline{I})}$. Then $C_\gamma(\overline{I})$ endowed with
the norm $\|\cdot\|_{C(\overline{I})}$ is a Banach space. It can be shown that any element $u \in C_\gamma(\overline{I})$ can
be written as $u(t) = t^{-\gamma}v(t)$, with $v(t) \in C(\overline{I})$.

The following mapping property of the operator $_0I_t^\alpha$ holds on the space $C_\gamma(\overline{I})$.

Lemma 4.1 Let $0 \leq \gamma < 1$, and $\alpha > 0$, then $_0I_t^\alpha$ is bounded on $C_\gamma(\overline{I})$:

$$\|_0I_t^\alpha v\|_{C_\gamma(\overline{I})} \leq \Gamma(1-\gamma)\Gamma(1+\alpha-\gamma)^{-1}T^\alpha\|v\|_{C_\gamma(\overline{I})}.$$

Proof By the definition of the norm $\|\cdot\|_{C_\gamma(\overline{I})}$,

$$\|_0I_t^\alpha v\|_{C_\gamma(\overline{I})} = \sup_{t\in\overline{I}} \frac{t^\gamma}{\Gamma(\alpha)} \int_0^t (t-s)^{\alpha-1}v(s)ds$$

$$\leq \sup_{t\in\overline{I}} \frac{t^\gamma}{\Gamma(\alpha)} \int_0^t (t-s)^{\alpha-1}s^{-\gamma}ds\|v\|_{C_\gamma(\overline{I})}.$$

This and the identity $\int_0^t (t-s)^{\alpha-1}s^{-\gamma}ds = B(\alpha, 1-\gamma)t^{\alpha-\gamma}$ imply the desired estimate.
$\square$

The next two results are $C_\gamma(\overline{I})$ analogues of Proposition 4.2 and Theorem 4.5.
The proof of the proposition is omitted since it is identical with Proposition 4.2.

Proposition 4.3 Let $n - 1 < \alpha < n$, $n \in \mathbb{N}$. Let $J \subset \mathbb{R}$ be an open set, and
$f : I \times J \to \mathbb{R}$ be a function such that $f(t, u(t)) \in C_{n-\alpha}(\overline{I})$ for any $u \in C_{n-\alpha}(\overline{I})$. Then
$u \in C_{n-\alpha}(\overline{I})$ with $(_0I_t^{n-\alpha})^{(n)}u \in C_{n-\alpha}(\overline{I})$ solves (4.10) if and only if $u \in C_{n-\alpha}(\overline{I})$
satisfies (4.11).

Theorem 4.6 Let $n - 1 < \alpha < n$, $n \in \mathbb{N}$, $J \subset \mathbb{R}$ an open set and the function
$f : \overline{I} \times J \to \mathbb{R}$ be such that $f(t, u) \in C_{n-\alpha}(\overline{I})$ for any $u \in J$ and satisfy the condition
(4.12). Then there exists a unique solution $u \in C_{n-\alpha}(\overline{I})$ to problem (4.10) with
$^R\partial_t^\alpha u \in C_{n-\alpha}(\overline{I})$.

Proof The proof is similar to Theorem 4.5, and employs Banach fixed point theorem. By Proposition 4.3, it suffices to prove the assertion for (4.11). Equation (4.11) makes sense in any interval $[0, t_1] \subset \overline{I}$. Choose t_1 such that $Lt_1^\alpha \max \left(\Gamma(\alpha - n + 1)\Gamma(2\alpha - n + 1)^{-1}, \Gamma(\alpha + 1)^{-1} \right) < 1$. We rewrite (4.11) as $u(t) = Au(t)$, with $A : C_{n-\alpha}(\overline{I}) \to C_{n-\alpha}(\overline{I})$ defined in (4.13). Clearly, if $u \in C_{n-\alpha}[0, t_1]$, since $f(t, u(t)) \in C_{n-\alpha}[0, t_1]$ by assumption, then by Lemma 4.1, $_0I_t^\alpha f(t, u(t)) \in C_{n-\alpha}[0, t_1]$ and also $Au \in C_{n-\alpha}[0, t_1]$. Next, by the Lipschitz continuity of f and Lemma 4.1, we deduce that for any $u_1, u_2 \in C_{n-\alpha}[0, t_1]$,

$$\|Au_1 - Au_2\|_{C_{n-\alpha}[0,t_1]} = \|_0I_t^\alpha (f(t, u_1(t)) - f(t, u_2(t)))\|_{C_{n-\alpha}[0,t_1]}$$
$$\leq L\|_0I_t^\alpha |u_1 - u_2|\|_{C_{n-\alpha}[0,t_1]} \leq Lt_1^\alpha \Gamma(\alpha - n + 1)\Gamma(2\alpha - n + 1)^{-1}\|u_1 - u_2\|_{C_{n-\alpha}[0,t_1]}.$$

By the choice of t_1, the operator A is contractive on $C_{n-\alpha}[0, t_1]$, and Banach fixed point theorem implies that there exists a unique solution $u \in C_{n-\alpha}[0, t_1]$ to (4.11) over $[0, t_1]$. Next we continue the solution $u \in C_{n-\alpha}[0, t_1]$ to $[t_1, 2t_1]$ so that $u \in C_{n-\alpha}[0, 2t_1]$ (assuming $2t_1 \leq T$). Then for any $t \in [t_1, 2t_1]$, (4.11) is equivalent to

$$u(t) = \frac{1}{\Gamma(\alpha)} \int_{t_1}^t (t - s)^{\alpha-1} f(s, u(s))ds + u_1(t)$$

with $u_1(t) = u_0(t) + \frac{1}{\Gamma(\alpha)} \int_0^{t_1} (t-s)^{\alpha-1} f(s, u(s))ds \in C_{n-\alpha}[0, t_1]$, which is known, since $u \in C_{n-\alpha}[0, t_1]$ is known. Further, it is continuous over the interval $[t_1, 2t_1]$. Then repeating the argument but with $C[t_1, t_2]$ shows that there exists a unique solution $u \in C[t_1, 2t_1]$ and further $u \in C_{n-\alpha}[0, 2t_1]$. By repeating the process for a finite number of steps, we deduce that there exists a solution $u \in C[t_1, T]$ with $u \in C_{n-\alpha}(\overline{I})$. The construction also shows that $u \in C_{n-\alpha}(\overline{I})$ satisfies $^R\partial_t^\alpha u \in C_{n-\alpha}(\overline{I})$. $\qquad \square$

In the preceding discussions, we give sufficient conditions for unique global existence in $L^1(I)$ and $C_\gamma(\overline{I})$. Now we discuss the existence of a local solution. There are several possible versions; See Exercise 4.9 for another result.

Theorem 4.7 *Let $n - 1 < \alpha \leq n$, $n \in \mathbb{N}$, let $K > 0$, $0 < \chi_* \leq T$ and $c_j \in \mathbb{R}$, $j = 1, \ldots, n$. Let $\mathcal{D}$ be the set given by*

$$\mathcal{D} = \left\{ (t, u) \in \mathbb{R}^2 : 0 < t \leq \chi_* : \left| t^{n-\alpha}u - \sum_{j=1}^n \frac{b_j t^{n-j}}{\Gamma(\alpha - j + 1)} \right| \leq K \right\}.$$

Let $f(t, u) : \mathcal{D} \to \mathbb{R}$ be such that $t^{n-\alpha} f(t, u) \in C(\overline{\mathcal{D}})$ and

$$|f(t, u_1) - f(t, u_2)| \leq L|u_1 - u_2|, \quad \forall (t, u_1), (t, u_2) \in \mathcal{D}.$$

Then there exists an $\chi \leq \chi_*$, such that problem (4.10) has a unique solution u in the space $C_{n-\alpha}[0, \chi]$ with $^R\partial_t^\alpha u \in C_{n-\alpha}[0, \chi]$.

Proof By Proposition 4.3, it suffices to prove it for the Volterra equation (4.11). Let $c_f = \max_{(t,u)\in\overline{\mathcal{D}}} |t^{n-\alpha} f(t, u)|$. We choose χ such that $\chi = \min(\chi_*, (\Gamma(2\alpha - n +$

$1)c_f^{-1}\Gamma(\alpha + 1 - n)^{-1})^{\frac{1}{\alpha}})$ and $L\Gamma(\alpha - n + 1)\Gamma(2\alpha - n + 1)^{-1}\chi^{\alpha} < 1$. Let

$$U = \left\{ u \in C(0, \chi] : \sup_{t \in [0, \chi]} \left| t^{n-\alpha}u(t) - \sum_{j=1}^{n} c_j \frac{t^{n-j}}{\Gamma(\alpha - j + 1)} \right| \le K \right\} \subset C_{n-\alpha}[0, \chi].$$

Then U is a complete subset of $C_{n-\alpha}[0, \chi]$. On the set U, we define an operator A by (4.13). Let $u \in U$. Then for any $t \in [0, \chi]$, we have

$$\left| t^{n-\alpha} Au(t) - \sum_{j=1}^{n} \frac{c_j t^{n-j}}{\Gamma(\alpha - j + 1)} \right| \le \frac{t^{n-\alpha}}{\Gamma(\alpha)} \int_0^t (t - s)^{\alpha-1} |f(s, u(s))| ds$$

$$\le \frac{t^{n-\alpha} c_f}{\Gamma(\alpha)} \int_0^t (t - s)^{\alpha-1} s^{\alpha-n} ds = \frac{t^{\alpha} c_f \Gamma(\alpha + 1 - n)}{\Gamma(2\alpha - n + 1)} \le c_f h^{\alpha} \frac{\Gamma(\alpha + 1 - n)}{\Gamma(2\alpha - n + 1)}.$$

By the choice of χ, $Au \in U$. Next, by the Lipschitz condition,

$$|t^{n-\alpha}(Au_1(t) - Au_2(t))| \le L \frac{t^{n-\alpha}}{\Gamma(\alpha)} \int_0^t (t - s)^{\alpha-1} |u_1(s) - u_2(s)| ds$$

$$\le \frac{L t^{n-\alpha}}{\Gamma(\alpha)} \int_0^t s^{\alpha-n}(t - s)^{\alpha-1} ds \|u_1 - u_2\|_{C_{n-\alpha}[0,h]}$$

$$\le L\chi^{\alpha}\Gamma(\alpha - n + 1)\Gamma(2\alpha - n + 1)^{-1}\|u_1 - u_2\|_{C_{n-\alpha}[0,h]}.$$

By the choice of χ, A is a contraction on $C_{n-\gamma}[0, \chi]$, and by Banach contraction mapping theorem, there exists a unique solution $u \in C_{n-\gamma}[0, \chi]$. Similarly, one can prove $^R\partial_t^{\alpha} u \in C_{n-\gamma}[0, \chi]$. $\qquad\square$

For linear Cauchy problems with constant coefficients, explicit solutions can be obtained using Mittag-Leffler functions and the method of successive approximations. This problem was first considered by Barrett [Bar54] in 1954, who showed the existence and uniqueness by essentially the argument given below.

Proposition 4.4 *Let $n - 1 < \alpha \le n$, $n \in \mathbb{N}$, $\lambda \in \mathbb{R}$ and $g \in C(\overline{I})$. Then the solution u to*

$$\begin{cases} ^R\partial_t^{\alpha} u = \lambda u + g, & t \in I, \\ (^R\partial_t^{\alpha-j} u)(0) = c_j, & j = 1, \dots, n, \end{cases} \qquad (4.14)$$

with $\{c_j\}_{j=1}^{n} \subset \mathbb{R}$, is given by

$$u(t) = \sum_{j=1}^{n} c_j t^{\alpha-j} E_{\alpha,\alpha-j+1}(\lambda t^{\alpha}) + \int_0^t (t - s)^{\alpha-1} E_{\alpha,\alpha}(\lambda(t - s)^{\alpha}) g(s) ds. \qquad (4.15)$$

Proof By the linearity, we can split u into $u = u_h + u_i$, with u_h and u_i solving

$$\begin{cases} ^R\partial_t^{\alpha} u_h = \lambda u_h, & t \in I, \\ (^R\partial_t^{\alpha-j} u_h)(0) = c_j, & j = 1, \dots, n, \end{cases} \quad \text{and} \quad \begin{cases} ^R\partial_t^{\alpha} u_i = \lambda u_i + g, & t \in I, \\ (^R\partial_t^{\alpha-j} u_i)(0) = 0, & j = 1, \dots, n, \end{cases}$$

respectively. By Proposition 4.2, it suffices to apply the method of successive approximations to (4.11) to find u_h and u_i. Let $u_h^0(t) = \sum_{j=1}^{n} \frac{c_j}{\Gamma(\alpha-j+1)} t^{\alpha-j}$, and $u_h^m(t) = u_h^0(t) + \lambda_0 I_t^\alpha u_h^{m-1}(t)$, $m = 1, 2, \ldots$, By the identity (2.19), we deduce

$$u_h^1(t) = \sum_{k=1}^{2} \sum_{j=1}^{n} c_j \frac{\lambda^{k-1} t^{\alpha k-j}}{\Gamma(\alpha k - j + 1)},$$

and by mathematical induction, we obtain

$$u_h^m(t) = \sum_{k=1}^{m+1} \sum_{j=1}^{n} c_j \frac{\lambda^{k-1} t^{\alpha k-j}}{\Gamma(\alpha k - j + 1)}.$$

Taking the limit as $m \to \infty$, we get

$$u_h(t) = \sum_{j=1}^{n} c_j \sum_{k=1}^{\infty} \frac{\lambda^{k-1} t^{\alpha k-j}}{\Gamma(\alpha k - j + 1)} = \sum_{j=1}^{n} c_j t^{\alpha-j} E_{\alpha,\alpha-j+1}(\lambda t^\alpha),$$

where, by definition, the inner series is $E_{\alpha,\alpha-j+1}(z)$, which converges uniformly over $\mathbb{C}$. To find u_i, for any $g \in C(\bar{I})$, let $u_i^1(t) = ({}_0 I_t^\alpha g)(t)$. Then by Theorem 2.1,

$$u_i^2(t) = \lambda({}_0 I_t^\alpha u_i^1)(t) + ({}_0 I_t^\alpha g)(t) = \sum_{k=1}^{2} \lambda^{k-1} {}_0 I_t^{k\alpha} g(t).$$

Like before, continuing the procedure, we obtain

$$u_i^m(t) = \sum_{k=1}^{m} \lambda^{k-1}({}_0 I_t^{k\alpha} g)(t) = \int_0^t \sum_{k=1}^{m} \frac{\lambda^{k-1}}{\Gamma(k\alpha)} (t - s)^{\alpha k-1} g(s) ds.$$

Taking the limit as $m \to \infty$ and using the definition of $E_{\alpha,\alpha}(\lambda t^\alpha)$ give

$$u_i(t) = \int_0^t \sum_{k=1}^{\infty} \frac{\lambda^{k-1}}{\Gamma(k\alpha)} (t - s)^{\alpha k-1} g(s) \, ds = \int_0^t (t - s)^{\alpha-1} E_{\alpha,\alpha}(\lambda(t - s)^\alpha) g(s) \, ds.$$

Combining the representations of $u_h(t)$ and $u_i(t)$ gives the desired result. $\square$

Example 4.2 When $g \equiv 0$, $\alpha \in (0, 1)$ and $c_0 = 1$, the solution $u(t)$ is given by $u(t) = t^{\alpha-1} E_{\alpha,1}(\lambda t^\alpha)$. When $\alpha = 1$, $E_{1,1}(\lambda t) = e^{\lambda t}$, which recovers the familiar result for classical ODEs. For $\lambda > 0$, u is exponentially growing like the case $\alpha = 1$, for large t, but at a faster rate. However, u is no longer monotone in t. For $\lambda < 0$, $u(t)$ is monotonically decreasing, by Theorem 3.5, and the smaller is α, the slower is the asymptotic decay, but the behavior at small time is quite the reverse. This behavior differs drastically from the case $\alpha = 1$. See Fig. 4.1 for a schematic illustration.

So far we have discussed only global/local existence and uniqueness of solutions using fixed point theorems. One important issue in the study of ODEs is

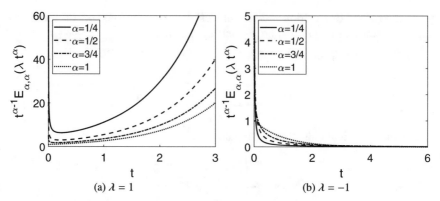

Fig. 4.1 The solution $t^{\alpha-1}E_{\alpha,\alpha}(\lambda t^{\alpha})$ to the homogeneous problem.

the smoothness of the solution with respect to the problem data, and for classical ODEs, the solution smoothness improves steadily with that of f: loosely speaking, $f \in C^k([0, \chi] \times \mathbb{R})$ implies $u \in C^{k+1}[0, \chi]$ for first-order ODEs. The solution $u_h(t)$ in Proposition 4.4 indicates that with nonzero c_n, $u_h(t) \approx c_n t^{\alpha-n}$ as $t \to 0^+$, which is singular around $t = 0$, and thus the classical counterpart generally does not hold. Of course this does not preclude a smooth solution for special data, e.g., $c_n = 0$. This observation holds also for problem (4.10), provided that f is smooth in u. It is conjectured that for $\alpha \in (0, 1)$, additional singular terms include $t^{\alpha-1+m\alpha}$, $m = 0, 1, \ldots$, even though a precise characterization seems missing from the literature.

Theorem 4.8 *For $n - 1 < \alpha < n$, $n \in \mathbb{N}$, and $f \in C(\bar{I} \times \mathbb{R})$ satisfy the condition $|f(t, u)| \le a(t) + b|u|$ for some $a(t) \in C(\bar{I})$ and $b \ge 0$. Then the solution u to problem (4.10) belongs to $C(I)$ and $|u(t)| \le c \sum_{j=1}^{n} t^{\alpha-j}$, as $t \to 0^+$. Further $u \in C(\bar{I})$ if and only if $c_n = 0$.*

Proof Under the given conditions, Proposition 4.2 holds, and there exists a unique solution u. It follows from (4.11) that

$$u(t) = u_0(t) + {}_0I_t^{\alpha} f(t, u(t)), \quad \text{with } u_0 = \sum_{j=1}^{n} \frac{c_j t^{\alpha-j}}{\Gamma(\alpha - j + 1)} \in C^{n-1}(I).$$

Under the given assumption on f, for $t \in I$,

$$|u(t)| \le \sum_{j=1}^{n} \frac{|c_j| t^{\alpha-j}}{\Gamma(\alpha - j + 1)} + \frac{1}{\Gamma(\alpha)} \int_0^t (t - s)^{\alpha-1}(a(s) + b|u(s)|)ds.$$

Then Gronwall's inequality implies the bound on $|u(t)|$. Further, for ${}_0I_t^{\alpha} f(t, u(t))$,

$$|({}_0I_t^{\alpha} f(\cdot, u(\cdot)))^{(n-1)}(t)| = |({}_0I_t^{\alpha-n+1} f(\cdot, u(\cdot)))(t)| \in C(\bar{I}).$$

Thus, $u \in C^{n-1}(I)$. The second assertion follows similarly. ☐

4.3 ODEs with a Djrbashian-Caputo Fractional Derivative

This section is devoted to the following fractional ODEs with a Djrbashian-Caputo fractional derivative (with $n - 1 < \alpha \leq n, n \in \mathbb{N}$):

$$\begin{cases} \partial_t^\alpha u(t) = f(t, u(t)), & t \in I, \\ u^{(j)}(0) = c_j, & j = 0, 1, \ldots, n - 1, \end{cases} \tag{4.16}$$

where $I \subset \mathbb{R}_+$ is an interval (to be determined), and f is a measurable function. Note that there are several different definitions of the Djrbashian-Caputo fractional derivative $\partial_t^\alpha u$, e.g., the classical version $\partial_t^\alpha u$ for $u \in AC^n(\bar{I})$, and also the regularized version $\partial_t^{\alpha*} u$, cf. Definitions 2.3 and 2.4 in Chapter 2. The concept of a "solution" will change accordingly. In spite of this, the difference is often not explicitly specified in the literature, and mostly we also do not distinguish between ∂_t^α and $\partial_t^{\alpha*}$ below.

First we give an explicit solution for linear problems with constant coefficients. This result was already obtained by Djrbashian and Nersesyan [DN68], who also generalized this to include coefficients $p_k(t) \in C(\bar{I})$ coupling each fractional derivative, which represents one of the earliest studies on fractional ODEs.

Proposition 4.5 *Let $n - 1 < \alpha < n, n \in \mathbb{N}, \lambda \in \mathbb{R}, g \in C(\bar{I})$. Then the solution u to*

$$\begin{cases} \partial_t^\alpha u = \lambda u + g, & t \in I, \\ u^{(j)}(0) = c_j, & j = 0, \ldots, n - 1 \end{cases} \tag{4.17}$$

is given by

$$u(t) = \sum_{j=0}^{n-1} c_j t^j E_{\alpha, j+1}(\lambda t^\alpha) + \int_0^t (t - s)^{\alpha-1} E_{\alpha, \alpha}(\lambda (t - s)^\alpha) g(s) \, \mathrm{d}s.$$

Proof Clearly, it can be equivalently converted into a Volterra integral equation as in Proposition 4.2, and then solved using the method of successive approximations. We employ Laplace transform instead. We apply Laplace transform in t, denoted by $\hat{\cdot}$. Then by Lemma 2.9, $(\lambda + z^\alpha)\hat{u}(z) = \sum_{j=0}^{n-1} c_j z^{\alpha-1-j} + \hat{g}(z)$, that is, $\hat{u}(z) = \sum_{j=0}^{n-1} c_j \frac{z^{\alpha-1-j}}{\lambda + z^\alpha} + \frac{1}{\lambda + z^\alpha} \hat{g}(z)$. It remains to invert the functions $\frac{z^\gamma}{z^\alpha + \lambda}$ with γ set to be $\gamma = 1, \alpha - 1, \ldots, \alpha - n$. These are available directly from Lemma 3.2. Together with the convolution rule for Laplace transform, we obtain the representation. $\square$

Example 4.3 When $g \equiv 0, \alpha \in (0, 1)$ and $c_0 = 1$, the solution u to problem (4.17) is given by $u(t) = E_{\alpha,1}(\lambda t^\alpha)$, cf. Fig. 4.2. With $\alpha = 1$, it recovers the well known result for ODEs. Depending on the sign of λ, $u(t)$ is either increasing or decreasing in t: for $\lambda > 0$, u is exponentially growing like the case $\alpha = 1$, but at a faster rate, whereas for $\lambda < 0$, it is only polynomially decaying, and the smaller is α, the slower is the asymptotic decay. This behavior is drastically different from the case $\alpha = 1$.

Example 4.4 Let $\alpha \in (0, 1)$. Consider the fractional ODE $\partial_t^\alpha u(t) = \lambda u + \Gamma(\gamma)^{-1} t^{\gamma-1}$ with $u(0) = 0$, and $\gamma > 0$. By Proposition 4.5, the solution $u(t)$, if there is one, is

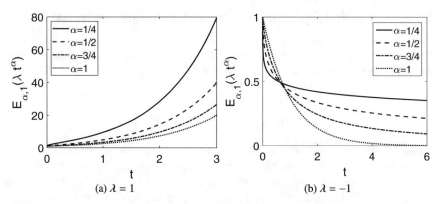

Fig. 4.2 The solution $E_{\alpha,1}(\lambda t^{\alpha})$ to the homogeneous problem.

given by $u(t) = t^{\alpha+\gamma-1}E_{\alpha,\alpha+\gamma}(\lambda t^{\alpha})$. Hence, $\lim_{t\to 0^+} u(t) = 0$ if and only $\alpha + \gamma > 1$. Thus $u(t)$ is indeed a solution to the ODE only if $\alpha + \gamma > 1$. However, for any $\gamma > 0$, $u(t)$ does satisfy the following ODE with a Riemann-Liouville fractional derivative: $^R\partial_t^{\alpha} u(t) = \lambda u + \Gamma(\gamma)^{-1}t^{\gamma-1}$ with $^R\partial_t^{\alpha-1}u(0) = 0$, since $\gamma > 0$. This shows the difference of the ODEs with the Riemann-Liouville and Djrbashian-Caputo cases; and the condition on f is more restrictive in the latter case.

Now we turn to the well-posedness of problem (4.16). The analysis strategy is similar to the Riemann-Liouville case, based on reduction to a Volterra integral equation and the application of fixed point theorems. The following result gives the equivalence of problem (4.18) with a Volterra integral equation, in either classical or regularized sense [Web19a, Theorem 4.6], where $T_n u$ denotes the Taylor expansion of u up to order n at $t = 0$. The factor $t^{-\gamma}$ allows weakly singular source f. It is frequently asserted that (when $\gamma = 0$) problem (4.18) is equivalent to (4.19), but the continuity of f is not enough to ensure $u \in AC^n(\bar{I})$ in (4.19), and thus the assertion is generally not valid then. The condition $_0I_t^{n-\alpha}(u - T_{n-1}u) \in AC^n(\bar{I})$ in (iii) ensures the validity of the integration. This is also implicit in the notion of "solution" for the problem, because if $\partial_t^{\alpha*}u$ exists and $\partial_t^{\alpha*}u(t) = t^{-\gamma}f(t, u(t))$, when u and f are continuous, then $t^{-\gamma}f(t, u(t)) \in L^1(I)$ so that $(_0I_t^{n-\alpha}(u - T_{n-1}u))^{(n)} \in L^1(I)$ i.e., $_0I_t^{n-\alpha}(u - T_{n-1}u) \in AC^n(\bar{I})$. When the term $t^{-\gamma}$ is absent, and (4.20) is satisfied, then $_0I_t^{n-\alpha}(u - T_{n-1}u) \in C^n(\bar{I})$.

Proposition 4.6 *Let f be continuous on $\bar{I} \times \mathbb{R}$, $n - 1 < \alpha < n$, $n \in \mathbb{N}$, and $0 \leq \gamma < \alpha - n + 1$. Then the following statements hold.*

(i) *If a function $u \in AC^n(\bar{I})$ satisfies*

$$\begin{cases} \partial_t^{\alpha}u(t) = t^{-\gamma}f(t, u(t)), & t \in I, \\ u^{(j)}(0) = c_j, & j = 0, \ldots, n-1, \end{cases} \quad (4.18)$$

then u satisfies the Volterra integral equation

$$u(t) = \sum_{j=0}^{n-1} \frac{c_j t^j}{j!} + {}_0I_t^\alpha(t^{-\gamma}f(t, u(t)))(t), \quad t \in \bar{I}. \tag{4.19}$$

(ii) If $u \in C(\bar{I})$ satisfies (4.19), then $u \in C^{n-1}(\bar{I})$ and $\partial_t^{\alpha*}u$ exists a.e. and u satisfies

$$\begin{cases} \partial_t^{\alpha*}u(t) = t^{-\gamma}f(t, u(t)), & a.e.\ t \in I, \\ u^{(j)}(0) = c_j, & j = 0, \dots, n-1. \end{cases} \tag{4.20}$$

Moreover, ${}_0I_t^{n-\alpha}(u - T_{n-1}u) \in AC^n(\bar{I})$.

(iii) If $u \in C^{n-1}(\bar{I})$ and ${}_0I_t^{n-\alpha}(u - T_{n-1}u) \in AC^n(\bar{I})$ and if u solves (4.20), then u solves (4.18).

Proof (i) Let $g(t) = t^{-\gamma}f(t, u(t)), t \in I$. Then $g \in L^1(I)$, since $0 \le \gamma \le \alpha - n + 1 < 1$ and f is continuous. Suppose $u \in AC^n(\bar{I})$, i.e., $u^{(n)} \in L^1(I)$, and

$$\partial_t^\alpha u(t) = t^{-\gamma}f(t, u(t)) = g(t) \quad a.e.\ t \in I.$$

Thus, $u^{(n)} \in L^1(I)$, and ${}_0I_t^{n-\alpha}(u^{(n)}) = g$ holds in $L^1(I)$, and ${}_0I_t^\alpha {}_0I_t^{n-\alpha}u^{(n)} = {}_0I_t^\alpha g$. By Theorem 2.1, ${}_0I_t^n u^{(n)} = {}_0I_t^\alpha g$, and integrating n times the term ${}_0I_t^n u^{(n)}$ leads to $u(t) - T_{n-1}u(t) = {}_0I_t^\alpha g$, i.e. (4.19) holds.

(ii) Let $u \in C^{n-1}(\bar{I})$ satisfy (4.19). Then for every $\beta \ge \alpha - n + 1 > \gamma$,

$$
{}_0I_t^\beta g(t) = \frac{t^{\beta-\gamma}}{\Gamma(\beta)} \int_0^1 (1-s)^{\beta-1} s^{-\gamma} f(ts, u(ts)) ds
$$

exists for every t, since $f(ts, u(ts))$ is bounded by the continuity of u and f. Thus, ${}_0I_t^\beta g(t) \in C(\bar{I})$, and in particular, ${}_0I_t^\beta g \in C^{n-1}(\bar{I})$ and also $u \in C^{n-1}(\bar{I})$. Further, for any $\beta > \gamma$, ${}_0I_t^\beta g(0) = 0$, and then taking $\beta = \alpha$ in (4.19) gives $u(0) = c_0$. By differentiating (4.19) and taking $\beta = \alpha - 1$, we obtain $u'(0) = c_1$. Similarly, we obtain $u^{(j)}(0) = c_j, j = 2, \dots, n-1$. Thus, we have

$$
\partial_t^{\alpha*}u = {}^R\partial_t^\alpha(u - T_{n-1}u) = {}^R\partial_t^\alpha\left(u - \sum_{j=0}^{n-1}\frac{c_j}{j!}t^j\right) = {}^R\partial_t^\alpha {}_0I_t^\alpha g = g.
$$

Thus, u satisfies (4.20). Moreover, from ${}_0I_t^{n-\alpha}(u - T_{n-1}u) = {}_0I_t^{n-\alpha}{}_0I_t^\alpha g = {}_0I_t^n g$, it follows that ${}_0I_t^{n-\alpha}(u - T_{n-1}u) \in AC^n(\bar{I})$.

(iii) Let $g(t) = t^{-\gamma}f(t, u(t))$. Then by the assumption on f, $g \in L^1(I)$. Since ${}_0I_t^{n-\alpha}(u - T_{n-1}u) \in AC^n(\bar{I})$, integrating the identity $({}_0I_t^{n-\alpha}(u - T_{n-1}u))^{(n)} = g(t)$ n times gives

$$
{}_0I_t^{n-\alpha}(u - T_{n-1}u) = {}_0I_t^n g(t) + \sum_{j=0}^{n-1} a_j t^{j-1},
$$

with $a_j \in \mathbb{R}$. Applying the operator ${}_0I_t^{\alpha-n+1}$ to both sides of the identity yields

$$_0I_t^1(u - T_{n-1}u) = _0I_t^{\alpha+1}g + \sum_{j=0}^{n-1} b_j t^{j+\alpha},$$

with $b_j \in \mathbb{R}$. It follows from the assumption $u \in C^{n-1}(\overline{I})$ that $_0I_t^1(u - T_{n-1}u) \in C^n(\overline{I})$ and $_0I_t^{\alpha+1}g \in C^n(\overline{I})$, both with its derivatives up to order $n - 1$ at $t = 0$ vanishing, and hence, there holds $b_j = 0$, $j = 0, 1, \dots, n - 1$. Then differentiating the identity gives $u - T_{n-1}u = _0I_t^\alpha g$. This completes the proof of the proposition. $\qquad\square$

Example 4.5 For $\alpha \in (1, 2)$, consider the initial value problem

$$\begin{cases} \partial_t^\alpha u(t) = (u(t) - 1)^{-1}, & t > 0, \\ u(0) = 1, & u'(0) = 0. \end{cases}$$

If we ignore the discontinuity of $f(t, u)$ at $(0, u(0)) = (0, 1)$ and blindly apply Proposition 4.6 with $\gamma = 0$ (which is valid for continuous f), we get $u(t) = 1 - _0I_t^\alpha(1 - u)^{-1}$. Clearly, it is satisfied by $u(t) = 1 \pm \Gamma(1 - \frac{\alpha}{2})^{\frac{1}{2}} \Gamma(1 + \frac{\alpha}{2})^{-\frac{1}{2}} t^{\frac{\alpha}{2}}$. However, either function has an unbounded first derivative at 0, and cannot be a solution of the ODE. This partly shows the necessity of the continuity on f in order to have the equivalence.

Now we establish the local unique existence for problem (4.16). Our discussions shall focus on the case $0 < \alpha < 1$, since the general case can be analyzed in a similar manner. The next result gives the local well-posedness for $\alpha \in (0, 1)$ and $c_0 \in \mathbb{R}$, when $f(t, u)$ is a continuous and locally Lipschitz function. It was first analyzed systematically by Diethelm and Ford [DF02, Section 2]. Kilbas and Marzan [KM04, KM05] also studied the problem via its integral formulation and proved existence and uniqueness of a global continuous solution in the case of continuous and global Lipschitz continuous function f.

Theorem 4.9 *If there exist $\chi^* > 0$ and $\gamma > 0$ such that $f(t, u)$ is continuous on $\mathcal{D} = [0, \chi^*] \times [c_0 - \gamma, c_0 + \gamma]$ such that there exists an $L > 0$,*

$$|f(t, u_1) - f(t, u_2)| \le L|u_1 - u_2|, \quad \forall (t, u_1), (t, u_2) \in \mathcal{D}.$$

Let $\chi = \min(\chi^*, \sup_{t \ge 0}\{\|f\|_{L^\infty(\mathcal{D})}\Gamma(1 + \alpha)^{-1}t^\alpha E_{\alpha,1}(Lt^\alpha) \le \gamma\})$. Then problem (4.16) has a unique solution $u \in C[0, \chi]$, which depends continuously on c_0.

Proof The proof employs Picard iteration, as for classical ODEs. In view of Proposition 4.6, it suffices to prove the result for the Volterra integral equation (4.19). We define a sequence of functions in $C[0, \chi]$ by $u^0(t) \equiv c_0$ and

$$u^n(t) = c_0 + (_0I_t^\alpha f(t, u^{n-1}))(t), \quad n = 1, 2, \dots.$$

Let $d^n = |u^n - u^{n-1}|$. Then for any $t \in [0, \chi]$,

$$d^1(t) = |_0I_t^\alpha f(t, u^0)| \le \|f\|_{L^\infty(\mathcal{D})}\Gamma(1 + \alpha)^{-1}\chi^\alpha =: c_f.$$

By the choice of χ, $c_f \le \gamma$, i.e., $(t, u^1) \in \mathcal{D}$. Now assume that $\sum_{m=1}^{n-1} d^m \le \gamma$ so that $|u^{n-1} - c_0| \le \gamma$. We show this for d^n. By the induction hypothesis, we have that for $t \in [0, \chi]$ and $m = 2, 3, \ldots, n$, there holds $d^m \le L_0 I_t^\alpha d^{m-1}$, and $d^m \le c_f L^{m-1} ({_0}I_t^{1+(m-1)\alpha} 1)(t)$, $m = 1, 2, \ldots, n$. It then follows that

$$\sum_{m=1}^{n} d^m < c_f \sum_{m=1}^{\infty} \frac{L^{m-1} t^{(m-1)\alpha}}{\Gamma((m-1)\alpha + 1)} = c_f E_{\alpha,1}(Lt^\alpha).$$

By the definition of χ,

$$\sum_{m=1}^{n} d^m \le \gamma, \quad \forall t \in [0, \chi].$$

Hence, $(t, u^n(t)) \in \mathcal{D}$ for all $t \in [0, \chi]$ and $n \ge 0$, and $\sum_n |u^n - u^{n-1}|$ converges uniformly on $[0, \chi]$. Thus, u^n converges to some u uniformly on $[0, \chi]$, and u is continuous. Passing to limit gives $u(t) = c_0 + {_0}I_t^\alpha f(t, u)$, for $t \in [0, \chi]$, i.e., u is indeed a solution. Next we show the uniqueness. Let u_1, u_2 be two solutions on the interval $[0, \chi]$, and $w = u_1 - u_2$. Then, both u_1 and u_2 fall into $[c_0 - \gamma, c_0 + \gamma]$ for $t \le \chi$, and $\partial_t^\alpha w = f(t, u_1) - f(t, u_2)$, with $w(0) = 0$. Since f is Lipschitz on $\mathcal{D}$ and $u_1, u_2 \in L^1(0, \chi)$, we have $f(t, u_1) - f(t, u_2) \in L^1(0, \chi)$. Thus, for $t \in (0, \chi)$, $w(t) = ({_0}I_t^\alpha (f(t, u_1) - f(t, u_2)))(t)$. For all $t \le \chi$, $|w(t)| \le L({_0}I_t^\alpha |w|)(t)$, and Theorem 4.1 implies $|w| = 0$ on $[0, \chi]$. The continuity in u_0 follows similarly. □

The next result gives a global existence.

Corollary 4.3 *Let $f \in C(\overline{\mathbb{R}}_+ \times \mathbb{R})$, and for any $T > 0$, there exists L_T such that $|f(t, u_1) - f(t, u_2)| \le L_T |u_1 - u_2|$ for any $t \in [0, T], u_1, u_2 \in \mathbb{R}$, then there exists a unique solution u on $\mathbb{R}_+$.*

Proof This is direct from the proof of Theorem 4.9: For any $\chi^* > 0$, γ can be chosen arbitrarily large so that $\chi = \chi^*$. Hence, the solution u exists and is unique on $[0, \chi]$. Since χ^* is arbitrary, the claim follows. Next we present an alternative proof using Bielecki norm [Bie56]. For any fixed $T > 0$ and any $u \in C(\overline{I})$, we define a family of weighted norms on $C(\overline{I})$ by $\|u\|_\lambda = \sup_{t \in \overline{I}} e^{-\lambda t} |u(t)|$, where $\lambda > 0$ is to be chosen. The norm $\|\cdot\|_\lambda$ is equivalent to the standard norm. For any $u \in C(\overline{I})$, let $Au = c_0 + {_0}I_t^\alpha f(t, u)$. Then $Au \in C(\overline{I})$. Next we claim that A is a contraction on $C(\overline{I})$. Indeed, for any $u, v \in C(\overline{I})$, then

$$Au(t) - Av(t) = {_0}I_t^\alpha [f(\cdot, u(\cdot)) - f(\cdot, v(\cdot))](t).$$

By the Lipschitz continuity of f in u, we deduce

$$e^{-\lambda t} |(Au - Av)(t)| \le \frac{e^{-\lambda t}}{\Gamma(\alpha)} \int_0^t (t - s)^{\alpha-1} e^{\lambda s} e^{-\lambda s} L_T |u(s) - v(s)| ds$$

$$\le \frac{e^{-\lambda t} L_T \|u - v\|_\lambda}{\Gamma(\alpha)} \int_0^t (t - s)^{\alpha-1} e^{-\lambda s} ds \le \lambda^{-\alpha} L_T \|u - v\|_\lambda.$$

Thus by choosing λ sufficiently large, A is contractive on $C(\bar{I})$ in the norm $\| \cdot \|_\lambda$, and by Theorem A.10, there exists a unique solution $u \in C(\bar{I})$ to $Au = u$, or equivalently problem (4.16). Since the choice of T is arbitrary, the desired assertion follows. $\square$

The next result gives the following global behavior of the solution [LL18a, Proposition 4.6]. Throughout t_* denotes the largest time of existence (or the blow-up time).

Proposition 4.7 *Let $J \equiv (a, b) \subset \mathbb{R}$, $f(t, u) \in C(\overline{\mathbb{R}}_+ \times J)$ be continuous and locally Lipschitz in u, i.e., for any $K \subset J$ compact and $\chi > 0$, there exists $L_{\chi,K} > 0$ such that $|f(t, u_1) - f(t, u_2)| \leq L_{\chi,K}|u_1 - u_2|$ for any $t \in [0, \chi], u_1, u_2 \in K$. Then, for any $c_0 \in J$, problem (4.16) has a unique solution $u \in C[0, t_*)$, with*

$$t_* = \sup\{h > 0 : \text{The solution } u \in C[0, h), u(t) \in D, \forall t \in [0, h]\}.$$

If $t_* < \infty$, then either $\liminf_{t \to t_*^-} u(t) = a$ or $\limsup_{t \to t_*^-} = b$.

Proof By Theorem 4.9, the solution with $[\liminf_{t \to 0^+} u(t), \limsup_{t \to 0^+} u(t)] \subset J$ is unique, and it exists locally and is continuous. Hence, $t_* > 0$. It suffices to show that if $t_* < \infty$ and $a < k_1 := \liminf_{t \to t_*^-} u(t) \leq k_2 := \limsup_{t \to t_*^-} < b$, then the solution can be extended to a larger interval. Then $[k_1, k_2] \subset J$ is compact. Pick $\delta > 0$ so that $[k_1 - \delta, k_2 + \delta] \subset J$, and define $\tilde{f}(t, u)$ on $\mathcal{D}_\delta = [0, t_* + \delta] \times [k_1 - \delta, k_2 + \delta]$ by

$$\tilde{f}(t, u) = \begin{cases} f(t, u), & (t, u) \in \mathcal{D}_\delta, \\ f(t, k_2 + \delta), & t \leq t_* + \delta, u \geq k_2 + \delta, \\ f(t_* + \delta, k_2 + \delta), & t \geq t_* + \delta, u \geq k_2 + \delta, \\ f(t_* + \delta, u), & t \geq t_* + \delta, u \in [k_1 - \delta, k_2 + \delta] \\ f(t, k_1 - \delta), & t \leq t_* + \delta, u \leq k_1 - \delta, \\ f(t_* + \delta, k_1 - \delta), & t \geq t_* + \delta, u \leq k_1 - \delta. \end{cases}$$

It agrees with $f(t, u)$ on $\mathcal{D}_\delta$ and is globally Lipschitz. Thus, by Corollary 4.3, there exists a unique continuous solution $\tilde{u}$ to $\partial_t^\alpha \tilde{u} = \tilde{f}(t, \tilde{u})$ on $\overline{\mathbb{R}}_+$. Since f and $\tilde{f}$ agree on $\mathcal{D}_\delta$, u also solves $\partial_t^\alpha \tilde{u} = \tilde{f}(t, \tilde{u})$ on $[0, t_*)$. Hence $\tilde{u} = u$ on $[0, t_*)$. It follows that there exists $\delta_1 \in (0, \delta)$ such that $(t, \tilde{u}(t)) \in \mathcal{D}_\delta$ for any $t \leq t_* + \delta_1$. On the interval $[0, t_* + \delta_1]$, $\tilde{u}$ also solves $\partial_t^\alpha u = f(t, u)$, contradicting with the definition of t_*. $\square$

The following example shows the necessity of Lipschitz continuity on f.

Example 4.6 In the absence of the Lipschitz assumption on f in u, the solution u is not necessarily unique. To see this, given $0 < \alpha < 1$, consider the problem $\partial_t^\alpha u = u^q$ for $t > 0$, with $u(0) = 0$. Consider $0 < q < 1$, so that the function on the right-hand side is continuous but not Lipschitz. Obviously, $u \equiv 0$ is a solution of the problem. However, since $\partial_t^\alpha t^\gamma = \Gamma(\gamma + 1)\Gamma(\gamma + 1 - \alpha)^{-1}t^{\gamma-\alpha}$, the function $u(t) = \Gamma(\gamma + 1 - \alpha)^{\frac{1}{1-q}}\Gamma(\gamma + 1)^{-\frac{1}{1-q}}t^\gamma$, with $\gamma = \frac{\alpha}{1-q}$ is also a solution, indicating nonuniqueness.

However, it is possible to recover the uniqueness if the Lipschitz condition is properly modified. One possible choice for $0 < \alpha < 1$ is a fractional analogue of the classial Nagumo's condition [Nag26]. The next result is taken from [Die12, Theorem 3.1] (see [Die17] for the erratum and [LL09] for the Riemann-Liouville case).

Theorem 4.10 *Let $T > 0$ and $c_0 \in \mathbb{R}$. If the function $f : \overline{I} \times \mathbb{R} \to \mathbb{R}$ is continuous at $(0, c_0)$ and satisfies the inequality $t^\alpha | f(t, u_1) - f(t, u_2)| \le \Gamma(\alpha + 1)|u_1 - u_2|$ for all $t \in \overline{I}$ and all $u_1, u_2 \in \mathbb{R}$, then problem (4.16) has at most one continuous solution $u \in C(\overline{I})$ with $\partial_t^\alpha u \in C(\overline{I})$.*

Proof The proof proceeds by contradiction, following the approach developed in [DW60], using the mean value theorem (cf. Theorem 2.14). Assume that equation (4.16) has two continuous solutions u and $\tilde{u}$ on $\overline{I}$. Let $w : \overline{I} \to \mathbb{R}$ be defined by

$$w(t) = \begin{cases} t^{-\alpha}|u(t) - \tilde{u}(t)|, & t \in I, \\ 0, & t = 0. \end{cases}$$

Then w is continuous on I. Since both u and $\tilde{u}$ solve (4.16) and by Theorem 2.14,

$$w(t) = t^{-\alpha}|(u(t) - u(0)) - (\tilde{u}(t) - \tilde{u}(0))| = t^{-\alpha}|(u(t) - \tilde{u}(t)) - (u(0) - \tilde{u}(0))|$$

$$= \Gamma(\alpha + 1)^{-1}|(\partial_t^\alpha(u - \tilde{u}))(\xi)| = \Gamma(\alpha + 1)^{-1}|f(\xi, u(\xi)) - f(\xi, \tilde{u}(\xi))|,$$

with some $\xi \in (0, t)$. Thus, as $t \to 0^+$, $\xi \to 0^+$, and since $u, \tilde{u}$ are continuous, we also have $u(\xi) \to u(0) = c_0$ and $\tilde{u}(\xi) \to \tilde{u}(0) = c_0$. These observations and the continuity of f at the point $(0, c_0)$ imply

$$w(t) = f(\xi, u(\xi)) - f(\xi, \tilde{u}(\xi)) \to f(0, c_0) - f(0, c_0) = 0 \quad \text{as } t \to 0^+.$$

Thus, $\lim_{t\to 0^+} w(t)$ exists and coincides with $w(0)$. Thus w is continuous at the origin. Now assume that w is not identically zero on $\overline{I}$, and let $\eta := \inf \{ t^* \in [0, T] : w(t^*) = \sup_{t \in \overline{I}} w(t) \}$. Since w is continuous and nonnegative and $w(0) = 0$, we have $w(\eta) = \sup_{t \in \overline{I}} w(t)$ and

$$w(s) < w(\eta), \quad \forall s \in [0, \eta). \tag{4.21}$$

By Theorem 2.14 and Nagumo's condition, we derive

$$w(\eta) = \eta^{-\alpha}|[u(\eta) - u(0)] - [\tilde{u}(\eta) - \tilde{u}(0)]| = \Gamma(\alpha + 1)^{-1}|\partial_t^\alpha u(s) - \partial_t^\alpha \tilde{u}(s)|$$

$$= \Gamma(\alpha + 1)^{-1}|f(s, u(s)) - f(s, \tilde{u}(s))| \le w(s)$$

with some $s \in (0, \eta)$, which contradicts (4.21). Thus w vanishes identically. $\qquad\square$

Example 4.7 The continuity of f at $(0, c_0)$ is essential. Let $c_0 = 0$ and define

$$f(t, u) = \begin{cases} \Gamma(2 - \alpha), & u > t, \\ \Gamma(2 - \alpha)t^{-\alpha}u, & 0 < u \le t \\ 0, & u \le 0. \end{cases}$$

Then Nagumo's condition is satisfied, since for $0 < \alpha \le 1$, $\Gamma(2 - \alpha) \le \Gamma(\alpha)$, but f is not continuous at $(0, c_0)$. Moreover, $u(t) = ct$ is a solution of the problem for all $c \in \mathbb{R}$, so the uniqueness does not hold. Theorem 4.10 is a pure uniqueness statement. This is caused by the fact that the hypotheses require f to be continuous only at the single point $(0, c_0)$ but not the entire domain of definition.

So far it has been assumed that f is continuous. However, when $u \in AC(\overline{I})$, $\partial_t^\alpha u(t)$ can be singular at $t = 0$. For example, for $u(t) = t^\rho \in AC(\overline{I})$, $\partial_t^\alpha u(t) = \Gamma(1 + \rho)\Gamma(1 + \rho - \alpha)^{-1}t^{\rho-\alpha}$, which belongs to $L^1(I)$ whenever $\rho > \alpha - 1$, and can be singular at $t = 0$. Hence, it makes sense to allow $\partial_t^\alpha u(t)$ to be singular at $t = 0$ in relevant ODEs. This leads to the following initial value problem (with $0 < \alpha < 1$):

$$\partial_t^\alpha u(t) = t^{-\gamma} f(t, u(t)), \quad t \in I, \quad \text{with } u(0) = c_0, \tag{4.22}$$

where $0 \leq \gamma < \alpha$, and $f \geq 0$ is continuous. By Proposition 4.6, it is equivalent to

$$u(t) = c_0 + \frac{1}{\Gamma(\alpha)} \int_0^t (t - s)^{\alpha-1} s^{-\gamma} f(s, u(s)) ds.$$

Below we show the existence of a nonnegative solution [Web19b, Theorem 4.8].

Theorem 4.11 Let $f : \overline{I} \times \mathbb{R}_+ \to \mathbb{R}_+$ be continuous, $0 \leq \gamma < \alpha < 1$ and $c_0 > 0$, and there is a $c_f > 0$ such that $f(t, u) \leq c_f(1 + u)$ for all $t \in \overline{I}$ and $u \geq 0$. Then problem (4.22) has a nonnegative solution. Further, if there exists $L > 0$ such that $|f(t, u) - f(t, v)| \leq L|u - v|$ for all $t \in \overline{I}, u, v \geq 0$, then the solution is unique.

Proof Let $P = \{u \in C(\overline{I}) : u(t) \geq 0, t \in \overline{I}\}$, and define $A : P \to P$ by

$$Au(t) := c_0 + \frac{1}{\Gamma(\alpha)} \int_0^t (t - s)^{\alpha-1} s^{-\gamma} f(s, u(s)) ds.$$

First, we prove that there is a bounded open ball U_R of radius R (centered at 0) containing 0 such that $Au \neq \lambda u$ for all $u \in \partial U \cap P$ and all $\lambda \geq 1$. In fact, if there exists $\lambda \geq 1$ and $u \neq 0$ such that $\lambda u(t) = Au(t)$, then

$$u(t) \leq \lambda u(t) = c_0 + \frac{1}{\Gamma(\alpha)} \int_0^t (t - s)^{\alpha-1} s^{-\gamma} f(s, u(s)) ds$$

$$\leq c_0 + \frac{c_f}{\Gamma(\alpha)} \int_0^t (t - s)^{\alpha-1} s^{-\gamma}(1 + u(s)) ds$$

$$\leq c_0 + \frac{c_f}{\Gamma(\alpha)} B(\alpha, 1 - \gamma)T^{\alpha-\gamma} + \frac{c_f}{\Gamma(\alpha)} \int_0^t (t - s)^{\alpha-1} s^{-\gamma} u(s) ds.$$

Since $1 - \alpha + \gamma < 1$, by Theorem 4.3, there is $c > 0$ such that $\|u\|_{C(\overline{I})} \leq c$. By choosing $R > c$, then $Au \neq \lambda u$ for all $u \in \partial U_R \cap P$ and all $\lambda \geq 1$. Next we prove A is compact. Let $c_R = \sup_{(t,u) \in \overline{I} \times [0,R]} f(t, u)$. Clearly, $A(\overline{U}_R)$ is bounded. For any $0 \leq t_1 < t_2 \leq T$,

$$\Gamma(\alpha)|Au(t_1) - Au(t_2)|$$

$$= |\int_0^{t_2} (t_2 - s)^{\alpha-1} s^{-\gamma} f(s, u(s)) ds - \int_0^{t_1} (t_1 - s)^{\alpha-1} s^{-\gamma} f(s, u(s)) ds|$$

$$\leq c_R \int_0^{t_1} ((t_1 - s)^{\alpha-1} - (t_2 - s)^{\alpha-1}) s^{-\gamma} ds + c_R \int_{t_1}^{t_2} (t_2 - s)^{\alpha-1} s^{-\gamma} ds.$$

Since the second term has an L^1 integrand, it is smaller than $\frac{\epsilon}{3}$ for $|t_1 - t_2| < \delta_1$. For any $0 < \eta < t_1$, we split the first term as $\int_0^\eta + \int_\eta^{t_1}$. By the integrability, we can fix η small so that $\int_0^\eta |(t_2 - s)^{\alpha-1} - (t_1 - s)^{\alpha-1}| s^{-\gamma} c_R ds < \frac{\epsilon}{3}$. Last,

$$\int_\eta^{t_1} |(t_2 - s)^{\alpha-1} - (t_1 - s)^{\alpha-1}| s^{-\gamma} ds \leq \eta^{-\gamma} \int_\eta^{t_1} ((t_1 - s)^{\alpha-1} - (t_2 - s)^{\alpha-1}) ds$$

$$= \alpha^{-1} \eta^{-\gamma} ((t_1 - \eta)^\alpha - (t_2 - \eta)^\alpha + (t_2 - t_1)^\alpha) \leq \alpha^{-1} \eta^{-\gamma}) (t_2 - t_1)^\alpha.$$

It is then smaller than $\frac{\epsilon}{3}$ for $|t_1 - t_2| < \delta_2$. This shows the equicontinuity, and by Arzela-Ascoli theorem, the operator $A : P \to P$ is compact. This and Theorem A.14 prove that A has a fixed point in P, i.e., the existence of a nonnegative solution. If f satisfies the Lipschitz condition and let u, v be two solutions. Then

$$u(t) - v(t) = \frac{1}{\Gamma(\alpha)} \int_0^t (t - s)^{\alpha-1} s^{-\gamma} (f(s, u(s)) - f(s, v(s))) ds,$$

and hence $|u(t) - v(t)| \leq L\Gamma(\alpha)^{-1} \int_0^t (t - s)^{\alpha-1} s^{-\gamma} |u(s) - v(s)| ds$. By Theorem 4.3, $|u(t) - v(t)| \equiv 0$. □

The next result states a comparison principle [FLLX18b, Theorem 2.2], which is an extension of the argument in [RV12, Theorem 2.3]. See also [LL18a, Theorem 4.10] and [VZ15, Lemma 2.6] for results under a monotonicity condition on $f(t, \cdot)$.

Proposition 4.8 *Let $f(t, u)$ be continuous and locally Lipschitz in u, and $v(t)$ be continuous. If $\partial_t^\alpha v \leq f(t, v)$, and $\partial_t^\alpha u = f(t, u)$, with $v_0 \leq u_0$. Then, $v \leq u$ on the common interval of existence.*

Proof By Proposition 4.7, problem (4.16) has a unique continuous solution u on $[0, t_*)$, with t_* being the blow-up time. Now fix $T \in (0, t_*)$, and $I = (0, T]$. Pick γ large enough so that $u(t)$ and $v(t)$ fall into $\bar{I} \times [-\gamma, \gamma]$. Let L be the Lipschitz constant of $f(t, \cdot)$ for the region $\bar{I} \times [-2\gamma, 2\gamma]$. Let $v^\epsilon = v - \epsilon w$, with $w = E_{\alpha,1}(2Lt^\alpha)$. If ϵ is sufficiently small, $v^\epsilon(t)$ falls into $\bar{I} \times [-2\gamma, 2\gamma]$. Thus,

$$\partial_t^\alpha v^\epsilon = \partial_t^\alpha v - \epsilon 2Lw \leq f(t, v) - \epsilon 2Lw \leq f(t, v^\epsilon) - \epsilon Lw.$$

We claim that for all small ϵ,

$$v^\epsilon(t) \leq u(t), \quad \forall t \in \bar{I}. \tag{4.23}$$

If not, define $t_1 = \sup\{t \in I : v^\epsilon(s) \leq u(s),\ \forall s \in [0, t]\}$. Since $v^\epsilon(0) = v_0 - \epsilon < u_0$, by continuity, we have $t_1 > 0$. Since (4.23) does not hold, $t_1 < T$. Hence, there exists $\delta_1 > 0$, such that $v^\epsilon(t_1) = u(t_1)$ and $v^\epsilon(t) > u(t)$ for $t \in (t_1, t_1 + \delta_1)$. Moreover,

$$\partial_t^\alpha(v^\epsilon - u) \leq f(t, v^\epsilon) - \epsilon L w - f(t, u).$$

By continuity, for some $\delta_2 \in (0, \delta_1)$, $\partial_t^\alpha(v^\epsilon - u) \leq 0$ on the interval $(t_1, t_1 + \delta_2)$. Thus, we have $v^\epsilon(t) \leq u(t)$ for $t \in (t_1, t_1 + \delta_2)$, which is a contradiction. This shows the claim. Taking $\epsilon \to 0$ yields the result on $\bar{I}$. Since T is arbitrary, the result follows. $\square$

Last we discuss solution regularity. One important issue in the study of classical ODEs is the smoothness of the solution u to problem (4.16) with $0 < \alpha < 1$, under suitable assumptions on f. For classical ODEs, the smoothness of u improves with that of f: if $f \in C^{k-1}([0, \chi] \times \mathbb{R})$, then the solution u to the ODE $u'(t) = f(t, u)$ with $u(0) = c_0$ belongs to $C^k[0, \chi]$. This generally does not hold in the fractional case. A first result gives local regularity of u to problem (4.16) (assuming the unique existence over $[0, \chi]$ and $\mathcal{D}$ being the domain for f). Thus, even for a smooth f, u can have a weak singularity at $t = 0$, but away from the origin, it is indeed smooth.

Theorem 4.12 *If $f \in C(\mathcal{D})$, then $u \in C[0, \chi]$. Further, if $f \in C^k(\mathcal{D})$, then $u \in C^k(0, \chi] \cap C[0, \chi]$, with $|u^{(k)}(t)| = O(t^{\alpha-k})$ as $t \to 0^+$.*

Proof By Proposition 4.6, the solution u satisfies $u(t) = c_0 + {}_0I_t^\alpha f(\cdot, u(\cdot))$. The first term is $C[0, \chi]$, and ${}_0I_t^\alpha f(\cdot, u(\cdot)) \in C[0, \chi]$, since $u \in C[0, \chi]$ and $f(t, u(t)) \in C[0, \chi]$. This shows the first assertion. The technical proof of the second assertion can be found in [BPV99, Theorem 2.1 and Section 4], using theory of Fredholm integral equations developed in [Vai89, PV94]. $\square$

The next result gives the asymptotic expansion of the solution u at the origin, when α is rational [Lub83, Section 2] (see also [MF71, Theorem 6] for related analyticity results). The proof below is taken from [Die10, Section 6.4].

Theorem 4.13 *Let $\alpha = \frac{p}{q} \in \mathbb{Q}$, with $p < q$ being relative prime, and f be of the form $f(t, u) = \tilde{f}(t^{\frac{1}{q}}, u)$, with $\tilde{f}$ analytic in a neighborhood of $(0, c_0)$. Then there exists a uniquely determined analytic function $\bar{u} : (-r, r)$ for some $r > 0$ such that*

$$u(x) = \bar{u}(t^{\frac{1}{q}}), \quad \forall t \in [0, r).$$

Proof The proof is based on constructing a formal solution using Puiseux series, and proving the convergence of the series using Lindelöf's majorant method. Let $u(t) = \sum_{i=0}^\infty u_i x^{\frac{i}{q}}$ with $u(0) = c_0$. Next we show that there are coefficients u_i such that the series converges and the function u solves problem

$$u(t) = c_0 + \frac{1}{\Gamma(\alpha)} \int_0^t (t-s)^{\alpha-1} f(s, u(s)) ds. \tag{4.24}$$

Substituting the series into (4.24) and using the series expansion of $\tilde{f}$ near $(0, c_0)$:

$$f(t, u(t)) = \tilde{f}(t^{\frac{1}{q}}, u(t)) = \sum_{\ell_1, \ell_2 = 0}^{\infty} f_{\ell_1 \ell_2} t^{\frac{\ell_1}{q}} (u - c_0)^{\ell_2} = \sum_{\ell_1, \ell_2 = 0}^{\infty} f_{\ell_1 \ell_2} t^{\frac{\ell_1}{q}} \left(\sum_{i=1}^{\infty} u_i t^{\frac{i}{q}} \right)^{\ell_2},$$

we deduce

$$\sum_{i=0}^{\infty} u_i t^{\frac{i}{q}} = c_0 + \frac{1}{\Gamma(\alpha)} \int_0^t (t - s)^{\alpha - 1} \sum_{\ell_1, \ell_2 = 0}^{\infty} f_{\ell_1 \ell_2} s^{\frac{\ell_1}{q}} \left[\sum_{i=1}^{\infty} u_i s^{\frac{i}{q}} \right]^{\ell_2} ds.$$

Next we rearrange the terms in the square bracket as

$$\left[\sum_{i=1}^{\infty} u_i s^{\frac{i}{q}} \right]^{\ell_2} = \sum_{\ell_3 = 0}^{\infty} \left(\sum_{i_1 + \ldots + i_{\ell_2} = \ell_3} u_{i_1} \ldots u_{i_{\ell_2}} \right) s^{\frac{\ell_3}{q}}.$$

Since $i \geq 1$, the case $\ell_3 = 0$ occurs in the first sum only occurs for $\ell_2 = 0$, for which we set the coefficient to 1, by convention. Assuming uniform convergence, we can exchange the order of summation and integration and integrate termwise, and obtain

$$\sum_{i=0}^{\infty} u_i t^{\frac{i}{q}} = c_0 + \sum_{\ell_1, \ell_2, \ell_3 = 0}^{\infty} \left(f_{\ell_1 \ell_2} c_{\ell_1, \ell_3} \sum_{i_1 + \ldots + i_{\ell_2} = \ell_3} u_{i_1} \ldots u_{i_{\ell_2}} \right) t^{\frac{\ell_1 + \ell_3 + p}{q}} \tag{4.25}$$

with $c_{\ell, \ell_3} = \Gamma(\frac{\ell + \ell_3}{q} + 1) \Gamma(\frac{\ell + \ell_3 + p}{q} + 1)^{-1}$. Comparing the coefficients of $t^{\frac{i}{q}}$ on both sides gives

$$u_i = \begin{cases} c_0, & i = 0, \\ 0, & i = 1, \ldots, p - 1, \\ \displaystyle\sum_{\ell_1 + \ell_3 + p = i} \sum_{\ell_2 = 0}^{\infty} f_{\ell_1 \ell_2} \left(c_{\ell_1, \ell_3} \sum_{i_1 + \ldots + i_{\ell_2} = \ell_3} u_{i_1} \ldots u_{i_{\ell_2}} \right), & i \geq p. \end{cases}$$

Thus, for $0 \leq i < p$, u_is are uniquely determined, and for $i \geq p$, it satisfies a recurrence relation: u_i depends only on coefficients u_{i_ℓ}, with $i_\ell \leq i_1 + \ldots + i_{\ell_2} = \ell_3 = i - p - \ell_1 < i$, since $p > 0$. Thus, there exists a unique formal solution, and it suffices to show the series converges locally absolutely and uniformly in a neighborhood of $t = 0$ using Lindelöf's majorant method. Let

$$F(t^{\frac{1}{q}}, u(t)) = \sum_{\ell_1, \ell_2 = 0}^{\infty} |f_{\ell_1 \ell_2}| t^{\frac{\ell_1}{q}} (u - |c_0|)^{\ell_2}.$$

This series converges since $\tilde{f}$ is analytic. Next, we study the Volterra integral equation

$$U(t) = |c_0| + \frac{1}{\Gamma(\alpha)} \int_0^t (t - s)^{\alpha - 1} F(s, U(s)) ds.$$

The formal solution $U(t)$ can be computed as $u(t)$. Then $U(t)$ is a majorant of u, and all coefficients of U are positive. Thus, it suffices to prove that the series expansion of U converges for some $r > 0$. The nonnegativity of the expansion coefficients of U implies that the series expansion of U converges uniformly over $[0, r]$. Let

$$P_{\ell+1}(t) = \sum_{i=0}^{\ell+1} U_i t^{\frac{i}{q}}.$$

Then the nonnegativity of the coefficients and the recurrent relation imply

$$P_{\ell+1}(t) \leq |c_0| + \frac{1}{\Gamma(\alpha)} \int_0^t (t-s)^{\alpha-1} F(s, P_\ell(s)) ds.$$

Choose $b > 0$, and let $c' = \sup_{(x,u)\in[0,b]\times[0,2c_0]} \Gamma(\alpha+1)^{-1} F(t, u)$, and $r := \min(b, |c_0|^{\frac{1}{\alpha}} (c')^{-\frac{1}{\alpha}})$. We claim $|P_\ell(t)| \leq 2|c_0|$ for all $\ell = 0, 1, \ldots$ and all $t \in [0, r]$, and prove it by mathematical induction. The case $\ell = 0$ is obvious by the choice of r. The induction step from ℓ to $\ell + 1$ follows by

$$|P_{\ell+1}(t)| \leq |c_0| + \frac{1}{\Gamma(\alpha)} \int_0^t (t-s)^{\alpha-1} F(s, P_\ell(s)) ds$$

$$\leq |c_0| + \Gamma(\alpha+1)^{-1} r^\alpha \sup_{t\in[0,r]} F(t, P_\ell(t)) \leq |c_0| + r^\alpha c' \leq 2|c_0|.$$

Thus, the sequence $\{P_\ell\}$ is uniformly bounded on $[0, r]$ and monotone, and hence uniformly convergent. Since it has the form of a power series, it converges uniformly on compact subsets of $[0, r)$. This justifies the interchange of summation and integration and completes the proof of the theorem. $\square$

When α is irrational, the result is slightly different.

Corollary 4.4 *Let* $\alpha \in (0, 1)$ *be irrational, and* $f(t, u) = \tilde{f}(t, t^\alpha, u)$, *where* $\tilde{f}$ *be analytic in a neighborhood of* $(0, 0, c_0)$. *Then there exists a uniquely determined analytic function* $\tilde{u} : (-r, r) \times (-r^\alpha, r^\alpha) \to \mathbb{R}$ *with some* $r > 0$ *such that* $u(t) = \tilde{u}(t, t^\alpha)$ *for* $t \in [0, r)$.

Proof The assumption implies the local existence of a unique continuous solution. Then we substitute the formal expansion $u(t) = \sum_{i_1 i_2=0}^{\infty} u_{i_1 i_2} t^{i_1+i_2\alpha}$ into (4.24) and repeat the argument in Theorem 4.13 to get $u_{0,0} = c_0$, and

$$u_{i_1 i_2} = \sum_{\substack{m_1+j_1=i_1 \\ m_2+j_2=i_2}} \gamma_{m_1 m_2 j_1 j_2} f_{m_1 m_2 \ell} \sum_{\substack{n_1+\ldots+n_\ell=j_1 \\ k_1+\ldots+k_\ell=j_2}} u_{n_1 k_1} \cdots u_{n_\ell k_\ell}$$

for the remaining cases, with the coefficients $\gamma_{m_1 m_2 j_1 j_2} = \Gamma(m_1 + j_1 + (m_2 + j_2)\alpha + 1)\Gamma(m_1 + j_1 + (m_2 + j_2 + 1)\alpha + 1)^{-1}$. The coefficient $\gamma_{m_1 m_2 j_1 j_2}$ is uniquely determined by the coefficients with smaller indices. The rest of the proof is analogous to Theorem 4.13, and thus it is omitted. $\square$

The proofs give also results under weaker assumptions on f.

Corollary 4.5 *Let $k \in \mathbb{N}$, and $0 < \alpha < 1$.*

(i) *If $\alpha = \frac{p}{q} \in \mathbb{Q}$, with $p < q$ being relative prime, and f be of the form $f(t, u) = \tilde{f}(t^{\frac{1}{q}}, u)$, with $\tilde{f} \in C^k(\overline{\mathcal{D}})$. Then the solution of problem (4.16) has an asymptotic expansion in powers of $t^{\frac{1}{q}}$ as $t \to 0$, and the smallest noninteger exponent is α.*

(ii) *If α is irrational, and f is of the form $f(t, u) = \tilde{f}(t, t^{\alpha}, u)$ with $\tilde{f} \in C^k([0, \chi] \times [0, \chi^{\alpha}] \times [c_0 - r, c_0 + r])$ for some $r > 0$. Then the solution u to problem (4.16) has an asymptotic expansion in mixed powers of t and t^{α} as $t \to 0$.*

In the rest of the section, we discuss autonomous fractional ODEs (with $0 < \alpha < 1$)

$$\partial_t^{\alpha} u = f(u), \quad \text{in } I, \quad \text{with } u(0) = c_0 \tag{4.26}$$

with $c_0 \in \mathbb{R}$. This problem was studied in detail in [FLLX18a], from which all the results stated below are taken, unless otherwise stated. Similar to Proposition 4.7, if $f(u)$ is locally Lipschitz on an interval $(a, b) \subset \mathbb{R}$, then for any $c_0 \in (a, b)$, there is a unique continuous solution u to problem (4.26) with $u(0) = c_0$. Either it exists globally on $\overline{\mathbb{R}}_+$ or there exists $t_* > 0$ such that either $\lim \inf_{t \to t_*^-} u(t) = a$ or $\lim \sup_{t \to t_*^-} u(t) = b$. The next result shows that $f(u(t))$ does not change sign.

Theorem 4.14 *Let f be locally Lipschitz. If $f(c_0) \neq 0$, then $f(u(t))f(c_0) \geq 0$, for all $t \in (0, t_*)$, and $f(u(t))f(c_0) = 0$ can hold only on a nowhere dense set.*

Proof We may assume $f(c_0) > 0$. Let $t^* = \inf\{t > 0 : \exists \delta > 0, \text{ s.t. } f(u(s)) \leq 0, \forall s \in [t, t + \delta]\}$. Since $f(c_0) > 0$, $t^* > 0$. It suffices to show $t^* = \infty$. We argue by contradiction. Suppose $t^* < \infty$, then there exists a $\delta > 0$ such that $f(u(t)) \leq 0$ for all $t \in [t^*, t^* + \delta]$. We claim that $u(t) < u(t^*)$ for any $t \in (t^*, t^* + \delta)$. Indeed,

$$u(t) - u(t^*) = ({}_0I_t^{\alpha} f(u))(t) - ({}_0I_t^{\alpha} f(u))(t^*)$$

$$= \frac{1}{\Gamma(\alpha)} \left(\int_0^{t^*} ((t - s)^{\alpha - 1} - (t^* - s)^{\alpha - 1}) f(u(s)) ds + \int_{t^*}^t (t - s)^{\alpha - 1} f(u(s)) ds \right).$$

Note that $f(u(t)) \geq 0$ for $t \in (0, t^*)$, and $f(u(t))$ is strictly positive when t is close to 0. In addition, $f(u(t)) \leq 0$ for $t \in [t^*, t^* + \delta]$. Thus, the right-hand side is strictly negative. Hence $u(t) < u(t^*)$ and the claim is proved. By the continuity of u, there exist $t_1, t_2, 0 \leq t_1 < t_2 \leq t^*$ such that for any $s \in [t_1, t_2]$, there exists $t_s \in [t^*, t^* + \delta]$ such that $u(s) = u(t_s)$. Then for any $s \in [t_1, t_2]$, $f(u(s)) = f(u(t_s)) \leq 0$, which contradicts with the definition of t^*. $\qquad\qquad\square$

For the usual derivative, the fact $f(u(t))$ has a definite sign implies that the solution u is monotone. For fractional derivatives, this is less obvious, however it does hold if f is close to C^2. First we prove the positivity of the solution to an integral equation, which is a slightly different version of [Wei75, Theorem 1].

Lemma 4.2 *Let $h \in L^1(I)$, $h > 0$ a.e. satisfy*

$$h(t) - \int_0^t r_\lambda(t - s)h(s)ds > 0 \quad a.e.\ t \in I$$

for any $\lambda > 0$, where r_λ is the resolvent for kernel $\lambda t^{\alpha-1}$ satisfying

$$r_\lambda(t) + \lambda \int_0^t (t - s)^{\alpha-1} r_\lambda(s)ds = \lambda t^{\alpha-1}. \tag{4.27}$$

Then for $v \in C(\bar{I})$, the integral equation

$$y(t) + \int_0^t (t - s)^{\alpha-1} v(s)y(s)ds = h(t) \tag{4.28}$$

has a unique solution $y(t) \in L^1(I)$, which satisfies $y(t) > 0$ a.e.

Proof It can be verified directly that

$$r_\lambda(t) = -\frac{d}{dt} E_{\alpha,1}(-\lambda \Gamma(\alpha)t^\alpha) \in L^1(I) \cap C(0, T]$$

and $r_\lambda > 0$. The existence and uniqueness of a solution $y \in L^1(I)$ to (4.28) follow similar as Theorem 4.5. We prove only $y > 0$ a.e. Convolving (4.28) with r_λ gives

$$\int_0^t r_\lambda(t - s)y(s)ds + \int_0^t \int_0^{t-s} \lambda(t - s - \xi)^{\gamma-1} r_\lambda(\xi)d\xi \frac{v(s)}{\lambda} y(s)ds$$

$$= \int_0^t r_\lambda(t - s)h(s)ds.$$

Subtracting this identity from (4.28) and using (4.27) yield

$$y(t) - \int_0^t r_\lambda(t - s)y(s)ds + \int_0^t r_\lambda(t - s)\frac{v}{\lambda}y(s)ds = h(t) - \int_0^t r_\lambda(t - s)h(s)ds.$$

Thus, y also solves

$$y(t) = \left(h(t) - \int_0^t r_\lambda(t - s)h(s)ds\right) + \int_0^t r_\lambda(t - s)(1 - \lambda^{-1}v(s))y(s)ds.$$

By assumption, $h(t) - \int_0^t r_\lambda(t - s)h(s)ds > 0$. Since $v \in C(\bar{I})$, there exists a $c_v > 0$ such that $|v| \le c_v$ on $\bar{I}$. Picking $\lambda > c_v$, $1 - \frac{v}{\lambda} > 0$ and then $y \ge 0$ a.e. on $[0, T)$ follows. $\qquad \square$

Theorem 4.15 Let $f \in C^1(a, b)$ for $(a, b) \subset \mathbb{R}$ and f' be locally Lipschitz on (a, b). Then, the solution u to problem (4.26) with $u(0) = c_0 \in (a, b)$ is monotone on the interval $(0, t_*)$ of existence. If $f(c_0) \ne 0$, the monotonicity is strict.

Proof Clearly, if $f(c_0) = 0$, then $u = c_0$ is the solution by the uniqueness and is monotone. Now, we assume $f(c_0) > 0$. Then by the regularity theory, $u \in C^1(0, t_*) \cap C[0, t_*]$. Now, we fix $T \in (0, t_*)$. The derivative $y = u'$ satisfies

$$y(t) = \frac{f(c_0)}{\Gamma(\alpha)}t^{\alpha-1} + \frac{1}{\Gamma(\alpha)}\int_0^t (t-s)^{\alpha-1}f'(u(s))y(s)\mathrm{d}s.$$

Since $f'(u(t))$ is continuous on $\bar{I}$ and $f(c_0) > 0$, Lemma 4.2 implies that y is positive on $(0, T)$. Since T is arbitrary, $y > 0$ on $(0, t_*)$, and u is increasing. The argument for $f(c_0) < 0$ is similar. $\qquad\square$

Example 4.8 Given $0 < \alpha < 1$ and $c_0 \in \mathbb{R}$, consider

$$\partial_t^\alpha u(t) = -u(t)(1 - u(t)), \quad t > 0, \quad \text{with } u(0) = c_0. \tag{4.29}$$

For $\alpha = 1$, it is known that (i) if $c_0 \in (0, 1)$, the solution u exists globally, and (ii) if $c_0 > 1$, u does not exist globally. For $0 < \alpha < 1$, by Theorem 4.9, locally, there exists a unique continuous solution u. If $0 < c_0 < 1$, by Proposition 4.5, u satisfies

$$u(t) = E_{\alpha,1}(-t^\alpha)c_0 + \int_0^t (t-s)^{\alpha-1}E_{\alpha,\alpha}(-(t-s)^\alpha)u^2(s)\mathrm{d}s.$$

Since $E_{\alpha,1}(-t^\alpha) > 0$ and $E_{\alpha,\alpha}(-t^\alpha) > 0$, we have $u(t) > 0$ for $u_0 > 0$. Let $\bar{u}(t) \equiv 1$ for $t > 0$. Since $0 < u_0 < 1$, then $u_0 < \bar{u}(0)$, and $\partial_t^\alpha \bar{u}(t) = 0 = -\bar{u}(t)(1 - \bar{u}(t))$. Meanwhile, $u(t) < \bar{u}(t) = 1$, and thus $0 < u(t) < 1$ for all $t > 0$. See Fig. 4.3 for an illustration. Note that in all cases, $u(t)$ decays to zero as t grows. However, the decay rate differs markedly with α.

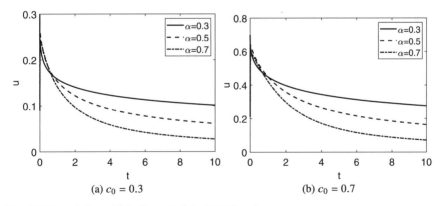

Fig. 4.3 The solution $u(t)$ for Example 4.8 with different c_0.

For classical ODEs, the solution curves do not intersect, which holds also for problem (4.26). (See also [DF12, Theorem 4.1] for problem (4.16) with $\alpha \in (0, 1)$.)

Proposition 4.9 *If $f(u)$ is locally Lipschitz and nondecreasing, then the solution curves of (4.26) do not intersect with each other.*

Now we present some results regarding the blow-up behavior.

Lemma 4.3 *Let $f(u)$ be locally Lipschitz and nondecreasing on $(0, \infty)$, $c_0 > 0$ and $f(c_0) > 0$. Then, the solution u to problem (4.26) is nondecreasing on $(0, t_*)$ and $\lim_{t \to t_*^-} u(t) = +\infty$, with $t_* \in (0, \infty]$.*

Proof Note that f is less regular than that in Theorem 4.15, and thus it cannot be applied directly. Let $u^0 = c_0$, and for $n \geq 1$, define $\partial_t^\alpha u^n = f(u^{n-1})$, with $u^n(0) = c_0$. Then u^n is continuous on $\overline{\mathbb{R}}_+$. Since $f(u^0) > 0$, we have

$$u^1(t) = c_0 + {}_0I_t^\alpha f(u^0)(t) \geq c_0 = u^0(t) \quad \text{for } t \in \overline{\mathbb{R}}_+.$$

Thus, $f(u^1(t)) \geq f(u^0(t))$ and

$$u^2(t) = c_0 + {}_0I_t^\alpha f(u^1)(t) \geq c_0 + {}_0I_t^\alpha f(u^0)(t) = u^1(t).$$

By induction, $u^n(t) \geq u^{n-1}(t)$ for all $n \geq 1$. Next, we claim $u(t) > u^0$, $\forall t \in (0, t_*)$. Let $t^* = \sup\{\bar{t} \in (0, t_*) : f(u(t)) > 0, \forall t \in (0, \bar{t})\}$, and we prove $t^* = t_*$. By the continuity of $u(t)$ and $f(u)$, the fact $f(c_0) > 0$ implies $t^* > 0$. Indeed, if $t^* < t_*$, then $f(u(t^*)) = 0$ by the continuity of f and u. In addition, by the definition of t^*,

$$u(t^*) = c_0 + ({}_0I_t^\alpha f(u))(t^*) > u^0.$$

Since f is nondecreasing, $f(u(t^*)) \geq f(u^0) > 0$, which is a contradiction. Using $u(t) \geq u^0$, we obtain

$$u(t) = c_0 + ({}_0I_t^\alpha f(u))(t) \geq c_0 + ({}_0I_t^\alpha f(u^0))(t) = u^1(t)$$

for $t \in [0, t_*)$, and by induction, $u(t) \geq u^2(t)$ and $u(t) \geq u^3(t)$, etc. Moreover, since f is nondecreasing and $f(u^n)$ is positive, for any $0 \leq t_1 < t_2 < \infty$:

$$u^1(t_2) = c_0 + ({}_0I_t^\alpha f(u^0))(t_2) \geq c_0 + ({}_0I_t^\alpha f(u^0))(t_1) = u^1(t_1).$$

Thus, u^1 is nondecreasing on $\overline{\mathbb{R}}_+$. Similarly,

$$u^2(t_2) = c_0 + ({}_0I_t^\alpha f(u^1))(t_2) \geq c_0 + \frac{1}{\Gamma(\alpha)} \int_{t_2-t_1}^{t_2} (t_2 - s)^{\alpha-1} f(u^1(s)) ds$$

$$\geq c_0 + \frac{1}{\Gamma(\alpha)} \int_{t_2-t_1}^{t_2} (t_2 - s)^{\alpha-1} f(u^1(s - (t_2 - t_1))) ds = u^2(t_1),$$

i.e., u^2 is nondecreasing on $\overline{\mathbb{R}}_+$. By induction, u^n is nondecreasing. Hence the sequence $\{u^n(t)\}$ converges to a nondecreasing function $\bar{u}(t)$ for any $t \in [0, t_*)$. By the monotone convergence theorem, $\bar{u}$ satisfies $\bar{u}(t) = c_0 + ({}_0I_t^\alpha f(\bar{u}))(t)$. This and the uniqueness shows $\bar{u} = u$. Hence, u is nondecreasing. If $t_* < \infty$, by the definition of t_* and the monotonicity, we have $\lim_{t \to t_*^-} u(t) = \infty$. If $t_* = \infty$, we find

$$u(t) = c_0 + ({}_0I_t^\alpha f(u))(t) \geq c_0 + f(u_0)({}_0I_t^\alpha 1)(t) \to \infty.$$

This completes the proof of the lemma. $\square$

Now we can give an Osgood type criterion for blow up.

Proposition 4.10 *Let $f(u)$ be locally Lipschitz, nondecreasing on $\mathbb{R}_+$, $c_0 > 0$ and $f(c_0) > 0$. Then, $t_* < \infty$ if and only if there exists $U > 0$ such that $\int_U^\infty (\frac{u}{f(u)})^{\frac{1}{\alpha}} \frac{du}{u} < \infty$.*

Proof By Lemma 4.3, u is increasing and $u(t) \to \infty$ as $t \to t_*^-$. Pick $r > \max(1, c_0^{\frac{1}{\alpha}})$. There exists $t_n < t_*$ so that $u(t_n) = r^{n\alpha}$ for $n = 1, 2, \ldots$. Then we have

$$u(t_n) = c_0 + ({}_0I_t^\alpha f(u))(t_n) \geq \frac{f(u(t_{n-1}))}{\Gamma(\alpha)} \int_{t_{n-1}}^{t_n} (t_n - s)^{\alpha-1} ds$$

$$= \Gamma(1+\alpha)^{-1}(t_n - t_{n-1})^\alpha f(u(t_{n-1})).$$

Thus, there exist constants $c_1(\alpha) > 0$ and $c_2(\alpha, r) > 0$ such that

$$t_n - t_{n-1} \leq c_1(\alpha) u(t_n)^{\frac{1}{\alpha}} f(u(t_{n-1}))^{-\frac{1}{\alpha}} = c_1(\alpha) r^2 (r-1)^{-1}(r^{n-1} - r^{n-2}) f(r^{(n-1)\alpha})^{-\frac{1}{\alpha}}$$

$$\leq c_2(\alpha, r) \int_{r^{n-2}}^{r^{n-1}} f(s^\alpha)^{-\frac{1}{\alpha}} ds.$$

Meanwhile,

$$u(t_n) = c_0 + ({}_0I_t^\alpha f(u))(t_n) \leq c_0 + \frac{1}{\Gamma(\alpha)} \int_0^{t_{n-1}} (t_{n-1} - s)^{\alpha-1} f(u(s)) ds$$

$$+ \frac{1}{\Gamma(\alpha)} \int_{t_{n-1}}^{t_n} (t_n - s)^{\alpha-1} f(u(s)) ds \leq u(t_{n-1}) + \frac{f(u(t_n))}{\Gamma(1+\alpha)}(t_n - t_{n-1})^\alpha.$$

Thus, there exist $\bar{c}_1(\alpha) > 0$ and $\bar{c}_2(\alpha, r) > 0$ such that

$$t_n - t_{n-1} \geq \bar{c}_1(\alpha)(r-1)^{\frac{1}{\alpha}}(r(r-1))^{-1}(r^{n+1} - r^n) f(r^{n\alpha})^{-\frac{1}{\alpha}} \geq \bar{c}_2(\alpha, r) \int_{r^n}^{r^{n+1}} f(s^\alpha)^{-\frac{1}{\alpha}} ds.$$

Hence, $t_* < \infty$ if and only if $\int^\infty f(\tau^\alpha)^{-\frac{1}{\alpha}} d\tau < \infty$, or equivalently there exists some $U > 0$ such that $\int_U^\infty u^{\frac{1}{\alpha}-1} f(u)^{-\frac{1}{\alpha}} du < \infty$. This completes the proof. $\square$

To gain further insights, below we analyze the power nonlinearity: given $c_0 > 0$, $\alpha \in (0, 1)$, $\gamma > 0$ and $\nu \in \mathbb{R}$,

$$\partial_t^\alpha u = \nu u^\gamma, \quad t > 0, \quad \text{with } u(0) = c_0. \tag{4.30}$$

This problem is representative among autonomous problems. For example, if $\gamma > 0$ and there exist $c_1, c_2 > 0$ such that $c_1 u^\gamma \leq f(u) \leq c_2 u^\gamma$, then the solution u is under control, according to the comparison principles. Several properties can be deducted directly from preceding discussions: the solution curves do not intersect; for $\gamma \in [0, 1]$ and $\nu > 0$, the solution u exists globally, but for $\nu > 0$ and $\gamma > 1$, the solution blows up in finite time. Below we discuss two cases, i.e., $\nu < 0$ and $\gamma > 0$, and $\nu > 0$ and $\gamma > 1$, and refer to the exercises for several other cases. The next result [VZ15, Theorem 7.1] analyzes the case $\nu < 0$ and $\gamma > 0$: the case $\alpha \in (0, 1)$

differs markedly from $\alpha = 1$, which has an algebraic decay $u(t) \sim ct^{-\frac{1}{\gamma-1}}$ for $\gamma > 1$, exponential decay for $\gamma = 1$, and extinction in finite time for $\gamma < 1$.

Proposition 4.11 *Let $\gamma > 0$ and $v < 0$, $c_0 > 0$, and $u \in W_{loc}^{1,1}(\mathbb{R}_+)$ be the solution to problem (4.30). Then there exist constants c_1, $c_2 > 0$ such that*

$$c_1(1 + t^{\frac{\alpha}{\gamma}})^{-1} \leq u(t) \leq c_2(1 + t^{\frac{\alpha}{\gamma}})^{-1}, \quad t \geq 0.$$

Proof The proof employs Proposition 4.8. Let $\omega_\alpha(t) = \Gamma(\alpha)^{-1}t^{\alpha-1}$. First we construct a subsolution. Let $\mu := -c_\alpha v c_0^\gamma$ and $\varepsilon := (c_0\Gamma(1 + \alpha))^{\frac{1}{\alpha}}(2\mu)^{-\frac{1}{\alpha}}$, with $c_\alpha = \Gamma(1 - \alpha)\Gamma(1 + \alpha) = \frac{\alpha\pi}{\sin\alpha\pi} > 1$, in view of (2.2) and (2.3). Now let $v(t) = c_0 - \mu\omega_{1+\alpha}(t)$ for $t \in [0, \varepsilon]$ and $v(t) = ct^{-\frac{\alpha}{\gamma}}$ for $t > \varepsilon$, with $c = \varepsilon^{\frac{\alpha}{\gamma}}\frac{c_0}{2}$. Then $v \in W_{loc}^{1,1}(\mathbb{R}_+)$, $v(0) = c_0$, $v(\varepsilon) = \frac{c_0}{2}$, v is nonincreasing, and $v(t) > 0$ for all $t \geq 0$. For any $t \in (0, \varepsilon)$, since $\partial_t^\alpha \omega_{1+\alpha}(t) = 1$, we have

$$\partial_t^\alpha v - vv^\gamma = -\mu\partial_t^\alpha \omega_{1+\alpha} - v(c_0 - \mu\omega_{1+\alpha})^\gamma$$
$$\leq -\mu - vc_0^\gamma = -(-c_\alpha + 1)vc_0^\gamma \leq 0,$$

by the definition of μ. Meanwhile, since $v'(t) \leq 0$ and $\omega_{1-\alpha}$ is nonnegative and decreasing, for $t > \varepsilon$, there holds

$$\partial_t^\alpha v(t) \leq \int_0^\varepsilon \omega_{1-\alpha}(t - s)v'(s)\,ds \leq \omega_{1-\alpha}(t)\int_0^\varepsilon v'(s)\,ds = -\omega_{1-\alpha}(t)\frac{c_0}{2}.$$

This, using the definitions of $\omega_{1-\alpha}$, μ, ε and c, we deduce

$$\partial_t^\alpha v - vv^\gamma \leq -\omega_{1-\alpha}(t)\frac{c_0}{2} - vc^\gamma t^{-\alpha} = -\omega_{1-\alpha}(t)\frac{c_0}{2}(1 - 2^{-\gamma}) \leq 0.$$

Hence v is a subsolution of problem (4.30). Next we construct a supersolution. Define $t_0 > 0$ by $vt_0^\alpha = c_0^{1-\gamma}(\omega_{1-\alpha}(\frac{1}{2}) + 2^{\alpha+\frac{\alpha}{\gamma}}c_\alpha')$, with $c_\alpha' = \alpha(\gamma\Gamma(2 - \alpha))^{-1}$. Let $w(t) = c_0$ for $t \in [0, t_0]$ and $w(t) = ct^{-\frac{\alpha}{\gamma}}$ for $t \geq t_0$ with $c = c_0 t_0^{\frac{\alpha}{\gamma}}$. For $t < t_0$, $\partial_t^\alpha w - vw^\gamma = -vw^\gamma \geq 0$. Note that for $t > t_0$,

$$\partial_t^\alpha w(t) = -c\frac{\alpha}{\gamma}\int_{t_0}^t \omega_{1-\alpha}(t - s)s^{-\frac{\alpha}{\gamma}-1}\,ds.$$

We denote the integral by $I(t)$, and further analyze the cases $t \in [t_0, 2t_0]$ and $t > 2t_0$ separately. For $t \in [t_0, 2t_0]$, we have

$$I(t) \leq t_0^{-\frac{\alpha}{\gamma}-1}\int_{t_0}^t \omega_{1-\alpha}(t - s)\,ds = t_0^{-\frac{\alpha}{\gamma}-1}\omega_{2-\alpha}(t - t_0) \leq t_0^{-\frac{\alpha}{\gamma}-1}\omega_{2-\alpha}(t_0).$$

This and the value of c imply $\partial_t^\alpha w(t) \geq -c_\alpha' c_0 t_0^{-\alpha} \geq -2^\alpha c_\alpha' c_0 t^{-\alpha}$. Now the definition of w and the choice of t_0 lead to

$$\partial_t^\alpha w(t) \geq vw(t)^\gamma(2^\alpha c_\alpha' c_0^{1-\gamma}(vt_0^\alpha)^{-1}) \geq vw(t)^\gamma.$$

Further, for $t > 2t_0$, by changing variables, we have

$$I(t) = t^{-\alpha - \frac{\alpha}{\gamma}} \int_{\frac{t_0}{t}}^{1} \omega_{1-\alpha}(1 - s)s^{-\frac{\alpha}{\gamma} - 1} \, ds.$$

Next we split the integral over $[\frac{t_0}{t}, \frac{1}{2}]$ and $[\frac{1}{2}, 1]$, denoted by $I_1(t)$ and I_2, respectively. Then direct computation gives $I_1 \leq \omega_{1-\alpha}(\frac{1}{2})\gamma\alpha^{-1}(\frac{t_0}{t})^{-\frac{\alpha}{\gamma}}$ and $I_2 \leq \omega_{2-\alpha}(\frac{1}{2})2^{\frac{\alpha}{\gamma}+1}$. These estimates and the choice of t_0 give

$$\partial_t^\alpha w(t) \geq vw(t)^\gamma c_0^{1-\gamma}(vt_0^\alpha)^{-1}(\omega_{1-\alpha}(\frac{1}{2}) + c_\alpha' 2^{\alpha + \frac{\alpha}{\gamma}}) = vw(t)^\gamma.$$

Thus, w is a supersolution of (4.30). The assertion follows from Proposition 4.8. $\square$

Now we derive suitable bounds on the blow-up time t_*, for $v > 0$ and $\gamma > 1$, and to understand the effects of the memory.

Theorem 4.16 *Let $\gamma > 1$, $v > 0$ and $c_0 > 0$. Then for the blow-up time t_* of the solution to problem* (4.30), *there holds*

$$\Gamma(1 + \alpha)^{\frac{1}{\alpha}}(vc_0^{\gamma-1}G(\gamma))^{-\frac{1}{\alpha}} \leq t_* \leq \Gamma(1 + \alpha)^{\frac{1}{\alpha}}(vc_0^{\gamma-1}H(\gamma, \alpha))^{-\frac{1}{\alpha}}$$

with $G(\gamma) = \min(2^\gamma, \gamma^\gamma(\gamma - 1)^{1-\gamma})$ and $H(\gamma, \alpha) = \max(\gamma - 1, 2^{-\frac{\gamma\alpha}{\gamma-1}})$. Hence, with $v > 0, \gamma > 1$ fixed, there exist $c_{02} > c_{01} > 0$ such that whenever $c_0 < c_{01}$, $\lim_{\alpha \to 0^+} t_* = \infty$, while $c_0 > c_{02}$ implies $\lim_{\alpha \to 0^+} t_* = 0$.

Proof Let $r > 1$, and choose t_n such that $u(t_n) = c_0 r^{n\alpha}$. Then $0 = t_0 < t_1 < t_2 \ldots$. Let $\omega_\alpha(t) = \Gamma(\alpha)^{-1} t^{\alpha-1}$. The following relation

$$u(t_n) = c_0 + \int_0^{t_{n-1}} \omega_\alpha(t_n - s)f(u(s))ds + \int_{t_{n-1}}^{t_n} \omega_\alpha(t_n - s)f(u(s))ds$$

$$\leq c_0 + \int_0^{t_{n-1}} \omega_\alpha(t_{n-1} - s)f(u(s))ds + \omega_{\alpha+1}(t_n - t_{n-1})f(u(t_n))$$

yields $\omega_{\alpha+1}(t_n - t_{n-1}) \geq c_0 r^{n\alpha}(1 - r^{-\alpha})f(c_0 r^{n\alpha})^{-1}$. Hence

$$t_n - t_{n-1} \geq \Gamma(1 + \alpha)^{\frac{1}{\alpha}}(vc_0^{\gamma-1})^{-\frac{1}{\alpha}}(1 - r^{-\alpha})^{\frac{1}{\alpha}}r^{-n(\gamma-1)}.$$

Then the lower bound follows by

$$t_* = \sum_{n=1}^{\infty} t_n - t_{n-1} \geq \Gamma(1 + \alpha)^{\frac{1}{\alpha}}(vc_0^{\gamma-1})^{-\frac{1}{\alpha}}(r^\alpha - 1)^{\frac{1}{\alpha}}r^{-1}(r^{\gamma-1} - 1)^{-1}.$$

To obtain the upper bound, we fix $m \geq 1$, and then deduce

$$u(t) \geq c_0 + \int_0^t \omega_\alpha(t_m - s)f(u(s))ds := v(t), \quad t \in (0, t_m).$$

Then $v(t_m) = u(t_m)$, and $v'(t) = \Gamma(\alpha)^{-1}(t_m - t)^{\alpha-1} f(u(t)) \geq \Gamma(\alpha)^{-1}(t_m - t)^{\alpha-1} f(v(t))$ and $\int_{c_0}^{u(t_m)} \frac{dv}{f(v)} \geq \Gamma(1+\alpha)^{-1} t_m^{\alpha}$, implying

$$t_m \leq \Gamma(1+\alpha)^{\frac{1}{\alpha}} (v(\gamma-1)c_0^{\gamma-1})^{-\frac{1}{\alpha}} (1 - r^{m\alpha(1-\gamma)})^{\frac{1}{\alpha}}. \tag{4.31}$$

For $n \geq m+1$, we have

$$u(t_n) \geq c_0 + \int_{t_{n-1}}^{t_n} \omega_\alpha(t_n - s) f(u(t_{n-1})) ds,$$

and thus

$$\Gamma(1+\alpha)^{-1}(t_n - t_{n-1})^{\alpha} f(u(t_{n-1})) \leq c_0 r^{n\alpha} - c_0 \leq u_0 r^{n\alpha}. \tag{4.32}$$

Combining (4.31) and (4.32) gives an upper bound

$$t_* = \sum_{n=m+1}^{\infty} (t_n - t_{n-1}) + t_m \leq \Gamma(1+\alpha)^{\frac{1}{\alpha}} (v c_0^{\gamma-1})^{-\frac{1}{\alpha}} r^{\gamma} (r^{(m+1)(\gamma-1)} - r^{m(\gamma-1)})^{-1} + t_m$$

$$= \Gamma(1+\alpha)^{\frac{1}{\alpha}} (v c_0^{\gamma-1})^{-\frac{1}{\alpha}} (r^{\gamma}(r^{(m+1)(\gamma-1)} - r^{m(\gamma-1)})^{-1} + (1 - r^{m\alpha(1-\gamma)})^{\frac{1}{\alpha}} (\gamma-1)^{-\frac{1}{\alpha}}).$$

Next we optimize the bounds with respect to the parameters r and m to obtain the desired bound. For the lower bound, picking $r = 2^{\frac{1}{\alpha}} > 1$ gives $\sup_{r>1}(r^{\alpha} - 1)^{\frac{1}{\alpha}}(r^{\gamma} - r)^{-1} \geq (2^{\frac{\gamma}{\alpha}} - 2^{\frac{1}{\alpha}})^{-1} \geq 2^{-\frac{\gamma}{\alpha}}$. Similarly, picking $r = \gamma^{\frac{1}{\alpha}}(\gamma-1)^{-\frac{1}{\alpha}}$ yields $\sup_{r>1}(r^{\alpha} - 1)^{\frac{1}{\alpha}}(r^{\gamma} - r)^{-1} \geq [(\gamma-1)^{\gamma-1}\gamma^{-\gamma}]^{\frac{1}{\alpha}}$. This shows the lower bound. For the upper bound, we fix $m > (\gamma-1)^{-1}$, and let $r \to \infty$:

$$r^{\gamma}(r^{(m+1)(\gamma-1)} - r^{m(\gamma-1)})^{-1} + [(1 - r^{m\alpha(1-\gamma)})(\gamma-1)^{-1}]^{\frac{1}{\alpha}} \to (\gamma-1)^{-\frac{1}{\alpha}}.$$

Instead choosing $m = 1$ and $r = 2^{\frac{1}{\gamma-1}} > 1$ gives

$$r^{\gamma}[r^{(m+1)(\gamma-1)} - r^{m(\gamma-1)}] + (1 - r^{m\alpha(1-\gamma)})^{\frac{1}{\alpha}}(\gamma-1)^{-\frac{1}{\alpha}} = 2^{\frac{\gamma}{\gamma-1}-1} + 2^{-1}(2^{\alpha} - 1)^{\frac{1}{\alpha}}(\gamma-1)^{-\frac{1}{\alpha}}.$$

Let $Q(\gamma) := 2^{\frac{\gamma}{\gamma-1}} (\frac{\gamma-1}{2^{\alpha}-1})^{\frac{1}{\alpha}}$. By elementary calculus, we have

$$Q'(\gamma) = Q(\gamma)(\ln Q(\gamma))' = Q(\gamma)(\gamma-1)^{-2}\alpha^{-1}(\gamma - 1 - \alpha \ln 2).$$

Hence,

$$Q(\gamma) \geq Q(\alpha \ln 2 + 1) = 2(\alpha e \log 2)^{\frac{1}{\alpha}}(2^{\alpha} - 1)^{-\frac{1}{\alpha}} \geq 2(e \ln 2)^{\frac{1}{\alpha}} \geq 2.$$

For the second inequality, since $\alpha - 2^{\alpha} + 1$ is concave on $(0, 1)$ and equals zero at $\alpha = 0, 1$, $\alpha > 2^{\alpha} - 1$ for $\alpha \in (0, 1)$. We find

$$t_* \leq 2^{-1}2^{\frac{\gamma}{\gamma-1}} + 2^{-1}(2^{\alpha} - 1)^{\frac{1}{\alpha}}(\gamma-1)^{-\frac{1}{\alpha}} < 2^{\frac{\gamma}{\gamma-1}},$$

and the upper bound follows. Then we can pick $c_{01} = v^{-\frac{1}{\gamma-1}} \max(2^{-\frac{\gamma}{\gamma-1}}, (\gamma-1)\gamma^{\frac{\gamma}{1-\gamma}})$ and $c_{02} = v^{-\frac{1}{\gamma-1}} \min(1, (\gamma-1)^{-\frac{1}{\gamma-1}})$ to obtain the last assertion.   □

Remark 4.2 Theorem 4.16 shows the role of memory, which is getting stronger as $\alpha \to 0^+$. When c_0 is very small, the memory defers the blow up. If c_0 is large, the memory accelerates the blow up. The critical value of c_0 might be determined by the limiting case $\alpha = 0$: $u - c_0 = vu^\alpha$. If $c_0 > \gamma^{-1}(\gamma-1)(v\gamma)^{-\frac{1}{\gamma-1}}$, this algebraic equation has no solution and it means that the blow-up time is zero. If $c_0 < \gamma^{-1}(\gamma-1)(v\gamma)^{-\frac{1}{\gamma-1}}$, there is a constant solution for $t > 0$, i.e., the blow-up time is infinity.

Exercises

Exercise 4.1 Prove the standard Gronwall's inequality in (4.1).

Exercise 4.2 Prove the following uniform version of Gronwall's inequality. Let g, h, $u \in L^1_{loc,+}(0, \infty)$ such that $u' \leq gu + h$, for all $t \geq 0$, and $\int_t^{t+r} g(s)ds \leq a_1$, $\int_t^{t+r} h(s)ds \leq a_2$, $\int_t^{t+r} u(s)ds \leq a_3$, for any $t \geq 0$, with $r, a_1, a_2, a_3 > 0$. Then

$$u(t + r) \leq (a_3 r^{-1} + a_2)e^{a_1}, \quad \forall t \geq 0.$$

Exercise 4.3 Let $a, b > 0$, and $u \in L^\infty_+(I)$ satisfies

$$u(t) \leq a + b \int_0^t (t - s)^{-\beta} u(s)ds, \quad \forall t \in I.$$

Then Theorems 4.2 and 4.3 can both be applied. Which theorem gives a sharper estimate?

Exercise 4.4 [LY20, Lemma 2.4] This exercise is about an extended Gronwall's inequality. Let $\alpha \in (0, 1)$ and $q > \alpha^{-1}$. Let $L \in L^q_+(I)$ and $a, u \in L^{\frac{q}{q-1}}_+(I)$. Suppose

$$u(t) \leq a(t) + \int_0^t (t - s)^{\alpha-1} L(s)u(s)ds, \quad \text{a.e. } t \in \bar{I}.$$

Then there exists a constant $c > 0$ such that

$$u(t) \leq a(t) + c \int_0^t (t - s)^{\alpha-1} L(s)a(s)ds, \quad \text{a.e. } t \in \bar{I}.$$

Exercise 4.5 For problem (4.14), it is very tempting to specify the initial conditions in the classical manner: $u^{(j)}(0) = c_j$, $j = 0, 1, \ldots, n - 1$. For the case $0 < \alpha < 1$ and $u \in C(\bar{I})$, show that $({}_0I_t^{1-\alpha}u)(0) = 0$. Discuss the implications of this observation.

Exercise 4.6 Prove that if the mapping $f \mapsto f(t, u(t)) \in L^1(I)$ holds for every $u \in L^1(I)$, then it is necessary and sufficient that there exist $a \in L^1(I)$ and some constant $b > 0$ such that $|f(t, u)| \leq a(t) + b|u|$ for all (t, u).

Exercise 4.7 In Theorem 4.5, can the Lipschitz continuity of $f(t, u)$ in u be relaxed to a Hölder continuity with exponent $\gamma \in (0, 1)$, i.e., $|f(t, u_1) - f(t, u_2)| \leq L_\gamma |u_1 - u_2|^\gamma$?

Exercise 4.8 Prove Theorem 4.6 using Bielecki's weighted norm.

Exercise 4.9 In this exercise, we consider a nonlinear problem

$$
\begin{cases}
{}^R\partial_t^\alpha u = f(t, u), & t \in I \\
({}^R\partial_t^{\alpha-j} u)(0) = c_j, & j = 1, \ldots, n
\end{cases}
$$

with $\{c_j\}_{j=1}^n \subset \mathbb{R}$. Suppose that f is continuous and uniformly bounded by $c \in \mathbb{R}$, and it is Lipschitz with respect to u with a constant L. Show that there exists a unique continuous solution to the problem in the region $\mathcal{D}$ defined by

$$
\mathcal{D} = \left\{ (t, u) : 0 < t \leq h, |t^{n-\alpha} u(t) - c_n \Gamma(\alpha - n + 1)^{-1}| \leq a \right\}
$$

with the constant $a > \sum_{j=0}^{n-1} \frac{h^{n-j} c_j}{\Gamma(\alpha - j + 1)}$.

Exercise 4.10 Prove Proposition 4.4 using Laplace transform.

Exercise 4.11 Determine the first three leading terms in the solution for problem (4.10) with $1 < \alpha < 2$.

Exercise 4.12 Under what conditions on f does the solution u to problem (4.10) with $0 < \alpha < 1$ belong to $u \in C^1(\bar{I})$?

Exercise 4.13 Consider the following singular Cauchy problem for the Riemann-Liouville fractional derivative with $0 < \gamma < \alpha < 1$:

$$
\begin{cases}
{}^R\partial_t^\alpha u(t) = f(t, u, {}^R\partial_t^\gamma u), & \text{in } I, \\
{}_0 I_t^{1-\alpha} u(0) = c_0
\end{cases}
$$

with $c_0 > 0$, and let $f : \bar{I} \times \mathbb{R} \times \mathbb{R} \to \mathbb{R}$ be continuous, for $t \in I$, with $f(t, u, v) \in L^1(I)$ for $u, v \in L^1(I)$. Prove the following results.

(i) $u \in L^1(I)$ with ${}_0 I_t^{1-\alpha} u \in AC(\bar{I})$ solves the problem if and only if u satisfies

$$
u(t) = \frac{c_0}{\Gamma(\alpha)} t^{\alpha-1} + \frac{1}{\Gamma(\alpha)} \int_0^t (t - s)^{\alpha-1} f(s, u(s), {}^R\partial_s^\gamma u(s)) ds.
$$

(ii) If, in addition, f satisfies the Lipschitz condition $|f(t, u, p) - f(t, v, q)| \leq L(|u - v| + |p - q|)$ for some $L > 0$ for all $t \in \bar{I}$ and all $u, v, p, q \in \mathbb{R}$, then the problem has a unique solution $u \in L^1(I)$.

Exercise 4.14 Consider the following Cauchy problem for the Riemann-Liouville fractional derivative with $0 < \alpha < 1$:

$$\begin{cases} {}^{R}\partial_t^\alpha u(t) = t^{-\gamma} f(t, u), & \text{in } I, \\ {}_0 I_t^{1-\alpha} u(0) = c_0 \end{cases}$$

with $0 < \gamma < 1 - \alpha$, $c_0 > 0$. Let $f : \bar{I} \times [0, \infty) \to [0, \infty)$ be continuous, for $t \in I$, and there is a constant c_f such that $f(t, u) \le c_f(1 + u)$ for all $t \in \bar{I}$ and $u \ge 0$. Prove the following results.

(i) The problem has a nonnegative fixed point in the set $X = \{v \in C(I) : v \ge 0, t^{1-\alpha} v \in C(\bar{I})\}$ with the norm $\|v\|_X = \sup_{t \in \bar{I}} t^{1-\alpha} |v(t)|$.

(ii) If, in addition, f satisfies the Lipschitz condition $|f(t, u) - f(t, v)| \le L|u - v|$ for some $L > 0$ for all $t \in \bar{I}$ and all $u, v \ge 0$, then the fixed point is unique.

Exercise 4.15 Prove Theorem 4.9 for general $\alpha > 0$.

Exercise 4.16 Let $0 < \alpha < 1$, and $\lambda \in \mathbb{R}$, with $g, \bar{g} \in C(\bar{I})$, and u and $\bar{u}$ solve, respectively,

$$\partial_t^\alpha u_1 = \lambda u_1 + g_1, \quad t \in I, \quad \text{with } u_1(0) = c_1,$$
$$\partial_t^\alpha u_2 = \lambda u_2 + g_2, \quad t \in I, \quad \text{with } u_2(0) = c_2.$$

(i) Using Proposition 4.5 and properties of the Mittag-Leffler function $E_{\alpha,\beta}(-x)$, prove that $c_1 \le c_2$ and $g_1 \le g_2$ imply $u_1(t) \le u_2(t)$ for all $t \in \bar{I}$.

(ii) Is there an analogue of (i) for $\alpha \in (1, 2)$?

Exercise 4.17 Consider the following Cauchy problem (with $1 < \alpha < 2$)

$$\begin{cases} \partial_t^\alpha u(t) = f(t), & \text{in } I, \\ u(0) = c_0, \\ u'(0) = c_1. \end{cases}$$

Prove the following regularity results.

(i) For $f \in AC(\bar{I})$, the solution u can be written as $u(t) = c_0 + c_1 t + f(0)\Gamma(\alpha + 1)^{-1} t^\alpha + v(t)$ with $v \in AC^2(\bar{I})$.

(ii) For $f \in C^1(\bar{I})$, the expansion remains valid, with $v \in C^2(\bar{I})$.

Exercise 4.18 Prove Corollary 4.5.

Exercise 4.19 Prove the existence and uniqueness of a solution $y \in L^1(I)$ to (4.28).

Exercise 4.20 This exercise is concerned with equation (4.29) with $c_0 > 1$. For $c_0 > 1$, let $w = u - 1$. Then the function w satisfies

$$\partial_t^\alpha w(t) = w(t)(1 + w(t)), \quad w(0) = w_0 := c_0 - 1 > 0.$$

Then by Proposition 4.6, we have the solution representation

$$w(t) = E_{\alpha,1}(t^\alpha)w_0 + \int_0^t (t-s)^{\alpha-1} E_{\alpha,\alpha}((t-s)^\alpha)w^2(s)ds.$$

Note that $t^{\alpha-1}E_{\alpha,\alpha}(t^\alpha)$ is now exponentially growing, indicating that the solution $w(t)$ may blow up in finite time. Establish this blow-up behavior, and derive the upper and lower bounds on the blow-up time.

Exercise 4.21 [FLLX18b] Prove the following bounds for the solution $u(t)$ to problem (4.30) with $v > 0$ and $0 < \gamma < 1$: there exist $c_1, c_2 > 0$ such that

$$c_1 t^{\frac{\alpha}{1-\gamma}} \le u(t) \le c_2 t^{\frac{\alpha}{1-\gamma}}, \quad t \ge 1.$$

Exercise 4.22 Let $0 < \alpha < \beta < 1$, and $\lambda \in \mathbb{R}$. Consider the following Cauchy problem with two Djrbashian-Caputo fractional derivatives

$$\begin{cases} \partial_t^\beta u + \lambda \partial_t^\alpha u = f, & t \in I, \\ u(0) = c_0. \end{cases}$$

(i) For $\lambda \ge 0$, derive the explicit series solution using Laplace transform and the method of successive approximations separately.

(ii) Discuss the asymptotics of the solution in (i) as $t \to 0^+$ and $t \to \infty$. [Hint: use Theorem A.9.]

(iii) Discuss the well-posedness of the problem when $\lambda < 0$.

Chapter 5
Boundary Value Problem for Fractional ODEs

This chapter describes the basic mathematical theory for two-point boundary value problems (BVPs) for one-dimensional stationary superdiffusion derived in Chapter 1:

$$_0D_x^\alpha u = f(x, u) \quad \text{in } D, \tag{5.1}$$

with $\alpha \in (1, 2)$ and $D = (0, 1)$, where the notation $_0D_x^\alpha$ denotes either Djrbashian-Caputo or Riemann-Liouville fractional derivative of order α. We describe the solution theory via Green's function and variational formulation, and also discuss the related Sturm-Liouville problem. These approaches resemble closely that for the classical two-point BVPs, but the analysis is more intricate, due to nonlocality of fractional derivatives.

5.1 Green's Function

First, we analyze problem (5.1) using the associated Green's function, which has been extensively studied with the focus on the existence of positive solutions of (nonlinear) fractional BVPs. Djrbashian [DN61] (see [Djr93] for the survey of some of these results) probably is the first researcher to study Dirichlet-type problems for FDES, and also related results for the fractional Sturm-Liouville problem in Section 5.3. Other more recent works include [Zha00, BL05, Zha06a, Zha06b]. We discuss the Riemann-Liouville and Djrbashian-Caputo cases separately, under a relatively strong condition on f.

5.1.1 Riemann-Liouville case

First, we comment on the boundary condition. Clearly, $-_0^R D_x^\alpha u = f$ is equivalent to $-(_0I_x^{2-\alpha}u)'' = f$, and the condition $_0I_x^{2-\alpha}u \in AC^2(\overline{D})$ is needed for the equation to

B. Jin, *Fractional Differential Equations*, Applied Mathematical Sciences 206, https://doi.org/10.1007/978-3-030-76043-4_5

make sense in $L^1(D)$. Upon letting $v = {}_0I_x^{2-\alpha}u$, v satisfies $-v'' = f$, for which the natural boundary conditions are given in terms of v or v' or their combinations at $x = 0, 1$. Thus, the most "natural" boundary conditions in the Riemann-Liouville case should specify ${}_0I_x^{2-\alpha}u$ and ${}_0^RD_x^{\alpha-1}u$ at $x = 0, 1$, whose physical interpretation is however not obvious and thus has not been extensively studied. Below, we employ a Dirichlet type boundary condition. The next result gives the equivalence between the BVP and its Fredholm integral reformulation via Green's function. The derivation employs Theorem 2.6 and (2.24), similar to Proposition 4.2 in Chapter 4.

Theorem 5.1 Let $f : D \times \mathbb{R} \to \mathbb{R}$ be such that $f(x, u(x)) \in L^1(D)$ for $u \in L^1(D)$. Then a function $u \in L^1(D)$ with ${}_0I_x^{2-\alpha}u \in AC^2(D)$ solves

$$\begin{cases} -{}_0^RD_x^\alpha u(x) = f(x, u(x)), & in\ D, \\ ({}_0^RD_x^{\alpha-2}u)(0) = u(1) = 0, \end{cases} \tag{5.2}$$

if and only if $u \in L^1(D)$ satisfies

$$u(x) = \int_0^1 G(x, s)f(s, u(s))ds, \tag{5.3}$$

where Green's function $G(x, s)$ is given by

$$G(x, s) = \frac{1}{\Gamma(\alpha)} \begin{cases} (x(1 - s))^{\alpha-1} - (x - s)^{\alpha-1}, & 0 \le s \le x \le 1, \\ x^{\alpha-1}(1 - s)^{\alpha-1}, & 0 \le x \le s \le 1. \end{cases}$$

Proof By Theorem 2.6, for $u \in L^1(D)$ with ${}_0I_x^{2-\alpha}u \in AC^2(\overline{D})$, by the assumption on f, $g(x) := f(x, u(x)) \in L^1(D)$. Integrating both sides of (5.2) gives

$$u(x) = c_0x^{\alpha-2} + c_1x^{\alpha-1} - {}_0I_x^\alpha g(x),$$

for some $c_0, c_1 \in \mathbb{R}$. The boundary conditions are fulfilled if $c_0 = 0$ and $c_1 = ({}_0I_x^\alpha g)(1)$ which, upon substitution, directly gives (5.3). Conversely, if $u \in L^1(D)$ satisfies (5.3), let $g = f(x, u(x))$ so $g \in L^1(D)$. Then u in (5.3) satisfies

$$u(x) = ({}_0I_x^\alpha g)(1)x^{\alpha-1} - ({}_0I_x^\alpha g)(x).$$

Clearly, ${}_0^RD_x^{\alpha-2}u(0) = 0$ and $u(1) = 0$. Furthermore, it follows that ${}_0I_x^{2-\alpha}u \in AC^2(\overline{D})$ and $({}_0I_x^{2-\alpha}u)''(x)$ exists almost everywhere with

$${}_0^RD_x^\alpha u(x) = {}_0^RD_x^\alpha(({}_0I_x^\alpha g)(1)x^{\alpha-1} - {}_0I_x^\alpha g) = g \quad \text{a.e. } D.$$

So the identity $-{}_0^RD_x^\alpha u = f(x, u(x))$ holds for a.e. $x \in D$. This proves the theorem. $\square$

It is evident from the expression of the general solution $u(x)$ that one cannot specify directly $u(0) = c_0 \neq 0$. The proper choice is to specify ${}_0^RD_x^{\alpha-2}u(0) = c_0$ or ${}_0^RD_x^{\alpha-1}u(0) = c_0$, which, respectively, resembles a Dirichlet/Neumann-type boundary condition in classical two-point BVPs; see [BKS18] for further discussions. Nonetheless, the condition ${}_0^R D_x^{\alpha-2} u(0) = 0$ may be replaced by $u(0) = 0$. Note also that a similar issue arises if one seeks a solution $u \in C(\overline{D})$. Since we focus on the zero "Dirichlet" boundary condition below, we will adopt this convention and write

$$\begin{cases} -{}_0^R D_x^\alpha u(x) = f(x, u(x)), & \text{in } D, \\ u(0) = u(1) = 0. \end{cases} \tag{5.4}$$

Formally, for any $x \in D$, the Green's function $G(x, s)$ satisfies

$$\begin{cases} -{}_0^R D_s^\alpha G(x, s) = \delta_x(s), & \text{in } D, \\ G(x, 0) = G(x, 1) = 0, \end{cases}$$

where $\delta_x(s)$ is the Dirac Delta function at $s = x$. Surely one has to make proper sense of the equation since fractional calculus in Chapter 2 is only defined for $L^1(D)$ functions. A rigorous treatment requires extending fractional operators to distributions. When $\alpha = 2$, it recovers classical two-point BVPs, and accordingly, the Green's function $G(x, s)$ also converges.

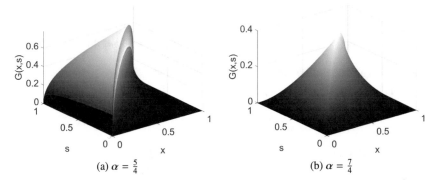

(a) $\alpha = \frac{5}{4}$ (b) $\alpha = \frac{7}{4}$

Fig. 5.1 Green's function $G(x, s)$ in the Riemann-Liouville case.

The Green's function $G(x, s)$ is shown in Fig. 5.1 for $\alpha = \frac{5}{4}$ and $\alpha = \frac{7}{4}$. Clearly, it is nonnegative and continuous, and positive in $D \times D$ and achieves its maximum along with the wedge $x = s$. For α close to unit, it has a sharp gradient along the boundary and the line $x = s$, indicating a possible limited smoothing property of the solution operator. When α tends to two, it becomes increasingly smoother.

The following lemma collects a few properties [BL05, Lemma 2.4].

Lemma 5.1 *The Green's function $G(x, s)$ has the following properties.*

(i) $G(x, s) > 0$ *for any $x, s \in D$.*

(ii) *The following bound holds:*

$$\min_{\frac{1}{4} \le x \le \frac{3}{4}} G(x, s) \ge \gamma(s) \max_{0 \le x \le 1} G(x, s) = \gamma(s) G(s, s), \quad s \in D,$$

where the positive function $\gamma \in C(D)$ *is given by*

$$\gamma(s) = \begin{cases} ((\frac{3}{4}(1-s))^{\alpha-1} - (\frac{3}{4} - s)^{\alpha-1})(s(1-s))^{1-\alpha}, & s \in (0, r], \\ (4s)^{1-\alpha}, & s \in [r, 1), \end{cases}$$

with r defined in the proof.

Proof Part (i) is obvious. For part (ii), note that for any fixed $s \in D$, $G(x, s)$ is decreasing in x on $[s, 1)$, and increasing on $(0, s]$. Let

$$g_1(x, s) = \Gamma(\alpha)^{-1}((x(1-s))^{\alpha-1} - (x-s)^{\alpha-1}),$$
$$g_2(x, s) = \Gamma(\alpha)^{-1}(x(1-s))^{\alpha-1}.$$

Then there holds

$$\min_{\frac{1}{4} \le x \le \frac{3}{4}} G(x, s) = \begin{cases} g_1(\frac{3}{4}, s), & s \in (0, \frac{1}{4}], \\ \min(g_1(\frac{3}{4}, s), g_2(\frac{1}{4}, s)), & s \in (\frac{1}{4}, \frac{3}{4}), \\ g_2(\frac{1}{4}, s), & s \in [\frac{3}{4}, 1) \end{cases} = \begin{cases} g_1(\frac{3}{4}, s), & s \in (0, r], \\ g_2(\frac{1}{4}, s), & s \in [r, 1), \end{cases}$$

where $r \in (\frac{1}{4}, \frac{3}{4})$ is the unique root to $\eta(s) := g_1(\frac{3}{4}, s) - g_2(\frac{1}{4}, s)$, i.e.,

$$\eta(s) := (\frac{3}{4}(1-s))^{\alpha-1} - (\frac{3}{4} - s)^{\alpha-1} - (\frac{1-s}{4})^{\alpha-1} = 0.$$

See Fig. 5.2(a) for an illustration. One can show that for $\alpha \to 1$, $r \to \frac{3}{4}$ and for $\alpha \to 2$, $r \to \frac{1}{2}$ (for $\alpha = 2$, $r = \frac{1}{2}$). Note that

$$\max_{0 \le x \le 1} G(x, s) = G(s, s) = \Gamma(\alpha)^{-1}(s(1-s))^{\alpha-1}, \quad s \in D.$$

Hence, (ii) follows by the choice of $\gamma(s)$. See also Fig. 5.2(b) for an illustration. □

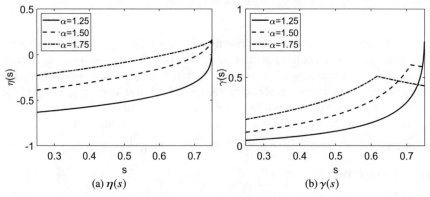

Fig. 5.2 The plots of the functions $\eta(s)$ and $\gamma(s)$ over the interval $[\frac{1}{4}, \frac{3}{4}]$.

The solution u is generally only Hölder continuous, and not Lipschitz. This lack of regularity is characteristic of BVPs involving $_0^R D_x^\alpha u$, similar to Cauchy problems, cf. Theorem 4.8.

Proposition 5.1 *If $f \in C(\overline{D} \times \mathbb{R})$, then the solution u to problem (5.4), if it exists, belongs to $C^{\alpha-1}(\overline{D})$.*

Proof It follows from Theorem 5.1 that the solution u, if it exists, is given by

$$u(x) = x^{\alpha-1}(_0I_x^\alpha f(s, u(s)))(1) - (_0I_x^\alpha f(s, u(s))(x).$$

It can be verified directly that $x^{\alpha-1} \in C^{\alpha-1}(\overline{D})$. Further, by the mapping property of $_0I_x^\alpha, _0I_x^\alpha f(s, u(s)) \in C^{1,\alpha-1}(\overline{D})$. Thus, the solution u belongs to $C^{\alpha-1}(\overline{D})$. □

Remark 5.1 When f is independent of u, u is Lipschitz continuous if and only if f satisfies the nonlocal condition $\int_0^1 (1-s)^{\alpha-1} f(s) ds = 0$, of which there is no analogue in the integer-order derivative case. Generally, it is unclear under what condition(s) that u is smooth.

Now we analyze the existence issue for problem (5.4). The basic strategy is to reformulate the problem into a Fredholm integral equation using Green's function, then apply a suitable version of fixed point theorem; see Appendix A.4 for several versions. If f is continuous, by Theorem 5.1, problem (5.4) is equivalent to the following Fredholm integral equation:

$$u(x) = \int_0^1 G(x, s) f(s, u(s)) ds := Tu(x). \tag{5.5}$$

Then problem (5.4) is equivalent to an operator equation $u = Tu$ in suitable function spaces. We below describe one existence result in the space $C(\overline{D})$ [BL05, Theorem 3.1]. Let $P = \{u \in C(\overline{D}) : u(x) \geq 0\} \subset C(\overline{D})$. Note that the constant M can be evaluated explicitly to be $M = \frac{\Gamma(2\alpha)}{\Gamma(\alpha)}$, in view of the identity

$$\int_0^1 G(s, s) ds = \frac{1}{\Gamma(\alpha)} \int_0^1 s^{\alpha-1}(1-s)^{\alpha-1} ds = \frac{B(\alpha, \alpha)}{\Gamma(\alpha)} = \frac{\Gamma(\alpha)}{\Gamma(2\alpha)}.$$

Assumption 5.1 Let $f(x, u) : \overline{D} \times [0, \infty) \to \mathbb{R}_+$ be continuous, and there exist two constants $r_2 > r_1 > 0$ such that

(i) $f(x, u) \leq Mr_2$ for all $(x, u) \in \overline{D} \times [0, r_2]$;
(ii) $f(x, u) \geq Nr_1$ for all $(x, u) \in \overline{D} \times [0, r_1]$,

with $M = \left(\int_0^1 G(s, s) ds \right)^{-1}$ and $N = \left(\int_{\frac{1}{4}}^{\frac{3}{4}} \gamma(s) G(s, s) ds \right)^{-1}$.

Theorem 5.2 *Let $f(x, u) : \overline{D} \times [0, \infty) \to \mathbb{R}_+$ be continuous. Then under Assumption 5.1, problem (5.4) has at least one positive solution u such that $r_1 \leq \|u\|_{C(\overline{D})} \leq r_2$.*

Proof By Theorem 5.1, problem (5.4) has a solution u if and only if it solves $u = Tu$. Next, we show that the operator $T : P \to P$ is compact. Since both $G(x, s)$ and $f(x, u)$ are nonnegative and continuous, T is continuous. Let $P' \subset P$ be a bounded subset with $\|u\|_{C(\overline{D})} \leq M$ for all $u \in P'$. Then for $u \in P'$, by Theorem 5.1, we have

$$|Tu(x)| \leq \int_0^1 G(x, s)|f(s, u(s))|\, ds \leq c_f \int_0^1 G(s, s)\, ds = \frac{\Gamma(\alpha)}{\Gamma(2\alpha)} c_f,$$

with $c_f = \|f\|_{L^\infty(D \times [0,M])}$. Hence, the set $T(P')$ is bounded. Next, we claim that for every $u \in P'$, $x_1, x_2 \in \overline{D}$ with $x_1 < x_2$, there holds

$$|Tu(x_2) - Tu(x_1)| \leq c_f \Gamma(\alpha + 1)^{-1}(x_2^{\alpha-1} - x_1^{\alpha-1} + x_2^\alpha - x_1^\alpha). \tag{5.6}$$

Indeed, using the expression of the Green's function $G(x, s)$ in Theorem 5.1, we have

$$|Tu(x_2) - Tu(x_1)| \leq |({}_0I_x^\alpha f(\cdot, u(\cdot)))(1)|(x_2^{\alpha-1} - x_1^{\alpha-1})$$
$$+ \frac{1}{\Gamma(\alpha)} \int_{x_1}^{x_2} (x_2 - s)^{\alpha-1}|f(s, u(s))|ds$$
$$+ \frac{1}{\Gamma(\alpha)} \int_0^{x_1} ((x_2 - s)^{\alpha-1} - (x_1 - s)^{\alpha-1})|f(s, u(s))|ds.$$

Next, we bound the three terms, denoted by I_i, $i = 1, 2, 3$. Actually,

$$I_1 \leq \frac{c_f}{\Gamma(\alpha)} \int_0^1 (1 - s)^{\alpha-1}ds(x_2^{\alpha-1} - x_1^{\alpha-1}) = \frac{c_f}{\Gamma(\alpha + 1)}(x_2^{\alpha-1} - x_1^{\alpha-1}),$$

$$I_2 \leq \frac{c_f}{\Gamma(\alpha)} \int_{x_1}^{x_2} (x_2 - s)^{\alpha-1}ds = \frac{c_f}{\Gamma(\alpha + 1)}(x_2 - x_1)^\alpha,$$

$$I_3 \leq \frac{c_f}{\Gamma(\alpha)} \int_0^{x_1} ((x_2 - s)^{\alpha-1} - (x_1 - s)^{\alpha-1})ds = \frac{c_f}{\Gamma(\alpha + 1)}(x_2^\alpha - (x_2 - x_1)^\alpha - x_1^\alpha).$$

Collecting these estimates yields (5.6). Thus, the set $T(P')$ is equicontinuous. By Arzelá-Ascoli theorem, $T : P \to P$ is compact. Below, we verify the two conditions in Theorem A.13. Let $P_1 := \{u \in P : \|u\|_{C(\overline{D})} < r_1\}$. For $u \in \partial P_1$, we have $0 \leq u(x) \leq r_1$ for all $x \in \overline{D}$. By Assumption 5.1(ii) and Lemma 5.2, for $x \in [\frac{1}{4}, \frac{3}{4}]$, there holds

$$Tu(x) = \int_0^1 G(x, s)f(s, u(s))\, ds \geq \int_0^1 \gamma(s)G(s, s)f(s, u(s))\, ds$$
$$\geq Nr_1 \int_{\frac{1}{4}}^{\frac{3}{4}} \gamma(s)G(s, s)\, ds = r_1 = \|u\|_{C(\overline{D})}.$$

So $\|Tu\|_{C(\overline{D})} \geq \|u\|_{C(\overline{D})}$ for $u \in \partial P_1$. Next, let $P_2 := \{u \in P : \|u\|_{C(\overline{D})} \leq r_2\}$. For $u \in \partial P_2$, we have $0 \leq u(x) \leq r_2$ for all $x \in \overline{D}$. By Assumption 5.1(i), for $x \in \overline{D}$

$$\|Tu\|_{C(\overline{D})} = \sup_{x \in D} \int_0^1 G(x, s) f(s, u(s)) \, ds \leq M r_2 \int_0^1 G(s, s) \, ds = r_2 = \|u\|_{C(\overline{D})}.$$

Thus, the two conditions in Theorem A.13 hold, and this completes the proof. $\square$

Example 5.1 Consider the following problem

$$\begin{cases} -{}_0^R D_x^{\frac{3}{2}} u = u^2 + \frac{\sin x}{4} + 1, & \text{in } D, \\ u(0) = u(1) = 0. \end{cases}$$

One can verify $M = \frac{4}{\sqrt{\pi}}$ and $N = 13.6649$. Choosing $r_1 = \frac{1}{14}$ and $r_2 = 1$ gives

$$f(x, u) = 1 + \frac{\sin x}{4} + u^2 \leq 2.2107 \leq M r_2, \quad \forall (x, u) \in \overline{D} \times [0, 1],$$
$$f(x, u) = 1 + \frac{\sin x}{4} + u^2 \geq 1 \geq N r_1, \quad \forall (x, u) \in \overline{D} \times [0, \tfrac{1}{14}].$$

Thus, Assumption 5.1 holds. Then by Theorem 5.2, this problem has at least one solution u such that $\frac{1}{14} \leq u \leq 1$.

Note that Theorem 5.2 only ensures the existence of a positive solution, but it does not say anything about the uniqueness or existence of other solutions, which have to be established by other methods. The next result gives the uniqueness under a Lipschitz assumption on $f(x, u)$ in u [CMSS18, Theorem 1]; see also [Bai10] for related results.

Theorem 5.3 *If $f : \overline{D} \times \mathbb{R} \to \mathbb{R}$ is continuous and there exists an $L > 0$ such that*

$$|f(x, u) - f(x, v)| \leq L|u - v|, \quad \forall x \in \overline{D}, u, v \in \mathbb{R},$$

with $L < \alpha^{1+\alpha}(\alpha - 1)^{1-\alpha}\Gamma(\alpha)$, then problem (5.4) has a unique solution $u \in C(\overline{D})$.

Proof Note that we can bound $\int_0^1 G(x, s)ds$ by

$$\Gamma(\alpha) \int_0^1 G(x, s)ds = x^{\alpha-1} \int_0^1 (1 - s)^{\alpha-1}ds - \int_0^x (x - s)^{\alpha-1}ds$$
$$= \alpha^{-1}(x^{\alpha-1} - x^\alpha) \leq \alpha^{-1}(x^{\alpha-1} - x^\alpha)|_{x = \frac{\alpha-1}{\alpha}} = \alpha^{-1-\alpha}(\alpha - 1)^{\alpha-1},$$

since the function $x^{\alpha-1} - x^\alpha$ attains its maximum over $\overline{D}$ at $x = \alpha^{-1}(\alpha - 1)$. This and the nonnegativity of $G(x, s)$ in Lemma 5.1(i) imply that for any $x \in \overline{D}$, there holds

$$|Tu(x) - Tv(x)| \leq \int_0^1 G(x, s)|f(s, u(s)) - f(s, v(s))|ds$$
$$\leq L \int_0^1 G(x, s)ds\|u - v\|_{C(\overline{D})} \leq L\Gamma(\alpha)^{-1}\alpha^{-1-\alpha}(\alpha - 1)^{\alpha-1}\|u - v\|_{C(\overline{D})}.$$

Under the condition on L, T is a contraction on $C(\overline{D})$, and Theorem A.10 implies the existence of a unique solution $u \in C(\overline{D})$ to $u = Tu$, or equivalently, problem (5.4). $\square$

5.1.2 Djrbashian-Caputo case

Now we turn to the Djrbashian-Caputo case, which is analytically more delicate, especially in the classical sense. First, we state the solution representation in the case of general separated boundary conditions (also known as Sturm-Liouville boundary conditions), when the Djrbashian-Caputo fractional derivative is in the regularized sense, cf. Definition 2.4. The classical version in Definition 2.3 requires the condition $u \in AC^2(\overline{D})$, which, however, is not direct from the integral representation (5.8).

Theorem 5.4 Let $a, b, c, d \in \mathbb{R}$ with $ad + bc + ac \neq 0$, and $f : \overline{D} \times \mathbb{R} \to \mathbb{R}$ be continuous. Then a function $u \in C^1(\overline{D})$ solves

$$\begin{cases} -{}_0^C D_x^\alpha u(x) = f(x, u(x)), & \text{in } D, \\ au(0) - bu'(0) = 0, \\ cu(1) + du'(1) = 0, \end{cases} \qquad (5.7)$$

if and only if $u \in C^1(\overline{D})$ satisfies

$$u(x) = \int_0^1 G(x, s) f(s, u(s)) ds, \qquad (5.8)$$

where Green's function $G(x, s) : D^2 \to \mathbb{R}$ is given by (with $\Lambda = ad + bc + ac$)

$$G(x, s) = \frac{1}{\Lambda \Gamma(\alpha)} \begin{cases} (b + ax)[(\alpha - 1) + c(1 - s)](1 - s)^{\alpha-2} - \Lambda(x - s)^{\alpha-1}, & s \leq x, \\ (b + ax)[(\alpha - 1) + c(1 - s)](1 - s)^{\alpha-2}, & x \leq s. \end{cases}$$

Proof By Theorem 2.13, for $u \in C^1(\overline{D})$ satisfying (5.7), we have $u(x) = c_0 + c_1 x - {}_0I_x^\alpha g(x)$, with $g = f(x, u(x))$ and $c_0, c_1 \in \mathbb{R}$ to be determined. The boundary conditions are fulfilled if

$$ac_0 - bc_1 = 0, \quad cc_0 + cc_1 - c({}_0I_x^\alpha g)(1) + dc_1 - d({}_0I_x^{\alpha-1}g)(1) = 0.$$

These two equations can uniquely determine c_0 and c_1 if and only if

$$\det \begin{bmatrix} a & -b \\ c & c + d \end{bmatrix} = ad + bc + ac \equiv \Lambda \neq 0.$$

Since $\Lambda = ad + bc + ac \neq 0$, by assumption, c_0 and c_1 is uniquely determined by

$$c_0 = \Lambda^{-1} a(c({}_0I_x^\alpha g)(1) + d({}_0I_x^{\alpha-1}g)(1)),$$
$$c_1 = \Lambda^{-1} b(c({}_0I_x^\alpha g)(1) + d({}_0I_x^{\alpha-1}g)(1)).$$

Substituting the values of c_0 and c_1 gives (5.8). Conversely, if $u \in C^1(\overline{D})$ satisfies (5.8), let $g(x) = f(x, u(x))$ so $g \in L^1(D)$. Then u in (5.8) satisfies

$$u(x) = \Lambda^{-1}(a + bx)(c({}_0I_x^\alpha g)(1) + d({}_0I_x^{\alpha-1}g)(1)) - ({}_0I_x^\alpha g)(x).$$

It follows that

$$u(0) = \Lambda^{-1} a(c({_0}I_x^\alpha g)(1) + d({_0}I_x^{\alpha-1} g)(1)),$$
$$u'(0) = \Lambda^{-1} b(c({_0}I_x^\alpha g)(1) + d({_0}I_x^{\alpha-1} g)(1)),$$
$$u(1) = \Lambda^{-1}(a + b)(c({_0}I_x^\alpha g)(1) + d({_0}I_x^{\alpha-1} g)(1)) - ({_0}I_x^\alpha g)(1),$$
$$u'(1) = \Lambda^{-1} b(c({_0}I_x^\alpha g)(1) + d({_0}I_x^{\alpha-1} g)(1)) - ({_0}I_x^{\alpha-1} g)(1).$$

Clearly, $au(0) - bu'(0) = 0$ and $cu(1) + du'(1) = 0$. Furthermore,

$$ {_0^C}D_x^\alpha u(x) = {_0^R}D_x^\alpha(u - u(0) - u'(0)x) = {_0^R}D_x^\alpha {_0}I_x^\alpha g = g \quad \text{a.e. } D.$$

So, the identity ${_0^C}D_x^\alpha u = f(x, u(x))$ holds for a.e. $x \in D$. This completes the proof. $\qquad\square$

The next result from [LL13, Theorems 2.4 and 2.7] discusses linear BVPs, i.e., $f = f(x)$, in the classical sense, cf. Definition 2.3. Note that in (iii), the condition $f \in AC(\overline{D})$ cannot be relaxed to $C(\overline{D})$, since T does not map $C(\overline{D})$ to $AC^2(\overline{D})$. But it can be relaxed to $f \in C(\overline{D})$ with ${_0}I_x^{\alpha-1} f \in AC(\overline{D})$, cf. Lemma 2.4.

Theorem 5.5 *Let $a, b, c, d \in \mathbb{R}$ satisfy $ad + bc + ac \neq 0$, and T be defined by*

$$Tf(x) = \int_0^1 G(x, s)f(s)ds,$$

where the Green's function $G(x, s)$ is given in Theorem 5.4. Then the following statements hold.

(i) *If $f \in L^\infty(D)$ and $u = Tf$, then $u'(x)$ exists for each $x \in \overline{D}$, and $\|u'\|_{L^\infty(D)} < \infty$.*

(ii) *The integral operator T maps $AC(\overline{D})$ into $AC^2(\overline{D})$.*

(iii) *If $f \in AC(\overline{D})$, then $u = Tf$ is a solution of*

$$\begin{cases} -{_0^C}D_x^\alpha u(x) = f(x), & \text{in } D, \\ au(0) - bu'(0) = 0, \\ cu(1) + du'(1) = 0. \end{cases} \qquad (5.9)$$

(iv) *If $u \in AC^2(\overline{D})$ and $f \in L^1(D)$ satisfy (5.9), then $u = Tf$.*

Proof (i) Let $\Lambda = ad + bc + ac$ and $c_1 = \Lambda^{-1} b(c({_0}I_x^\alpha f)(1) + d({_0}I_x^{\alpha-1} f)(1))$. If $f \in L^\infty(D)$, then by Theorem 2.2(i),

$$|c_1| \leq |\Lambda|^{-1}|b|(|c|({_0}I_x^\alpha f)(1)| + |d|({_0}I_x^{\alpha-1} f)(1)|)$$
$$\leq (|\Lambda|\Gamma(\alpha))^{-1}|b|(|c|\alpha^{-1} + |d|)\|f\|_{L^\infty(D)} < \infty.$$

Now by the identity

$$u'(x) = c_1 - {_0}I_t^{\alpha-1} f(x), \qquad (5.10)$$

we deduce

$$|u'(x)| \le (|\Lambda||\Gamma(\alpha)|)^{-1}(|b|(|c|\alpha^{-1} + |d|) + |\Lambda|)\|f\|_{L^\infty(D)} < \infty, \quad \forall x \in \overline{D}.$$

Since $u'(x)$ exists for every $x \in \overline{D}$, $u \in C(\overline{D})$.

(ii) This is direct from (5.10) and Theorem 2.2(iv).

(iii) Since $f \in AC(\overline{D})$, Theorem 2.2(iv) indicates $_0I_x^{\alpha-1} f \in AC(\overline{D})$ with $_0I_x^{\alpha-1} f(0) = 0$. Then by (5.10), $u''(x) = -(_0I_x^{\alpha-1} f)'$. Applying $_0I_x^{2-\alpha}$ to both sides leads to

$$_0^C D_x^\alpha u = -_0I_x^{2-\alpha}(_0I_x^{\alpha-1} f)' = -f.$$

The boundary conditions can be verified directly. Thus, $u = Tf$ solves (5.9).

(iv) Since $-_0^C D_x^\alpha u = f$ for a.e. $x \in D$, we have $-_0I_x^\alpha(_0^C D_x^\alpha u) = _0I_x^\alpha f(x)$ for a.e. $x \in D$. Since $u \in AC^2(\overline{D})$, by Theorem 2.13,

$$_0I_x^\alpha\, _0^C D_x^\alpha u(x) = u(x) - u(0) - u'(0)x \in C(\overline{D}), \quad \text{a.e. } x \in D.$$

Since $f \in L^1(D)$ and $\alpha \in (1, 2)$, by Theorem 2.2(iii), $_0I_x^\alpha f \in C(\overline{D})$. Thus, we have

$$u(x) = u(0) + u'(0)x - _0I_x^\alpha f(x), \quad \forall x \in \overline{D}.$$

The expression of Green's function follows as Theorem 5.4, and thus it is omitted. □

The next result [LL13, Proposition 3.1] gives the equivalence for $_0^C D_x^\alpha u$ in the classical sense, under a Lipschitz condition on f, thereby extending Theorem 5.4.

Theorem 5.6 *Let the conditions in Theorem 5.4 be fulfilled, and $f(x, u)$ be Lipschitz, i.e., $|f(x_2, u_2) - f(x_1, u_1)| \le L_r \max(|x_2 - x_1|, |u_2 - u_1|)$, for all $x_1, x_2 \in \overline{D}$ and $u_1, u_2 \in \mathbb{R}$, with $|u_1| < r$, $|u_2| < r$. If $u \in C(\overline{D})$ satisfies (5.8), then $u \in AC^2(\overline{D})$ and it solves (5.7).*

Proof If $u \in C(\overline{D})$ solves (5.8), i.e., $u = \int_0^1 G(x, s) f(s, u(s)) ds$ for $x \in \overline{D}$. Let $g = f(x, u(x))$. Then $g \in C(\overline{D})$, and by Theorem 5.5(i), u' exists for every $x \in \overline{D}$, and $c := \|u'\|_{L^\infty(D)} < \infty$. It follows that $|u(x_2) - u(x_1)| \le c|x_2 - x_1|$ for any $x_1, x_2 \in \overline{D}$. This and the Lipschitz continuity of f imply that for $x_1, x_2 \in \overline{D}$

$$|g(x_2) - g(x_1)| = |f(x_2, u(x_2)) - f(x_1, u(x_1))|$$
$$\le L \max(|x_2 - x_1|, |u(x_2) - u(x_1)|) \le L(1 + c)|x_2 - x_1|.$$

Hence, $g \in AC(\overline{D})$. By Theorem 5.5(iii), $u \in AC^2(\overline{D})$, and $_0^C D_x^\alpha u(x) = g(x) = f(x, u(x))$ a.e. $x \in \overline{D}$, i.e., u solves problem (5.7). □

The next result is an immediate corollary of Theorem 5.4 in the Dirichlet case.

Corollary 5.1 *In the Djrbashian-Caputo case, let $f : D \times \mathbb{R} \to \mathbb{R}$ be continuous. Then a function $u \in C^1(\overline{D})$ solves*

$$\begin{cases} -_0^C D_x^\alpha u(x) = f(x, u(x)), & \text{in } D, \\ u(0) = u(1) = 0, \end{cases} \tag{5.11}$$

if and only if $u \in C^1(\overline{D})$ *satisfies* $u(x) = \int_0^1 G(x, s) f(s, u(s)) \, ds$, *where the Green's function* $G(x, s)$ *is given by*

$$G(x, s) = \frac{1}{\Gamma(\alpha)} \begin{cases} x(1 - s)^{\alpha-1} - (x - s)^{\alpha-1}, & 0 \leq s \leq x \leq 1, \\ x(1 - s)^{\alpha-1}, & 0 \leq x \leq s \leq 1. \end{cases}$$

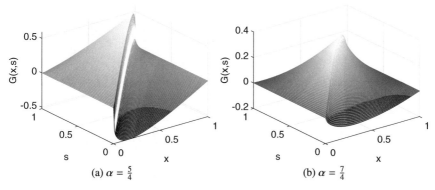

(a) $\alpha = \frac{5}{4}$ (b) $\alpha = \frac{7}{4}$

Fig. 5.3 Green's function $G(x, s)$ in the Djrbashian-Caputo case.

The Green's function $G(x, s)$ in Corollary 5.1 is shown in Fig. 5.3. It is not necessarily positive in $D \times D$, which implies that the solution operator is not positivity preserving, and lacks the comparison principle, etc. It is continuous, but there is a very steep change around the diagonal $x = s$ for α close to unity, similar to the Riemann-Liouville case. The magnitude of the negative part decreases to zero as α approaches two. One possible strategy to regain the positivity is to use a Robin-type boundary condition at $x = 0$; see [LL13] for detailed discussions.

Example 5.2 Let $\alpha = \frac{6}{5}$, and $u(x) = x^3 - \frac{5}{3}x^2 + \frac{2}{3}x$ on D. Clearly, $u(0) = u(1) = 0$. Further, for any $x \in D$, $-{_0^C}D_x^\alpha u(x) > 0$. However, $u(\frac{4}{5}) = -\frac{8}{375} < 0$, so the BVP does not preserve the positivity, and lacks a comparison principle. See Fig. 5.4 for a graphical illustration. This contrasts sharply with the Riemann-Liouville case.

The next result gives the solution regularity, which is better than the Riemann-Liouville case. Generally, it cannot be smoother than $C^{1,\alpha-1}(\overline{D})$, since the term ${_0}I_x^\alpha f$ has limited regularity, even when f is smooth (e.g., $f \equiv 1$). Higher regularity can only be obtained under extra compatibility conditions on f. More generally, the regularity may be derived from theory of Fredholm integral equations [Gra82, PV06], but a systematic study seems still missing (for either fractional derivative).

Theorem 5.7 *If* $f \in C(\overline{D} \times \mathbb{R})$, *then the solution* u *to problem* (5.11), *if it exists, belongs to* $C^{1,\alpha-1}(\overline{D})$.

Proof It follows from Corollary 5.1 that the solution u is given by

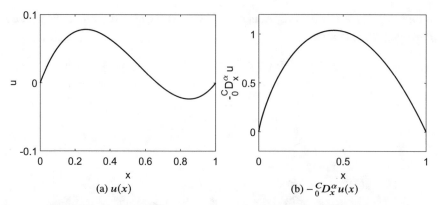

Fig. 5.4 The solution u and its Djrbashian-Caputo fractional derivative $-{}_0^C D_x^\alpha u$ with $\alpha = \frac{6}{5}$.

$$u(x) = \frac{1}{\Gamma(\alpha)} x \int_0^1 (1-s)^{\alpha-1} f(s, u(s)) ds - \frac{1}{\Gamma(\alpha)} \int_0^x (x-s)^{\alpha-1} f(s, u(s)) ds.$$

For $f \in C(\overline{D} \times \mathbb{R})$, ${}_0 I_x^\alpha f(s, u(s)) \in C^{1,\alpha-1}(\overline{D})$. Thus, u belongs to $C^{1,\alpha-1}(\overline{D})$.   □

Now we give an existence result to problem (5.11) [Zha06a, Theorem 4.1]. The proof relies on the Leray-Schauder alternative in Appendix A.4.

Theorem 5.8 *Let* $f : \overline{D} \times \mathbb{R} \to \mathbb{R}$ *be continuous, and* $|f(x,u)| \le c_\mu |u|^\mu + c$, *for some* $c_\mu \in (0, \frac{\Gamma(1+\alpha)}{6})$, $c > 0$ *and* $0 < \mu \le 1$. *Then problem (5.11) has a solution.*

Proof We define an operator $T : C(\overline{D}) \to C(\overline{D})$ by $Tu = \int_D G(x,s) f(s, u(s)) ds$. It suffices to show that T has a fixed point. It is continuous. For $u \in Z := \{u \in C(\overline{D}) : \|u\|_{C(\overline{D})} \le M\}$, with $c_f = \|f\|_{L^\infty(\overline{D} \times [0,M])}$, we have

$$|Tu(x)| \le \frac{c_f}{\Gamma(\alpha)} \left(\int_0^x (x-s)^{\alpha-1} ds + x \int_0^1 (1-s)^{\alpha-1} ds \right) \le \frac{c_f(x + x^\alpha)}{\Gamma(\alpha+1)}.$$

Thus, the set $T(Z)$ is bounded. Next, we show the equicontinuity of $T(Z)$. For any $u \in Z$, $x_1, x_2 \in \overline{D}$ with $x_1 < x_2$, by the triangle inequality,

$$
\begin{aligned}
|Tu(x_2) - Tu(x_1)| \le &\frac{1}{\Gamma(\alpha)} \int_0^{x_1} |(x_2-s)^{\alpha-1} - (x_1-s)^{\alpha-1}| |f(s, u(s))| ds \\
&+ \frac{1}{\Gamma(\alpha)} \int_{x_1}^{x_2} (x_2-s)^{\alpha-1} |f(s, u(s))| ds \\
&+ \frac{x_2 - x_1}{\Gamma(\alpha)} \int_0^1 (1-s)^{\alpha-1} |f(s, u(s))| ds \\
\le &c_f \Gamma(\alpha+1)^{-1} ((x_2 - x_1) + (x_2^\alpha - x_1^\alpha)).
\end{aligned}
$$

Thus, the set $T(Z)$ is equicontinuous, and by Arzelà-Ascoli theorem, T is compact. Let $P = \{u \in C(\overline{\Omega}), \|u\|_{C(\overline{D})} < R\} \subset C(\overline{D})$, with $R > \max(1, 6c\Gamma(\alpha+1)^{-1})$. Suppose that there is $u \in \partial P$ and $\lambda \in (0, 1)$ such that $u = \lambda Tu$. Then for any $u \in \partial P$, we have

$$\lambda |Tu(x)| \leq \frac{1}{\Gamma(\alpha)} \left[\int_0^x (x - s)^{\alpha-1} (c_u |u|^\mu + c) ds + x \int_0^1 (1 - s)^{\alpha-1} (c_u |u(s)|^\mu + c) ds \right]$$
$$\leq (x^\alpha + x) \Gamma(\alpha + 1)^{-1} (c_\mu R^\mu + c) \leq 2\Gamma(\alpha + 1)^{-1} (c_\mu R + c) < R,$$

since $\mu \in (0, 1]$, which implies $\lambda \|Tu\|_{C(\overline{D})} \neq R = \|u\|_{C(\overline{D})}$, contradicting the assumption. Then Theorem A.15 implies that T has a fixed point in $C(\overline{D})$. $\square$

Example 5.3 For the BVP (with $\gamma > 0$ and $\mu \in (0, 1)$),

$$\begin{cases} {}_0^C D_x^\alpha u = (1 + u^2)^{-1} u^\mu + x^\gamma, & \text{in } D, \\ u(0) = u(1) = 0, \end{cases}$$

the conditions in Theorem 5.8 hold, and thus the problem has a solution.

The uniqueness of the solution holds under a suitable Lipschitz condition on f.

Theorem 5.9 *If $f : \overline{D} \times \mathbb{R} \to \mathbb{R}$ is continuous and satisfies $|f(x, u) - f(x, v)| \leq L|u - v|$ for any $x \in \overline{D}, u, v \in \mathbb{R}$, with L satisfying $L \leq (\sup_{x \in \overline{D}} \int_0^1 |G(x, s)| ds)^{-1}$, then problem* (5.4) *has a unique solution in $C(\overline{D})$.*

Proof The proof is identical with that for Theorem 5.3, except that $G(x, s)$ in the Djrbashian-Caputo case is not always nonnegative. $\square$

5.2 Variational Formulation

Now we develop the variational solution theory for

$$\begin{cases} -{}_0 D_x^\alpha u + qu = f, & \text{in } D, \\ u(0) = u(1) = 0, \end{cases} \tag{5.12}$$

with $f \in L^2(D)$ or suitable Sobolev space, and the potential $q \in L^\infty(D)$. Variational formulations are useful for constructing numerical approximations. The study on variational formulations for BVPs with a fractional derivative was initiated by [ER06], where the Riemann-Liouville case (involving both left-sided and right-sided versions) was studied. The well-posedness of the variational problem was shown. We follow the description in [JLPR15], where both Riemann-Liouville and Djrbashian-Caputo cases were analyzed, with proven regularity estimates. We refer to [WY13] for the case of variable coefficients, and [JLZ16b] for Petrov-Galerkin formulations, which are useful for convection-diffusion problems. These investigations mostly focus on homogeneous Dirichlet boundary conditions, which greatly facilitates the analysis. It is not always trivial to extend to nonzero boundary conditions or other types of boundary conditions, e.g., Neumann type or Robin type.

5.2.1 One-sided fractional derivatives

5.2.1.1 Variational formulations

First, we derive the strong solution representation for the case $q = 0$. The procedure is similar to the derivation of Green's function in Section 5.1, but with weaker regularity assumptions on f. First, consider the Riemann-Liouville case. Fix $f \in L^2(D)$ and set $g = {}_0I_x^\alpha f \in H_{0,L}^\alpha(D)$. By Corollary 2.2, the fractional derivative ${}_0^R D_x^\alpha g$ is well defined. Now by Theorem 2.1, we deduce

$$
{}_0I_x^{2-\alpha} g = {}_0I_x^{2-\alpha} {}_0I_x^\alpha f = {}_0I_x^2 f \in H_{0,L}^2(D).
$$

Clearly, $({}_0I_x^2 f)'' = f$ holds for $f \in L^2(D)$. This yields the fundamental relation, cf. Theorem 2.6, ${}_0^R D_x^\alpha g = f$. Consequently, the representation

$$
u = -{}_0I_x^\alpha f + ({}_0I_x^\alpha f)(1)x^{\alpha-1} \tag{5.13}
$$

is a solution of problem (5.12) in the Riemann-Liouville case when $q = 0$ since u satisfies the correct boundary condition and ${}_0^R D_x^\alpha x^{\alpha-1} = (c_\alpha x)'' = 0$.

Next consider the Djrbashian-Caputo case. To this end, we choose $s \geq 0$ so that $\alpha + s \in (\frac{3}{2}, 2)$. For smooth u and $\alpha \in (1, 2)$, the Djrbashian-Caputo and Riemann-Liouville derivatives are related by (cf. Theorem 2.12)

$$
{}_0^C D_x^\alpha u = {}_0^R D_x^\alpha u - \frac{u(0)}{\Gamma(1-\alpha)} x^{-\alpha} - \frac{u'(0)}{\Gamma(2-\alpha)} x^{1-\alpha}.
$$

Applying it to $g = {}_0I_x^\alpha f \in H_{0,L}^{\alpha+s}(D)$, which by Theorem 2.9 satisfies $g(0) = g'(0) = 0$ for $f \in H_0^s(D)$, shows that ${}_0^C D_x^\alpha g$ makes sense and equals ${}_0^R D_x^\alpha g = f$. Thus, a solution u of problem (5.12) in the Djrbashian-Caputo case with $q = 0$ is given by

$$
u = -{}_0I_x^\alpha f + ({}_0I_x^\alpha f)(1) x. \tag{5.14}
$$

Remark 5.2 The representation (5.14) of the model (5.12) with ${}_0^C D_x^\alpha u$ and $f \in H_0^s(D)$ such that $\alpha + s \leq \frac{3}{2}$ remains unclear. The expressions (5.13) and (5.14) are suggestive: their difference in regularity stems from the kernel of the corresponding differential operator. In the Riemann-Liouville case, the kernel consists of the weakly singular functions $cx^{\alpha-1}$, whereas in the Djrbashian-Caputo case, the kernel consists of smooth functions cx.

Now we develop variational formulations of problem (5.12) and establish regularity pickup in Sobolev spaces for the variational solution. We shall first consider the case $q = 0$, where we have an explicit representation of the solutions, and then the case of a general $q \neq 0$. We begin with a useful lemma.

Lemma 5.2 *For $u \in H_{0,L}^1(D)$ and $\beta \in (0, 1)$, ${}_0^R D_x^\beta u = {}_0I_x^{1-\beta} u'$. Similarly, for $u \in H_{0,R}^1(D)$ and $\beta \in (0, 1)$, ${}_x^R D_1^\beta u = -{}_x I_1^{1-\beta} u'$.*

Proof It suffices to prove the result for the left-sided derivative ${}^R_0D^\beta_x$. For $u \in C^1(\overline{D})$ with $u(0) = 0$, Theorem 2.12 implies that

$$
{}^R_0D^\beta_x u = {}^C_0D^\beta_x u = {}_0I^{1-\beta}_x(u'). \tag{5.15}
$$

Corollary 2.2 implies that the left-hand side extends to a continuous operator on $H^\beta_{0,L}(D)$ and hence $H^1_{0,L}(D)$ into $L^2(D)$. Meanwhile, Theorem 2.9 implies that the right-hand side of (5.15) extends to a bounded operator from $H^1(D)$ into $L^2(D)$. The lemma now follows by a density argument. $\qquad\square$

Next, we derive the variational formulation in the Riemann-Liouville case. Upon taking u as in (5.13), $g = {}_0I^\alpha_x f \in \widetilde{H}^\alpha_L(D)$ and $v \in C^\infty_0(D)$, since ${}^R_0D^\alpha_x x^{\alpha-1} = 0$, Lemma 5.2 implies

$$
\begin{aligned}
({}^R_0D^\alpha_x u, v) &= -\big(({}_0I^{2-\alpha}_x g)'', v\big) = \big(({}_0I^{2-\alpha}_x g)', v'\big) \\
&= \big({}^R_0D^{\alpha-1}_x g, v'\big) = ({}_0I^{2-\alpha}_x g', v').
\end{aligned} \tag{5.16}
$$

Now Theorem 2.1 and Lemma 2.3 yield

$$
({}^R_0D^\alpha_x u, v) = \big({}_0I^{1-\frac{\alpha}{2}}_x g', {}_xI^{1-\frac{\alpha}{2}}_1 v'\big). \tag{5.17}
$$

Since $g \in H^\alpha_{0,L}(D)$ and $v \in H^1_0(D)$, we can apply Lemma 5.2 again to conclude

$$
({}^R_0D^\alpha_x u, v) = -({}^R_0D^{\frac{\alpha}{2}}_x g, {}^R_xD^{\frac{\alpha}{2}}_1 v).
$$

Further, by noting the identity ${}^R_0D^{\frac{\alpha}{2}}_x x^{\alpha-1} = c_\alpha x^{\frac{\alpha}{2}-1} \in L^2(D)$ and $v' \in L^2(D)$ from the assumption $v \in H^1_0(D)$, since $v \in H^1_0(D)$, we apply Theorem 2.1 to deduce

$$
({}^R_0D^{\frac{\alpha}{2}}_x x^{\alpha-1}, {}^R_xD^{\frac{\alpha}{2}}_1 v) = (c_\alpha x^{\frac{\alpha}{2}-1}, {}_xI^{1-\frac{\alpha}{2}}_1 v') = (c_{\alpha 0}I^{1-\frac{\alpha}{2}}_x x^{\frac{\alpha}{2}-1}, v') = c'_\alpha(1, v') = 0.
$$

Consequently,

$$
A(u, v) := -({}^R_0D^\alpha_x u, v) = -({}^R_0D^{\frac{\alpha}{2}}_x u, {}^R_xD^{\frac{\alpha}{2}}_1 v). \tag{5.18}
$$

Thus, the variational formulation of problem (5.12) (with $q = 0$) in the Riemann-Liouville case is to find $u \in U := H^{\frac{\alpha}{2}}_0(D)$ satisfying

$$
A(u, v) = (f, v), \quad \forall v \in U.
$$

We next consider the Djrbashian-Caputo case. Following the definition of the solution in (5.14), we choose $\beta \geq 0$ so that $\alpha + \beta \in (\frac{3}{2}, 2)$ and $f \in H^\beta_0(D)$. By Theorem 2.9, the function $g = {}_0I^\alpha_x f$ lies in $H^{\alpha+\beta}_{0,L}(D)$ and hence $g'(0) = 0$. Differentiating both sides of (5.14) and setting $x = 0$ yields the identity $u'(0) = ({}_0I^\alpha_x f)(1)$, and thus, the solution representation (5.14) can be rewritten as

$$
u = -{}_0I^\alpha_x f + x u'(0) := -g(x) + x u'(0).
$$

Meanwhile, for $v \in C^{\frac{\alpha}{2}}(\overline{D})$ with $v(1) = 0$, (5.16) implies

$$-({}_0^R D_x^\alpha g, v) = (({}_0 I_x^{2-\alpha} u)'', v) = (({}_0 I_x^{2-\alpha} g)', v'),$$

where we have again used the relation $g'(0) = 0$ in the last step. Analogous to the derivation of (5.17) and the identities ${}_0^C D_x^\alpha g = {}_0^R D_x^\alpha g$ and ${}_0^C D_x^\alpha x = 0$, we then have

$$-({}_0^C D_x^\alpha u, v) = ({}_0^R D_x^\alpha g, v) = -A(g, v) = A(u, v) - u'(0)A(x, v). \qquad (5.19)$$

The term involving $u'(0)$ cannot appear in a variational formulation in $H_0^{\frac{\alpha}{2}}(D)$. To circumvent this issue, we reverse the preceding steps and arrive at

$$A(x, v) = ({}_0 I_x^{2-\alpha} 1, v') = (\Gamma(3-\alpha))^{-1}(x^{2-\alpha}, v') = -(\Gamma(2-\alpha))^{-1}(x^{1-\alpha}, v).$$

Hence, in order to get rid of the troublesome term $-u'(0)A(x, v)$ in (5.19), we require our test functions v to satisfy $(x^{1-\alpha}, v) = 0$. Thus, the variational formulation of (5.12) in the Djrbashian-Caputo case (with $q = 0$) is to find $u \in U$ satisfying

$$A(u, v) = (f, v), \quad \forall v \in V,$$

with the test space

$$V = \left\{ \phi \in H_{0,R}^{\frac{\alpha}{2}}(D) \ : \ (x^{1-\alpha}, \phi) = 0 \right\}. \qquad (5.20)$$

Below we discuss the stability of the variational formulations. Throughout, we denote by U^* (respectively, V^*) the set of bounded linear functionals on U (respectively, V), and slightly abuse $\langle \cdot, \cdot \rangle$ for duality pairing between U and U^* (or between V and V^*). Further, we will denote by $\| \cdot \|_U$ the norm on the space U, etc.

Remark 5.3 We have seen that when f is in $H_0^s(D)$, with $\alpha + s > \frac{3}{2}$, then the solution u constructed by (5.14) satisfies the variational equation and hence coincides with the unique solution to the variational equation. This may not be the case when f is only in $L^2(D)$. Indeed, for $\alpha \in (1, \frac{3}{2})$, the function $f = x^{1-\alpha}$ is in $L^2(D)$. However, the variational solution (with $q = 0$) in this case is $u = 0$ and clearly does not satisfy the strong form of the differential equation ${}_0^C D_x^\alpha u = f$.

5.2.1.2 Variational stability in the Riemann-Liouville case

We now establish the stability of the variational formulations. The following lemma implies the variational stability in the Riemann-Liouville case with $q \equiv 0$.

Lemma 5.3 *For $\gamma \in (\frac{1}{2}, 1)$, there exists $c = c(\gamma) > 0$ satisfying*

$$c\|u\|_{H_0^\gamma(D)}^2 \leq -({}_0^R D_x^\gamma u, {}_x^R D_1^\gamma u) = A(u, u), \quad \forall u \in H_0^\gamma(D). \qquad (5.21)$$

Proof For $u \in C_0^\infty(D)$ (with $\tilde{u}$ being the zero extension of u to $\mathbb{R}$), by Plancherel's identity (A.9), there holds

$$
-({}_0^R D_x^\gamma u, {}_x^R D_1^\gamma u) = -\frac{1}{2\pi} \int_{-\infty}^\infty (i\xi)^{2\gamma} |\mathcal{F}(\tilde{u})|^2 \, d\xi
$$

$$
= \frac{\cos((1-\gamma)\pi)}{\pi} \int_0^\infty \xi^{2\gamma} |\mathcal{F}(\tilde{u})|^2 \, d\xi = \frac{\cos((1-\gamma)\pi)}{2\pi} \int_{-\infty}^\infty |\xi|^{2\gamma} |\mathcal{F}(\tilde{u})|^2 \, d\xi.
$$

Now suppose that there does not exist a constant satisfying (5.21). Then by the compactness of the embedding from $H_0^\gamma(D)$ into $L^2(D)$, there is a sequence $\{u_j\} \subset H_0^\gamma(D)$ with $\|u_j\|_{H_0^\gamma(D)} = 1$ convergent to some $u \in L^2(D)$ and satisfying

$$
\|u_j\|_{H_0^\gamma(D)}^2 > -j({}_0^R D_x^\gamma u_j, {}_x^R D_1^\gamma u_j).
$$

Thus, the sequence $\{\mathcal{F}(\tilde{u}_j)\}$ converges to zero in $\int_{-\infty}^\infty |\xi|^{2\gamma} | \cdot |^2 d\xi$, and by the convergence of the sequence $\{u_j\}$ in $L^2(D)$ and Plancherel's theorem, $\{\mathcal{F}(\tilde{u}_j)\}$ converges in $\int_{-\infty}^\infty | \cdot |^2 d\xi$. Hence, $\{\mathcal{F}(\tilde{u}_j)\}$ is a Cauchy sequence and hence converges to $\mathcal{F}(\tilde{u})$ in the norm $(\int_{-\infty}^\infty (1+\xi^2)^\gamma | \cdot |^2 d\xi)^{\frac{1}{2}}$. This implies that u_j converges to u in $H_0^\gamma(D)$. Theorem 2.10 implies $-({}_0^R D_x^\gamma u, {}_x^R D_1^\gamma u) = 0$, from which it follows that $\mathcal{F}(\tilde{u}) = \tilde{u} = 0$. This contradicts the assumption $\|u_j\|_{H_0^\gamma(D)} = 1$, completing the proof. $\square$

We now return to problem (5.12) with $q \neq 0$ and define

$$
a(u, v) = A(u, v) + (qu, v).
$$

We make the following uniqueness assumption on the bilinear form.

Assumption 5.2 Let the bilinear form $a(u, v)$ with $u, v \in U$ satisfy

(a) The problem of finding $u \in U$ such that $a(u, v) = 0$ for all $v \in U$ has only the trivial solution $u \equiv 0$.

(a*) The problem of finding $v \in U$ such that $a(u, v) = 0$ for all $u \in U$ has only the trivial solution $v \equiv 0$.

The following result, the Petree-Tartar lemma, is a useful tool; the proof can be found in [EG04, pp. 469, Lemma A.38].

Lemma 5.4 *Let X, Y and Z be Banach spaces with A a bounded linear, injective operator from $X \to Y$ and B a compact linear operator from $X \to Z$. If there exists a constant $\gamma > 0$ such that*

$$
\gamma \|x\|_X \leq \|Ax\|_Y + \|Bx\|_Z,
$$

then for some $c > 0$,

$$
c\|x\|_X \leq \|Ax\|_Y, \quad \forall x \in X.
$$

We then have the following existence result.

Theorem 5.10 *Let Assumption 5.2 hold and $q \in L^\infty(D)$. Then for any given $F \in U^*$, there exists a unique solution $u \in U$ solving*

$$a(u, v) = \langle F, v \rangle, \quad \forall v \in U. \tag{5.22}$$

Proof We define, respectively, $S : U \to U^*$ and $T : U \to U^*$ by

$$\langle Su, v \rangle = a(u, v), \quad \text{and} \quad \langle Tu, v \rangle = -(qu, v), \quad \forall v \in U.$$

Assumption 5.2(a) implies that S is injective. Further, Lemma 5.3 implies

$$\|u\|_U^2 \le cA(u, u) = c(\langle Su, u \rangle + \langle Tu, u \rangle)$$
$$\le c(\|Su\|_{U^*} + \|Tu\|_{U^*})\|u\|_U, \quad \forall u \in U.$$

Meanwhile, the compactness of T follows from the fact $q \in L^\infty(D)$ and the compactness of U in $L^2(D)$. Now Lemma 5.4 implies that there exists $\gamma > 0$ satisfying

$$\gamma\|u\|_U \le \sup_{v \in U} \frac{a(u, v)}{\|v\|_U}. \tag{5.23}$$

This together with Assumption 5.2(a*) shows that the operator $S : U \to U^*$ is bijective, i.e., there is a unique solution of (5.22), by Lemma 5.4. $\square$

We now show that the variational solution u in Theorem 5.10, in fact, is a strong solution when $\langle F, v \rangle = (f, v)$ for some $f \in L^2(D)$. We consider the problem

$$-{}_0^R D_x^\alpha w = f - qu. \tag{5.24}$$

A strong solution is given by (5.13) with a right-hand side $\widetilde{f} = f - qu$. It satisfies the variational equation and hence coincides with the unique variational solution.

Theorem 5.11 *Let Assumption 5.2 hold, and $q \in L^\infty(D)$. Then with a right-hand side $\langle F, v \rangle = (f, v)$ for some $f \in L^2(D)$, the solution u to (5.22) is in $H^{\alpha-1+\beta}(D) \cap H_0^{\frac{\alpha}{2}}(D)$ for any $\beta \in (1 - \frac{\alpha}{2}, \frac{1}{2})$, and it satisfies*

$$\|u\|_{H^{\alpha-1+\beta}} \le c\|f\|_{L^2(D)}.$$

Proof It follows from Theorem 5.10 and Assumption 5.2 that there exists a solution $u \in H_0^{\frac{\alpha}{2}}(D)$. Next we rewrite into (5.24) with $\widetilde{f} = f - qu$. Since $q \in L^\infty(D)$ and $u \in H_0^{\frac{\alpha}{2}}(D)$, we have $qu \in L^2(D)$, and hence $\widetilde{f} \in L^2(D)$. Now the desired assertion follows from (5.13), Theorem 2.9 and Corollary 2.1. $\square$

Remark 5.4 In general, the best possible regularity of the solution u to problem (5.12) with a Riemann-Liouville fractional derivative is $H^{\alpha-1+\beta}(D)$ for any $\beta \in (1 - \frac{\alpha}{2}, \frac{1}{2})$, due to the presence of the singular term $x^{\alpha-1}$. The only possibility of an improved regularity is the case $({}_0I_x^\alpha f)(1) = 0$ (for $q \equiv 0$).

Now we turn to the adjoint problem: given $F \in U^*$, find $w \in U$ such that

$$a(v, w) = \langle v, F \rangle, \quad \forall v \in U, \tag{5.25}$$

there exists a unique solution $w \in U$ to the adjoint problem. Indeed, Assumption 5.2 and (5.23) imply that the inf-sup condition for the adjoint problem holds with the same constant. Now, by proceeding as in (5.13), for $q = 0$ and $\langle F, v \rangle \equiv (f, v)$ with $f \in L^2(D)$, we have

$$w = -_x I_1^\alpha f + (_x I_1^\alpha f)(0)(1 - x)^{\alpha - 1}.$$

This implies a similar regularity pickup, i.e., $w \in H^{\alpha - 1 + \beta}(D)$, for (5.25), provided that $q \in L^\infty(D)$. Further, we can repeat the arguments in the proof of Theorem 5.11 for a general q to deduce the regularity of the adjoint solution w.

Theorem 5.12 *Let Assumption 5.2 hold, and $q \in L^\infty(D)$. Then with $\langle F, v \rangle = (f, v)$ for some $f \in L^2(D)$, the solution w to (5.25) is in $H^{\alpha - 1 + \beta}(D) \cap H_0^{\frac{\alpha}{2}}(D)$ for any $\beta \in (1 - \frac{\alpha}{2}, \frac{1}{2})$, and it satisfies*

$$\|w\|_{H^{\alpha - 1 + \beta}(D)} \leq c \|f\|_{L^2(D)}.$$

The time-dependent counterpart of problem (5.12) with a Riemann-Liouville fractional derivative was studied in [JLPZ14], where the existence of a weak solution and the solution regularity was established, and also numerical schemes were developed.

5.2.1.3 Variational stability in the Djrbashian-Caputo case

Now we consider the Djrbashian-Caputo case, which, unlike the Riemann-Liouville case, involves a test space V different from the solution space U. We set $\phi_0 = (1 - x)^{\alpha - 1}$. By Corollary 2.1, ϕ_0 is in the space $H_{0, R}^{\frac{\alpha}{2}}(D)$. Further, we observe

$$A(u, \phi_0) = 0, \quad \forall u \in U. \tag{5.26}$$

Indeed, for $u \in H_0^1(D)$, by Theorem 2.1, we have

$$A(u, (1 - x)^{\alpha - 1}) = -(_0 I_x^{1 - \frac{\alpha}{2}} u', {}_x^R D_1^{\frac{\alpha}{2}} (1 - x)^{\alpha - 1})$$
$$= c_\alpha(u', {}_x I_1^{1 - \frac{\alpha}{2}} (1 - x)^{\frac{\alpha}{2} - 1}) = c'_\alpha(u', 1) = 0.$$

Now for a given $u \in U$, we set $v = u - \gamma_u \phi_0$, where γ_u is defined by

$$\gamma_u = \frac{(x^{1 - \alpha}, u)}{(x^{1 - \alpha}, \phi_0)}. \tag{5.27}$$

Clearly, the norms $\|\cdot\|_{H^{\frac{\alpha}{2}}(D)}$ and $\|\cdot\|_{H^{\frac{\alpha}{2}}_{0,R}(D)}$ for $\alpha \in (1,2)$ are equivalent on the space $H^{\frac{\alpha}{2}}_{0,R}(D)$. Thus,

$$|\gamma_u| \leq c|(x^{1-\alpha}, u)| \leq c\|u\|_{L^\infty(D)}\|x^{1-\alpha}\|_{L^1(D)} \leq c\|u\|_{H^{\frac{\alpha}{2}}_0(D)},$$

and the function $v \in H^{\frac{\alpha}{2}}(D)$ and $v(1) = 0$, i.e., it is in the space V. By Lemma 5.3,

$$A(u,v) = A(u,u) \geq c\|u\|^2_{H^{\frac{\alpha}{2}}_0(D)}.$$

Finally, there also holds

$$\|v\|_{H^{\frac{\alpha}{2}}_{0,R}(D)} \leq \|u\|_{H^{\frac{\alpha}{2}}_0(D)} + c|\gamma_u| \leq c\|u\|_{H^{\frac{\alpha}{2}}_0(D)},$$

and thus the inf-sup condition follows immediately

$$\|u\|_{H^{\frac{\alpha}{2}}_0(D)} \leq c \sup_{v \in V} \frac{A(u,v)}{\|v\|_{H^{\frac{\alpha}{2}}_{0,R}(D)}} \quad \forall u \in U. \tag{5.28}$$

Now given any $v \in V$, we set $u = v - v(0)\phi_0$. Obviously, u is nonzero whenever $v \neq 0$ and by Lemma 5.3, we get $A(u,v) = A(u,u) > 0$. Thus, if $A(u,v) = 0$ for all $u \in U$, then $v = 0$. This and (5.28) imply that the variational problem is stable.

We next consider the case $q \not\equiv 0$ and make the following assumption.

Assumption 5.3 Let the bilinear form $a(u,v)$ with $u \in U$, $v \in V$ satisfy

(b) The problem of finding $u \in U$ such that $a(u,v) = 0$ for all $v \in V$ has only the trivial solution $u \equiv 0$.

(b*) The problem of finding $v \in V$ such that $a(u,v) = 0$ for all $u \in U$ has only the trivial solution $v \equiv 0$.

We then have the following existence result.

Theorem 5.13 *Let Assumption 5.3 hold and $q \in L^\infty(D)$. Then for any given $F \in V^*$, there exists a unique solution $u \in U$ to*

$$a(u,v) = \langle F, v \rangle, \quad \forall v \in V. \tag{5.29}$$

Proof The proof is similar to that of Theorem 5.10. In this case, we define $S : U \to V^*$ and $T : U \to V^*$ by

$$\langle Su, v \rangle = a(u,v) \quad \text{and} \quad \langle Tu, v \rangle = -(qu, v), \quad \forall v \in V.$$

By Assumption 5.3(b), S is injective. By applying (5.28), we get for any $u \in U$

$$\|u\|_U \leq c \sup_{v \in V} \frac{A(u,v)}{\|v\|_V} \leq c \sup_{v \in V} \frac{a(u,v)}{\|v\|_V} + c \sup_{v \in V} \frac{-(qu, v)}{\|v\|_V} = c(\|Su\|_{V^*} + \|Tu\|_{V^*}).$$

The rest of the proof, including verifying the inf-sup condition,

$$\gamma\|u\|_U \le \sup_{v \in V} \frac{a(u,v)}{\|v\|_U}, \quad \forall u \in U, \tag{5.30}$$

is essentially identical with that of Theorem 5.10. □

Theorem 5.14 *Let $s \in [0, \frac{1}{2})$ and Assumption 5.3 be fulfilled. Suppose that $\langle F, v \rangle = (f, v)$ for some $f \in H_0^s(D)$ with $\alpha + s > \frac{3}{2}$, and $q \in L^\infty(D) \cap H^s(D)$. Then the variational solution $u \in U$ of (5.29) is in $H_0^{\frac{\alpha}{2}}(D) \cap H^{\alpha+s}(D)$ and it is a solution of (5.12). Further, it satisfies*

$$\|u\|_{H^{\alpha+s}(D)} \le c\|f\|_{H_0^s(D)}.$$

Proof Let u be the solution of (5.29). We consider the problem

$$-{}_0^C D_x^\alpha w = f - qu. \tag{5.31}$$

Under the given assumption, $qu \in H_0^s(D)$ (exercise). A strong solution of (5.31) is given by (5.14) with a right-hand side $\tilde{f} = f - qu \in H_0^s(D)$. Since it satisfies problem (5.29), it coincides with u. The regularity $u \in H^{\alpha+s}(D)$ is direct from Theorem 2.9. □

Remark 5.5 The solution to problem (5.12) in the Djrbashian-Caputo case can achieve full regularity, which contrasts sharply with the Riemann-Liouville case since for the latter, generally the best possible regularity is $H^{\alpha-1+\beta}(D)$, for any $\beta \in (1 - \frac{\alpha}{2}, \frac{1}{2})$, due to the presence of the singular term $x^{\alpha-1}$.

Last we discuss the adjoint problem: find $w \in V$ such that

$$a(v, w) = \langle v, F \rangle, \quad \forall v \in U,$$

for some $F \in U^*$. If $\langle F, v \rangle = (f, v)$ for some $f \in L^2(D)$, the strong form reads

$$-{}_x^R D_1^\alpha w + qw = f,$$

with $w(1) = 0$ and $(x^{1-\alpha}, w) = 0$. By repeating the steps leading to (5.14), we deduce that for $q \equiv 0$, the solution w can be expressed as $w = c_f(1-x)^{\alpha-1} - {}_xI_1^\alpha f$, with the prefactor c_f given by

$$c_f = \frac{(x^{1-\alpha}, {}_xI_1^\alpha f)}{(x^{1-\alpha}, (1-x)^{\alpha-1})} = \frac{({}_0I_x^\alpha x^{1-\alpha}, f)}{B(2 - \alpha, \alpha)} = \frac{1}{\Gamma(\alpha)}(x, f).$$

Clearly, there holds $|c_f| \le c|(x^{1-\alpha}, {}_xI_1^\alpha f)| \le c\|f\|_{L^2(D)}$. Hence, $w \in H_0^{\frac{\alpha}{2}}(D) \cap H^{\alpha-1+\beta}(D)$, for any $\beta \in (1-\frac{\alpha}{2}, \frac{1}{2})$. The case of a general q can be deduced analogously to the proof of Theorem 5.11. Hence, we have the following improved regularity estimate for the adjoint solution w.

Theorem 5.15 *Let Assumption 5.3 hold, and $q \in L^\infty(D)$. Then with a right-hand side $\langle F, v \rangle = (f, v)$ for some $f \in L^2(D)$, the solution w to (5.29) is in $H_0^{\frac{\alpha}{2}}(D) \cap H^{\alpha-1+\beta}(D)$ for any $\beta \in (1 - \frac{\alpha}{2}, \frac{1}{2})$, and further there holds*

$$\|w\|_{H^{\alpha-1+\beta}(D)} \leq c\|f\|_{L^2(D)}.$$

Remark 5.6 The adjoint problem for both Djrbashian-Caputo and Riemann-Liouville cases is of Riemann-Liouville type, albeit with slightly different boundary conditions, and thus shares the same singularity.

Example 5.4 Consider problem (5.12) with $q(x) = x$ and $f \equiv 1$. The solution u is shown in Fig. 5.5 for the Riemann-Liouville and Djrbashian-Caputo case separately. Clearly, in the Riemann-Liouville case, the solution u has a weak singularity at $x = 0$, which contrasts sharply with the Djrbashian-Caputo case. Interestingly, the change of the solution with respect to the fractional-order α also differs markedly: in the Riemann-Liouville case, the magnitude of the solution u decreases with α, and this trend is reversed in the Djrbashian-Caputo case.

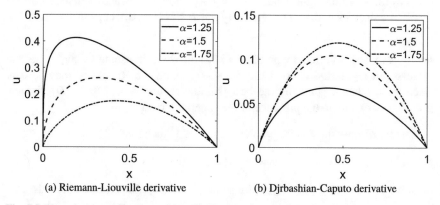

(a) Riemann-Liouville derivative   (b) Djrbashian-Caputo derivative

Fig. 5.5 The solution profiles for problem (5.12) with $q = x$ and $f = 1$.

Last, the nonstationary counterpart of problem (5.12) with a Djrbashian-Caputo fractional derivative is less studied: the existence and uniqueness of the solutions was studied in [NR20] via the concept of viscosity solutions.

5.2.2 Two-sided mixed fractional derivatives

In this part, we describe some results for the two-sided problem

$$\begin{cases} L_r^\alpha u = f, & \text{in } D, \\ u(0) = u(1) = 0. \end{cases} \tag{5.32}$$

The associated differential operator L_r^α is defined by

$$L_r^\alpha u(x) = D(r_0 I_x^{2-\alpha} + (1-r)_x I_1^{2-\alpha}) Du(x),$$

where $r \in [0, 1]$, and D denotes taking the first-order derivative. The operator L_r^α is closely related to but not equivalent to a linear combination of the Riemann-Liouville or Djrbashian-Caputo fractional derivative. Putting D inside the fractional integral, introduced by Patie and Simon [PS12, p. 570], changes the kernel of the operator, and thus also the regularity of the solution. However, when considering a zero Dirichlet boundary condition, which we discuss below, the equivalence to the Riemann-Liouville fractional derivative does hold. Problems of this type were first studied in [ER06], where the variational formulation was derived and finite element approximation was developed. The behavior of the solution of this problem is drastically different from either Riemann-Liouville or Djrbashian-Caputo case, in the sense, the solution generally exhibits weak singularity at both end points, a fact only established very recently [EHR18], which establishes the kernel of the operator. Two-sided BVPs were studied in [HHK16] in a probabilistic framework.

Following the preceding procedure, we can derive the weak formulation. With $U = H_0^{\frac{\alpha}{2}}(D)$, the weak formulation of the Dirichlet BVP reads: given $f \in H^{-\frac{\alpha}{2}}(D)$, find $u \in U$ such that

$$B(u, v) = \langle f, v \rangle, \quad \forall v \in U,$$

where the bilinear form $B(\cdot, \cdot) : U \times U \to \mathbb{R}$ is defined by

$$B(u, v) := r({}_0^R D_x^{\frac{\alpha}{2}} u, {}_x^R D_1^{\frac{\alpha}{2}} v) + (1-r)({}_x^R D_1^{\frac{\alpha}{2}} u, {}_0^R D_x^{\frac{\alpha}{2}} v).$$

By Lemma 5.3, the bilinear form B is coercive and continuous on U, and thus the weak formulation has a unique solution $u \in U$. Below we analyze the kernel of the operator L_r^α, following [EHR18], but replacing some computer verification with direct evaluation. The analysis uses the three-parameter hypergeometric function.

Definition 5.1 The Gauss three-parameter hypergeometric function $_2F_1(a, b; c; x)$ is defined by an integral and series as follows:

$$_2F_1(a, b; c; x) = \frac{\Gamma(c)}{\Gamma(b)\Gamma(c-b)} \int_0^1 s^{b-1}(1-s)^{c-b-1}(1-sx)^{-a} ds$$

$$= \sum_{n=0}^{\infty} \frac{(a)_n (b)_n}{(c)_n} \frac{x^n}{n!},$$

with convergence only if $\mathfrak{R}(c) > \mathfrak{R}(b) > 0$, where $(a)_n$ denotes the rising Pochhammer symbol, i.e., $(a)_n = \frac{\Gamma(a+n)}{\Gamma(a)}$.

It is clear from the series definition that the interchange property holds.

Lemma 5.5 For $\mathfrak{R}(c) > \mathfrak{R}(b) > 0$ and $\mathfrak{R}(c) > \mathfrak{R}(a) > 0$, there holds

$$_2F_1(a, b; c; x) = {}_2F_1(b, a; c; x).$$

Further, the following useful identities hold: the first Bolz formula (see, e.g., [AS65, p. 559, (15.3.3)])

$$_2F_1(a, b; c; x) = (1 - x)^{c-a-b} {}_2F_1(c - a, c - b; c; x), \tag{5.33}$$

for $a + b - c \neq 0, \pm 1, \pm 2, \ldots$, and reflection-type formula (see, e.g., [AS65, p. 559. (15.3.6)], [WW96, p. 291])

$$
\begin{aligned}
_2F_1(a, b; c; 1 - x) &= \frac{\Gamma(c)\Gamma(c - a - b)}{\Gamma(c - a)\Gamma(c - b)} {}_2F_1(a, b; a + b - c + 1; x) \\
&+ x^{c-a-b} \frac{\Gamma(c)\Gamma(a + b - c)}{\Gamma(a)\Gamma(b)} {}_2F_1(c - a, c - b; 1 + c - a - b; x).
\end{aligned}
\tag{5.34}
$$

Next, we give the kernel of the operator with a general weight $r \in (0, 1)$.

Lemma 5.6 *The function $k(x) = x^p (1 - x)^q$, where $p, q \in [\alpha - 2, 0)$ satisfy*

$$p + q = \alpha - 2 \quad and \quad r \sin q\pi = (1 - r) \sin p\pi. \tag{5.35}$$

is in the kernel of the operator $D(r {}_0 I_x^{2-\alpha} + (1 - r)_x I_1^{2-\alpha})$.

Proof By changing variables $s = zx$, we have

$$
\begin{aligned}
{}_0 I_x^{2-\alpha} k(x) &= \frac{1}{\Gamma(2 - \alpha)} \int_0^x (x - s)^{1-\alpha} s^p (1 - s)^q ds \\
&= \frac{1}{\Gamma(2 - \alpha)} x^{2-\alpha+p} \int_0^1 (1 - z)^{1-\alpha} z^p (1 - zx)^q dz \\
&= x^{2-\alpha+p} \Gamma(p + 1) \Gamma(3 - \alpha + p)^{-1} {}_2F_1(-q, p + 1; 3 - \alpha + p; x),
\end{aligned}
$$

where the last line follows from the fact $3 - \alpha + p > p + 1$, since $\alpha \in (1, 2)$. By Lemma 5.5, under the condition $3 - \alpha + p > -q > 0$, we have

$$
\begin{aligned}
{}_0 I_x^{2-\alpha} k(x) &= \frac{\Gamma(p + 1)}{\Gamma(3 - \alpha + p)} x^{2-\alpha+p} {}_2F_1(p + 1, -q; 3 - \alpha + p; x) \\
&= \frac{\Gamma(p + 1)}{\Gamma(-q)\Gamma(3 - \alpha + p + q)} x^{2-\alpha+p} \int_0^1 (1 - z)^{2-\alpha+p+q} z^{-q-1} (1 - zx)^{-p-1} dz \\
&= \frac{\Gamma(p + 1)}{\Gamma(-q)\Gamma(3 - \alpha + p + q)} \int_0^x (x - s)^{2-\alpha+p+q} s^{-q-1} (1 - s)^{-p-1} ds \\
&= \Gamma(p + 1)\Gamma(-q)^{-1} {}_0 I_x^{3-\alpha+p+q} x^{-q-1} (1 - x)^{-p-1}.
\end{aligned}
$$

Similarly, one has

$$_x I_1^{2-\alpha} k(x) = \Gamma(q + 1)\Gamma(-p)^{-1} {}_x I_1^{3-\alpha+p+q} x^{-q-1} (1 - x)^{-p-1}. \tag{5.36}$$

Comparing the last two identities with $D(r {}_0 I_x^{2-\alpha} + (1 - r)_x I_1^{2-\alpha}) k(x) = 0$ implies $2 - \alpha + p + q = 0$ and $r\Gamma(p + 1)\Gamma(-q)^{-1} = (1 - r)\Gamma(q + 1)\Gamma(-p)^{-1} \Leftrightarrow r\Gamma(-p)\Gamma(1 -$

$(-p)) = (1-r)\Gamma(-q)\Gamma(1-(-q))$. By the reflection formula (2.3) for $\Gamma(z)$, $r\frac{\pi}{\sin(-p\pi)} = (1-r)\frac{\pi}{\sin(-q\pi)}$. This shows the desired assertion. $\qquad\square$

Remark 5.7 Given $0 \leq r < 1$ and $1 < \alpha < 2$, the exponents p and q satisfying (5.35) can be equivalently expressed as $q = \alpha - 2 - p$ and $\alpha - 2 \leq p \leq 0$ such that $h(p) := r - \sin p\pi(\sin(\alpha - p)\pi + \sin p\pi)^{-1} = 0$. The existence and uniqueness of p follows from the fact that $h(\alpha - 2) = r - 1 \leq 0$, $h(0) = r \geq 0$ and $h'(p) > 0$.

Remark 5.8 When $r = \frac{1}{2}$, $p = q = \frac{\alpha}{2} - 1$, and when $r = 1$, $q = 0$ and $p = \alpha - 2$. Then $k(x)$ is given by $k(x) = x^{\alpha-2}$ and thus $K(x) = x^{\alpha-1}$, which indeed belongs to the kernel of ${}_0^R D_x^{\alpha-1}$.

The next result characterizes the kernel of the operator L_r^α.

Corollary 5.2 *The kernel* $\ker(L_r^\alpha)$ *of* L_r^α *is given by* $\mathrm{span}\{1, K(x)\}$, *with* $K(x) = \int_0^x k(s)ds = (p+1)^{-1}x^{p+1}{}_2F_1(-q, p+1; p+2; x)$, *where* $k(s)$ *is given in Lemma 5.6.*

Proof By Lemma 5.6, $\mathrm{span}\{1, K(x)\} \subset \ker(L_r^\alpha)$. It suffices to prove $\dim(\ker(L_r^\alpha)) = 2$. With $z(x) = 1 + x$ and $f(x) = \Gamma(2 - \alpha)^{-1}(-rx^{1-\alpha} + (1 - r)(1 - x)^{1-\alpha})$, then there holds $L_r^\alpha x = f(x)$ in D. Since $K(1) \neq 0$, we can choose c_1, c_2 such that $\hat{z}(x) = z(x) + c_1 + c_2 K(x)$ satisfies $\hat{z}(0) = \hat{z}(1) = 0$ and $L_r^\alpha \hat{z}(x) = f(x)$. Now suppose there is a third linearly independent function $s(x) \in \ker(L_r^\alpha)$, for which we may assume $s(0) = s(1) = 0$. Then $\tilde{z}(x) := \hat{z}(x) + s(x)$ satisfies $L_r^\alpha \tilde{z}(x) = f(x)$. However, the existence of $\tilde{z}(x) \neq \hat{z}(x)$ contradicts the uniqueness of the solution to $L_r^\alpha u(x) = f(x)$ with $u(0) = u(1) = 0$ (from Lax-Milgram theorem). $\qquad\square$

Remark 5.9 The regularity of $K(x) \in H^{\min(p,\alpha-p)+\frac{1}{2}-\epsilon}(D)$ (with small $\epsilon > 0$), which can be less regular than the Riemann-Liouville case, i.e., a worsening regularity as the ratio r decreases from 1 to $\frac{1}{2}$. However, one may obtain better regularity in weighted Sobolev spaces; see the work [Erv21], which discuss also the influence of a convective term.

Lemma 5.7 *For* $1 \leq \alpha < \frac{3}{2}$, $D_0 I_x^{2-\alpha} D$ *maps from* $H^\alpha(D)$ *onto* $L^2(D)$.

Proof We have $D : H^\alpha(D) \to H^{\alpha-1}(D)$. For $\alpha \in [1, \frac{3}{2})$, $H^{\alpha-1}(D) = H_0^{\alpha-1}(D)$. Since $D_0 I_x^{2-\alpha} = {}_0^R D_x^{\alpha-1}$, by Theorem 2.10, $D_0 I_x^{2-\alpha} : H_0^{\alpha-1}(D) \to L^2(D)$. To show "onto", note that for $f \in L^2(D)$, $D_0 I_x^{2-\alpha} D u = f$, with $u = {}_0 I_x^\alpha f$. $\qquad\square$

The next result gives a concise description of the range of L_r^α, with domain $H^\alpha(D)$. Note that for $\alpha \in [1, \frac{3}{2})$, $\mathrm{span}(\{x^{1-\alpha}\}) \subset L^2(D)$, and $\mathrm{span}(\{(1 - x)^{1-\alpha}\}) \subset L^2(D)$.

Proposition 5.2 *For* $1 < \alpha < 2$, L_r^α *maps from* $H^\alpha(D)$ *into* $L^2(D) \oplus \mathrm{span}(\{x^{1-\alpha}\}) \oplus \mathrm{span}(\{(1 - x)^{1-\alpha}\})$.

Proof The case $1 < \alpha < \frac{3}{2}$ is covered by Lemma 5.7. For $f \in H^\alpha(D)$, $\alpha \geq \frac{3}{2}$, let $p(x)$ denote the Hermite cubit interpolant of f, and $\tilde{f}(x) = f(x) - p(x) \in H_0^\alpha(D)$. By Theorem 2.10, $L_r^\alpha \tilde{f} \in L^2(D)$. Further, direct computation shows $L_r^\alpha p(x) \in L^2(D) \oplus \mathrm{span}(\{x^{1-\alpha}\}) \oplus \mathrm{span}(\{(1 - x)^{1-\alpha}\})$. This shows the desired assertion. $\qquad\square$

Example 5.5 Consider problem (5.32) with L_r^α, $q \equiv 0$ and $f \equiv 1$. The solution u is shown in Fig. 5.6. The shape is similar to the Riemann-Liouville case in that it has weak singularity at one end point, depending on the weight r. This partly confirms the analysis of the kernel of L_r^α, but the precise regularity is to be ascertained.

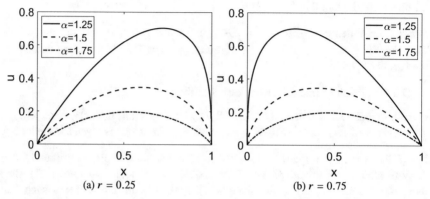

(a) $r = 0.25$ (b) $r = 0.75$

Fig. 5.6 The solutions for problem (5.32) with $q = 0$ and $f = 1$.

Last, we provide an eigen-type expansion.

Proposition 5.3 *For $1 < \alpha < 2$, $0 < \beta < \alpha$, and $r = \frac{\sin\beta\pi}{\sin\beta\pi+\sin(\alpha-\beta)\pi} \in [0,1]$,*

$$L_r^\alpha x^\beta (1-x)^{\alpha-\beta} x^n = \sum_{j=0}^{n} a_{n,j} x^j, \quad n = 0, 1, 2, \ldots,$$

with the coefficients $a_{n,j}$ given by

$$a_{n,j} = (-1)^{(n+1)}(1-r)\frac{\sin\alpha\pi}{\sin(\alpha-\beta)\pi}\frac{(-1)^j\Gamma(1+\alpha-\beta)\Gamma(1+\alpha+j)}{\Gamma(1+\alpha-\beta-n+j)\Gamma(1+n-j)\Gamma(j+1)}.$$

Proof Let $u(x) = x^{\beta+n}(1-x)^{\alpha-\beta}$. By the definition of $_2F_1(a,b;c;x)$,

$$_0I_x^{2-\alpha}u(x) = \frac{1}{\Gamma(2-\alpha)}\int_0^x (x-s)^{1-\alpha}s^{\beta+n}(1-s)^{\alpha-\beta}ds$$

$$= \frac{1}{\Gamma(2-\alpha)}\int_0^1 (x-xt)^{1-\alpha}(xt)^{\beta+n}(1-xt)^{\alpha-\beta}x\,dt \quad [\text{with } s = tx]$$

$$= \frac{1}{\Gamma(2-\alpha)}x^{2+n+\beta-\alpha}\int_0^1 t^{\beta+n}(1-t)^{1-\alpha}(1-xt)^{\alpha-\beta}dt$$

$$= \frac{\Gamma(n+\beta+1)}{\Gamma(n+3+\beta-\alpha)}x^{n+2+\beta-\alpha}{}_2F_1(\beta-\alpha,\beta+n+1;n+3+\beta-\alpha;x).$$

Next, we evaluate the integral $_xI_1^{2-\alpha}u(x)$. By changing variables $s = 1 - (1-x)t$,

$$_xI_1^{2-\alpha}u(x) = \frac{1}{\Gamma(2-\alpha)}\int_x^1 (s-x)^{1-\alpha}s^{\beta+n}(1-s)^{\alpha-\beta}ds$$

$$= \frac{1}{\Gamma(2-\alpha)}\int_0^1 [(1-x)(1-t)]^{1-\alpha}(1-(1-x)t)^{\beta+n}((1-x)t)^{\alpha-\beta}(1-x)dt$$

$$= \frac{1}{\Gamma(2-\alpha)}(1-x)^{2-\beta}\int_0^1 t^{\alpha-\beta}(1-t)^{1-\alpha}(1-(1-x)t)^{\beta+n}dt$$

$$= \frac{\Gamma(\alpha-\beta+1)}{\Gamma(3-\beta)}(1-x)^{2-\beta}{}_2F_1(-\beta-n,\alpha-\beta+1;3-\beta;1-x).$$

Below we apply (5.33) and (5.34) to rewrite the formula. By (5.34), we have

$$_2F_1(-\beta-n,\alpha-\beta+1;3-\beta;1-x)$$
$$= \frac{\Gamma(3-\beta)\Gamma(2+n-\alpha+\beta)}{\Gamma(3+n)\Gamma(2-\alpha)}{}_2F_1(-\beta-n,\alpha-\beta+1;-n-1+\alpha-\beta;x)$$
$$+ x^{n+2-\alpha+\beta}\frac{\Gamma(3-\beta)\Gamma(-n-2+\alpha-\beta)}{\Gamma(-\beta-n)\Gamma(\alpha-\beta+1)}{}_2F_1(n+3,2-\alpha;n+3-\alpha+\beta;x).$$

Now, by the first Bolz formula (5.33), we have

$$_2F_1(3+n,2-\alpha,3+n-\alpha+\beta;x)$$
$$= (1-x)^{-2+\beta}{}_2F_1(-\alpha+\beta,n+1+\beta,n+3-\alpha+\beta;x),$$

and

$$_2F_1(-n-\beta,\alpha-\beta+1,-n-1+\alpha-\beta;x)$$
$$= (1-x)^{-2+\beta}{}_2F_1(\alpha-1,-n-2,-n-1+\alpha-\beta;x).$$

Combining the preceding identities, we obtain

$$_xI_1^{2-\alpha}u(x) = \frac{\Gamma(\alpha-\beta+1)\Gamma(n+2-\alpha+\beta)}{\Gamma(2-\alpha)\Gamma(n+3)}{}_2F_1(\alpha-1,-n-2;-n-1+\alpha-\beta;x)$$
$$+ \frac{\Gamma(-n-2+\alpha-\beta)}{\Gamma(-\beta-n)}x^{n+2+\beta-\alpha}{}_2F_1(-\alpha+\beta,n+1+\beta;n+3-\alpha+\beta;x).$$

Consequently,

$$(r_0I_x^{2-\alpha}+(1-r)_xI_1^{2-\alpha})u(x)$$
$$= \left[r\frac{\Gamma(n+\beta+1)}{\Gamma(n+3+\beta-\alpha)}+(1-r)\frac{\Gamma(-n-2+\alpha-\beta)}{\Gamma(-\beta-n)}\right]$$
$$\times x^{n+2+\beta-\alpha}{}_2F_1(\beta-\alpha,\beta+n+1;n+3+\beta-\alpha;x)$$
$$+ (1-r)\frac{\Gamma(\alpha-\beta+1)\Gamma(n+2-\alpha+\beta)}{\Gamma(2-\alpha)\Gamma(n+3)}{}_2F_1(\alpha-1,-n-2;-n-1+\alpha-\beta;x).$$

Now we simplify the term in the square bracket, denoted by I:

$$I = \frac{r\Gamma(n + \beta + 1)\Gamma(-\beta - n) + (1 - r)\Gamma(-n - 2 + \alpha - \beta)\Gamma(n + 3 + \beta - \alpha)}{\Gamma(n + 3 + \beta - \alpha)\Gamma(-\beta - n)}.$$

By the reflection formula (2.3) for the Gamma function,

$$\Gamma(-\beta - n)\Gamma(n + \beta + 1) = \frac{\pi}{\sin(-\beta - n)\pi} = \frac{(-1)^{n+1}\pi}{\sin \beta\pi},$$

$$\Gamma(-2 - n + \alpha - \beta)\Gamma(n + 3 + \beta - \alpha) = \frac{\pi}{\sin(-2 - n + \alpha - \beta)\pi} = \frac{(-1)^{n+2}\pi}{\sin(\alpha - \beta)\pi}.$$

It follows from these two identities and the choice of β that

$$I = \frac{1}{\Gamma(n + 3 + \beta - \alpha)\Gamma(-\beta - n)}\left[r\frac{(-1)^{n+1}\pi}{\sin \beta\pi} + (1 - r)\frac{(-1)^{n+2}\pi}{\sin(\alpha - \beta)\pi}\right]$$

$$= \frac{(-1)^{n+1}\pi}{\Gamma(n + 3 + \beta - \alpha)\Gamma(-\beta - n)}\left[\frac{r}{\sin \beta\pi} - \frac{1 - r}{\sin(\alpha - \beta)\pi}\right] = 0.$$

Meanwhile,

$$\frac{\Gamma(1 + \alpha - \beta)\Gamma(n + 2 - \alpha + \beta)}{\Gamma(2 - \alpha)\Gamma(n + 3)}{}_2F_1(\alpha - 1, -n - 2; -n - 1 + \alpha - \beta; x)$$

$$= \frac{\Gamma(1 + \alpha - \beta)\Gamma(n + 2 - \alpha + \beta)}{\Gamma(2 - \alpha)\Gamma(n + 3)}\sum_{k=0}^{n+2}\frac{(\alpha - 1)_k(-n - 2)_k}{(-n - 1 + \alpha - \beta)_k}x^k$$

$$= \frac{\Gamma(1 + \alpha - \beta)\Gamma(n + 2 - \alpha + \beta)\Gamma(-n - 1 + \alpha - \beta)}{\Gamma(2 - \alpha)\Gamma(\alpha - 1)\Gamma(n + 3)}\sum_{k=0}^{n+2}\frac{\Gamma(\alpha - 1 + k)(-n - 2)_k}{\Gamma(-n - 1 + \alpha - \beta + k)}x^k.$$

Now using the reflection formula for the Gamma function,

$$\Gamma(n + 2 - \alpha + \beta)\Gamma(-n - 1 + \alpha - \beta) = \frac{\pi}{\sin(-n - 1 + \alpha - \beta)\pi} = \frac{(-1)^{n+1}\pi}{\sin(\alpha - \beta)\pi},$$

$$\Gamma(2 - \alpha)\Gamma(\alpha - 1) = \frac{\pi}{\sin(\alpha - 1)\pi} = \frac{-\pi}{\sin \alpha\pi}.$$

Collecting the identities leads to

$$\frac{\Gamma(1 + \alpha - \beta)\Gamma(n + 2 - \alpha + \beta)}{\Gamma(2 - \alpha)\Gamma(n + 3)}{}_2F_1(\alpha - 1, -n - 2; -n - 1 + \alpha - \beta; x)$$

$$= \frac{\Gamma(1 + \alpha - \beta)(-1)^n \sin \alpha\pi}{\sin(\alpha - \beta)\pi}\sum_{k=0}^{n+2}\frac{\Gamma(\alpha - 1 + k)(-n - 2)_k}{\Gamma(-n - 1 + \alpha - \beta + k)\Gamma(n + 3)}x^k.$$

Differentiating both sides gives the desired assertion.   □

5.3 Fractional Sturm-Liouville Problem

The classical Sturm-Liouville problem plays a fundamental role in many areas. A lot is known about Sturm-Liouville theory and indeed the picture was almost complete by the middle of the nineteenth century. The eigenvalues are real, the eigenfunctions are simple and those corresponding to distinct eigenvalues are mutually orthogonal in $L^2(D)$. The zeros of eigenfunctions arising from successive eigenvalues strictly interlace, and there are various monotonicity theorems relating the spectrum to the coefficients and boundary conditions. None of these results in the above generality are known in the case of Riemann-Liouville /Djrbashian-Caputo fractional derivatives, and in many instances the conclusions are false. In this section, we discuss the fractional case:

$$\begin{cases} -{}_0D_x^\alpha u + qu = \lambda u, & \text{in } D, \\ u(0) = u(1) = 0, \end{cases} \tag{5.37}$$

where the potential $q \in L^\infty(D)$. The two fractional derivatives not only share a lot of similarities but also some big differences, and thus we discuss them separately. Since the operator ${}_0D_x^\alpha$ is not self-adjoint, generally, the eigenvalues and the eigenfunctions are genuinely complex. In particular, if $\lambda \in \mathbb{C}$ is an eigenvalue of $-{}_0D_x^\alpha + q$, then its complex conjugate $\bar{\lambda}$ is also an eigenvalue. We collect several known analytical results, for the case of $q \equiv 0$, using the Green's function in Section 5.1. The eigenvalue problem (5.37) with a nonzero q appears analytically unexplored, although extensive numerical investigations were presented in [JR12].

Djrbashian [Džr70] probably first considered Dirichlet-type problems for FDES. The first such problem is to find the solution $u(x)$ (in $L^1(D)$ or $L^2(D)$) of the BVP

$$\begin{cases} {}_0D_x^{\sigma_2}u(x) - [\lambda + q(x)]u(x) = 0, & \text{in } D, \\ \cos\phi_0({}_0D_x^{\sigma_0}u)(0) + \sin\phi_0({}_0D_x^{\sigma_1}u)(0) = 0, \\ \cos\phi_1({}_0D_x^{\sigma_0}u)(1) + \sin\phi_1({}_0D_x^{\sigma_1}u)(1) = 0, \end{cases}$$

with a Lipschitz potential $q(x)$, and $\phi_0, \phi_1 \in [0, \pi)$. For the set $\{\gamma_i\}_{i=0}^2 \subset (0, 1]$, the Djrbashian fractional derivative ${}_0D_x^{\sigma_2}$ is defined in Exercise 2.19. Let $\sigma_k = \sum_{j=0}^k \gamma_k - 1$, $k = 0, 1, 2$, and assume $\sigma_2 > 1$. Then the operators ${}_0D_x^{\sigma_i}$ are defined by

$$_0D_x^{\sigma_0}u = {}_0I_x^{1-\gamma_0}u, \quad {}_0D_x^{\sigma_1}u = {}_0I_x^{1-\gamma_1R}{}_0D_x^{\gamma_0}u, \quad {}_0D_x^{\sigma_2}u = {}_0I_x^{1-\gamma_2R}{}_0D_x^{\gamma_1R}{}_0D_x^{\gamma_0}u.$$

Djrbashian defined the function $\omega(\lambda) = \cos\phi_1({}_0D_x^{\sigma_0}u)(1; \lambda) + \sin\phi_1({}_0D_x^{\sigma_1}u)(1; \lambda)$, where $u(x; \lambda)$ solves the Cauchy-type problem

$$\begin{cases} {}_0D_x^{\sigma_2}u(x) - [\lambda + q(x)]u(x) = 0, & \text{in } D, \\ ({}_0D_x^{\sigma_0}u)(0) = \sin\phi_0, \\ ({}_0D_x^{\sigma_1}u)(0) = -\cos\phi_0. \end{cases}$$

Djrbashian showed that $\omega(\lambda)$ is an entire function in λ of order σ_2^{-1}, and proved that if $\lambda = \lambda_0$ is a zero of the function $\omega(\lambda)$, then $u(x; \lambda_0)$ is an eigenfunction of the Dirichlet problem corresponding to this eigenvalue. The work [Gal80] proved the completeness of the system of the eigen and associated functions in the case $q \equiv 0$.

Note that for the fractional Sturm-Liouville problem, several researchers [KA13, KOM14, ZK13, TT16] have focused on retrieving as much as possible from the results of the classical case, in particular seeking conditions on the types of fractional operators (and boundary conditions) that lead to real eigenvalues and mutually orthogonal eigenfunctions. The one-sided case as (5.37) is far less understood.

5.3.1 Riemann-Liouville case

The following theorem summarizes several results on problem (5.37) with $q \equiv 0$ in the Riemann-Liouville case. (iii) and (iv) are taken from [AKT13, Theorems 2.1 and 2.3] and (ii) from [JLPR15, appendix]. The work [JLPR15] also gives a variational approach for efficiently approximating the eigenvalues and eigenfunctions.

Theorem 5.16 *In the Riemann-Liouville case, the following statements hold for problem (5.37) with $q \equiv 0$.*

(i) *The eigenvalues $\{\lambda_n\}_{n\geq 1}$ are zeros of the function $E_{\alpha,\alpha}(-\lambda)$, and the corresponding eigenfunctions are given by $x^{\alpha-1}E_{\alpha,\alpha}(-\lambda_n x^\alpha)$.*

(ii) *The lowest Dirichlet eigenvalue is real and positive, and the associated eigenfunction can be taken to be strictly positive in D.*

(iii) *All eigenvalues lie in the sector $\{z \in \mathbb{C} : |\arg z| < \frac{2-\alpha}{2}\pi\}$.*

(iv) *There is no eigenvalue inside the circle with a radius $\frac{\Gamma(2\alpha)}{\Gamma(\alpha)}$ centered at the origin.*

Proof (i). Consider the following ODE for $\mu \in \mathbb{C}$:

$$\begin{cases} -{}_0^R D_x^\alpha v = \mu v, & \text{in } D, \\ v(0) = 0, & {}_0^R D_x^{\alpha-1} v(0) = 1. \end{cases} \tag{5.38}$$

In view of the Dirichlet boundary condition $u(0) = u(1) = 0$ in problem (5.37), we choose $v(0) = 0$, and ${}_0^R D_x^{\alpha-1} v(0) = 1$ (but other normalizing conditions are possible). Then according to the theory of fractional ODEs in Proposition 4.4, for any $\mu \in \mathbb{C}$, problem (5.38) has a unique solution $v_\mu(x)$, given by $v_\mu(x) = x^{\alpha-1}E_{\alpha,\alpha}(-\mu x^\alpha)$. $v_\mu(1) = 0$ if and only if μ is a zero of the Mittag-Leffler function $E_{\alpha,\alpha}(-\lambda)$, and the corresponding solution $x^{\alpha-1}E_{\alpha,\alpha}(-\mu x^\alpha)$ is precisely an eigenfunction.

(ii). Let the operator $T : C(\overline{D}) \to C(\overline{D})$ be defined by

$$Tf(x) = \int_0^1 G(x, s)f(s)ds = ({}_0I_x^\alpha f)(1)x^{\alpha-1} - {}_0I_x^\alpha f(x), \tag{5.39}$$

where $G(x, s)$ is the Green's function given in Theorem 5.1. Clearly, T is linear and, further it is compact. Let $P = \{v \in C(\overline{D}) : v \geq 0 \text{ in } D\}$. We claim that T is positive on P. Indeed, let $f \in C(\overline{D})$, and $f \geq 0$. Then

$$Tf(x) = \frac{x^{\alpha-1}}{\Gamma(\alpha)} \int_x^1 (1-s)^{\alpha-1} f(s)ds + \frac{1}{\Gamma(\alpha)} \int_0^x ((x-xs)^{\alpha-1} - (x-s)^{\alpha-1}) f(s)ds.$$

For any $x \in D$, the first term is nonnegative. Similarly, since $(x-xs)^{\alpha-1} > (x-s)^{\alpha-1}$ for $s \in (0, x)$, the second term is also nonnegative. Hence, $Tf \in P$, i.e., the operator T is positive. Now by Krein-Rutman theorem [Dei85, Theorem 19.2], the spectral radius of T is an eigenvalue of T, and there exists an eigenfunction $u \in P \setminus \{0\}$.

(iii). It suffices to analyze the expression $-({}_0^R D_x^\alpha u, u)$. Note that $-({}_0^R D_x^\alpha u, u) = ({}_0 I_x^{2-\alpha} u', u')$. By the Macaev-Palant theorem [MP62] (see also [GK70, p. 402, footnote (46)]), the values of the quadratic form $({}_0 I_x^{2-\alpha} u', u')$ lies in the sector $\{z \in \mathbb{C} : |\arg z| < \frac{2-\alpha}{2}\pi\}$.

(iv). By Theorem 5.1, problem (5.37) is equivalent to the following integral equation $u(x) = \lambda Tu$, with T defined in (5.39), and by part (i), a number $\lambda \in \mathbb{C}$ is an eigenvalue of problem (5.37) if and only if it is a zero of $E_{\alpha,\alpha}(-\lambda)$. Now we decompose the operator Tu into $Tu = -T_0 u + T_1 u$ with $T_0 u = {}_0 I_x^\alpha u$ and $T_1 u = x^{\alpha-1}({}_0 I_x^\alpha u)(1)$. For any $\alpha \in (1, 2)$, the operators T_0 and T_1 are trace-class. Hence, the trace of $\mathrm{tr}(T)$ the operator T is given by $\mathrm{tr}(T) = \mathrm{tr}(-T_0 + T_1) = -\mathrm{tr}(T_0) + \mathrm{tr}(T_1)$. Moreover, we have

$$\mathrm{tr}(T_0) = 0 \quad \text{and} \quad \mathrm{tr}(T_1) = \frac{1}{\Gamma(\alpha)} \int_0^1 s^{\alpha-1}(1-s)^{\alpha-1}ds = \frac{\Gamma(\alpha)}{\Gamma(2\alpha)}.$$

Hence, $\mathrm{tr}(T) = \frac{\Gamma(\alpha)}{\Gamma(2\alpha)}$, i.e., $\lambda_1^{-1} + \sum_{i=2}^\infty \lambda_i^{-1} = \frac{\Gamma(\alpha)}{\Gamma(2\alpha)}$. By (ii), λ_1^{-1} is a positive number. By the sector property of the roots of $E_{\alpha,\alpha}(-x)$ (cf. (iii)), $\sum_{i=2}^\infty \lambda_i^{-1}$ is also a positive number. Therefore, $\lambda_1^{-1} < \frac{\Gamma(\alpha)}{\Gamma(2\alpha)}$. That is, the function $E_{\alpha,\alpha}(-\lambda)$ has no zero inside the circle with radius $\frac{\Gamma(2\alpha)}{\Gamma(\alpha)}$ centered at the origin. $\qquad\square$

Remark 5.10 The first result connects eigenvalues of problem (5.37) with the function $E_{\alpha,\alpha}(-\lambda)$. Since a number $\lambda \in \mathbb{C}$ is an eigenvalue if and only if it is a zero of $E_{\alpha,\alpha}(-\lambda)$, (ii) implies that $E_{\alpha,\alpha}(-\lambda)$ has at least one positive root.

Theorem 5.16(i) indicates that the eigenfunctions $\{u_n\}$ are given by $u_n(x) = x^{\alpha-1} E_{\alpha,\alpha}(-\lambda_n x^\alpha)$. Since $\lim_{x\to 0} E_{\alpha,\alpha}(-\lambda x^\alpha) = \frac{1}{\Gamma(\alpha)}$, it follows that $u_n(x) \sim \frac{x^{\alpha-1}}{\Gamma(\alpha)}$, as $x \to 0$. Thus, for $1 < \alpha < 2$, $u_n(x) \sim \frac{x^{\alpha-1}}{\Gamma(\alpha)}$ and is not Lipschitz at $x = 0$. Hence, it is impossible to normalize by setting its derivative at $x = 0$ to be unity. Theorem 5.16(ii) indicates that there is always at least one real eigenvalue and the asociated eigenfunction is strictly positive in D. Fig. 5.7 shows the lowest eigenvalues (which are positive and real, by Theorem 5.16(ii)) and the lower bound $\frac{\Gamma(2\alpha)}{\Gamma(\alpha)}$ given by Theorem 5.16(iii). The eigenvalues and eigenfunctions are computed using a numerical method developed in [JLPR15]. Interestingly, the lowest eigenvalue is not monotone with respect to the fractional-order α: It first decreases with α, and then increases with α. Further, the lower bound $\frac{\Gamma(2\alpha)}{\Gamma(\alpha)}$ is fairly loose. The corresponding eigenfunctions are also presented in Fig. 5.7, with the normalization ${}_0^R D_x^{\alpha-1} u_1(0) = 1$.

The eigenfunctions become more symmetric about $x = \frac{1}{2}$ as α increases. This is to be expected since as $\alpha \to 2$, the eigenfunction of $-u''$, $\sin n\pi x$, must be recovered.

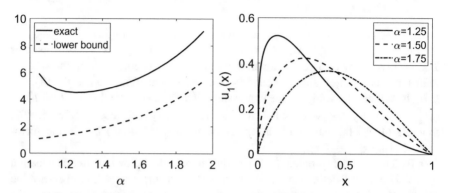

Fig. 5.7 The lowest Dirichlet eigenvalues of the operator $-{}_0^R D_x^\alpha$ versus its lower bound $\frac{\Gamma(2\alpha)}{\Gamma(\alpha)}$, and the corresponding eigenfunction.

For α sufficiently close to 1, there is only a single real eigenvalue and corresponding eigenfunction (actually this holds for α less than approximately 1.34, after which there are three real eigenvalues with linearly independent eigenfunctions). With increasing α from $\alpha = 1$, each subsequent real eigenvalue added has an eigenfunction with one more zero. When subsequently added eigenvalues become complex, these occur in complex conjugate pairs, and then the added eigenfunctions have two more zeros in D than the previous one (but there is no proof of this observation). Fig. 5.8 shows a few higher eigenfunctions for the case $\alpha = \frac{4}{3}$, where the index excludes the complex conjugate. Note both the singular behavior at the origin (asymptotically $\frac{x^{\alpha-1}}{\Gamma(\alpha)}$) and the sinusoidal with rapidly decreasing amplitude as $x \to 1$. However, these computational observations are purely empirical, and no proof is available. It is unclear whether the eigenvalue(s) at the bifurcation point is geometrically/algebraically simple. Naturally, the bifurcation is not stable under the perturbation of a potential term and then the eigenfunctions can be noticeably different.

5.3.2 Djrbashian-Caputo case

The Djrbashian-Caputo fractional Sturm-Liouville problem is more complex than the Riemann-Liouville case, and even fewer results are known. Nakhushev [Nak77] investigated the spectrum of the Dirichlet-type problem

$$\begin{cases} u''(x) + \lambda({}_0^R D_x^\beta u)(x) = 0, & \text{in } D, \\ u(0) = u(1) = 0, \end{cases}$$

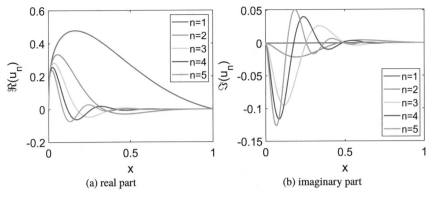

Fig. 5.8 The Dirichlet eigenfunctions $\{u_n\}_{n=1}^5$ of the operator $-{}_0^R D_x^\alpha$ with $\alpha = \frac{4}{3}$.

with $\beta \in (0, 1)$, and showed that λ is an eigenvalue if and only if it is a zero of the Mittag-Leffler function $E_{2-\beta,2}(-\lambda)$ (and thus identical with problem (5.37) with a Djrbashian-Caputo fractional derivative: This is no coincidence, since Nakhushev's formulation can be obtained from problem (5.37) by applying the operator ${}_0^C D_x^\beta$, with $\beta = 2 - \alpha$). Further, Nakhshev showed that all the zeros whose moduli are sufficiently large are simple zeros, and they have the asymptotic estimates $|\lambda_k| = O(k^{2-\beta})$ as $k \to \infty$. These investigations were continued by several researchers, e.g., Malamud [Mal94, MO01] and Aleroev [AKT13].

Theorem 5.17 *In the Djrbashian-Caputo case, the following statements hold for problem (5.37) with $q \equiv 0$.*

(i) *The eigenvalues $\{\lambda_n\}_{n \geq 1}$ are zeros of the function $E_{\alpha,2}(-\lambda)$, and the corresponding eigenfunctions are given by $xE_{\alpha,2}(-\lambda_n x^\alpha)$.*

(ii) *If $\alpha \in (\frac{5}{3}, 2)$, then the smallest eigenvalue is real and positive, and the associated eigenfunction is strictly positive in D.*

Proof The proof of assertion (i) is identical with that in Theorem 5.16. (ii). The existence of a real positive eigenvalue for $\alpha > \frac{5}{3}$ was proved by [Pop06] by a careful analysis of the asymptotics of $E_{\alpha,2}(-\lambda)$, and we refer to [Pop06] for details. The corresponding eigenfunctions is given by $u(x) = xE_{\alpha,2}(-\lambda x^\alpha)$. It suffices to show that $u_1(x) = xE_{\alpha,2}(-\lambda_1 x^\alpha)$ does not vanish in D. Let $x_0 \in D$ such that $x_0 E_{\alpha,2}(-\lambda_1 x_0^\alpha) = 0$. Then $\lambda_1 x_0^\alpha$ is a zero of $E_{\alpha,2}(-\lambda)$, and moreover, $\lambda_1 x_0^\alpha < \lambda_1$. This contradicts the assumption that λ_1 is the first zero of $E_{\alpha,2}(-\lambda)$. $\square$

There are only finitely many real eigenvalues for any $\alpha < 2$. In fact, the existence of real eigenvalues is only guaranteed for α sufficiently close to 2. It has been shown [Pop06] that there exists real eigenvalues provided $\alpha > \frac{5}{3}$. Careful numerical experiments indicate that the first real zeros appear for $\alpha \in (1.5991152, 1.5991153)$ and they occur in pairs [JR12]. By Proposition 3.2, we have the following asymptotic:

$$z_n^{\frac{1}{\alpha}} = 2n\pi i - (\alpha - 1)\left(\ln 2\pi|n| + \frac{\pi}{2}\text{sign}(n)\,i\right) + \ln \frac{\alpha}{\Gamma(2-\alpha)} + O(|n|^{-1}\ln|n|)$$

This leads to the following asymptotics for the magnitude and phase:

$$|\lambda_n| \sim \left((2n + \tfrac{1-\alpha}{2})^2\pi^2 + ((1-\alpha)\ln 2\pi n + \ln \tfrac{\alpha}{\Gamma(2-\alpha)})^2\right)^{\frac{\alpha}{2}} \sim (2\pi n)^\alpha,$$

$$\arg(\lambda_n) \sim \pi - \alpha \arctan\left(\frac{2\pi n + (1-\alpha)\tfrac{\pi}{2}}{(\alpha-1)\ln 2\pi n - \ln \tfrac{\alpha}{\Gamma(2-\alpha)}}\right) \sim \frac{(2-\alpha)\pi}{2}. \tag{5.40}$$

This asymptotic behavior is illustrated in Fig. 5.9, where the index n excludes the conjugate eigenvalues. It is observed that the magnitude prediction in (5.40) is fairly accurate except for the first few eigenvalues, but the phase is accurate only for very large eigenvalue numbers and this is particularly evident as α approaches 2, when the real eigenvalues kick in. Thus, these formulas are accurate indeed only in the "asymptotic" regime.

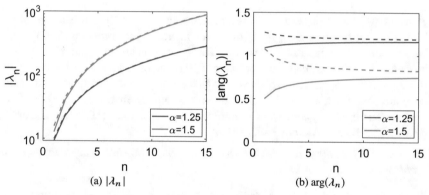

Fig. 5.9 The magnitude $|\lambda_n|$ and angle $\arg(\lambda_n)$ of the eigenvalues λ_ns for problem (5.37) with $-\,_0^C D_x^\alpha$. The dashed and solid lines denote the estimate and true value, respectively.

The eigenfunctions u_n for the operator $-\,_0^C D_x^\alpha$ are given by $xE_{\alpha,2}(-\lambda_n x^\alpha)$. These functions are sinusoidal in nature but significantly attenuated near $x = 1$ with the degree of attenuation strongly depending on α. Fig. 5.10 shows the first five eigenfunctions (excluding the complex conjugate ones) for $\alpha = \tfrac{3}{2}$, where we have the normalization. With complex-valued eigenfunctions, one can always multiply by any complex number so that multiplying by the imaginary unit i interchanges the real to imaginary ones. The choice $u'_n(0) = 1$ sets a consistent choice for selecting the real part of the eigenfunction. A close inspection of the plots indicates that the number of interior zeros of both real and imaginary parts increases by two with consecutive complex eigenfunctions. Also, the number of interior zeros of the real and imaginary parts of the respective eigenfunction u_n always differs by one. It should be stressed that this pattern is purely a numerical observation.

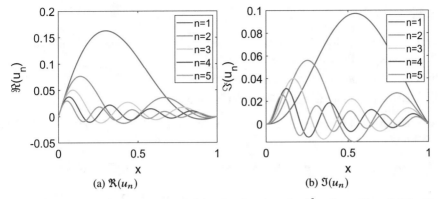

Fig. 5.10 The real and imaginary parts of the eigenfunctions $\{u_n\}_{n=1}^{5}$ for problem (5.37) with $-{}_0^C D_x^\alpha$ and $\alpha = \frac{3}{2}$.

Exercises

Exercise 5.1 In Lemma 5.1, examine the dependence of the value of r on the fractional-order α.

Exercise 5.2 Prove the comparison principle for the two-point BVP in the Riemann-Liouville case. Let u_1 and u_2 be the solutions to

$$\begin{cases} -{}_0^R D_x^\alpha u_1 = f_1, & \text{in } D, \\ u_1(0) = u_1(1) = 0, \end{cases} \quad \text{and} \quad \begin{cases} -{}_0^R D_x^\alpha u_2 = f_2, & \text{in } D, \\ u_2(0) = u_2(1) = 0. \end{cases}$$

Then $f_1 \leq f_2$ implies $u_1 \leq u_2$.

Exercise 5.3 This exercise studies the "natural" Dirichlet boundary condition in the Riemann-Liouville case for $c_0, c_1 \in \mathbb{R}$,

$$\begin{cases} -{}_0^R D_x^\alpha u = f(x, u), & \text{in } D, \\ {}_0^R D_x^{\alpha-2} u(0) = c_0, \\ {}_0^R D_x^{\alpha-2} u(1) = c_1. \end{cases}$$

(i) Derive the solution representation via Green's function.

(ii) Discuss the positivity of Green's function.

(iii) Discuss the solution regularity, if $f(x, u)$ is smooth.

Exercise 5.4 This exercise studies the "natural" Neumann boundary condition in the Riemann-Liouville case

$$\begin{cases} -{}_0^R D_x^\alpha u = f(x, u), & \text{in } D, \\ {}_0^R D_x^{\alpha-1} u(0) = c_0, \\ {}_0^R D_x^{\alpha-1} u(1) = c_1, \end{cases}$$

with $c_0, c_1 \in \mathbb{R}$.

(i) Derive the solution representation via Green's function. Is there any compatibility condition as the classical case ?

(ii) Discuss the positivity of Green's function.

(iii) Discuss the solution regularity, if $f(x, u)$ is smooth.

Exercise 5.5 Let $0 < \alpha < 1$, $a, b, c \in \mathbb{R}$ with $a + b \neq 0$, and $f : \overline{D} \times \mathbb{R} \to \mathbb{R}$ be continuous.

(i) Prove that a function $u \in C(\overline{D})$ with ${}_0I_t^{1-\alpha}(u - u(0)) \in AC(\overline{D})$ solves

$$\begin{cases} {}_0^C D_x^\alpha u(x) = f(x, u(x)), & \text{in } D, \\ au(0) + bu(1) = c, \end{cases}$$

if and only if $u \in C(\overline{D})$ satisfies

$$u(x) = \frac{c}{a + b} + ({}_0I_x^\alpha f(\cdot, u(\cdot)))(x) - \frac{b}{a + b}({}_aI_x^\alpha f(\cdot, u(\cdot)))(1).$$

(ii) Now suppose that f is globally Lipschitz in u with a constant L. Prove that if L is sufficiently small, then the problem has a unique solution.

(iii) Derive a sharp bound on the required Lipschitz constant on f.

Exercise 5.6 Prove Theorem 5.9.

Exercise 5.7 Let $\gamma > 0$, $1 < \alpha < 2$, and $f \in C(\overline{D})$. Consider the following BVP

$$\begin{cases} -{}_0^C D_x^\alpha u = f, & \text{in } D, \\ u(0) - \gamma u'(0) = u(1) = 0. \end{cases}$$

(i) Derive Green's function $G(x, s)$ for the problem.

(ii) Discuss the positivity of $G(x, s)$ in relation to the parameter γ.

(iii) Derive an explicit expression for the critical value γ as a function of α (so as to ensure positivity).

Exercise 5.8 Show that the operator T defined in (5.39) is compact on $C(\overline{D})$.

Exercise 5.9 Discuss the asymptotically behaviour of the "Dirichlet"/"Neumann" eigenvalues with a Riemann-Liouville fractional derivative.

Exercise 5.10 Investigate numerically and analytically the properties of the following eigenvalue problem:

$$\begin{cases} -{}_0^C D_x^\alpha u = \lambda u, & \text{in } D, \\ u'(0) = u(1) = 0. \end{cases}$$

Part III
Time-Fractional Diffusion

Chapter 6
Subdiffusion: Hilbert Space Theory

In this chapter, we describe the solution theory in Hilbert spaces for the time-fractional diffusion or subdiffusion model derived in Chapter 1, which involves a (regularized) Djrbashian-Caputo fractional derivative $\partial_t^\alpha u$ of order $\alpha \in (0,1)$ in time. We will discuss the following aspects: unique existence of a (weak) solution, solution representations via Laplace transform, Hilbert space regularity, time-dependent diffusion coefficients, nonlinear problems, (weak) maximum principle, inverse problems, and numerical methods for discretizing $\partial_t^\alpha u$. Most regularity results are described using the space $\dot{H}^s(\Omega)$ defined in Section A.2.4 in the appendix. Throughout, the problem is defined on an open bounded domain $\Omega \subset \mathbb{R}^d$ ($d = 1, 2, 3$) with a boundary $\partial\Omega$, and $T > 0$ the final time. The notation $Q = \Omega \times (0, T]$ denotes the parabolic cylinder, $\partial_L Q = \partial\Omega \times (0, T]$ and $\partial_p Q = (\partial\Omega \times (0, T]) \cup (\Omega \times \{0\})$ the lateral and parabolic boundary, respectively, and $I = (0, T]$. For a function $v : \Omega \times I \rightarrow \mathbb{R}$, we denote by $v(t) = v(\cdot, t)$ for any $t \in I$.

6.1 Existence and Uniqueness in an Abstract Hilbert Space

First, we analyze the subdiffusion problem in an abstract Hilbert space setting and give the uniqueness, existence and basic energy estimates, following discussions in [Zac09] (which covers general completely positive kernels). See also the work [Aka19] for differential inclusions, which discusses also the concept of strong solutions, and gives sufficient conditions for the classical initial condition. Let V and H be real Hilbert spaces such that V is densely and continuously embedded into H, with $\|v\|_H \leq c_{V \to H}\|v\|_V$, for any $v \in V$, where $c_{V \to H}$ denotes the constant of the embedding $V \hookrightarrow H$. By identifying H with its dual H', we have

$$V \hookrightarrow H \hookrightarrow V', \tag{6.1}$$

$$(h, v)_H = \langle h, v \rangle_{V' \times V}, \quad \forall h \in H, v \in V, \tag{6.2}$$

© The Author(s), under exclusive license to Springer Nature Switzerland AG 2021
B. Jin, *Fractional Differential Equations*, Applied Mathematical Sciences 206,
https://doi.org/10.1007/978-3-030-76043-4_6

where $(\cdot, \cdot)_H$ and $\langle \cdot, \cdot \rangle_{V' \times V}$ denote the inner product in H and duality pairing between V' (the dual space of V) and V, respectively. Consider the abstract evolution problem

$$\partial_t((k_\alpha * (u - u_0))(t), v)_H + a(t, u(t), v) = \langle f(t), v \rangle_{V' \times V}, \quad \forall v \in V, \qquad (6.3)$$

where $u_0 \in H$ and $f \in L^2(I; V')$ (see Section A.2.5 in the appendix for the definition of the space) are given data, and $a : I \times V \times V \to \mathbb{R}$ is a time-dependent bounded V-coercive bilinear form; see Assumption 6.1 below for the precise assumptions. Note that the bilinear form a is only assumed to be measurable in time t, which allows, e.g., treating time-fractional PDEs in divergence form with merely bounded and measurable coefficients. The term $\partial_t^\alpha u$ denotes the (regularized) Djrbashian-Caputo fractional derivative in time t

$$\partial_t^\alpha u = {}^R\partial_t^\alpha (u - u_0), \quad \text{with } k_\alpha(t) = \Gamma(1 - \alpha)^{-1} t^{-\alpha};$$

see Definition 2.4 in Chapter 2. We look for a solution of problem (6.3) in the space

$$W(u_0, V, H) = \{u \in L^2(I; V) : k_\alpha * (u - u_0) \in {}_0H^1(I; V')\},$$

where the left subscript 0 in the notation ${}_0H^1(I; V')$ means a vanishing trace at $t = 0$. The vector u_0 can be regarded as the initial data for u, at least in a weak sense.

Note that problem (6.3) is equivalent to the following operator equation:

$$\partial_t(k_\alpha * (u - u_0))(t) + A(t)u(t) = f(t), \quad \text{a.e. } t \in I,$$

in V', where the operator $A(t) : V \to V'$ is defined by

$$\langle A(t)u, v \rangle_{V' \times V} = a(t, u, v), \quad \forall u, v \in V. \qquad (6.4)$$

We need one preliminary result. The sequence $\{k_n\}$ is often called Yosida approximations of the singular kernel k_α, in view of its construction, and allows overcoming the singularity of k_α with an approximation argument. (iii) can be viewed as a version of Alikhanov's inequality in Lemma 2.11 for the Djrbashian-Caputo fractional derivative ∂_t^α.

Lemma 6.1 *For the kernel k_α, the following statements hold.*

(i) $k_\alpha * k_{1-\alpha} \equiv 1$ *on I.*
(ii) *There exists a sequence of nonnegative and nonincreasing functions $\{k_n\}_{n=1}^\infty \subset W^{1,1}(I)$, such that $k_n \to k_\alpha$ in $L^1(I)$.*
(iii) *Any k_n from (ii) satisfies $u\partial_t(k_n * u) \geq \frac{1}{2}\partial_t(k_n * u^2)$.*

Proof (i) can be verified directly using the identity (2.10) as

$$(k_\alpha * k_{1-\alpha})(t) = \frac{1}{\Gamma(\alpha)\Gamma(1 - \alpha)} \int_0^t (t - s)^{\alpha-1} s^{-\alpha} ds$$

$$= \frac{1}{\Gamma(\alpha)\Gamma(1 - \alpha)} \int_0^1 (1 - s)^{\alpha-1} s^{-\alpha} ds = \frac{B(\alpha, 1 - \alpha)}{\Gamma(\alpha)\Gamma(1 - \alpha)} = 1.$$

For an arbitrary completely positive kernel, to which the kernel k_α belongs to, the required approximating sequence can be constructed via the so-called Yosida approximation. Specifically, Let $k_n = ns_n$, with s_n being the unique solution of the scalar Volterra equation

$$s_n(t) + n(s_n * k_{1-\alpha})(t) = 1, \quad t > 0, n \in \mathbb{N}.$$

Then it can be verified that $k_n = ns_n$ satisfies the desired properties using the general theory for completely positive kernels (cf. [Prü93, Proposition 4.5] and [CN81, Proposition 2.1]). This can be verified directly for the kernel k_α:

$$k_n = ns_n = nE_{1-\alpha,1}(-nt^{1-\alpha}) \in W^{1,1}(I),$$
$$k_n \to k_\alpha \quad \text{in } L^1(I), \quad \text{as } n \to \infty.$$

The latter convergence is direct from the exponential asymptotic in Theorem 3.2. Now by the complete monotonicity of the function $E_{\alpha,1}(-x)$ in Theorem 3.5, k_n is nonnegative and decreasing. (iii) is a direct consequence of (ii), i.e., k_n is decreasing and nonnegative.   □

We begin with an embedding result for the space $W(u_0, V, H)$. In the theory of abstract parabolic equations, the following continuous embedding

$$H^1(I; V') \cap L^2(I; V) \hookrightarrow C(\overline{I}; H) \tag{6.5}$$

is well known [LM72, Chapter 1, Proposition 2.1 and Theorem 3.1]. The following theorem provides an analogue for the space $W(u_0, V, H)$. When $u_0 = 0$, the property $k_\alpha * u \in C(\overline{I}; H)$ follows directly from the embedding (6.5): $u \in L^2(I; V)$ implies $k_\alpha * u \in L^2(I; V)$, by Young's inequality, and by the definition of $W(u_0, V, H)$, we have $k_\alpha * u \in H^1(I; V') \cap L^2(I; V) \hookrightarrow C(\overline{I}; H)$. However, for $u_0 \neq 0$, this simple reduction is not feasible.

Theorem 6.1 *Let V and H be real Hilbert spaces satisfying (6.1), $u_0 \in H$ and $u \in W(u_0, V, H)$. Then $k_\alpha * (u - u_0)$ and $k_\alpha * u$ belong to $C(\overline{I}; H)$ (possibly after redefinition on a set of measure zero), and*

$$\|k_\alpha * u\|_{C(\overline{I};H)} \leq c(\|k_\alpha\|_{L^1(I)}, T, c_{V \to H})(\|\partial_t[k_\alpha * (u - u_0)]\|_{L^2(I;V')}$$
$$+ \|u\|_{L^2(I;V)} + \|u_0\|_H).$$

Further, $\|k_\alpha * u\|_H^2(t) \in W^{1,1}(I)$, and for a.e. $t \in \overline{I}$,

$$\partial_t \|k_\alpha * u\|_H^2(t) = 2\langle[k_\alpha * (u - u_0)]'(t), [k_\alpha * u](t)\rangle_{V' \times V} + 2k_\alpha(t)(u_0, (k_\alpha * u)(t))_H.$$

Proof Since

$$(k_\alpha * u_0) = (1 * k_\alpha)u_0 \in W^{1,1}(I; H) \hookrightarrow C(\overline{I}; H),$$

$k_\alpha * (u - u_0) \in C(\overline{I}; H)$ if and only if $k_\alpha * u \in C(\overline{I}; H)$. Let $k_n \in W^{1,1}(I)$, $n \in \mathbb{N}$, be given by Lemma 6.1(ii). Then $k_n * u \in H^1(I; V)$. For $n \in \mathbb{N}$, let $v_n = k_n * u$. Then

$$\partial_t \|v_n(t) - v_m(t)\|_H^2 = 2(v_n'(t) - v_m'(t), v_n(t) - v_m(t))_H.$$

Thus, for all $s, t \in \bar{I}$, with $w_{n,m} = v_n - v_m$, there holds

$$\|w_{n,m}(t)\|_H^2 = \|w_{n,m}(s)\|_H^2 + 2 \int_s^t [k_n(\xi) - k_m(\xi)](u_0, w_{n,m}(\xi))_H \, d\xi$$

$$+ 2 \int_s^t \langle [k_n * (u - u_0)]'(\xi) - [k_m * (u - u_0)]'(\xi), w_{n,m}(\xi) \rangle_{V' \times V} \, d\xi.$$

This and Young's inequality imply

$$\|w_{n,m}(t)\|_H^2 \le \|w_{n,m}(s)\|_H^2 + \|[k_n * (u - u_0)]' - [k_m * (u - u_0)]'\|_{L^2(0,T;V')}^2$$

$$+ \|w_{n,m}\|_{L^2(0,T;V)}^2 + 2\|u_0\|_H^2 \|k_n - k_m\|_{L^1(0,T)}^2 + \tfrac{1}{2} \|w_{n,m}\|_{C([0,T];H)}^2. \quad (6.6)$$

Since $k_n \to k_\alpha$ in $L^1(I)$ as $n \to \infty$ and $k_\alpha * (u - u_0) \in {}_0H^1(I; V')$, we have

$$v_n \to k_\alpha * u := v \text{ in } L^2(I; H) \text{ and } L^2(I; V), \quad (6.7)$$

$$\partial_t(k_n * (u - u_0)) \to \partial_t(k_\alpha * (u - u_0)) \text{ in } L^2(I; V').$$

Now we fix a point $s \in (0, T)$ for which $v_n(s) \to v(s)$ in H as $n \to \infty$. Taking then in (6.6) the maximum over all $t \in \bar{I}$ and absorbing the last term imply that $\{v_n\}$ is a Cauchy sequence in $C(\bar{I}; H)$ and converges to some $\tilde{v} \in C(\bar{I}; H)$. By the convergence $v_n \to v$ in $L^2(I; H)$ in (6.7), the sequence $\{v_n\}$ actually converges to v a.e. in $\bar{I}$. This proves the first assertion. Similarly, for all $s, t \in \bar{I}$, and $n \in \mathbb{N}$,

$$\|v_n(t)\|_H^2 = \|v_n(s)\|_H^2 + 2 \int_s^t k_n(\xi)(u_0, v_n(\xi))_H \, d\xi$$

$$+ 2 \int_s^t \langle [k_n * (u - u_0)]'(\xi), v_n(\xi) \rangle_{V' \times V} \, d\xi.$$

Taking the limit as $n \to \infty$ gives (recall $v = k_\alpha * u$),

$$\|v(t)\|_H^2 = \|v(s)\|_H^2 + 2 \int_s^t k_\alpha(\xi)(u_0, v(\xi))_H \, d\xi$$

$$+ 2 \int_s^t \langle [k_\alpha * (u - u_0)]'(\xi), v(\xi) \rangle_{V' \times V} \, d\xi. \quad (6.8)$$

Hence, the mapping $\{t \mapsto \|v\|_H^2(t)\}$ is absolutely continuous on $\bar{I}$, and differentiating the identity (6.8) shows the third assertion. Next, integrating (6.8) with respect to s yields for all $t \in \bar{I}$

$$\|v(t)\|_H^2 \le T^{-1} \|v\|_{L^2(I;H)}^2 + \|\partial_t(k_\alpha * (u - u_0))\|_{L^2(I;V')}^2$$

$$+ \|v\|_{L^2(I;V)}^2 + 2\|u_0\|_H^2 \|k_\alpha\|_{L^1(I)}^2 + \tfrac{1}{2} \|v\|_{C(\bar{I};H)}^2.$$

Then taking the maximum over all $t \in \bar{I}$ and Young's inequality for convolutions imply the desired estimate, thereby completing the proof.   □

Next, we turn to the existence and uniqueness of a solution u to problem (6.3). We will make the following assumption on the bilinear form $a(t, \cdot, \cdot)$. The condition $a(t, u, u) \geq c_1 \|u\|_V^2 - c_2 \|u\|_H^2$ is an abstract form of Gärding's inequality.

Assumption 6.1 For a.e. $t \in I$, for the bilinear form $a(t, \cdot, \cdot) : V \times V \to \mathbb{R}$, there exist constants $c_0, c_1 > 0$ and $c_2 \geq 0$ independent of t such that for all $u, v \in V$

$$|a(t, u, v)| \leq c_0 \|u\|_V \|v\|_V \quad \text{and} \quad a(t, u, u) \geq c_1 \|u\|_V^2 - c_2 \|u\|_H^2.$$

Moreover, the function $\{t \mapsto a(t, u, v)\}$ is measurable on I for all $u, v \in V$.

Example 6.1 Let $\Omega \subset \mathbb{R}^d$, $d = 1, 2, 3$, be an open bounded domain. Consider

$$\begin{cases} \partial_t^\alpha u - \nabla \cdot (a\nabla u) + b \cdot \nabla u + cu = f, & \text{in } Q, \\ u = 0, & \text{on } \partial_L Q, \\ u(0) = u_0, & \text{in } \Omega. \end{cases}$$

Assume $u_0 \in L^2(\Omega)$, $g \in L^2(I; H^{-1}(\Omega))$, $a \in L^\infty(Q; \mathbb{R}^{d \times d})$ and there exists $\lambda > 0$ such that for a.e. $(x, t) \in Q$, $a(x, t)\xi \cdot \xi \geq \lambda |\xi|^2$ for any $\xi \in \mathbb{R}^d$, where $\cdot$ and $|\cdot|$ denote the Euclidean inner product and norm on $\mathbb{R}^d$, respectively. Further, $b \in L^\infty(Q; \mathbb{R}^d)$, $c \in L^\infty(Q)$. To apply the abstract theory, let $V = H_0^1(\Omega)$, $H = L^2(\Omega)$, and

$$a(t, u, v) = \int_\Omega (a(x, t)\nabla u, \nabla v) + (b(x, t), \nabla u)v + c(x, t)uv \, dx$$

and

$$\langle f(t), v \rangle_{V' \times V} = \int_\Omega f(x, t)v(x) dx.$$

Then the weak formulation of the problem reads

$$\partial_t ((k_\alpha * (u - u_0))(t), v)_H + a(t, u(t), v) = \langle f(t), v \rangle_{V' \times V}, \quad \forall v \in V, \text{ a.e. } t \in I,$$

and we seek a solution in the space

$$W(u_0, H_0^1(\Omega), L^2(\Omega)) = \{u \in L^2(I; H_0^1(\Omega)) : k_\alpha * (u - u_0) \in {}_0 H^1(I; H^{-1}(\Omega))\}.$$

It is folklore that Assumption 6.1 holds under these conditions on the coefficients.

To construct a solution, we use the standard Galerkin method and derive suitable *a priori* estimates for the weak solution, in a manner analogous to the standard parabolic case. Let $\{w_m\}_{m=1}^\infty$ be a basis of V, and $\{u_{0,m}\}_{m=1}^\infty \subset H$ be a sequence such that $u_{0,m} \in \text{span}\{w_1, \ldots, w_m\}$ such that $u_{0,m} \to u_0$ in H as $m \to \infty$. Let

$$u_m(t) = \sum_{j=1}^m c_{jm}(t)w_j, \quad u_{0,m} = \sum_{j=1}^m b_{jm}w_j.$$

Then formally, for every $m \in \mathbb{N}$, the system of Galerkin equations reads

$$\sum_{j=1}^{m} \partial_t(k_\alpha * (c_{jm} - b_{jm}))(t)(w_j, w_i)_H + \sum_{j=1}^{m} c_{jm}(t)a(t, w_j, w_i) = \langle f(t), w_i \rangle_{V' \times V}, \quad (6.9)$$

for a.e. $t \in I$, $i = 1, \ldots, m$.

Then we have the following existence and uniqueness result.

Theorem 6.2 *Under Assumption 6.1, for $u_0 \in H$ and $f \in L^2(I; V')$, problem (6.3) has exactly one solution $u \in W(u_0, V, H)$ and*

$$\|k_\alpha * (u - u_0)\|_{H^1(I;V')} + \|u\|_{L^2(I;V)} \leq c(\|u_0\|_H + \|f\|_{L^2(I;V')}).$$

Moreover, for every $m \in \mathbb{N}$, the Galerkin system (6.9) has a unique solution $u_m \in W(u_{0,m}, V, H)$, and the sequence $\{u_m\}$ converges weakly to u in $L^2(I; V)$ as $m \to \infty$.

Proof First, we prove the uniqueness. Suppose that $u_1, u_2 \in W(u_0, V, H)$ are two solutions of problem (6.3). Then $u = u_1 - u_2$ belongs to $W(0, V, H)$ and satisfies

$$\langle (k_\alpha * u)'(t), v \rangle_{V' \times V} + a(t, u(t), v) = 0, \quad \forall v \in V.$$

Taking $v = u(t)$ gives

$$\langle (k_\alpha * u)'(t), u(t) \rangle_{V' \times V} + a(t, u(t), u(t)) = 0.$$

Let $k_n \in W^{1,1}(I)$ be the regularized kernel given in Lemma 6.1(ii). Then the last identity is equivalent to

$$\langle (k_n * u)'(t), u(t) \rangle_{V' \times V} + a(t, u(t), u(t)) = \zeta_n(t),$$

with $\zeta_n(t) = \langle (k_n * u)'(t) - (k_\alpha * u)'(t), u(t) \rangle_{V' \times V}$. It is easy to prove that $\zeta_n \to 0$ in $L^1(I)$. Since $k_n * u \in H^1(I; H)$, by (6.2),

$$(\partial_t(k_n * u)(t), u(t))_H + a(t, u(t), u(t)) = \zeta_n(t).$$

k_n is increasing and nonnegative, and Lemma 6.1(iii) implies

$$\tfrac{1}{2}\partial_t(k_n * \|u(\cdot)\|_H^2)(t) \leq (\partial_t(k_n * u)(t), u(t))_H.$$

This and Gärding's inequality in Assumption 6.1 yield

$$\partial_t(k_n * \|u(\cdot)\|_H^2)(t) \leq 2c_2\|u(t)\|_H^2 + 2\zeta_n(t).$$

It can be verified that all terms as functions of t belong to $L^1(I)$. Then convolving the inequality with $k_{1-\alpha}$ and passing to limit (possibly on a subsequence) give

$$\|u(t)\|_H^2 \leq 2c_2(k_{1-\alpha} * \|u\|_H^2)(t), \quad (6.10)$$

where we have used the fact that since $\zeta_n \to 0$ in $L^1(I)$, Young's inequality for convolution entails $k_{1-\alpha} * \zeta_n \to 0$ in $L^1(I)$, and that

$$k_{1-\alpha} * \partial_t(k_n * \|u\|_H^2) = \partial_t(k_n * k_{1-\alpha} * \|u\|_H^2)$$
$$\rightarrow \quad \partial_t(k_\alpha * k_{1-\alpha} * \|u\|_H^2) = \|u\|_H^2, \quad \text{in } L^1(I),$$

since $k_n \rightarrow k_\alpha$ in $L^1(I)$. Since $k_{1-\alpha}$ is nonnegative, now by Gronwall's inequality, (6.10) implies $\|u(t)\|_H^2 = 0$, i.e., $u = 0$. To prove the existence, first, we show that for every $m \in \mathbb{N}$, the Galerkin system (6.9) has a unique solution $\mathbf{c}_m := (c_{1m}, \dots, c_{mm})^T \in \mathbb{R}^m$ on $\overline{I}$ in the class $W(\xi_m, \mathbb{R}^m, \mathbb{R}^m)$, where $\mathbf{b}_m := (b_{1m}, \dots, b_{mm})^T \in \mathbb{R}^m$. Since $w_1, \dots, w_m$ are linearly independent, the matrix $((w_j, w_i)_H) \in \mathbb{R}^{m \times m}$ is invertible. Hence, (6.9) can be solved for $\partial_t(k_\alpha * (c_{jm} - b_{jm}))$,

$$\partial_t[k_\alpha * (\mathbf{c}_m - \mathbf{b}_m)](t) = \mathbf{B}(t)\mathbf{c}_m(t) + \mathbf{f}(t), \quad \text{a.e. } t \in I,$$

where $\mathbf{B} \in L^\infty(I; \mathbb{R}^{m \times m})$, and $\mathbf{f} \in L^2(I; \mathbb{R}^m)$, by the assumptions on a and f. Next, we transform it into a system of Volterra equations

$$\mathbf{c}_m(t) = \mathbf{b}_m + k_{1-\alpha} * [\mathbf{B}(\cdot)\mathbf{c}_m(\cdot)](t) + (k_{1-\alpha} * \mathbf{f})(t),$$

which has a unique solution $\mathbf{c}_m \in L^2(I; \mathbb{R}^m)$. Then $\mathbf{c}_m \in W(\mathbf{b}_m, \mathbb{R}^m, \mathbb{R}^m)$, and hence it is also a solution of (6.9). This shows that for every $m \in \mathbb{N}$, the Galerkin system (6.9) has exactly one solution $u_m \in W(u_{0,m}, V, H)$. Next, we derive *a priori* estimates on the Galerkin solutions $\{u_m\}$. The Galerkin system is equivalent to

$$(\partial_t(k_\alpha * (u_m - u_{0,m}))(t), w_i)_H + a(t, u_m(t), w_i) = \langle f(t), w_i \rangle_{V' \times V}, i = 1, \dots, m. \quad (6.11)$$

By multiplying by c_{im} and then summing over i, we obtain

$$(\partial_t(k_\alpha * (u_m - u_{0,m}))(t), u_m(t))_H + a(t, u_m(t), u_m(t)) = \langle f(t), u_m(t) \rangle_{V' \times V}.$$

Then it can be equivalently written as

$$(\partial_t(k_n * u_m)(t), u_m(t))_H + a(t, u_m(t), u_m(t))$$
$$= k_n(t)(u_{0,m}, u_m(t)) + \langle f(t), u_m(t) \rangle_{V' \times V} + \zeta_{mn}(t),$$

with $\zeta_{mn}(t) = \langle [k_n * (u_m - u_{0,m})]'(t) - [k_\alpha * (u_m - u_{0,m})]'(t), u_m \rangle_{V' \times V}$. Using Lemma 6.1(iii) and Assumption 6.1, we obtain

$$\frac{1}{2}\partial_t(k_n * \|u_m\|_H^2)(t) + \frac{1}{2}k_n(t)\|u_m(t)\|_H^2 + c_1\|u_m(t)\|_V^2$$
$$\leq c_2\|u_m(t)\|_H^2 + k_n(t)(u_{0,m}, u_m(t)) + \langle f(t), u_m(t) \rangle_{V' \times V} + \zeta_{mn}(t),$$

which, together with Young's inequality, yields

$$\partial_t(k_n * \|u_m(t)\|_H^2)(t) + c_1\|u_m(t)\|_V^2$$
$$\leq 2c_2\|u_m\|_H^2 + k_n(t)\|u_{0,m}\|_H^2 + c_1^{-1}\|f(t)\|_{V'}^2 + 2\zeta_{mn}(t). \quad (6.12)$$

Similarly, as $n \rightarrow \infty$,

$$k_{1-\alpha} * \zeta_{mn} \to 0 \quad \text{in } L^1(I),$$

$$k_{1-\alpha} * \partial_t(k_n * \|u_m(\cdot)\|_H^2) \to \|u_m(\cdot)\|_H^2 \quad \text{in } L^1(I).$$

Thus, convolving (6.12) with $k_{1-\alpha}$ and passing to limit (on a subsequence) give

$$\|u_m(t)\|_H^2 \le 2c_2(k_{1-\alpha} * \|u_m(\cdot)\|_H^2)(t) + \|u_{0,m}\|_H^2 + c_1^{-1}(k_{1-\alpha} * \|f(\cdot)\|_{V'}^2)(t),$$

for all $m \in \mathbb{N}$. By the positivity of $k_{1-\alpha}$, it follows that

$$\|u_m\|_{L^2(I;H)} \le c(c_1, c_2, \|k_{1-\alpha}\|_{L^1(I)}, T)(\|u_{0,m}\|_H + \|f\|_{L^2(I;V')}).$$

We integrate (6.12) over I, since $(k_n * \|u_m(\cdot)\|_H^2)(0) = 0$ and let $n \to \infty$ to obtain

$$c_1\|u_m\|_{L^2(I;V)}^2 \le 2c_2\|u_m(t)\|_{L^2(I;H)}^2 + \|k_\alpha\|_{L^1(I)}\|u_{0,m}\|_H^2 + c_1^{-1}\|f(t)\|_{L^2(I;V')}^2.$$

The last two estimates and the assumption $u_{0,m} \to u_0$ in H yields

$$\|u_m\|_{L^2(I;V)} \le c(c_1, c_2, \|k_\alpha\|_{L^1(I)}, T)(\|u_0\|_H + \|f\|_{L^2(I;V')}). \qquad (6.13)$$

Thus, there exists a subsequence of $\{u_m\}$, still denoted by $\{u_m\}$, and some $u \in L^2(I;V)$ such that

$$u_m \rightharpoonup u \text{ in } L^2(I;V). \qquad (6.14)$$

We claim that $u \in W(u_0, V, H)$ and it is a solution of problem (6.3). Let $\varphi \in C^1(\overline{I})$ with $\varphi(T) = 0$. By multiplying (6.11) by φ and integration by parts, we obtain

$$-\int_0^T \varphi'(t)([k_\alpha * (u_m - u_{0,m})](t), w_i)_H dt + \int_0^T \varphi(t)a(t, u_m(t), w_i)dt$$

$$= \int_0^T \varphi(t)\langle f(t), w_i \rangle_{V' \times V} dt, \quad i = 1, \ldots, m,$$

since $(k_\alpha * (u_m - u_{0,m}))(0) = 0$. Then by (6.14) and $u_{0,m} \to u_0$ in H, Assumption 6.1, and Young's and Hölder's inequalities, this leads to

$$-\int_0^T \varphi'(t)([k_\alpha * (u - u_0)](t), w_i)_H dt + \int_0^T \varphi(t)a(t, u(t), w_i)dt$$

$$= \int_0^T \varphi(t)\langle f(t), w_i \rangle_{V' \times V} dt, \quad \forall i \in \mathbb{N}.$$

By (6.2), $(k_\alpha * (u - u_0), w_i)_H = \langle k_\alpha * (u - u_0), w_i \rangle_{V' \times V}$. Thus, all the terms in this identity represent bounded linear functionals on V. Since $\{w_i\}$ is a basis of V,

$$-\int_0^T \varphi'(t)([k_\alpha * (u - u_0)](t), v)_{V' \times V} dt + \int_0^T \varphi(t)a(t, u(t), v)dt$$

$$= \int_0^T \varphi(t)\langle f(t), v \rangle_{V' \times V} dt, \quad \forall v \in V.$$

Since the identity holds for all $\varphi \in C_0^\infty(I)$, $k_\alpha * (u - u_0)$ has a weak derivative on I given by

$$\partial_t[k_\alpha * (u - u_0)](t) + A(t)u(t) = f(t), \tag{6.15}$$

where $A(t)$ is defined in (6.4). It follows from $u \in L^2(I; V)$ and $\|A(t)u(t)\|_{V'} \leq c_0\|u(t)\|_V$ that $A(t)u(t) \in L^2(I; V')$. Since $f \in L^2(I; V')$, $(k_\alpha * (u - u_0))' \in L^2(I; V')$. To show $u \in W(u_0, V, H)$, it suffices to show $(k_\alpha * (u - u_0))(0) = 0$. Let $z := k_\alpha * (u - u_0)$. The preceding discussion indicates $z \in H^1(I; V') \hookrightarrow C(\bar{I}; V')$, and the last two identities yield

$$-\int_0^T \varphi'(t)\langle z(t), v\rangle_{V' \times V} dt = \int_0^T \varphi(t)\langle z'(t), v\rangle_{V' \times V} dt, \quad \forall v \in V,$$

for all $\varphi \in C^1(\bar{I})$ with $\varphi(T) = 0$. By choosing φ such that $\varphi(0) = 1$, and approximating z in $H^1(I; V')$ by a sequence of functions $C^1(\bar{I}; V')$, by integration by parts, $\langle z(0), v\rangle_{V' \times V} = 0$ for all $v \in V$, i.e., $z(0) = 0$. In sum, the function $u \in W(u_0, V, H)$ solves (6.15), or equivalently (6.3). Moreover, (6.3) has exactly one solution u in $W(u_0, V, H)$. Thus, all weakly convergent subsequences of $\{u_m\}$ (in $L^2(I; V)$) have the same limit u. Hence, the whole sequence $\{u_m\}$ converges weakly to u in $L^2(I; V)$. Last, we prove continuous dependence on the data. From $u_m \rightharpoonup u$ in $L^2(I; V)$ and (6.13), by the weak lower semi-continuity of norms, we deduce

$$\|u\|_{L^2(I;V)} \leq \liminf_{m \to \infty} \|u_m\|_{L^2(I;V)} \leq c(\|u_0\|_H + \|f\|_{L^2(I;V')}).$$

This estimate, the inequality $\|Au\|_{L^2(I;V')} \leq c_0\|u\|_{L^2(I;V)}$, (6.15) and the assumption $f \in L^2(I; V')$ yield the desired stability estimate. $\square$

Next, we interpret the role of u_0. It plays the role of the initial data for u, at least in a weak sense. For example, if u and $\partial_t(k_\alpha * (u - u_0)) = \tilde{f} \in C(\bar{I}; V')$, then $(k_\alpha * (u - u_0))(0) = 0$ implies $u(0) = u_0$. Indeed,

$$u - u_0 = \partial_t(k_{1-\alpha} * k_\alpha * (u - u_0)) = k_{1-\alpha} * \tilde{f}, \quad \text{in } C(\bar{I}; V').$$

This implies $u(0) = u_0$ in V'. Under additional conditions, e.g., $k_{1-\alpha} \in L^2(I)$, $u(0) = u_0$ actually holds in a classical sense [Aka19, Proposition 2.5]. The kernel $k_{1-\alpha}(t) = \Gamma(\alpha)^{-1}t^{\alpha-1}$ belongs to $L^2(I)$ for any $T > 0$ if and only if $\alpha > \frac{1}{2}$.

Theorem 6.2 gives basic energy estimates. Next, we give a stronger result by exploiting the better integrability of $k_{1-\alpha}$ rather than merely $k_{1-\alpha} \in L^1(I)$. It contains an analogue of the parabolic counterpart: $\int_0^T t^{-1}\|u(t)\|_H^2 dt < \infty$ for any $u \in {_0}H^1(I; V') \cap L^2(I; V)$; see [LM72, Chapter 3, Propositions 5.3 and 5.4], for the space $W(u_0, V, H)$.

Theorem 6.3 *Let V and H be real Hilbert spaces with (6.1), and $u_0 \in H$, and $u \in W(u_0, V, H)$ solve problem (6.3). Then the following statements hold.*

(i) *If $k_{1-\alpha} \in L^p(I)$, $p > 1$, then $u \in L^{2p}(I; H)$ and with $c = c(\|k_{1-\alpha}\|_{L^p(I)}, T)$*

$$\|u\|_{L^{2p}(I;H)} \le c\big(\|\partial_t(k_\alpha * (u - u_0))\|_{L^2(I;V')} + \|u\|_{L^2(I;V)} + \|u_0\|_H\big).$$

(ii) *The following estimate holds with $c = c(\|k_\alpha\|_{L^1(I)})$*

$$\left(\int_0^T k_\alpha(t)\|u(t)\|_H^2 dt\right)^{\frac{1}{2}} \le c\big(\|\partial_t(k_\alpha * (u - u_0))\|_{L^2(I;V')} + \|u\|_{L^2(I;V)} + \|u_0\|_H\big).$$

Proof The proof proceeds similarly as Theorem 6.2. Let $\{k_n\}$ be the sequence of Yosida approximations in Lemma 6.1(ii), and for $t \in I$, let $g(t) = \langle(k_\alpha * (u - u_0))'(t), u(t)\rangle_{V' \times V}$ and $\zeta_n(t) = \langle(k_n * (u - u_0))'(t) - (k_\alpha * (u - u_0))'(t), u(t)\rangle_{V' \times V}$. Then

$$(\partial_t(k_n * u), u(t))_H = k_n(t)(u_0, u(t))_H + g(t) + \zeta_n(t),$$

with each term being in $L^1(I)$. By Lemma 6.1(iii) and Young's inequality,

$$\tfrac{1}{2}\partial_t(k_n * \|u\|_H^2)(t) - \tfrac{1}{4}k_n(t)\|u(t)\|_H^2 \le k_n(t)\|u_0\|_H^2 + g(t) + \zeta_n(t). \tag{6.16}$$

To prove part (i), convolving with $k_{1-\alpha}$, and passing to limit lead to $\|u(t)\|_H^2 \le 2(\|u_0\|_H^2 + (k_{1-\alpha} * g)(t))$. Young's inequality then gives

$$\|u\|_{L^{2p}(I;H)} = \|\|u\|_H^2\|_{L^p(I)} \le 2(\|k_{1-\alpha}\|_{L^p(I)}\|g\|_{L^1(I)} + T^{\frac{1}{p}}\|u_0\|_H^2),$$

which together with the inequality $2\|g\|_{L^1(I)} \le \|\partial_t(k_\alpha * (u - u_0))\|_{L^2(I;V')}^2 + \|u\|_{L^2(I;V)}^2$ implies the desired estimate in (i). To show (ii), by integrating (6.16) over I and dropping the term $k_n * \|u(\cdot)\|_H^2$, we obtain

$$\int_0^T k_n(t)\|u(t)\|_H^2 dt \le 4\big((1 * k_n)(T)\|u_0\|_H^2 + \|g\|_{L^1(I)} + \|\zeta_n\|_{L^1(I)}\big).$$

Since $(1 * k_n)(T) \to (1 * k_\alpha)(T)$ and $\|\zeta_n\|_{L^1(I)} \to 0$ as $n \to \infty$, for sufficiently large n, there holds

$$\int_0^T k_n(t)\|u(t)\|_H^2 dt \le 8\big((1 * k_\alpha)(T)\|u_0\|_H^2 + \|g\|_{L^1(I)}\big).$$

Since $k_n \to k_\alpha$ in $L^1(I)$, the desired assertion then follows from Fatou's lemma. $\square$

We have the following corollary. The space $L^{p,\infty}(I)$ denotes the weak $L^p(I)$ space; see Section A.2.1 in the appendix for the definition.

Corollary 6.1 *Let Assumption 6.1 be fulfilled, $u_0 \in H$ and $f \in L^2(I;V')$, Then problem (6.3) has exactly one solution u in the space $W_\alpha(u_0, V, H) := \{u \in L^2(I;V) : u - u_0 \in {}_0W^{\alpha,2}(I;V')\}$. Furthermore, $k_\alpha * u \in C(\bar{I};H)$, $u \in L^{\frac{2}{1-\alpha},\infty}(I;H)$ and $\int_0^T t^{-\alpha}\|u(t)\|_H^2 dt < \infty$, and*

$$\|u - u_0\|_{W^{\alpha,2}(I;V')} + \|u\|_{L^2(I;V)} + \|k_\alpha * u\|_{C(\overline{I};H)} + \|u\|_{L^{\frac{2}{1-\alpha},\infty}(I;H)}$$

$$+ \left(\int_0^T t^{-\alpha} \|u(t)\|_H^2 \, dt \right)^{\frac{1}{2}} \le c(\alpha, T)(\|u_0\|_H + \|f\|_{L^2(I;V')}).$$

Note that formally letting $\alpha \to 1^-$ in the estimate (with $\mu = 0$) in the corollary recovers the well-known estimate for the solutions of abstract parabolic problems.

The energy argument (by using suitable test functions in the weak formulation) can handle integrodifferential equations involving a completely positive kernel under low regularity condition on the coefficients, and extends to certain semilinear/quasilinear problems. Many further results have been obtained using the approach. These include L^∞ bounds [Zac08, VZ10], weak Harnack's inequality [Zac13b], interior Hölder estimates using De Giorgi-Nash-type argument [Zac13a], and asymptotic decay [VZ15]. Sobolev regularity results were also derived recently [MMAK19, MMAK20] for the time-fractional advection-diffusion equation using an energy argument.

6.2 Linear Problems with Time-Independent Coefficients

Now we study linear subdiffusion with time-independent coefficients. This is a relatively simple case that allows more direct treatment via separation of variables and Laplace transform. Consider the following subdiffusion problem for u

$$\begin{cases} \partial_t^\alpha u = \mathcal{L}u + f, & \text{in } Q, \\ \quad u = 0, & \text{on } \partial_L Q, \\ u(\cdot, 0) = u_0, & \text{in } \Omega. \end{cases} \tag{6.17}$$

In the model, $\alpha \in (0, 1)$ is the fractional order, and $\partial_t^\alpha u$ denotes the left-sided Djrbashian-Caputo fractional derivative of the function u of order $\alpha \in (0, 1)$ with respect to time t (based at zero), i.e.,

$$\partial_t^\alpha u(x, t) = \frac{1}{\Gamma(1 - \alpha)} \int_0^t (t - s)^{-\alpha} \partial_s u(x, s) \, ds,$$

cf. Definition 2.3, and $\mathcal{L}$ is a time-independent strongly elliptic second-order differential operator, defined by

$$\mathcal{L}u = \sum_{i,j=1}^d \partial_{x_i}(a_{ij}(x)\partial_{x_j}u) - q(x)u(x), \quad x \in \overline{\Omega}, \tag{6.18}$$

where $q(x) \ge 0$ is smooth, $a_{ij} = a_{ji}, i, j = 1, \ldots, d$, and the symmetric matrix-valued function $a = [a_{ij}(x)] : \overline{\Omega} \to \mathbb{R}^{d \times d}$ is smooth and satisfies the uniform ellipticity

$$\lambda|\xi|^2 \le \xi \cdot a(x)\xi \le \lambda^{-1}|\xi|^2 \quad \forall \xi \in \mathbb{R}^d, x \in \overline{\Omega},$$

for some $\lambda \in (0, 1)$, where $\cdot$ and $|\cdot|$ denote the Euclidean inner product and norm, respectively. The precise regularity condition on f and u_0 will be specified later. We shall derive explicit solution representations using separation of variables and Laplace transform, and then establish the existence, uniqueness, and regularity theory in a Hilbert space setting.

Much of the material of this section can be traced back to the paper [SY11]: it gives a first rigorous treatment of the solution theory in the Hilbert space $\dot{H}^s(\Omega)$, regularity estimates and applications in inverse problems including backward subdiffusion and inverse source problem, etc.) and covering both cases of subdiffusion and diffusion wave problems. The analysis uses the standard separation of variables technique and decays property of $E_{\alpha,\beta}(z)$ in Theorem 3.2 and complete monotonicity of $E_{\alpha,1}(-t)$ in Theorem 3.5. Another early work is [McL10], which treats a slightly different mathematical model $\partial_t u + {}^R\partial_t^{1-\alpha} Au = f$. Our presentation is more operator theoretic (see e.g., the monograph [Prü93]). The current presentation using Laplace transform follows largely [JLZ18, Sections 2 and 3].

Most of our discussions are only for a zero Dirichlet boundary condition. The extension to a zero Neumann boundary condition is analogous. The case of a nonzero Dirichlet boundary in $L^2(\partial_L Q)$ was analyzed in a weak setting by the transposition method in [Yam18] and then applied to Dirichlet boundary control and [KY21] for further results (including Neumann one); see the monograph [KRY20] for a detailed treatment of FDES with the Djrbashian-Caputo fractional derivative in Sobolev spaces.

6.2.1 Solution representation

First, we derive a solution representation to problem (6.17) using the separation of variables technique. Let $\Omega \subset \mathbb{R}^d$ be an open bounded smooth domain with a boundary $\partial\Omega$. Let $A : H^2(\Omega) \cap H_0^1(\Omega) \to L^2(\Omega)$ be the realization of the time-independent second-order symmetric elliptic operator $-\mathcal{L}$ defined in (6.18) in the space $L^2(\Omega)$ with its domain $D(A) = \{v \in H_0^1(\Omega) : Av \in L^2(\Omega)\}$, i.e., with a zero Dirichlet boundary condition. It is unbounded, closed and, by elliptic regularity theory and Sobolev embedding theorem, its inverse $A^{-1} : L^2(\Omega) \to L^2(\Omega)$ is compact. Thus, by the standard spectral theory for compact operators, its spectrum is discrete, positive and accumulates at infinity. We repeat each eigenvalue of A according to its (finite) multiplicity: $0 < \lambda_1 < \lambda_2 \le \cdots \le \lambda_j \le \cdots \to \infty$, as $j \to \infty$, and we denote by $\varphi_j \in H^2(\Omega) \cap H_0^1(\Omega)$ an eigenfunction corresponding to λ_j, i.e.,

$$\begin{cases} -\mathcal{L}\varphi_j = \lambda_j\varphi_j, & \text{in } \Omega, \\ \varphi_j = 0, & \text{on } \partial\Omega. \end{cases}$$

The eigenfunctions $\{\varphi_j\}_{k=1}^\infty$ can be taken to form an orthonormal basis of $L^2(\Omega)$.

Then by multiplying both sides of problem (6.17) by φ_j, integrating over the domain Ω, and applying integration by parts twice, we obtain

$$\partial_t^\alpha (u(\cdot, t), \varphi_j) = (\mathcal{L}u(\cdot, t), \varphi_j) + (f(\cdot, t), \varphi_j)$$
$$= (u(\cdot, t), \mathcal{L}\varphi_j) + (f(\cdot, t), \varphi_j) = -\lambda_j (u(\cdot, t), \varphi_j) + (f(\cdot, t), \varphi_j),$$

with $(\cdot, \cdot)$ being the $L^2(\Omega)$ inner product. Let $u_j(t) = (u(\cdot, t), \varphi_j)$, $f_j(t) = (f(\cdot, t), \varphi_j)$ and $u_{0j} = (u_0, \varphi_j)$. Then we arrive at a system of fractional ODEs

$$\partial_t^\alpha u_j(t) = -\lambda_j u_j(t) + f_j(t), \quad \forall t > 0, \quad \text{with } u_j(0) = u_{0j},$$

for $j = 1, 2, \ldots$. It remains to find the scalar functions $u_j(t)$, $j = 1, 2, \ldots$. To this end, consider the following fractional ODE for $\lambda > 0$:

$$\partial_t^\alpha u_\lambda(t) = -\lambda u_\lambda(t) + f(t), \quad t > 0, \quad \text{with } u_\lambda(0) = c_0. \tag{6.19}$$

By means of Laplace transform, the unique solution $u_\lambda(t)$ is given by (cf. Proposition 4.5 in Chapter 4)

$$u_\lambda(t) = c_0 E_{\alpha,1}(-\lambda t^\alpha) + \int_0^t (t - s)^{\alpha-1} E_{\alpha,\alpha}(-\lambda(t - s)^\alpha) f(s) \, ds, \tag{6.20}$$

where $E_{\alpha,\beta}(z)$ is the Mittag-Leffler function defined in (3.1) in Chapter 3. Hence, the solution $u(t)$ to problem (6.17) can be formally represented by

$$u(x, t) = \sum_{j=1}^\infty (u_0, \varphi_j) \varphi_j(x) E_{\alpha,1}(-\lambda_j t^\alpha)$$
$$+ \sum_{j=1}^\infty \int_0^t (t - s)^{\alpha-1} E_{\alpha,\alpha}(-\lambda_j (t - s)^\alpha)(f(\cdot, s), \varphi_j) \, ds \varphi_j(x).$$

The solution $u(t)$ can be succinctly represented by

$$u(t) = F(t)u_0 + \int_0^t E(t - s) f(s) \, ds,$$

where the solution operators F and E are defined by

$$F(t)v = \sum_{j=1}^\infty E_{\alpha,1}(-\lambda_j t^\alpha)(v, \varphi_j) \varphi_j, \tag{6.21}$$

$$E(t)v = \sum_{j=1}^\infty t^{\alpha-1} E_{\alpha,\alpha}(-\lambda_j t^\alpha)(v, \varphi_j) \varphi_j. \tag{6.22}$$

respectively. The operators F and E denote the solution operator for problem (6.17) with $f \equiv 0$ and $u_0 \equiv 0$, respectively. Note that as $\alpha \to 1^-$, the two operators $F(t)$ and $E(t)$ coincide and recover that for the standard parabolic case.

Next, we re-derive the representation by means of (vector-valued) Laplace transform (cf. Section A.3.1). We begin with the necessary functional analytic framework. We define an operator A in $L^2(\Omega)$ by

$$(Au)(x) = (-\mathcal{L}u)(x), \quad x \in \Omega,$$

with its domain $D(A) = H^2(\Omega) \cap H_0^1(\Omega)$. Then it is known that the operator A satisfies the following resolvent estimate (The notation $\|\cdot\|$ denotes the operator norm from $L^2(\Omega)$ to $L^2(\Omega)$)

$$\|(z+A)^{-1}\| \le c_\theta |z|^{-1}, \quad \forall z \in \Sigma_\theta, \forall \theta \in (0, \pi), \tag{6.23}$$

with $\Sigma_\theta := \{0 \ne z \in \mathbb{C} : |\arg(z)| \le \theta\}$. Denote by $\widehat{u}$ the Laplace transform of u. We extend f from the interval I to $(0, \infty)$ by zero extension. By Lemma 2.9, the Laplace transform $\widehat{\partial_t^\alpha u}(z)$ of the Djrbashian-Caputo derivative $\partial_t^\alpha u$ is given by $\widehat{\partial_t^\alpha u}(z) = z^\alpha \widehat{u}(z) - z^{\alpha-1} u(0)$. Thus, applying Laplace transform to (6.17) leads to

$$z^\alpha \widehat{u}(z) + A\widehat{u}(z) = \widehat{f}(z) + z^{\alpha-1} u_0.$$

Thus, the Laplace transform $\widehat{u}$ of the solution u is given by

$$\widehat{u}(z) = (z^\alpha + A)^{-1}(\widehat{f}(z) + z^{\alpha-1} u_0).$$

By inverse Laplace transform and the convolution rule, we have

$$
\begin{aligned}
u(t) &= \frac{1}{2\pi i} \int_C e^{zt} z^{\alpha-1} (z^\alpha + A)^{-1} u_0 dz + \frac{1}{2\pi i} \int_C e^{zt} (z^\alpha + A)^{-1} \widehat{f}(z) dz \\
&= \frac{1}{2\pi i} \int_C e^{zt} z^{\alpha-1} (z^\alpha + A)^{-1} u_0 dz + \frac{1}{2\pi i} \int_0^t \left(\int_C e^{zs} (z^\alpha + A)^{-1} dz \right) f(t-s) ds,
\end{aligned}
$$

where $C \subset \mathbb{C}$ is a Hankel contour, oriented with an increasing imaginary part. Therefore, we obtain the following representation:

$$u(t) = F(t)u_0 + \int_0^t E(t-s) f(s) ds, \tag{6.24}$$

where the solution operators $F(t)$ and $E(t)$ are, respectively, defined by

$$F(t) := \frac{1}{2\pi i} \int_{\Gamma_{\theta,\delta}} e^{zt} z^{\alpha-1} (z^\alpha + A)^{-1} \, dz, \tag{6.25}$$

$$E(t) := \frac{1}{2\pi i} \int_{\Gamma_{\theta,\delta}} e^{zt} (z^\alpha + A)^{-1} \, dz, \tag{6.26}$$

with integral over a contour $\Gamma_{\theta,\delta} \subset \mathbb{C}$ (oriented with an increasing imaginary part), deformed from the contour C:

$$\Gamma_{\theta,\delta} = \{z \in \mathbb{C} : |z| = \delta, |\arg z| \le \theta\} \cup \{z \in \mathbb{C} : z = \rho e^{\pm i\theta}, \rho \ge \delta\}. \qquad (6.27)$$

Throughout, we fix $\theta \in (\frac{\pi}{2}, \pi)$ so that $z^\alpha \in \Sigma_{\alpha\theta}$ for all $z \in \Sigma_\theta$. One may deform the contour $\Gamma_{\theta,\delta}$ to obtain an explicit representation in terms of the eigenexpansion $\{(\lambda_j, \varphi_j)\}_{j \ge 1}$, and then recover (6.21) and (6.22). The details are left to an exercise. Below we employ both representations for the analysis.

Now we give several useful results. The first connects the operators E and F, where I denotes the identity operator.

Lemma 6.2 *The following identity holds* $AE(t) = -\frac{d}{dt}F(t)$.

Proof It follows from the identities $z^\alpha(z^\alpha + A)^{-1} = I - A(z^\alpha + A)^{-1}$ and $\int_{\Gamma_{\theta,\delta}} e^{zt} dz = 0$ that for any $t > 0$,

$$-\frac{d}{dt}F(t) = -\frac{1}{2\pi i} \int_{\Gamma_{\theta,\delta}} e^{zt} z^\alpha (z^\alpha + A)^{-1} dz$$

$$= -\frac{1}{2\pi i} \int_{\Gamma_{\theta,\delta}} e^{zt} (I - A(z^\alpha + A)^{-1}) dz$$

$$= A \frac{1}{2\pi i} \int_{\Gamma_{\theta,\delta}} e^{zt} (z^\alpha + A)^{-1} dz = AE(t).$$

This shows the assertion. Alternatively, this can be seen from (6.21) and (6.22)

$$\frac{d}{dt}F(t)v = \sum_{j=1}^{\infty} \frac{d}{dt} E_{\alpha,1}(-\lambda_j t^\alpha)(v, \varphi_j)\varphi_j$$

$$= \sum_{j=1}^{\infty} -\lambda_j t^{\alpha-1} E_{\alpha,\alpha}(-\lambda_j t^\alpha)(v, \varphi_j)\varphi_j = -AE(t)v.$$

This shows also the identity.   □

The next result gives the continuity of the operator $F(t)$ at $t = 0$.

Lemma 6.3 *For the operator F defined in (6.25),* $\lim_{t \to 0^+} \|I - F(t)\| = 0$.

Proof Note that

$$\frac{1}{2\pi i} \int_{\Gamma_{\theta,\delta}} z^{\alpha-1}(z^\alpha + A)^{-1} dz = \frac{1}{2\pi i} \int_{\Gamma_{\theta,\delta}} (z^{-1} - A(z^\alpha + A)^{-1}) dz = I.$$

This and the definition of $F(t)$ imply that for any $v \in L^2(\Omega)$ with $\|v\|_{L^2(\Omega)} \le 1$,

$$\lim_{t \to 0^+} \|F(t)v - v\|_{L^2(\Omega)} = \lim_{t \to 0^+} \frac{1}{2\pi} \left\| \int_{\Gamma_{\theta,\delta}} (e^{zt} - 1) z^{\alpha-1}(z^\alpha + A)^{-1} v dz \right\|_{L^2(\Omega)}.$$

Then Lebesgue's dominated convergence theorem yields the desired assertion. Alternatively, by representation (6.21),

$$\|F(t)v - v\|^2_{L^2(\Omega)} = \sum_{j=1}^{\infty} |(v, \varphi_j)|^2 (E_{\alpha,1}(-\lambda_j t^\alpha) - 1)^2$$

and $\lim_{t\to 0^+}(E_{\alpha,1}(-\lambda_j t^\alpha) - 1) = 0$, $j \in \mathbb{N}$. Meanwhile, by the complete monotonicity of the function $E_{\alpha,1}(-t)$ on $\mathbb{R}_+$ in Theorem 3.5, $E_{\alpha,1}(-t) \in [0, 1]$ for $t \geq 0$. Consequently, for any $0 \leq t \leq T$,

$$\sum_{j=1}^{\infty} |(v, \varphi_j)|^2 |E_{\alpha,1}(-\lambda_j t^\alpha) - 1|^2 \leq \sum_{j=1}^{\infty} |(v, \varphi_j)|^2 = \|v\|^2_{L^2(\Omega)} \leq 1.$$

Now Lebesgue's dominated convergence theorem yields the identity.   □

The next theorem summarizes smoothing properties of $F(t)$ and $E(t)$. The notation $F^{(k)}(t) = \frac{d^k}{dt^k} F(t)$ denotes the kth derivative of $F(t)$ in t, etc. (ii) indicates that unlike F, E can absorb A^2, which agrees with the asymptotic of $E_{\alpha,\alpha}(-x)$ in Theorem 3.2.

Theorem 6.4 *For any $k \in \mathbb{N}_0$, the operators F and E defined in (6.25)–(6.26) satisfy for any $t \in I$, the following estimates hold.*

(i) $t^{-\alpha}\|A^{-1}(I - F(t))\| + t^{1-\alpha}\|A^{-1}F'(t)\| \leq c$;
(ii) $t^{k+1-\alpha}\|E^{(k)}(t)\| + t^{k+1}\|AE^{(k)}(t)\| + t^{k+1+\alpha}\|A^2 E^{(k)}(t)\| \leq c$;
(iii) $t^k\|F^{(k)}(t)\| + t^{k+\alpha}\|AF^{(k)}(t)\| \leq c$;
(iv) $\|F(t)\| \leq E_{\alpha,1}(-\lambda_1 t^\alpha)$, $\|E(t)\| \leq t^{\alpha-1}E_{\alpha,\alpha}(-\lambda_1 t^\alpha)$.

Proof The proof employs the resolvent estimate (6.23). Obviously the following identity holds:

$$A(z^\alpha + A)^{-1} = I - z^\alpha(z^\alpha + A)^{-1},$$

and by the estimate (6.23),

$$\|A(z^\alpha + A)^{-1}\| \leq c, \quad \forall z \in \Gamma_{\theta,\delta}. \tag{6.28}$$

In part (i), by Lemma 6.2 and choosing $\delta = t^{-1}$ in the contour $\Gamma_{\theta,\delta}$ and letting $\hat{z} = tz$ (with $|dz|$ being the arc length element of $\Gamma_{\theta,\delta}$):

$$\|A^{-1}F'(t)\| = \|E(t)\| \leq \frac{1}{2\pi} \int_{\Gamma_{\theta,\delta}} e^{\Re(z)t}\|(z^\alpha + A)^{-1}\|\,|dz|$$

$$\leq ct^{\alpha-1}\int_{\Gamma_{\theta,1}} e^{\Re(\hat{z})}|\hat{z}|^{-\alpha}|d\hat{z}| \leq ct^{\alpha-1}\int_{\Gamma_{\theta,1}} e^{\cos(\theta)|\hat{z}|}(1 + |\hat{z}|^{-1})|d\hat{z}| \leq ct^{\alpha-1}.$$

Now for any $k \in \mathbb{N}_0$ and $m = 0, 1$, by choosing $\delta = t^{-1}$ in $\Gamma_{\theta,\delta}$ and changing variables $z = s\cos\varphi + is\sin\varphi$, we have

$$\|A^m F^{(k)}(t)\| \leq c\left\|\int_{\Gamma_{\theta,\delta}} e^{zt} z^{k+\alpha-1} A^m(z^\alpha + A)^{-1}\,dz\right\| \leq c\int_{\Gamma_{\theta,\delta}} e^{\Re(z)t}|z|^{k-1+m\alpha}\,|dz|$$

$$\leq c\int_{\delta}^{\infty} e^{st\cos\theta} s^{k-1+m\alpha}\,ds + c\int_{-\theta}^{\theta} e^{\cos\varphi} \delta^{k+m\alpha}\,d\varphi \leq ct^{-m\alpha-k}.$$

This implies

$$t^k\|F^{(k)}(t)\| + t^{k+\alpha}\|AF^{(k)}(t)\| \le c, \quad \forall t > 0,$$

showing (iii). Similarly, we can show (ii) by

$$\|A^m E^{(k)}(t)\| = \left\| \frac{1}{2\pi i} \int_{\Gamma_{\theta,\delta}} e^{zt} z^k A^m (z^\alpha + A)^{-1} dz \right\|$$

$$\le c \int_{\Gamma_{\theta,\delta}} e^{\Re(z)t} |z|^{k+(m-1)\alpha} |dz| \le ct^{(1-m)\alpha-k-1}.$$

Next, the proof of Lemma 6.2 gives

$$AE(t) = -\frac{1}{2\pi i} \int_{\Gamma_{\theta,\delta}} e^{zt} z^\alpha (z^\alpha + A)^{-1} dz.$$

It follows from this identity and (6.28) that

$$\|A^2 E^{(k)}(t)\| \le \frac{1}{2\pi} \int_{\Gamma_{\theta,\delta}} e^{\Re(z)t} |z|^{k+\alpha} |dz| \le ct^{-k-1-\alpha}.$$

Last, in view of the identity $F'(t) = -AE(t)$ from Lemma 6.2, $A^{-1}F'(t) = -E(t)$. Thus, (ii) implies the estimate $t^{\alpha-1}\|A^{-1}F'(t)\| \le c$ in (i). Now the bound on $\|A^{-1}(I - F(t))\|$ follows from Lemmas 6.2 and 6.3,

$$I - F(t) = \int_0^t \frac{d}{ds}(I - F(s))\, ds = \int_0^t AE(s)\, ds,$$

from which and (ii) it follows directly that

$$\|A^{-1}(I - F(t))\| \le \int_0^t \|E(s)\|\, ds \le c \int_0^t s^{\alpha-1}\, ds = ct^\alpha.$$

In (iv), the first estimate follows from Theorem 3.5

$$\|F(t)v\|^2_{L^2(\Omega)} = \sum_{j=1}^\infty E_{\alpha,1}(-\lambda_j t^\alpha)^2 (\varphi_j, v)^2$$

$$\le E_{\alpha,1}(-\lambda_1 t^\alpha)^2 \sum_{j=1}^\infty (\varphi_j, v)^2 = E_{\alpha,1}(-\lambda_1 t^\alpha)^2 \|v\|^2_{L^2(\Omega)}.$$

The second also holds since $E_{\alpha,\alpha}(-t)$ is completely monotone, cf. Corollary 3.2. $\square$

The following result is immediate from Theorem 6.4. The space $\dot{H}^s(\Omega)$ is defined in Section A.2.4 in the appendix.

Corollary 6.2 *The operators* $F(t)$ *and* $E(t)$ *defined in* (6.25)–(6.26) *satisfy*

$$\|F(t)v\|_{\dot{H}^\beta(\Omega)} \le ct^{\frac{\gamma-\beta}{2}\alpha}\|v\|_{\dot{H}^\gamma(\Omega)} \quad \text{and} \quad \|E(t)v\|_{\dot{H}^\beta(\Omega)} \le ct^{(1-\frac{\beta-\gamma}{2})\alpha-1}\|v\|_{\dot{H}^\gamma(\Omega)},$$

where $\beta, \gamma \in \mathbb{R}$ and $\gamma \leq \beta \leq \gamma + 2$, and the constant c depends only on α and $\beta - \gamma$.

Proof The estimates are direct from Theorem 6.4. We provide an alternative proof using the properties of $E_{\alpha,\beta}(z)$. Indeed, by the definition of $F(t)$,

$$\|F(t)v\|_{\dot{H}^\beta(\Omega)}^2 = \sum_{j=1}^\infty \lambda_j^\beta |E_{\alpha,1}(-\lambda_j t^\alpha)|^2 (v, \varphi_j)^2$$

$$= t^{(\gamma-\beta)\alpha} \sum_{j=1}^\infty \lambda_j^{\beta-\gamma} t^{(\beta-\gamma)\alpha} |E_{\alpha,1}(-\lambda_j t^\alpha)|^2 \lambda_j^\gamma (v, \varphi_j)^2.$$

By the decay property of the function $E_{\alpha,1}(-t)$ in Corollary 3.1, $|E_{\alpha,1}(-\lambda_j t^\alpha)| \leq c(1 + \lambda_j t^\alpha)^{-1}$, with $c = c(\alpha)$. For $0 \leq \beta - \gamma \leq 2$, there holds $\sup_j \lambda_j^{\beta-\gamma} t^{(\beta-\gamma)\alpha} (1 + \lambda_j t^\alpha)^{-2} \leq c$, with $c > 0$ depending only on α and $\beta - \gamma$. Thus,

$$\|F(t)v\|_{\dot{H}^\beta(\Omega)}^2 \leq ct^{(\gamma-\beta)\alpha} \sup_j \frac{\lambda_j^{\beta-\gamma} t^{(\beta-\gamma)\alpha}}{(1 + \lambda_j t^\alpha)^2} \sum_{j=1}^\infty \lambda_j^\gamma (v, \varphi_j)^2 \leq ct^{(\gamma-\beta)\alpha} \|v\|_{\dot{H}^\gamma(\Omega)}^2.$$

The other estimate can be derived similarly. $\square$

Remark 6.1 The restriction $\beta \leq \gamma + 2$ in Corollary 6.2 indicates that $F(t)$ has at best an order two smoothing in space, which contrasts sharply with the classical parabolic case, for which there holds

$$\|F(t)v\|_{\dot{H}^\beta(\Omega)} \leq ct^{\frac{\gamma-\beta}{2}} \|v\|_{\dot{H}^\gamma(\Omega)}, \quad \forall \beta \geq \gamma.$$

This restriction is due to the sublinear decay of $E_{\alpha,1}(-\lambda_j t^\alpha)$ on $\mathbb{R}_+$ for $0 < \alpha < 1$, instead of the exponential decay of $e^{-\lambda_j t}$ in the standard diffusion case. Limited smoothing is characteristic of many nonlocal models.

Before turning to the regularity issue, we note that a fractional analogue of the classical Duhamel principle holds for subdiffusion, which allows representing the solution of an inhomogeneous problem in terms of solutions of the associated homogeneous problems [US07]. We use the notation

$$^R\partial_t^\alpha v(\cdot, t; s) = \frac{d}{dt} \frac{1}{\Gamma(1-\alpha)} \int_s^t (t-r)^{-\alpha} v(\cdot, r; s) dr,$$

$$\partial_t^\alpha v(\cdot, t; s) = \frac{1}{\Gamma(1-\alpha)} \int_s^t (t-r)^{-\alpha} \partial_r v(\cdot, r; s) dr.$$

Theorem 6.5 Let $f : Q \to \mathbb{R}$ be a smooth function. If a smooth function u satisfies

$$\begin{cases} \partial_t^\alpha u + Au = f, & \text{in } Q, \\ u(\cdot, 0) = 0, & \text{in } \Omega, \end{cases} \tag{6.29}$$

then u can be represented by

$$u(\cdot, t) = \int_0^t {}^R\partial_t^{1-\alpha} v(\cdot, t; s) ds, \quad t \in I, \tag{6.30}$$

where $v(\cdot, \cdot; s)$ satisfies (with a parameter $s \in (0, T)$)

$$\begin{cases} \partial_t^\alpha v + Av = 0, & \text{in } \Omega \times (s, T], \\ v(\cdot, s) = f(\cdot, s), & \text{in } \Omega. \end{cases}$$

Proof It suffices to prove that the function u indeed satisfies (6.29). For smooth f, u and v, we may take any derivatives when needed, and further, we omit the dependence on x by writing $u(\cdot, t)$ as $u(t)$ etc. In the analysis, we employ the following identity

$$ {}^R\partial_t^{1-\alpha} f(t) = \frac{f(0)t^{\alpha-1}}{\Gamma(\alpha)} + \partial_t^{1-\alpha} f(t), \tag{6.31}$$

cf. Theorem 2.12. Since v is sufficiently smooth, we pass to $t \to 0^+$ in the definition of u in (6.30) and obtain $u(0) = 0$ by the absolute continuity of Lebesgue integral. Next, the definition of u in (6.30) and the initial condition for v give

$$\partial_t u(t) = {}^R\partial_t^{1-\alpha} v(t; t) + \int_0^t \partial_t {}^R\partial_t^{1-\alpha} v(t; s) ds$$

$$= {}^R\partial_t^{1-\alpha} f(t) + \int_0^t \partial_t {}^R\partial_t^{1-\alpha} v(t; s) ds.$$

Meanwhile, substituting it into the definition $\partial_t^\alpha u = \frac{1}{\Gamma(1-\alpha)} \int_0^t (t-s)^{-\alpha} \partial_s u(s) ds$ gives

$$\partial_t^\alpha u(t) = \frac{1}{\Gamma(1-\alpha)} \Big(\int_0^t (t-s)^{-\alpha} {}^R\partial_s^{1-\alpha} f(s) ds$$

$$+ \int_0^t (t-s)^{-\alpha} \int_0^s {}^R\partial_s^{1-\alpha} \partial_s v(s; r) dr ds \Big).$$

Next, we simplify the two terms in the bracket, denoted by I_1 and I_2. By the identities (6.31) and $\int_r^t (t-s)^{-\alpha} (s-r)^{\alpha-1} ds = \Gamma(\alpha)\Gamma(1-\alpha)$, for $r < t$, we deduce

$$I_1 = \frac{1}{\Gamma(\alpha)} \int_0^t (t-s)^{-\alpha} \Big(f(0)s^{\alpha-1} + \int_0^s (s-r)^{\alpha-1} \partial_r f(r) dr \Big) ds$$

$$= \frac{1}{\Gamma(\alpha)} \Big(f(0) \int_0^t (t-s)^{-\alpha} s^{\alpha-1} ds + \int_0^t \partial_r f(r) \int_r^t (t-s)^{-\alpha} (s-r)^{\alpha-1} ds dr \Big)$$

$$= \Gamma(1-\alpha) \Big(f(0) + \int_0^t \partial_r f(r) dr \Big) = \Gamma(1-\alpha) f(t).$$

For the term I_2, the identity (6.31) and changing integration order give

$$I_2 = \frac{1}{\Gamma(\alpha)} \int_0^t (t-s)^{-\alpha} \int_0^s (s-r)^{\alpha-1} \partial_r v(r;r) dr ds$$

$$+ \frac{1}{\Gamma(\alpha)} \int_0^t (t-s)^{-\alpha} \int_0^s \int_r^s (s-\xi)^{\alpha-1} \partial_\xi^2 v(\xi;r) d\xi dr ds$$

$$= \frac{1}{\Gamma(\alpha)} \int_0^t \partial_r v(r;r) \left(\int_r^t (t-s)^{-\alpha}(s-r)^{\alpha-1} ds \right) dr$$

$$+ \frac{1}{\Gamma(\alpha)} \int_0^t \int_0^\xi \partial_\xi^2 v(\xi;r) \left(\int_\xi^t (t-s)^{-\alpha}(s-\xi)^{\alpha-1} ds \right) dr d\xi$$

$$= \Gamma(1-\alpha) \left(\int_0^t \partial_r v(r;r) dr + \int_0^t \int_0^\xi \partial_\xi^2 v(\xi;r) dr d\xi \right).$$

Meanwhile, since the operator A is independent of t and v is smooth, $(\partial_t^\alpha v)(s,s) = 0$, cf. Lemma 2.13, and thus there holds

$$Au(t) = \int_0^t {}^R \partial_t^{1-\alpha} Av(t;s) ds = - \int_0^t {}^R \partial_t^{1-\alpha} \partial_t^\alpha v(t;s) ds = - \int_0^t \partial_t^{1-\alpha} \partial_t^\alpha v(t;s) ds.$$

Substituting the definition of $\partial_t^{1-\alpha} v(t;s)$ and applying the identity (6.31) lead to

$$-Au(t) = \frac{1}{\Gamma(\alpha)\Gamma(1-\alpha)} \int_0^t \int_s^t (t-r)^{\alpha-1} \partial_r \int_s^r (r-\xi)^{-\alpha} \partial_\xi v(\xi;s) d\xi dr ds$$

$$= \frac{1}{\Gamma(\alpha)\Gamma(1-\alpha)} \left(\int_0^t \partial_s v(s,s) \left(\int_s^t (t-r)^{\alpha-1}(r-s)^{-\alpha} dr \right) ds + I_3 \right),$$

where the term I_3 is given by

$$I_3 = \int_0^t \int_s^t (t-r)^{\alpha-1} \int_s^r (r-\xi)^{-\alpha} \partial_\xi^2 v(\xi;s) d\xi dr ds$$

$$= \int_0^t \int_s^t \left(\int_\xi^t (t-r)^{\alpha-1}(r-\xi)^{-\alpha} dr \right) \partial_\xi^2 v(\xi;s) d\xi ds$$

$$= \Gamma(\alpha)\Gamma(1-\alpha) \int_0^t \int_s^t \partial_\xi^2 v(\xi;s) d\xi ds.$$

The preceding identities together imply

$$(\partial_t^\alpha u + Au)(t) = \frac{I_1 + I_2}{\Gamma(1-\alpha)} - \int_0^t \partial_s v(s;s) ds - \frac{I_3}{\Gamma(\alpha)\Gamma(1-\alpha)} = f(t).$$

Thus, the function u defined in (6.30) satisfies (6.29).   □

6.2.2 Existence, uniqueness and regularity

Below we show the existence, uniqueness and regularity of a solution to problem (6.17), using the solution representation (6.24). First, we introduce the concept of weak solutions for problem (6.17).

Definition 6.1 We call u a weak solution to problem (6.17) if the equation in (6.17) holds in $L^2(\Omega)$ and $u(t) \in H_0^1(\Omega)$ for almost all $t \in I$ and $u \in C(\bar{I}; \dot{H}^{-\gamma}(\Omega))$, with

$$\lim_{t \to 0^+} \|u(\cdot, t) - u_0\|_{\dot{H}^{-\gamma}(\Omega)} = 0,$$

for some $\gamma \geq 0$, which may depend on α.

The following existence and uniqueness result holds for problem (6.17) with $f \equiv 0$. The inhomogeneous case can be analyzed similarly. Alternatively, one may employ the standard Galerkin approximation and energy estimates as in Section 6.1.

Theorem 6.6 *If $u_0 \in L^2(\Omega)$ and $f \equiv 0$, then there exists a unique weak solution $u \in C(\bar{I}; L^2(\Omega)) \cap C(I; \dot{H}^2(\Omega))$ in the sense of Definition 6.1.*

Proof First, we show that the representation $u(t) = F(t)u_0$ gives a weak solution to problem (6.17). By Corollary 6.2 (with $\beta = \gamma = 0$), we have

$$\|u(t)\|_{L^2(\Omega)} = \|F(t)u_0\|_{L^2(\Omega)} \leq c\|u_0\|_{L^2(\Omega)}, \tag{6.32}$$

and by Corollary 6.2 (with $\beta = 2$ and $\gamma = 0$), for any $t > 0$,

$$\|u(t)\|_{\dot{H}^2(\Omega)} = \|F(t)u_0\|_{\dot{H}^2(\Omega)} \leq ct^{-\alpha}\|u_0\|_{L^2(\Omega)}. \tag{6.33}$$

Further, by Theorem 6.4, $u \in C(\bar{I}; L^2(\Omega)) \cap C(I; \dot{H}^2(\Omega))$. Now using the governing equation $\partial_t^\alpha u = Au$ in (6.17), $\partial_t^\alpha u \in C(I; L^2(\Omega))$, and thus the equation is satisfied in $L^2(\Omega)$ a.e. Further, by Lemma 6.3, $u(t) = F(t)u_0$ satisfies

$$\lim_{t \to 0^+} \|u(t) - u_0\|_{L^2(\Omega)} \leq \lim_{t \to 0^+} \|F(t) - I\|\|u_0\|_{L^2(\Omega)} = 0. \tag{6.34}$$

Thus, $u(t) = F(t)u_0$ is indeed a solution to problem (6.17) in the sense of Definition 6.1. Next, we show the uniqueness of the weak solution. It suffices to show that problem (6.17) with $u_0 \equiv 0$ and $f \equiv 0$ has only a trivial solution. By taking inner product with φ_j and setting $u_j(t) = (u(t), \varphi_j)$ give

$$\partial_t^\alpha u_j(t) = -\lambda_j u_j(t), \quad \forall t \in I.$$

Since $u(t) \in L^2(\Omega)$ for $t \in I$, it follows from (6.34) that $u_j(0) = 0$. Due to the existence and uniqueness of solutions to fractional ODEs in Theorem 4.9 in Chapter 4, we deduce $u_j(t) = 0, j = 1, 2, \ldots$. Since $\{\varphi_j\}_{j=1}^\infty$ is an orthonormal basis in $L^2(\Omega)$, we have $u \equiv 0$ in Q. This shows the uniqueness, and completes the proof. $\square$

Now we derive the $\dot{H}^s(\Omega)$ regularity for $f \equiv 0$.

Theorem 6.7 *If $u_0 \in \dot{H}^\gamma(\Omega)$, with $0 \leq \gamma \leq 2$, and $f \equiv 0$. Then the solution u to problem (6.17) belongs to $C(\bar{I}; \dot{H}^\gamma(\Omega)) \cap C(I; \dot{H}^\beta(\Omega))$ for $\gamma \leq \beta \leq \gamma + 2$, and $\partial_t^\alpha u \in C(I; \dot{H}^\gamma(\Omega))$ and satisfies for any $k \in \mathbb{N}$,*

$$\left\| u^{(k)}(t) \right\|_{\dot{H}^\beta(\Omega)} \leq ct^{-\frac{(\beta-\gamma)\alpha}{2}-k} \|u_0\|_{\dot{H}^\gamma(\Omega)}, \quad \gamma \leq \beta \leq \gamma + 2, \qquad (6.35)$$

$$\|\partial_t^\alpha u(t)\|_{\dot{H}^\beta(\Omega)} \leq ct^{-\frac{\beta-\gamma+2}{2}\alpha} \|u_0\|_{\dot{H}^\gamma(\Omega)}, \quad \gamma - 2 \leq \beta \leq \gamma. \qquad (6.36)$$

Proof By (6.24), the solution u is given by $u(t) = F(t)u_0$. Then the bound (6.35) follows from Theorem 6.4(iii), and the bound (6.36) from the governing equation $\partial_t^\alpha u = Au$. The continuity of $u(t)$ in $\dot{H}^\gamma(\Omega)$ up to $t = 0$ is proven in Lemma 6.3. $\square$

Remark 6.2 Theorem 6.7 indicates that for any $t > 0$, $\gamma \geq 0$, $0 \leq \beta \leq \gamma + 2$ and $k = 1, 2, \ldots,$

$$\|u^{(k)}(t)\|_{\dot{H}^\beta(\Omega)} \leq ct^{-k-\frac{\beta-\gamma}{2}\alpha} \|u_0\|_{\dot{H}^\gamma(\Omega)}, \quad \text{for } t > 0. \qquad (6.37)$$

In contrast, for classical diffusion, we have

$$\|u^{(k)}(t)\|_{\dot{H}^\beta(\Omega)} \leq ct^{-k-\frac{\beta-\gamma}{2}} \|u_0\|_{\dot{H}^\gamma(\Omega)}$$

for every $\beta > \gamma$. The slow decay of $E_{\alpha,1}(-\lambda_j t^\alpha)$ accounts for the restriction $\beta \leq \gamma + 2$ in Corollary 6.2. Further, for any $u_0 \in L^2(\Omega)$ and $f \equiv 0$, the unique weak solution $u \in C([0, \infty); L^2(\Omega)) \cap C((0, \infty); \dot{H}^2(\Omega))$ to problem (6.17) satisfies

$$\|u(t)\|_{L^2(\Omega)} \leq c(1 + \lambda_1 t^\alpha)^{-1} \|u_0\|_{L^2(\Omega)}, \quad \forall t \geq 0.$$

This is direct from Theorem 6.4(iv) and Corollary 3.1.

We shall need the $L^p(I; X)$-norm and $L^{p,\infty}(I; X)$-norm (cf. [BL76, section 1.3])

$$\|u\|_{L^p(I;X)} := \left(\int_I \|u(t)\|_X^p dt \right)^{\frac{1}{p}},$$

$$\|u\|_{L^{p,\infty}(I;X)} := \sup_{\lambda>0} \lambda |\{t \in I : \|u(t)\|_X > \lambda\}|^{\frac{1}{p}}.$$

The next result is direct from Theorem 6.7. It can be viewed as the maximal L^p regularity for homogeneous subdiffusion. The notation $D(A)$ denotes the domain of the operator A, equipped with the graph norm.

Theorem 6.8 *The following maximal L^p-regularity estimates hold*

$$\|\partial_t^\alpha u\|_{L^p(I;L^2(\Omega))} + \|Au\|_{L^p(I;L^2(\Omega))} \leq c_p \|u_0\|_{(L^2(\Omega),D(A))_{1-\frac{1}{p\alpha},p}}, \quad p \in (\tfrac{1}{\alpha}, \infty],$$

$$\|\partial_t^\alpha u\|_{L^{p,\infty}(I;L^2(\Omega))} + \|Au\|_{L^{p,\infty}(I;L^2(\Omega))} \leq c_p \|u_0\|_{L^2(\Omega)}, \quad p = \tfrac{1}{\alpha},$$

$$\|\partial_t^\alpha u\|_{L^p(I;L^2(\Omega))} + \|Au\|_{L^p(I;L^2(\Omega))} \leq c_p \|u_0\|_{L^2(\Omega)}, \quad p \in [1, \tfrac{1}{\alpha}),$$

where the constant c_p depends on p.

Proof The mapping properties in Theorem 6.4 imply $\|Au\|_{L^2(\Omega)} \le c\|Au_0\|_{L^2(\Omega)}$. Similarly, one can show

$$\|u\|_{L^2(\Omega)} \le c\|u_0\|_{L^2(\Omega)} \quad \text{and} \quad \|Au\|_{L^2(\Omega)} \le ct^{-\alpha}\|u_0\|_{L^2(\Omega)}.$$

This last estimate immediately implies the third assertion. Now for $p \in (\frac{1}{\alpha}, \infty]$, the last two estimates imply

$$\|u\|_{L^\infty(I;D(A))} \le c\|u_0\|_{D(A)} \quad \text{and} \quad \|u\|_{L^{\frac{1}{\alpha},\infty}(I;D(A))} \le c\|u_0\|_{L^2(\Omega)},$$

which imply the first assertion with $p = \infty$ and the second assertion, respectively. The real interpolation of the last two estimates yields

$$\|u\|_{(L^{\frac{1}{\alpha},\infty}(I;D(A)),L^\infty(I;D(A)))_{1-\frac{1}{\alpha p},p}} \le c\|u_0\|_{(L^2(\Omega),D(A))_{1-\frac{1}{\alpha p},p}}, \quad \forall p \in (\alpha^{-1}, \infty).$$

Since $(L^{\frac{1}{\alpha},\infty}(I;, D(A)), L^\infty(I; D(A)))_{1-\frac{1}{\alpha p},p} = L^p(I; D(A))$ [BL76, Theorem 5.2.1], this implies the first assertion in the case $p \in (\frac{1}{\alpha}, \infty)$. $\qquad \square$

The next result gives the analyticity of the solution $u(t)$. It will play a role in investigating maximum principle and inverse problems.

Proposition 6.1 *If $u_0 \in L^2(\Omega)$, then the solution $u : I \to L^2(\Omega)$ to problem (6.17) with $f \equiv 0$ is analytic in the sector $\Sigma =: \{z \in \mathbb{C} : z \ne 0, |\arg(z)| \le \frac{\pi}{2}\}$.*

Proof Since $E_{\alpha,1}(-z)$ is entire, cf. Proposition 3.1 in Chapter 3, $E_{\alpha,1}(-\lambda_n t^\alpha)$ is analytic in $\Sigma \subset \mathbb{C}$. Hence, the finite sum

$$u_N(t) = \sum_{j=1}^N (u_0, \varphi_j) E_{\alpha,1}(-\lambda_j t^\alpha)\varphi_j$$

is analytic in Σ. Further, by Corollary 3.1, for any $z \in \overline{\Sigma}$,

$$\|u_N(z) - u(z)\|_{L^2(\Omega)}^2 = \sum_{j=N+1}^\infty (u_0, \varphi_j)^2 |E_{\alpha,1}(-\lambda_j z^\alpha)|^2 \le c \sum_{j=N+1}^\infty |(u_0, \varphi_j)|^2,$$

i.e., $\lim_{N\to\infty} \|u_N(z) - u(z)\|_{L^\infty(\overline{\Sigma};L^2(\Omega))} = 0$. Hence, u is actually analytic in Σ. $\qquad \square$

Example 6.2 Consider problem (6.17) on the unit interval $\Omega = (0, 1)$, i.e.,

$$\begin{cases} \partial_t^\alpha u = \partial_{xx}^2 u, & \text{in } Q, \\ u(\cdot, t) = 0, & \text{on } \partial_L Q, \\ u(\cdot, 0) = u_0, & \text{in } \Omega, \end{cases}$$

with (i) $u_0(x) = \sin \pi x$ and (ii) $u_0(x) = \delta_{\frac{1}{2}}(x)$, the Dirac delta function at $x = \frac{1}{2}$. In case (i), u_0 belongs to $\dot{H}^\gamma(\Omega)$ for any $\gamma > 0$, and $u(x, t) = E_{\alpha,1}(-\pi^2 t^\alpha) \sin \pi x$.

Despite the smoothness of u_0, the temporal regularity of u is limited for any $\alpha \in (0, 1)$: $E_{\alpha,1}(-\pi^2 t^\alpha) \sim 1 - \frac{\pi^2}{\Gamma(\alpha+1)} t^\alpha$, as $t \to 0^+$, which is continuous at $t = 0$ but $u'(t)$ is unbounded. It contrasts with the case $\alpha = 1$, for which $u(x, t) = e^{-\pi^2 t} \sin(\pi x)$ and is $C^\infty[0, T]$. Thus, the temporal regularity in Theorem 6.7 is sharp. In case (ii), by Sobolev embedding in Theorem A.3, u_0 belongs to $\dot{H}^\gamma(\Omega)$ for any $\gamma < -(\frac{1}{2} + \epsilon)$, with $\epsilon > 0$. In Fig. 6.1, we show the solution profiles for $\alpha = 0.5$ and $\alpha = 1$. For any $t > 0$, u is very smooth in x for $\alpha = 1$, but it remains nonsmooth for $\alpha = 0.5$. Actually, the kink at $x = 0.5$ remains no matter how long the problem evolves, showing the limited spatial smoothing property of the operator $F(t)$.

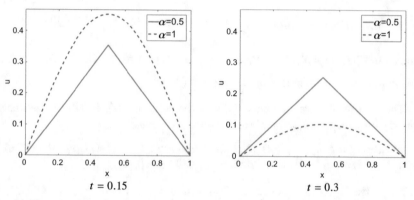

Fig. 6.1 The solution profiles for Example 6.2(ii) at two time instances for $\alpha = 0.5$ and 1.

Now we state two further results on the uniqueness of the solutions.

Theorem 6.9 Let $u_0 \in \dot{H}^\gamma(\Omega)$ with $\gamma > d$. Let $u \in C(\bar{I}; L^2(\Omega)) \cap C(I; \dot{H}^2(\Omega))$ satisfy (6.17) with $f \equiv 0$. Let $\omega \subset \Omega$ be an arbitrary subdomain. Then $u(x, t) = 0$ for $(x, t) \in \omega \times I$ implies $u \equiv 0$ in Q.

Proof By Sobolev embedding in Theorem A.3, since $\|\varphi_j\|_{L^2(\Omega)} = 1$,

$$\|\varphi_j\|_{L^\infty(\Omega)} \le c \|\varphi_j\|_{\dot{H}^{\frac{\gamma}{2}}(\Omega)} \le c \|A^{\frac{\gamma}{4}} \varphi_j\|_{L^2(\Omega)} \le c |\lambda_j|^{\frac{\gamma}{4}}.$$

Then in view of the well-known asymptotic $\lambda_j = O(j^{\frac{2}{d}})$ [Wey12] and the condition $u_0 \in \dot{H}^\gamma(\Omega)$ with $\gamma > d$, the Cauchy-Schwarz inequality, we deduce

$$\sum_{j=1}^\infty |(u_0, \varphi_j)| \|\varphi_j\|_{L^\infty(\Omega)} \le c \sum_{j=1}^\infty |(u_0, \varphi_j)| |\lambda_j|^{\frac{\gamma}{4}} = c \sum_{j=1}^\infty |(u_0, \varphi_j)| |\lambda_j|^{\frac{\gamma}{2}} \lambda_j^{-\frac{\gamma}{4}}$$

$$\le c \Big(\sum_{j=1}^\infty (u_0, \varphi_j)^2 |\lambda_j|^\gamma \Big)^{\frac{1}{2}} \Big(\sum_{j=1}^\infty \lambda_j^{-\frac{\gamma}{2}} \Big)^{\frac{1}{2}} < \infty. \qquad (6.38)$$

By Proposition 6.1, the function $\sum_{j=1}^{\infty}(u_0, \varphi_j)E_{\alpha,1}(-\lambda_j t^{\alpha})\varphi_j(x)$ can be extended analytically in t to a sector $\{z \in \mathbb{C} : z \neq 0, |\arg z| \leq \theta_0\}$ for some $\theta_0 > 0$. Therefore,

$$u(x, t) = \sum_{j=1}^{\infty}(u_0, \varphi_j)E_{\alpha,1}(-\lambda_j t^{\alpha})\varphi_j(x) = 0, \quad \forall(x, t) \in \omega \times (0, T)$$

implies

$$u(x, t) = \sum_{j=1}^{\infty}(u_0, \varphi_j)E_{\alpha,1}(-\lambda_j t^{\alpha})\varphi_j(x) = 0, \quad \forall(x, t) \in \omega \times (0, \infty).$$

Let $\sigma(A) = \{\mu_k\}_{k=1}^{\infty}$ be the spectrum of A and m_k the multiplicity of the kth eigenvalue μ_k, and denote by $\{\varphi_{kj}\}_{j=1}^{m_k}$ an $L^2(\Omega)$ orthonormal basis of $\mathrm{Ker}(\mu_k - A)$ (i.e., the spectrum $\sigma(A)$ is a set instead of a sequence with multiplicities). Then we can rewrite the series into

$$\sum_{k=1}^{\infty}\left(\sum_{j=1}^{m_k}(u_0, \varphi_{kj})\varphi_{kj}\right)E_{\alpha,1}(-\mu_k t^{\alpha}) = 0, \quad (x, t) \in \omega \times (0, \infty).$$

By the complete monotonicity of $E_{\alpha,1}(-x)$ in Theorem 3.5 and (6.38), we deduce

$$\sum_{k=1}^{\infty}\sum_{j=1}^{m_k}|(u_0, \varphi_{kj})||\varphi_{kj}(x)||E_{\alpha,1}(-\mu_k t^{\alpha})| \leq \sum_{k=1}^{\infty}\sum_{j=1}^{m_k}|(u_0, \varphi_{kj})|\|\varphi_{kj}\|_{L^{\infty}(\Omega)} < \infty.$$

Thus, Lebesgue's dominated convergence theorem yields

$$\int_0^{\infty}e^{-zt}\left(\sum_{k=1}^{\infty}\sum_{j=1}^{m_k}(u_0, \varphi_{kj})\varphi_{kj}(x)E_{\alpha,1}(-\mu_k t^{\alpha})\right)dt$$

$$= \sum_{k=1}^{\infty}\sum_{j=1}^{m_k}(u_0, \varphi_{kj})\left(\int_0^{\infty}e^{-zt}E_{\alpha,1}(-\mu_k t^{\alpha})\right)dt\varphi_{kj}(x), \quad x \in \omega, \mathfrak{R}(z) > 0.$$

By Lemma 3.2, $\int_0^{\infty}e^{-zt}E_{\alpha,1}(-\lambda t^{\alpha})dt = \frac{z^{\alpha-1}}{z^{\alpha}+\lambda}$, $\mathfrak{R}(z) > 0$, and thus

$$\sum_{k=1}^{\infty}\sum_{j=1}^{m_k}(u_0, \varphi_{kj})\frac{z^{\alpha-1}}{z^{\alpha}+\mu_k}\varphi_{kj}(x) = 0, \quad x \in \omega, \mathfrak{R}(z) > 0,$$

i.e.,

$$\sum_{k=1}^{\infty}\sum_{j=1}^{m_k}(u_0, \varphi_{kj})\frac{1}{\eta+\mu_k}\varphi_{kj}(x) = 0, \quad x \in \omega, \mathfrak{R}(\eta) > 0. \tag{6.39}$$

By (6.38), we can analytically continue both sides of (6.39) in η so that the identity (6.39) holds for $\eta \in \mathbb{C}\setminus\{-\mu_k\}_{k=1}^{\infty}$. Next, we take a suitable small disk which includes $-\mu_\ell$ and does not include $\{-\mu_k\}_{k\neq\ell}$. Integrating (6.39) along this disk gives

$$u_\ell(x) = \sum_{j=1}^{m_\ell} (u_0, \varphi_{\ell j}) \varphi_{\ell j}(x) = 0, \quad x \in \omega.$$

Since $(\mu_\ell - A)u_\ell = 0$ in Ω, and $u_\ell = 0$ in ω. The unique continuation principle for elliptic PDEs [Isa06, Section 3.3] implies $u_\ell = 0$ in Ω for each $\ell \in \mathbb{N}$. Since the set of functions $\{\varphi_{\ell j}\}_{j=1}^{m_\ell}$ is linearly independent in Ω, $(u_0, \varphi_{\ell j}) = 0$ for $j = 1, \ldots, m_\ell$, $\ell \in \mathbb{N}$. Therefore, $u \equiv 0$ in Q. $\square$

Remark 6.3 Theorem 6.9 corresponds to [SW78, Corollary 2.3]. For $\alpha = 1$, the uniqueness holds without the boundary condition (i.e., unique continuation). However, for $\alpha \in (0, 1)$, it is unclear whether the uniqueness holds in that case.

The next result gives a converse statement to the asymptotic decay (6.37). It asserts that the solution to the homogeneous problem cannot decay faster than t^{-m} with any $m \in \mathbb{N}$, if the solution does not vanish identically. This is distinct for subdiffusion because the classical diffusion (i.e., with $\alpha = 1$) admits nonzero solutions decaying exponentially fast. This is one description of the slow diffusion modeled by subdiffusion when compared with the classical case.

Theorem 6.10 *Let $u_0 \in \dot{H}^\beta(\Omega)$ with $\beta > d$ and $\omega \subset \Omega$ be an arbitrary subdomain. Let $u \in C(\bar{I}; L^2(\Omega)) \cap C(I; \dot{H}^2(\Omega))$ satisfy problem (6.17) with $f \equiv 0$. If for any $m \in \mathbb{N}$, there exists a constant $c(m) > 0$ such that*

$$\|u(\cdot, t)\|_{L^\infty(\omega)} \le c(m)t^{-m} \quad \text{as } t \to \infty, \tag{6.40}$$

then $u \equiv 0$ in $\Omega \times (0, \infty)$.

Proof By (6.38) and since $0 \le E_{\alpha,1}(-\eta) \le 1$ on $\mathbb{R}_+$, cf. Theorem 3.5, we deduce

$$u(x, t) = \sum_{k=1}^\infty \sum_{j=1}^{m_k} (u_0, \varphi_{k_j}) E_{\alpha,1}(-\mu_k t^\alpha) \varphi_{k_j}(x)$$

converges uniformly for $x \in \bar{\Omega}$ and $\delta \le t \le T$ with any $\delta, T > 0$. Hence, by Theorem 3.2, for any $p \in \mathbb{N}$, we have

$$u(x, t) = -\sum_{k=1}^\infty \sum_{j=1}^{m_k} \sum_{\ell=1}^p \frac{(-1)^\ell}{\Gamma(1 - \alpha\ell)\mu_k^\ell t^{\alpha\ell}} (u_0, \varphi_{k_j}) \varphi_{k_j}(x)$$

$$+ \sum_{k=1}^\infty \sum_{j=1}^{m_k} O\left(\frac{1}{\mu_k^{p+1} t^{\alpha(p+1)}}\right)(u_0, \varphi_{k_j}) \varphi_{k_j}(x) \quad \text{as } t \to \infty. \tag{6.41}$$

Since $\alpha \in (0, 1)$, $\Gamma(1 - \alpha) \ne 0$ by $1 - \alpha > 0$. By setting $m = 1$ in (6.40) and $p = 1$ in the identity (6.41), multiplying by t^α and letting $t \to \infty$, we deduce

$$\sum_{k=1}^\infty \sum_{j=1}^{m_k} \frac{1}{\Gamma(1 - \alpha)\mu_k} (u_0, \varphi_{k_j}) \varphi_{k_j}(x) = 0 \quad x \in \omega.$$

By $0 < \alpha < 1$, there exists $\{\ell_i\}_{i \in \mathbb{N}}$ such that $\lim_{i \to \infty} \ell_i = \infty$ and $\alpha \ell_i \notin \mathbb{N}$. In fact, let $\alpha \notin \mathbb{Q}$. Then $\ell \alpha \notin \mathbb{N}$ for any $\ell \in \mathbb{N}$. Meanwhile for $\alpha \in \mathbb{Q}$, i.e., $\alpha = \frac{n_1}{m_1}$, where $m_1, n_1 \in \mathbb{N}$ have no common divisors except for 1. There exist infinitely many $\ell \in \mathbb{N}$ with no common divisors with m_1, and $\ell \alpha \in \mathbb{Q} \setminus \mathbb{N}$. Then $\frac{1}{\Gamma(1-\alpha \ell_i)} \neq 0$. Therefore, by setting $p = 2, 3, \ldots$ and repeating the preceding argument, we obtain

$$\sum_{k=1}^{\infty} \frac{1}{\mu_k^{\ell_i}} \sum_{j=1}^{m_k} (u_0, \varphi_{k_j}) \varphi_{k_j}(x) = 0, \quad x \in \omega, \ell_i \in \mathbb{N}.$$

Hence,

$$\sum_{j=1}^{m_1} (u_0, \varphi_{1_j}) \varphi_{1_j}(x) + \sum_{k=2}^{\infty} \left(\frac{\mu_1}{\mu_k}\right)^{\ell_i} \sum_{j=1}^{m_k} (u_0, \varphi_{k_j}) \varphi_{k_j}(x) = 0, \quad x \in \omega, \ell_i \in \mathbb{N}.$$

Now using (6.38) and $0 < \mu_1 < \mu_2 < \ldots$, we have

$$\left\| \sum_{k=2}^{\infty} \sum_{j=1}^{m_k} \left(\frac{\mu_1}{\mu_k}\right)^{\ell_i} (u_0, \varphi_{k_j}) \varphi_{k_j} \right\|_{L^{\infty}(\Omega)} \leq \sum_{k=2}^{\infty} \sum_{j=1}^{m_k} \left|\frac{\mu_1}{\mu_k}\right|^{\ell_i} |(u_0, \varphi_{k_j})| \|\varphi_{k_j}\|_{L^{\infty}(\Omega)}$$

$$\leq \left|\frac{\mu_1}{\mu_2}\right|^{\ell_i} \sum_{k=2}^{\infty} \sum_{j=1}^{m_k} |(u_0, \varphi_{k_j})| \|\varphi_{k_j}\|_{L^{\infty}(\Omega)} \leq c \left|\frac{\mu_1}{\mu_2}\right|^{\ell_i}.$$

Letting $\ell_i \to \infty$ and noting $|\frac{\mu_1}{\mu_2}| < 1$ yield

$$\sum_{j=1}^{m_1} (u_0, \varphi_{1_j}) \varphi_{1_j}(x) = 0, \quad \forall x \in \omega.$$

Similarly, we obtain

$$\sum_{j=1}^{m_k} (u_0, \varphi_{k_j}) \varphi_{k_j}(x) = 0, \quad x \in \omega, k \in \mathbb{N}.$$

Since

$$u_0 = \sum_{k=1}^{\infty} \left(\sum_{j=1}^{m_k} (u_0, \varphi_{k_j}) \varphi_{k_j} \right), \quad \text{in } L^2(\Omega),$$

we deduce $u \equiv 0$ in $\Omega \times (0, \infty)$. $\qquad \square$

Next, we turn to problem (6.17) with $u_0 \equiv 0$. We begin with the L^p estimate in time. Given a Banach space X and a closed linear operator A with domain $D(A) \subset X$, the time fractional evolution equation (with $\alpha \in (0, 1)$)

$$\begin{cases} \partial_t^{\alpha} u(t) + Au(t) = f(t), & t \in I, \\ u(0) = 0, \end{cases} \qquad (6.42)$$

is said to have the property of maximal L^p regularity, if for each $f \in L^p(I; X)$, problem (6.42) possesses a unique solution u in the space $W_0^{\alpha,p}(I; X) \cap L^p(I; D(A))$ (see Appendix A.2.5 for the space $W_0^{s,p}(I; X)$, in the sense of complex interpolation, where the subscript 0 indicates a zero trace at $t = 0$). That is, both terms on the left-hand side belong to $L^p(I; X)$.

The following maximal L^p-regularity holds. It recovers the classical maximal regularity estimates for standard parabolic problems as $\alpha \to 1^-$. It can be found in the PhD thesis [Baj01, Chapters 4 and 5]; see also [Zac05] for relevant results for Volterra evolution equations. The proof below is taken from [JLZ20a, Theorem 2.2] and is a straightforward application of the now classical operator-valued Fourier multiplier theorem due to Weis [Wei01, Theorem 3.4].

Theorem 6.11 *If $u_0 \equiv 0$ and $f \in L^p(I; L^2(\Omega))$ with $1 < p < \infty$, then problem* (6.17) *has a unique solution $u \in L^p(I; \dot{H}^2(\Omega))$ such that $\partial_t^\alpha u \in L^p(I; L^2(\Omega))$ and*

$$\|u\|_{L^p(I;\dot{H}^2(\Omega))} + \|\partial_t^\alpha u\|_{L^p(I;L^2(\Omega))} \le c\|f\|_{L^p(I;L^2(\Omega))},$$

where the constant c does not depend on f and T.

Proof For $f \in L^p(I; L^2(\Omega))$, extending f to be zero on $\Omega \times [(-\infty, 0) \cup (T, \infty)]$ yields $f \in L^p(\mathbb{R}; L^2(\Omega))$ and

$$\|f\|_{L^p(\mathbb{R};L^2(\Omega))} = \|f\|_{L^p(I;L^2(\Omega))}. \tag{6.43}$$

Further, we have

$$\partial_t^\alpha f(t) = {}_{-\infty}^{R}\partial_t^\alpha f(t), \quad \forall t \in \bar{I} \quad \text{and} \quad \widetilde{{}_{-\infty}^{R}\partial_t^\alpha f} = (i\xi)^\alpha \widetilde{f}(\xi)$$

cf. Theorem 2.8, where $\sim$ denotes taking Fourier transform in t (i.e., $\widetilde{f} \equiv \mathcal{F}[f]$ the Fourier transform of f). Then, $\widetilde{u}(\xi) = ((i\xi)^\alpha + A)^{-1}\widetilde{f}(\xi)$ is a solution of (6.17) and

$$(i\xi)^\alpha \widetilde{u}(\xi) = (i\xi)^\alpha((i\xi)^\alpha + A)^{-1}\widetilde{f}(\xi).$$

The self-adjoint operator $A : D(A) \to L^2(\Omega)$ is invertible from $L^2(\Omega)$ to $D(A)$, and generates a bounded analytic semigroup. Thus, the operator

$$(i\xi)^\alpha((i\xi)^\alpha + A)^{-1} \tag{6.44}$$

is bounded from $L^2(\Omega)$ to $D(A)$ in a small neighborhood $\mathcal{N}$ of $\xi = 0$. Further, in $\mathcal{N}$, the operator

$$\xi\frac{d}{d\xi}[(i\xi)^\alpha((i\xi)^\alpha + A)^{-1}] = \alpha(i\xi)^\alpha((i\xi)^\alpha + A)^{-1} - \alpha(i\xi)^{2\alpha}((i\xi)^\alpha + A)^{-2} \tag{6.45}$$

is also bounded. By the resolvent estimate (6.23), for ξ away from zero, the following inequality

$$\|(i\xi)^\alpha((i\xi)^\alpha + A)^{-1}\| \le c$$

implies the boundedness of (6.44) and (6.45). Since boundedness of operators is equivalent to R-boundedness of operators in $L^2(\Omega)$ (see [KW04, p. 75] for the concept of R-boundedness), the boundedness of (6.44) and (6.45) implies that (6.44) is an operator-valued Fourier multiplier [Wei01, Theorem 3.4], and thus

$$
\begin{aligned}
\|\partial_t^\alpha u\|_{L^p(\mathbb{R};L^2(\Omega))} &\le \|\mathcal{F}^{-1}[(i\xi)^\alpha \widetilde{u}(\xi)]\|_{L^p(\mathbb{R};L^2(\Omega))} \\
&= \|\mathcal{F}^{-1}[(i\xi)^\alpha((i\xi)^\alpha + A)^{-1}\widetilde{f}(\xi)]\|_{L^p(\mathbb{R};L^2(\Omega))} \\
&\le c\|\mathcal{F}^{-1}[\widetilde{f}(\xi)]\|_{L^p(\mathbb{R};L^2(\Omega))} \le c\|f\|_{L^p(\mathbb{R};L^2(\Omega))}.
\end{aligned}
$$

This and (6.43) imply the bound on $\|\partial_t^\alpha u\|_{L^p(I;L^2(\Omega))}$. The bound on $\|Au\|_{L^p(I;L^2(\Omega))}$ follows similarly by replacing $(i\xi)^\alpha((i\xi)^\alpha + A)^{-1}$ with $A((i\xi)^\alpha + A)^{-1}$ in the proof. This completes the proof of the theorem. $\qquad\square$

Next, we derive pointwise in time regularity.

Theorem 6.12 *Let u be the solution to problem* (6.17) *with $u_0 = 0$. If $f \in C^{k-1}(\overline{I}; \dot{H}^\gamma(\Omega))$ and $\int_0^t (t-s)^{\alpha-1}\|f^{(k)}(s)\|_{\dot{H}^\gamma(\Omega)}ds < \infty$, for any $t \in \overline{I}$, then for any $\gamma \le \beta < \gamma + 2$ and $k \ge 0$, there holds*

$$
\begin{aligned}
\left\|u^{(k)}(t)\right\|_{\dot{H}^\beta(\Omega)} &\le c\sum_{j=0}^{k-1} t^{(1-\frac{\beta-\gamma}{2})\alpha-j-1}\|f^{(k-j-1)}(0)\|_{\dot{H}^\gamma(\Omega)} \\
&\quad + \int_0^t (t-s)^{(1-\frac{\beta-\gamma}{2})\alpha-1}\|f^{(k)}(s)\|_{\dot{H}^\gamma(\Omega)}ds.
\end{aligned}
$$

Similarly, if $f \in C^k(\overline{I}; \dot{H}^\gamma(\Omega))$ and $\int_0^t \|f^{(k+1)}(s)\|_{\dot{H}^\gamma(\Omega)}ds < \infty$, for any $t \in \overline{I}$, then for any $\beta = \gamma + 2$ and $k \ge 0$, there holds

$$
\left\|u^{(k)}(t)\right\|_{\dot{H}^{\gamma+2}(\Omega)} \le c\sum_{j=0}^{k} t^{-j}\|f^{(k-j)}(0)\|_{\dot{H}^\gamma(\Omega)} + \int_0^t \|f^{(k+1)}(s)\|_{\dot{H}^\gamma(\Omega)}ds.
$$

Proof In view of the representation (6.24), the solution $u(t)$ is given by

$$
u(t) = \int_0^t E(t-s)f(s)ds = \int_0^t E(s)f(t-s)ds.
$$

Differentiating the representation k times yields (with the convention summation with lower index greater than upper index being zero)

$$
u^{(k)}(t) = \sum_{j=0}^{k-1} E^{(j)}(t)f^{(k-j-1)}(0) + \int_0^t E(s)f^{(k)}(t-s)ds,
$$

and thus for $\gamma \le \beta < \gamma + 2$, by Theorem 6.4, we obtain

$$\left\|u^{(k)}(t)\right\|_{\dot{H}^\beta(\Omega)} \le \sum_{j=0}^{k-1} \left\|E^{(j)}(t)f^{(k-j-1)}(0)\right\|_{\dot{H}^\beta(\Omega)} + \int_0^t \left\|E(s)f^{(k)}(t-s)\right\|_{\dot{H}^\beta(\Omega)} ds$$

$$\le \sum_{j=0}^{k-1} \left\|A^{\frac{\beta-\gamma}{2}}E^{(j)}(t)\right\|\left\|f^{(k-j-1)}(0)\right\|_{\dot{H}^\gamma(\Omega)} + \int_0^t \left\|A^{\frac{\beta-\gamma}{2}}E(s)\right\|\left\|f^{(k)}(t-s)\right\|_{\dot{H}^\gamma(\Omega)} ds$$

$$\le c \sum_{j=0}^{k-1} t^{(1-\frac{\beta-\gamma}{2})\alpha-j-1}\left\|f^{(k-j-1)}(0)\right\|_{\dot{H}^\gamma(\Omega)} + c \int_0^t s^{(1-\frac{\beta-\gamma}{2})\alpha-1}\left\|f^{(k)}(t-s)\right\|_{\dot{H}^\gamma(\Omega)} ds.$$

This shows the first estimate. In the event $\beta = \gamma + 2$, by the identity $\frac{d}{dt}(I - F(t)) = AE(t)$ from Lemma 6.2, and integration by parts, we obtain

$$Au^{(k)}(t) = \sum_{j=0}^{k-1} AE^{(j)}(t)f^{(k-j-1)}(0) + \int_0^t AE(s)f^{(k)}(t-s)ds$$

$$= \sum_{j=0}^{k-1} AE^{(j)}(t)f^{(k-j-1)}(0) + \int_0^t \frac{d}{ds}(I-F(s))f^{(k)}(t-s)ds$$

$$= \sum_{j=0}^{k-1} AE^{(j)}(t)f^{(k-j-1)}(0) + (I-F(t))f^{(k)}(0) + \int_0^t (I-F(s))f^{(k+1)}(t-s)ds,$$

since $I - F(0) = 0$, cf. Lemma 6.3. Then by Theorem 6.4 and repeating the preceding argument, we obtain the second assertion.   □

Remark 6.4 In Theorem 6.12, setting $k = 0$ and $\gamma = 0$ gives the following estimate

$$\|u(t)\|_{\dot{H}^2(\Omega)} \le c\|f(0)\|_{L^2(\Omega)} + \int_0^t \|f'(s)\|_{L^2(\Omega)} ds.$$

Thus, with $f \in L^\infty(I; L^2(\Omega))$ only, the solution u generally does not belong to $L^\infty(0,T; \dot{H}^2(\Omega))$. Indeed, if $u_0 = 0$, and $f \in L^\infty(I; \dot{H}^\gamma(\Omega))$, $-1 \le \gamma \le 1$, then the solution $u \in L^\infty(I; \dot{H}^{\gamma+2-\epsilon}(\Omega))$ for any $0 < \epsilon < 1$, and

$$\|u(t)\|_{\dot{H}^{\gamma+2-\epsilon}(\Omega)} \le c\epsilon^{-1}t^{\frac{\epsilon}{2}\alpha}\|f\|_{L^\infty(0,t;\dot{H}^\gamma(\Omega))}.$$

Actually, by (6.24) and Theorem 6.4(ii),

$$\|u(t)\|_{\dot{H}^{\gamma+2-\epsilon}(\Omega)} = \left\|\int_0^t E(t-s)f(s)\,ds\right\|_{\dot{H}^{\gamma+2-\epsilon}(\Omega)} \le \int_0^t \|E(t-s)f(s)\|_{\dot{H}^{\gamma+2-\epsilon}(\Omega)}\,ds$$

$$\le c \int_0^t (t-s)^{\frac{\epsilon}{2}\alpha-1}\|f(s)\|_{\dot{H}^\gamma(\Omega)}\,ds \le c\epsilon^{-1}t^{\frac{\epsilon}{2}\alpha}\|f\|_{L^\infty(0,t;\dot{H}^\gamma(\Omega))},$$

which shows the desired estimate. The ϵ factor in the estimate reflects the limited smoothing property of the subdiffusion operator. The condition $f \in L^\infty(I; \dot{H}^\gamma(\Omega))$ can be further weakened to $f \in L^r(I; \dot{H}^\gamma(\Omega))$ with $r > \alpha^{-1}$. This follows from

Theorem 6.4 and the Cauchy-Schwarz inequality with the conjugate exponent r':

$$\|u(t)\|_{\dot{H}^\gamma(\Omega)} \le \int_0^t \|E(t-s)f(s)\|_{\dot{H}^\gamma(\Omega)}\,ds \le c\int_0^t (t-s)^{\alpha-1}\|f(s)\|_{\dot{H}^\gamma(\Omega)}\,ds$$

$$\le \frac{c}{1+r'(\alpha-1)}t^{1+r'(\alpha-1)}\|f\|_{L^r(0,t;\dot{H}^\gamma(\Omega))},$$

where $1 + r'(\alpha - 1) > 0$ by the condition $r > \alpha^{-1}$. It follows from this that the initial condition $u(0) = 0$ holds in a weak sense: $\lim_{t\to 0^+} \|u(t)\|_{\dot{H}^\gamma(\Omega)} = 0$. Hence, for any $\alpha \in (\frac{1}{2}, 1)$ the representation formula (6.24) remains a legitimate solution for $f \in L^2(I; \dot{H}^\gamma(\Omega))$. For a detailed treatise on FDEs in Sobolev spaces, we refer to the work [GLY15] and the recent monograph [KRY20].

Last we give a Hölder in time regularity estimate for problem (6.17). First, we give a lemma on the Hölder regularity of convolution with $E(t)$. Hölder estimates for the subdiffusion model will be discussed in more detail in Chapter 7.

Lemma 6.4 *For* $f \in C^\theta(\bar{I}; L^2(\Omega))$, *the function* $v(t) = \int_0^t E(t-s)(f(s) - f(t))\,ds$ *belongs to* $C^\theta(\bar{I}; L^2(\Omega))$ *and*

$$\|Av\|_{C^\theta(\bar{I};L^2(\Omega))} \le c\|f\|_{C^\theta(\bar{I};L^2(\Omega))}.$$

Proof By taking $0 \le t < t + \tau := \tilde{t} \le T$, then

$$v(\tilde{t}) - v(t) = \int_0^{\tilde{t}} E(\tilde{t}-s)(f(s) - f(\tilde{t}))\,ds - \int_0^t E(t-s)(f(s) - f(t))\,ds$$

$$= \int_0^t (E(\tilde{t}-s) - E(t-s))(f(s) - f(t))\,ds$$

$$+ \int_0^t E(\tilde{t}-s)(f(t) - f(\tilde{t}))\,ds + \int_t^{\tilde{t}} E(\tilde{t}-s)(f(s) - f(\tilde{t}))\,ds$$

$$:= I + II + III.$$

Now we bound the three terms separately. First, for the term I, by Theorem 6.4(ii),

$$\|AI(t)\|_{L^2(\Omega)} = \left\| \int_0^t \int_{t-s}^{\tilde{t}-s} AE'(\zeta)(f(s) - f(t))\,d\zeta ds \right\|_{L^2(\Omega)}$$

$$\le \int_0^t \int_{t-s}^{\tilde{t}-s} \|AE'(\zeta)\|\,d\zeta\|f(s) - f(t)\|_{L^2(\Omega)}\,ds$$

$$\le c\int_0^t \int_{t-s}^{\tilde{t}-s} \zeta^{-2}d\zeta(t-s)^\theta ds\|f\|_{C^\theta([0,t];L^2(\Omega))}$$

$$\le c\tau\int_0^t (t-s)^{-1+\theta}(\tilde{t}-s)^{-1}ds\|f\|_{C^\theta([0,t];L^2(\Omega))}.$$

Meanwhile, for $0 < \theta < 1$, using the following identity [GR15, 3.194, p. 318]

$$\int_0^\infty \frac{\eta^{\theta-1}}{\eta + \tau} d\eta = \frac{\pi \tau^{\theta-1}}{\sin \theta \pi}, \tag{6.46}$$

we deduce

$$\int_0^t (\tilde{t} - s)^{-1}(t - s)^{-1+\theta} \, ds = \int_0^t \frac{\eta^{\theta-1}}{\eta + \tau} d\eta \leq \int_0^\infty \frac{\eta^{\theta-1}}{\eta + \tau} d\eta \leq c\tau^{\theta-1},$$

and hence

$$\|AI(t)\|_{L^2(\Omega)} \leq c\tau^\theta \|f\|_{C^\theta(\bar{I}; L^2(\Omega))}.$$

For the second term II,

$$\|AII(t)\|_{L^2(\Omega)} \leq \left\| \int_0^t AE(\tilde{t} - s)(f(t) - f(\tilde{t})) \, ds \right\|_{L^2(\Omega)}$$

$$\leq \left\| \int_0^t AE(\tilde{t} - s) ds \right\| \|f(t) - f(\tilde{t})\|_{L^2(\Omega)}.$$

Now by Lemma 6.2, $AE(t) = \frac{d}{dt}(I - F(t))$, and thus by Theorem 6.4,

$$\left\| \int_0^t AE(\tilde{t} - s) ds \right\| = \|F(\tilde{t} - t) - F(\tilde{t})\| \leq c.$$

This and Hölder continuity of f imply

$$\|AII(t)\|_{L^2(\Omega)} \leq c\tau^\theta \|f\|_{C^\theta(\bar{I}; L^2(\Omega))}.$$

Last, by Cauchy-Schwarz inequality and Young's inequality, we deduce

$$\|AIII(t)\|_{L^2(\Omega)} \leq \int_t^{\tilde{t}} \|AE(\tilde{t} - s)\| \|f(s) - f(\tilde{t})\|_{L^2(\Omega)} ds$$

$$\leq c \int_t^{\tilde{t}} (\tilde{t} - s)^{-1+\theta} \|f\|_{C^\theta(\bar{I}; L^2(\Omega))} ds \leq c\tau^\theta \|f\|_{C^\theta(\bar{I}; L^2(\Omega))}.$$

Combining the bounds on I, II and III completes the proof of the lemma. □

Theorem 6.13 *For $0 < \alpha < 1$, $u_0 \in \dot{H}^2(\Omega)$, $f \in C^\theta(\bar{I}; L^2(\Omega))$, with $\theta \in (0, 1)$. Then for the solution u given by (6.24), there holds for every $\delta > 0$*

$$\|u\|_{C^\theta([\delta,T]; \dot{H}^2(\Omega))} + \|\partial_t^\alpha u\|_{C^\theta([\delta,T]; L^2(\Omega))} \leq c\big(\delta^{-1}\|f\|_{C^\theta([\delta,T]; L^2(\Omega))} + \|u_0\|_{H^2(\Omega)}\big),$$

$$\|u\|_{C(\bar{I}; \dot{H}^2(\Omega))} + \|\partial_t^\alpha u\|_{C(\bar{I}; L^2(\Omega))} \leq c\big(\|f\|_{C^\theta(\bar{I}; L^2(\Omega))} + \|u_0\|_{H^2(\Omega)}\big).$$

Further, if $u_0 = 0$ and $f(0) = 0$, then

$$\|Au\|_{C^\theta(\bar{I}; L^2(\Omega))} + \|\partial_t^\alpha u\|_{C^\theta(\bar{I}; L^2(\Omega))} \leq c\|f\|_{C^\theta(\bar{I}; L^2(\Omega))}.$$

Proof By the solution representation (6.24), we have

$$\partial_t^\alpha u(t) = -Au(t) + f(t) = -AF(t)u_0 + f(t) - \int_0^t AE(t-s)f(s)ds$$

$$= -AF(t)u_0 + \left(f(t) - \int_0^t AE(t-s)f(t)ds\right) - \int_0^t AE(t-s)(f(s) - f(t))ds.$$

The three terms are denoted by I, II and III. It follows from Lemma 6.4 that

$$\|\mathrm{III}(t)\|_{C^\theta(\bar{I};L^2(\Omega))} \le c\|f\|_{C^\theta(\bar{I};L^2(\Omega))}.$$

So it suffices to bound the first two terms. Let $\tilde{t} = t + \tau$, with $\tau > 0$. Then for the term I(t), Theorem 6.4(i) implies

$$\|\mathrm{I}(\tilde{t}) - \mathrm{I}(t)\|_{L^2(\Omega)} = \|A(F(\tilde{t})u_0 - F(t))u_0\|_{L^2(\Omega)} \le \|F(\tilde{t}) - F(t)\|\|Au_0\|_{L^2(\Omega)}$$

$$\le ct^{-1}\tau\|Au_0\|_{L^2(\Omega)} \le c\delta^{-1}\tau^\theta\|Au_0\|_{L^2(\Omega)}.$$

To bound the term II, using the identity $\frac{d}{dt}(I - F(t)) = AE(t)$, cf. Lemma 6.2,

$$\int_0^t AE(t-s)ds = \int_0^t AE(s)ds = \int_0^t \frac{d}{ds}(I - F(s))ds = I - F(t),$$

since $I - F(0) = 0$, cf. Lemma 6.3. Consequently, $\mathrm{II}(t) = F(t)f(t)$ and

$$\|\mathrm{II}(t)(\tilde{t}) - \mathrm{II}(t)\|_{L^2(\Omega)} \le \|F(\tilde{t}) - F(t)\|\|f(t)\|_{L^2(\Omega)} + \|F(\tilde{t})\|\|f(\tilde{t}) - f(t)\|_{L^2(\Omega)}$$

$$\le c(t^{-1}\tau + \tau^\theta)\|f\|_{C^\theta(\bar{I};L^2(\Omega))} \le c(\delta^{-1} + 1)\tau^\theta\|f\|_{C^\theta(\bar{I};L^2(\Omega))}.$$

This proves the first estimate. The second assertion follows from Lemma 6.4 and preceding discussions. Finally, we turn to the last assertion. It suffices to show $(F(\tilde{t}) - F(t))f(t) \in C^\theta(\bar{I}; L^2(\Omega))$. Since $f(0) = 0$, $f \in C^\theta(\bar{I}; L^2(\Omega))$ implies

$$\|f(t)\|_{L^2(\Omega)} \le ct^\theta\|f\|_{C^\theta(\bar{I};L^2(\Omega))},$$

and consequently,

$$\|(F(\tilde{t}) - F(t))f(t)\|_{L^2(\Omega)} \le ct^\theta \int_t^{\tilde{t}} s^{-1}ds\|f\|_{C^\theta(\bar{I};L^2(\Omega))}.$$

Now straightforward computation shows

$$t^\theta \int_t^{\tilde{t}} s^{-1}ds \le \int_t^{\tilde{t}} t^\theta s^{-1}ds \le \int_t^{\tilde{t}} s^{\theta-1}ds = \frac{(t + \tau)^\theta - t^\theta}{\theta} \le c\theta^{-1}\tau^\theta.$$

This completes the proof of the theorem. $\qquad\square$

Remark 6.5 The condition $u_0 = 0$ and $f(0) = 0$ is one sufficient compatibility condition for the Hölder continuity up to $t = 0$. Without suitable continuity condition, the Hölder regularity for an arbitrary exponent $\theta \in (0, 1)$ is generally false; see Example 6.2 for an illustration.

6.3 Linear Problems with Time-Dependent Coefficients

Now we consider subdiffusion with a time-dependent diffusion coefficient, following the works [JLZ19b, JLZ20b]; see also [KY18] for an alternative treatment via
approximating the coefficients by smooth functions. Due to the time-dependence of
the elliptic operator, the separation of variable and Laplace transform techniques
from Section 6.2 are no longer directly applicable, and the analysis requires different
techniques. In this section, we describe a perturbation argument to handle this.

Consider the following fractional-order parabolic problem:

$$\begin{cases} \partial_t^\alpha u - \nabla \cdot (a\nabla u) = f, & \text{in } Q, \\ \qquad\qquad\qquad\quad u = 0, & \text{on } \partial_L Q, \\ \qquad\qquad\quad u(\cdot, 0) = u_0, & \text{in } \Omega, \end{cases} \tag{6.47}$$

where $f \in L^\infty(I; L^2(\Omega))$ and $u_0 \in L^2(\Omega)$ are given source and initial data, respectively, and $a(x, t) : \overline{Q} \to \mathbb{R}^{d \times d}$ is a symmetric matrix-valued diffusion coefficient
such that for some constant $\lambda \in (0, 1)$, integer $K \geq 2$ and all $i, j = 1, \ldots, d$:

$$\lambda|\xi|^2 \leq a(x, t)\xi \cdot \xi \leq \lambda^{-1}|\xi|^2, \quad \forall \xi \in \mathbb{R}^d, \forall (x, t) \in \overline{Q}, \tag{6.48}$$

$$\left|\tfrac{\partial}{\partial t} a_{ij}(x, t)\right| + \left|\nabla_x \tfrac{\partial^k}{\partial t^k} a_{ij}(x, t)\right| \leq c, \forall (x, t) \in \overline{Q}, k = 0, \ldots, K + 1. \tag{6.49}$$

To derive regularity estimates, it is convenient to define a time-dependent elliptic
operator $A(t) : \dot{H}^2(\Omega) \to L^2(\Omega)$ by $A(t)\phi = -\nabla \cdot (a(x, t)\nabla\phi)$, for all $\phi \in \dot{H}^2(\Omega)$.
Under condition (6.49) with $k = 1$, direct computation gives

$$\|(A(t) - A(s))v\|_{L^2(\Omega)} \leq c|t - s|\,\|v\|_{H^2(\Omega)}. \tag{6.50}$$

Indeed, direct computation leads to

$$(A(t) - A(s))v(x) = (\nabla \cdot a(x, t) - \nabla \cdot a(x, s)) \cdot \nabla v(x) + (a(x, t) - a(x, s)) : \nabla^2 v,$$

where : denotes the Frobenius inner product for matrices. This and the differentia-
bility assumption in (6.49) imply the estimate (6.50).

First, we state a version of Gronwall's inequality for fractional ODEs.

Proposition 6.2 *Let X be a Banach space. For $\alpha \in (0, 1)$ and $p \in (\frac{1}{\alpha}, \infty)$, if a
function $u \in C(\overline{I}; X)$ satisfies $\partial_t^\alpha u \in L^p(I; X)$, $u(0) = 0$ and*

$$\|\partial_t^\alpha u\|_{L^p(0,s;X)} \leq \kappa\|u\|_{L^p(0,s;X)} + \sigma, \quad \forall s \in I, \tag{6.51}$$

for some positive constants κ and σ, then

$$\|u\|_{C(\overline{I};X)} + \|\partial_t^\alpha u\|_{L^p(I;X)} \leq c\sigma, \tag{6.52}$$

where the constant c is independent of σ, u and X, but depends on α, p, κ and T.

Proof Since $u(0) = 0$, the Riemann–Liouville and Djrbashian-Caputo fractional derivatives coincide. Thus, by the fundamental theorem of calculus, $u(t) = \frac{1}{\Gamma(\alpha)} \int_0^t (t - \xi)^{\alpha-1} \partial_\xi^\alpha u(\xi) \, d\xi$. Since $p > \frac{1}{\alpha}$, Hölder's inequality implies

$$\|u(t)\|_X \le c \left(\int_0^t (t - \xi)^{\frac{(\alpha-1)p}{p-1}} \, d\xi \right)^{\frac{p-1}{p}} \|\partial_\xi^\alpha u\|_{L^p(0,t;X)} \le c \|\partial_\xi^\alpha u\|_{L^p(0,t;X)}.$$

Upon taking the supremum with respect to $t \in (0, s)$ for any $s \in I$, we obtain

$$\|u\|_{L^\infty(0,s;X)} \le c\|\partial_\xi^\alpha u\|_{L^p(0,s;X)} \le c\kappa \|u\|_{L^p(0,s;X)} + c\sigma$$

$$\le \epsilon\kappa\|u\|_{L^\infty(0,s;X)} + c_\epsilon\kappa\|u\|_{L^1(0,s;X)} + c\sigma,$$

where $\epsilon > 0$ can be arbitrary. By choosing $\epsilon = \frac{1}{2\kappa}$, we have

$$\|u\|_{L^\infty(0,s;X)} \le c_\kappa\|u\|_{L^1(0,s;X)} + c\sigma, \quad \forall s \in I.$$

That is,

$$\|u(s)\|_X \le c_\kappa \int_0^s \|u(\xi)\|_X d\xi + c\sigma, \quad \text{for } s \in I.$$

Now the standard Gronwall's inequality yields $\sup_{s \in \bar{I}} \|u(s)\|_X \le e^{c_\kappa T} c\sigma$. Substituting it into (6.51) yields (6.52), completing the proof. $\qquad\square$

Now we can give the existence, uniqueness and regularity of solutions to problem (6.47) with $u_0 = 0$.

Theorem 6.14 *Under conditions* (6.48)–(6.49), *if* $u_0 = 0$ *and* $f \in L^p(I; L^2(\Omega))$, *with* $\frac{1}{\alpha} < p < \infty$, *then problem* (6.47) *has a unique solution* $u \in C(\bar{I}; L^2(\Omega)) \cap L^p(I; \dot{H}^2(\Omega))$ *such that* $\partial_t^\alpha u \in L^p(I; L^2(\Omega))$.

Proof For any $\theta \in [0, 1]$, consider the following subdiffusion problem

$$\partial_t^\alpha u(t) + A(\theta t)u(t) = f(t), \quad t \in I, \quad \text{with } u(0) = 0, \tag{6.53}$$

and define a set $D = \{\theta \in [0, 1] : (6.53) \text{ has a solution } u \in L^p(I; \dot{H}^2(\Omega)) \text{ and } \partial_t^\alpha u \in L^p(I; L^2(\Omega))\}$. Theorem 6.11 implies $0 \in D$ and so $D \ne \emptyset$. For any $\theta \in D$, by rewriting (6.53) as

$$\partial_t^\alpha u(t) + A(\theta t_0)u(t) = f(t) + (A(\theta t_0) - A(\theta t))u(t), \quad t \in I, \tag{6.54}$$

with $u(0) = 0$, and by applying Theorem 6.11 in the time interval $(0, t_0)$, we obtain

$$\|\partial_t^\alpha u\|_{L^p(0,t_0;L^2(\Omega))} + \|u\|_{L^p(0,t_0;H^2(\Omega))}$$
$$\le c\|f\|_{L^p(0,t_0;L^2(\Omega))} + c\|(A(\theta t_0) - A(\theta t))u(t)\|_{L^p(0,t_0;L^2(\Omega))}$$
$$\le c\|f\|_{L^p(0,t_0;L^2(\Omega))} + c\|(t_0 - t)u(t)\|_{L^p(0,t_0;H^2(\Omega))}, \tag{6.55}$$

where the last line follows from (6.50). Let $g(t) = \|u\|^p_{L^p(0,t;H^2(\Omega))}$, which satisfies $g'(t) = \|u(t)\|^p_{H^2(\Omega)}$. Then (6.55) and integration by parts imply

$$g(t_0) \leq c\|f\|^p_{L^p(0,t_0;L^2(\Omega))} + c \int_0^{t_0} (t_0 - t)^p g'(t)dt$$

$$= c\|f\|^p_{L^p(0,t_0;L^2(\Omega))} + cp \int_0^{t_0} (t_0 - t)^{p-1} g(t)dt$$

$$\leq c\|f\|^p_{L^p(0,t_0;L^2(\Omega))} + c \int_0^{t_0} g(t)dt,$$

which in turn implies (via the standard Gronwall's inequality in Theorem 4.1) $g(t_0) \leq c\|f\|^p_{L^p(0,t_0;L^2(\Omega))}$, i.e., $\|u\|_{L^p(0,t_0;H^2(\Omega))} \leq c\|f\|_{L^p(0,t_0;L^2(\Omega))}$. Substituting the last inequality into (6.55) yields

$$\|\partial_t^\alpha u\|_{L^p(0,t_0;L^2(\Omega))} + \|u\|_{L^p(0,t_0;H^2(\Omega))} \leq c\|f\|_{L^p(0,t_0;L^2(\Omega))}. \tag{6.56}$$

Since the estimate (6.56) is independent of $\theta \in D$, D is a closed subset of $[0, 1]$. Now we show that D is also open with respect to the subset topology of $[0, 1]$. In fact, if $\theta_0 \in D$, then problem (6.53) can be rewritten as

$$\partial_t^\alpha u(t) + A(\theta_0 t)u(t) + (A(\theta t) - A(\theta_0 t))u(t) = f(t), \quad t \in I, \quad \text{with } u(0) = 0,$$

which is equivalent to

$$\left[1 + (\partial_t^\alpha + A(\theta_0 t))^{-1}(A(\theta t) - A(\theta_0 t))\right]u(t) = (\partial_t^\alpha + A(\theta_0 t))^{-1} f(t).$$

It follows from (6.56) that the operator $(\partial_t^\alpha + A(\theta_0 t))^{-1}(A(\theta t) - A(\theta_0 t))$ is small in the sense that

$$\|(\partial_t^\alpha + A(\theta_0 t))^{-1}(A(\theta t) - A(\theta_0 t))\|_{L^p(I;H^2(\Omega)) \to L^p(I;H^2(\Omega))} \leq c|\theta - \theta_0|.$$

Thus, for θ sufficiently close to θ_0, the operator $1 + (\partial_t^\alpha + A(\theta_0 t))^{-1}(A(\theta t) - A(\theta_0 t))$ is invertible on $L^p(I; \dot{H}^2(\Omega))$, which implies $\theta \in D$. Thus, D is open with respect to the subset topology of $[0, 1]$. Since D is both closed and open respect to the subset topology of $[0, 1]$, $D = [0, 1]$. Further, note that for $\frac{1}{\alpha} < p < \infty$, the inequality (6.56) and the condition $u(0) = 0$ directly imply $u \in C(\bar{I}; L^2(\Omega))$, by Proposition 6.2, which completes the proof of the theorem.   □

Now we provide some regularity results in $\dot{H}^s(\Omega)$ on solutions to problem (6.47). The overall analysis strategy is to employ a perturbation argument and then to properly resolve the singularity. Specifically, for any fixed $t_* \in I$, we rewrite problem (6.47) into (with the shorthand $A_* = A(t_*)$)

$$\begin{cases} \partial_t^\alpha u(t) + A_* u(t) = (A_* - A(t))u(t) + f(t), & \forall t \in I, \\ u(0) = u_0. \end{cases} \tag{6.57}$$

By the representation (6.24), the solution $u(t)$ of problem (6.57) is given by

$$u(t) = F_*(t)u_0 + \int_0^t E_*(t-s)(f(s) + (A_* - A(s))u(s))ds, \qquad (6.58)$$

with the operators $F_*(t)$ and $E_*(t)$ defined by

$$F_*(t) = \frac{1}{2\pi i}\int_{\Gamma_{\theta,\delta}} e^{zt}z^{\alpha-1}(z^\alpha + A_*)^{-1}dz \text{ and } E_*(t) = \frac{1}{2\pi i}\int_{\Gamma_{\theta,\delta}} e^{zt}(z^\alpha + A_*)^{-1}dz,$$

respectively, with integrals over the contour $\Gamma_{\theta,\delta}$ defined in (6.27). The objective is to estimate the kth temporal derivative $u^{(k)}(t) := \frac{d^k}{dt^k}u(t)$ in $\dot{H}^\beta(\Omega)$ for $\beta \in [0,2]$ using (6.58). However, direct differentiation of $u(t)$ in (6.58) with respect to t leads to strong singularity that precludes the use of Gronwall's inequality in Theorem 4.2 directly, in order to handle the perturbation term. To overcome the difficulty, we instead estimate $\|(t^{k+1}u(t))^{(k)}\|_{\dot{H}^\beta(\Omega)}$ using the expansion of $t^{k+1} = [(t-s)+s]^{k+1}$ in the following expression:

$$t^{k+1}u(t) = t^{k+1}F_*(t)u_0 + t^{k+1}\int_0^t E_*(t-s)f(s)ds$$

$$+ \sum_{m=0}^{k+1} C(k+1,m)\int_0^t (t-s)^m E_*(t-s)(A_* - A(s))s^{k+1-m}u(s)ds, \quad (6.59)$$

where $C(k+1,m)$ denotes binomial coefficients. One crucial part in the proof is to bound kth-order time derivatives of the summands in (6.59).

In the analysis, the following perturbation estimate will be used extensively. It implies that $\|A(s)^{-1}A(t)\| \le c$ for any $s,t \in \bar{I}$.

Lemma 6.5 *Under conditions* (6.48)–(6.49), *for any* $\beta \in [\frac{1}{2},1]$, *there holds*

$$\|A^\beta(I - A(t)^{-1}A(s))v\|_{L^2(\Omega)} \le c|t-s|\|A^\beta v\|_{L^2(\Omega)}, \quad \forall A^\beta v \in L^2(\Omega).$$

Proof For any given $v \in H_0^1(\Omega)$, let $\varphi = A(s)v$ and $w = A(t)^{-1}\varphi$. Then

$$(A(t)w, \chi) = (\varphi, \chi) = (A(s)v, \chi), \quad \forall \chi \in H_0^1(\Omega),$$

which implies

$$(a(\cdot,t)\nabla w, \nabla\chi) = (a(\cdot,s)\nabla v, \nabla\chi), \quad \forall \chi \in H_0^1(\Omega).$$

Consequently,

$$(a(\cdot,t)\nabla(w-v), \nabla\chi) = ((a(\cdot,s) - a(\cdot,t))\nabla v, \nabla\chi), \quad \forall \chi \in H_0^1(\Omega).$$

Let $\phi = w - v \in H_0^1(\Omega)$ be the weak solution of the elliptic problem

$$(a(\cdot, t)\nabla\phi, \nabla\xi) = ((a(\cdot, s) - a(\cdot, t))\nabla v, \nabla\xi), \quad \forall \xi \in H_0^1(\Omega). \tag{6.60}$$

By Lax-Milgram theorem, ϕ satisfies the following *a priori* estimate:

$$\|\phi\|_{H^1(\Omega)} \le c\|(a(\cdot, s) - a(\cdot, t))\nabla v\|_{L^2(\Omega)} \le c|t - s|\|v\|_{H^1(\Omega)}.$$

Since the operator $A(t)$ is self-adjoint, the preceding estimate together with a duality argument yields

$$\|(I - A(s)A(t)^{-1})v\|_{L^2(\Omega)} \le c|t - s|\|v\|_{L^2(\Omega)}, \quad \forall v \in \dot{H}^2(\Omega).$$

Consequently,

$$\|(A(t) - A(s))v\|_{L^2(\Omega)} \le \|(I - A(s)A(t)^{-1})A(t)v\|_{L^2(\Omega)} \le c|t - s|\|A(t)v\|_{L^2(\Omega)}.$$

Further, the interpolation between $\beta = \frac{1}{2}, 1$ yields

$$\|A^\beta(t)(I - A(t)^{-1}A(s))v\|_{L^2(\Omega)} \le c|t - s|\|A^\beta(t)v\|_{L^2(\Omega)}.$$

This completes the proof of the lemma.   □

Now we can give the regularity for the homogeneous problem.

Theorem 6.15 *If $a(x, t)$ satisfies (6.48)–(6.49), $u_0 \in \dot{H}^\gamma(\Omega)$ with $\gamma \in [0, 2]$ and $f \equiv 0$, then for all $t \in I$ and $k = 0, \ldots, K$, the solution $u(t)$ to problem (6.47) satisfies*

$$\left\|A^{\frac{\beta}{2}}\frac{d^k}{dt^k}(t^k u(t))\right\|_{L^2(\Omega)} \le ct^{-\frac{\beta-\gamma}{2}\alpha}\|A^{\frac{\gamma}{2}}u_0\|_{L^2(\Omega)}, \quad \forall \beta \in [\gamma, 2].$$

Proof When $k = 0$, setting $f = 0$ and $t = t_*$ in (6.58) yields

$$A_*^{\frac{\beta}{2}}u(t_*) = A_*^{\frac{\beta}{2}}F_*(t_*)u_0 + \int_0^{t_*} A_*^{\frac{\beta}{2}}E_*(t_* - s)(A_* - A(s))u(s)ds,$$

where $\beta \in [\gamma, 2]$. By Theorem 6.4 and Lemma 6.5,

$$\|A_*^{\frac{\beta}{2}}u(t_*)\|_{L^2(\Omega)} \le \|A_*^{\frac{\beta-\gamma}{2}}F_*(t_*)A_*^{\frac{\gamma}{2}}u_0\|_{L^2(\Omega)}$$

$$+ \int_0^{t_*}\|A_*E_*(t_* - s)\|\|A_*^{\frac{\beta}{2}}(I - A_*^{-1}A(s))u(s)\|_{L^2(\Omega)}ds$$

$$\le ct_*^{-(\beta-\gamma)\alpha}\|A_*^{\frac{\gamma}{2}}u_0\|_{L^2(\Omega)} + c\int_0^{t_*}(t_* - s)\|A_*E_*(t_* - s)\|\|A_*^{\frac{\beta}{2}}u(s)\|_{L^2(\Omega)}ds$$

$$\le ct_*^{-\frac{\beta-\gamma}{2}\alpha}\|u_0\|_{\dot{H}^\gamma(\Omega)} + c\int_0^{t_*}\|A_*^{\frac{\beta}{2}}u(s)\|_{L^2(\Omega)}ds.$$

This and Gronwall's inequality in Theorem 4.2 yield

$$\|A_*^{\frac{\beta}{2}}u(t_*)\|_{L^2(\Omega)} \le c(1 - \frac{\beta-\gamma}{2}\alpha)^{-1}t_*^{-\frac{\beta-\gamma}{2}\alpha}\|u_0\|_{\dot{H}^\gamma(\Omega)}.$$

In particular, we have

$$\|A_*^{\frac{\beta+\gamma}{4}} u(t_*)\|_{L^2(\Omega)} \le ct_*^{-\frac{\beta-\gamma}{4}\alpha} \|u_0\|_{\dot{H}^\gamma(\Omega)},$$

with c being bounded as $\alpha \to 1^-$. This estimate, Theorem 6.4(ii) and Lemma 6.5 then imply

$$\|A_*^{\frac{\beta}{2}} u(t_*)\|_{L^2(\Omega)} \le \|A_*^{\frac{\beta-\gamma}{2}} F_*(t_*) A_*^{\frac{\gamma}{2}} u_0\|_{L^2(\Omega)}$$

$$+ \int_0^{t_*} \|A_*^{\frac{\beta-\gamma}{4}} A_* E_*(t_* - s)\| \|A_*^{\frac{\beta+\gamma}{4}} (I - A_*^{-1} A(s)) u(s)\|_{L^2(\Omega)} ds$$

$$\le ct_*^{-\frac{\beta-\gamma}{2}\alpha} \|A_*^{\frac{\gamma}{2}} u_0\|_{L^2(\Omega)} + c \int_0^{t_*} (t_* - s) \|A_*^{\frac{\beta-\gamma}{4}} A_* E_*(t_* - s)\| \|A_*^{\frac{\beta+\gamma}{4}} u(s)\|_{L^2(\Omega)} ds$$

$$\le c \left(t_*^{-\frac{\beta-\gamma}{2}\alpha} + \int_0^{t_*} (t_* - s)^{-\frac{\beta-\gamma}{4}\alpha} s^{-\frac{\beta-\gamma}{4}\alpha} ds \right) \|u_0\|_{\dot{H}^\gamma(\Omega)} \le ct_*^{-\frac{\beta-\gamma}{2}\alpha} \|u_0\|_{\dot{H}^\gamma(\Omega)}.$$

Equivalently, we have

$$\|A_*^{\frac{\beta}{2}} t_* u(t_*)\|_{L^2(\Omega)} \le ct_*^{1-\frac{\beta-\gamma}{2}\alpha} \|u_0\|_{\dot{H}^\gamma(\Omega)},$$

where c is bounded as $\alpha \to 1^-$. This proves the assertion for $k = 0$.

Next, we prove the case $1 \le k \le K$ using mathematical induction. Suppose that the assertion holds up to $k - 1 < K$, and we prove it for $k \le K$. Indeed, by Lemma 6.6 below,

$$\left\| A_*^{\frac{\beta}{2}} \frac{d^k}{dt^k} \int_0^t (t - s)^m E_*(t - s)(A_* - A(s)) s^{k+1-m} u(s) ds \big|_{t=t_*} \right\|_{L^2(\Omega)}$$

$$\le ct_*^{-\frac{\beta-\gamma}{2}\alpha+1} \|u_0\|_{\dot{H}^\gamma(\Omega)} + c \int_0^{t_*} \|A_*^{\frac{\beta}{2}} (s^{k+1} u(s))^{(k)}\|_{L^2(\Omega)} ds,$$

where $m = 0, 1, \ldots, k + 1$. Meanwhile, the estimates in Theorem 6.4 imply

$$\left\| A_*^{\frac{\beta}{2}} (t^{k+1} F_*(t) u_0)^{(k)} \right\|_{L^2(\Omega)} \le ct^{-\frac{\beta-\gamma}{2}\alpha+1} \|u_0\|_{\dot{H}^\gamma(\Omega)}.$$

By applying $A_*^{\frac{\beta}{2}} \frac{d^k}{dt^k}$ to (6.59) and using the last two estimates, we obtain

$$\left\| A_*^{\frac{\beta}{2}} (t^{k+1} u(t))^{(k)} \big|_{t=t_*} \right\|_{L^2(\Omega)} \le ct_*^{-\frac{\beta-\gamma}{2}\alpha+1} \|u_0\|_{\dot{H}^\gamma(\Omega)}$$

$$+ c \int_0^{t_*} \|A_*^{\frac{\beta}{2}} (s^{k+1} u(s))^{(k)}\|_{L^2(\Omega)} ds.$$

Last, applying Gronwall's inequality in Theorem 4.2 completes the induction step. $\square$

In the proof of Theorem 6.15, we have used the following result.

Lemma 6.6 *Under the conditions of Theorem 6.15, for $m = 0, \ldots, k + 1$, there holds*

$$\left\|A_*^{\frac{\beta}{2}}\frac{\mathrm{d}^k}{\mathrm{d}t^k}\int_0^t (t-s)^m E_*(t-s)(A_* - A(s))s^{k+1-m}u(s)\mathrm{d}s|_{t=t_*}\right\|_{L^2(\Omega)}$$

$$\leq ct_*^{-\frac{\beta-\gamma}{2}\alpha+1}\|u_0\|_{\dot{H}^\gamma(\Omega)} + c\int_0^{t_*}\left\|A_*^{\frac{\beta}{2}}(s^{k+1}u(s))^{(k)}\right\|_{L^2(\Omega)}\mathrm{d}s.$$

Proof Denote the integral on the left-hand side by $I_m(t)$, and let $v_m = t^m u(t)$ and $W_m(t) = t^m E_*(t)$. Direct computation using product rule and changing variables gives that for any $0 \leq m \leq k$, there holds

$$I_m^{(k)}(t) = \frac{\mathrm{d}^{k-m}}{\mathrm{d}t^{k-m}}\int_0^t W_m^{(m)}(t-s)(A_* - A(s))v_{k-m+1}(s)\mathrm{d}s$$

$$= \frac{\mathrm{d}^{k-m}}{\mathrm{d}t^{k-m}}\int_0^t W_m^{(m)}(s)(A_* - A(t-s))v_{k-m+1}(t-s)\mathrm{d}s$$

$$= \int_0^t W_m^{(m)}(s)\frac{\mathrm{d}^{k-m}}{\mathrm{d}t^{k-m}}\big((A_* - A(t-s))v_{k-m+1}(t-s)\big)\mathrm{d}s$$

$$= \sum_{\ell=0}^{k-m} C(k-m,\ell)\underbrace{\int_0^t W_m^{(m)}(s)(A_* - A(t-s))^{(k-m-\ell)}v_{k-m+1}^{(\ell)}(t-s)\mathrm{d}s}_{I_{m,\ell}(t)}.$$

Next, we bound the integrand

$$\widetilde{I}_{m,\ell}(s) := W_m^{(m)}(A_* - A(t_*-s))^{(k-m-\ell)}v_{k-m+1}^{(\ell)}(t_*-s)$$

of the integral $I_{m,\ell}(t_*)$. We shall distinguish between $\beta \in [\gamma, 2)$ and $\beta = 2$. First, we analyze the case $\beta \in [\gamma, 2)$. When $\ell < k$, by Theorem 6.4(ii) and 6.5 and the induction hypothesis, we bound the integrand $\widetilde{I}_{m,\ell}(s)$ by

$$\|A_*^{\frac{\beta}{2}}\widetilde{I}_{m,\ell}(s)\|_{L^2(\Omega)} \leq \|A_*^{\frac{\beta}{2}}W_m^{(m)}(s)\|\|(A_* - A(t_*-s))^{(k-m-\ell)}v_{k-m+1}^{(\ell)}(t_*-s)\|_{L^2(\Omega)}$$

$$\leq \begin{cases} cs^{(1-\frac{\beta}{2})\alpha-1}s\|A_*v_{k-m+1}^{(k-m)}(t_*-s)\|_{L^2(\Omega)}, & \ell = k-m, \\ cs^{(1-\frac{\beta}{2})\alpha-1}\|A_*v_{k-m+1}^{(\ell)}(t_*-s)\|_{L^2(\Omega)}, & \ell < k-m, \end{cases}$$

$$\leq \begin{cases} cs^{(1-\frac{\beta}{2})\alpha}(t_*-s)^{1-(1-\frac{\gamma}{2})\alpha}\|A_*^{\frac{\gamma}{2}}u_0\|_{L^2(\Omega)}, & \ell = k-m, \\ cs^{(1-\frac{\beta}{2})\alpha-1}(t_*-s)^{k-m-\ell+1-(1-\frac{\gamma}{2})\alpha}\|A_*^{\frac{\gamma}{2}}u_0\|_{L^2(\Omega)}, & \ell < k-m. \end{cases}$$

Similarly for the case $\ell = k$ (and thus $m = 0$), there holds

$$\|A_*^{\frac{\beta}{2}}\widetilde{I}_{0,k}(s)\|_{L^2(\Omega)} \leq \|A_*E_*(s)\|\|A_*^{\frac{\beta}{2}}(I - A_*^{-1}A(t_*-s))v_{k+1}^{(k)}\|_{L^2(\Omega)}$$

$$\leq c\|A_*^{\frac{\beta}{2}}v_{k+1}^{(k)}(t_*-s)\|_{L^2(\Omega)}.$$

Thus, for $0 \leq m \leq k$ and $\ell = k - m$, upon integrating from 0 to t_*, we obtain

$$\|A_*^{\frac{\beta}{2}}I_m^{(k)}(t_*)\|_{L^2(\Omega)} \le ct_*^{2+\frac{\gamma-\beta}{2}\alpha}\|A_*^{\frac{\gamma}{2}}u_0\|_{L^2(\Omega)} + c\int_0^{t_*}\|A_*^{\frac{\beta}{2}}v_{k+1}^{(k)}(s)\|_{L^2(\Omega)}ds,$$

and similarly for $0 \le m \le k$ and $\ell < k - m$,

$$\|A_*^{\frac{\beta}{2}}I_m^{(k)}(t_*)\|_{L^2(\Omega)} \le c((1-\frac{\beta}{2})\alpha)^{-1}t_*^{1+\frac{\gamma-\beta}{2}\alpha}\|A_*^{\frac{\gamma}{2}}u_0\|_{L^2(\Omega)}$$
$$+ c\int_0^{t_*}\|A_*^{\frac{\beta}{2}}v_{k+1}^{(k)}(s)\|_{L^2(\Omega)}ds.$$

Meanwhile, for $m = k + 1$, we have

$$A_*^{\frac{\beta}{2}}I_{k+1}^{(k)}(t_*) = \int_0^{t_*}A_*^{\frac{\beta}{2}+1-\frac{\gamma}{2}}W_{k+1}^{(k)}(t_* - s)A_*^{\frac{\gamma}{2}}(I - A_*^{-1}A(s))u(s)ds,$$

and consequently, by Theorem 6.4(ii) and Lemma 6.5 and the induction hypothesis,

$$\|A_*^{\frac{\beta}{2}}I_{k+1}^{(k)}(t_*)\|_{L^2(\Omega)} \le \int_0^{t_*}\|A_*^{\frac{\beta}{2}+1-\frac{\gamma}{2}}W_{k+1}^{(k)}(t_* - s)\|\|A_*^{\frac{\gamma}{2}}(I - A_*^{-1}A(s))u(s)\|_{L^2(\Omega)}ds$$
$$\le c\int_0^{t_*}(t_* - s)^{1-\frac{\beta-\gamma}{2}\alpha}\|A_*^{\frac{\gamma}{2}}u(s)\|_{L^2(\Omega)}ds$$
$$\le ct_*^{2+\frac{\gamma-\beta}{2}\alpha}\|A_*^{\frac{\gamma}{2}}u_0\|_{L^2(\Omega)}.$$

In the case $0 \le m \le k$ and $\ell < k - m$, the last estimates require $\beta \in [0, 2)$. When $0 \le m \le k$, $\ell < k - m$ and $\beta = 2$, we apply Lemma 6.2 and rewrite $A_*I_{m,\ell}(t_*)$ as

$$A_*I_{m,\ell}(t_*) = \int_0^{t_*}(s^m(I - F_*(s))')^{(m)}(A_* - A(t_* - s))^{(k-m-\ell)}v_{k-m+1}^{(\ell)}(t_* - s)ds.$$

Then integration by parts and product rule yield

$$A_*I_{m,\ell}(t_*) = -\int_0^{t_*}D(s)(A_* - A(t_* - s))^{(k-m-\ell+1)}v_{k-m+1}^{(\ell)}(t_* - s)ds$$
$$- \int_0^{t_*}D(s)(A_* - A(t_* - s))^{(k-m-\ell)}v_{k-m+1}^{(\ell+1)}(t_* - s)ds$$
$$- D(0)(A_* - A(t_* - s))^{(k-m-\ell)}|_{s=0}v_{k-m+1}^{(\ell)}(t_*), \qquad (6.61)$$

with

$$D(s) = \begin{cases} I - F_*(s), & m = 0, \\ (s^m(I - F_*(s))')^{(m-1)}, & m > 0. \end{cases}$$

By Theorem 6.4(i) and (iii), $\|D(s)\| \le c$, and thus the preceding argument with Theorem 6.4 and Lemma 6.5 and the induction hypothesis allows bounding the integrand $A_*\widetilde{I}_{m,\ell}(s)$ of (6.61) by

$$\|A_* \widetilde{I}_{m,\ell}(s)\|_{L^2(\Omega)} \le c(t_* - s)^{k-\ell-(1-\frac{\gamma}{2})\alpha}\|u_0\|_{\dot{H}^\gamma(\Omega)}$$

$$+ \begin{cases} c\|A_* v_{k+1}^{(k)}(t_* - s)\|_{L^2(\Omega)}, & \ell = k - 1, \\ c(t_* - s)^{k-1-\ell-(1-\frac{\gamma}{2})\alpha}\|u_0\|_{\dot{H}^\gamma(\Omega)}, & \ell < k - 1, \end{cases}$$

where for $\ell = k - 1$, we have $m = 0$ and hence $D(0) = 0$. Combining the last estimates and then integrating from 0 to t_* in s, we obtain the desired assertion. $\square$

Next, we analyze the case $f \not\equiv 0$. Now we consider $(t^k u(t))^{(k)}$, instead of $(t^{k+1} u(t))^{(k)}$ for the proof of Theorem 6.15. We begin with a bound on $\frac{d^k}{dt^k}(t^k \int_0^t E(t-s; t_*)f(s)ds)$, which follows from straightforward but lengthy computation.

Lemma 6.7 *Let $k \ge 1$. Then for any $\beta \in [0, 2)$, there holds*

$$\left\|A_*^{\frac{\beta}{2}} \frac{d^k}{dt^k}\left(t^k \int_0^t E_*(t-s)f(s)ds\right)\big|_{t=t_*}\right\|_{L^2(\Omega)}$$

$$\le c \sum_{m=0}^{k-1} t_*^{(1-\frac{\beta}{2})\alpha+m}\|f^{(m)}(0)\|_{L^2(\Omega)} + ct_*^k \int_0^{t_*} (t_* - s)^{(1-\frac{\beta}{2})\alpha-1}\|f^{(k)}(s)\|_{L^2(\Omega)}ds,$$

and further,

$$\left\|A_* \frac{d^k}{dt^k}\left(t^k \int_0^t E_*(t-s)f(s)ds\right)\big|_{t=t_*}\right\|_{L^2(\Omega)}$$

$$\le c \sum_{m=0}^{k} t_*^m\|f^{(m)}(0)\|_{L^2(\Omega)} + ct_*^k \int_0^{t_*} \|f^{(k+1)}(s)\|_{L^2(\Omega)}ds.$$

Proof Let $I(t) = \frac{d^k}{dt^k}(t^k \int_0^t E_*(t-s)f(s)ds)$. It follows from the elementary identity

$$\frac{d^m}{dt^m} \int_0^t E_*(s)f(t-s)ds = \sum_{\ell=0}^{m-1} E_*^{(\ell)}(t)f^{(m-1-\ell)}(0) + \int_0^t E_*(s)f^{(m)}(t-s)ds$$

and direct computation that

$$I(t) = \sum_{m=0}^{k} C(k,m)^2 t^m \left(\sum_{\ell=0}^{m-1} E_*^{(\ell)}(t)f^{(m-1-\ell)}(0) + \int_0^t E_*(s)f^{(m)}(t-s)ds\right).$$

Consequently, by Theorem 6.4, for $\beta \in [0, 2)$,

$$\|A_*^{\frac{\beta}{2}} I(t^*)\|_{L^2(\Omega)} \le c \sum_{m=0}^{k} t_*^m \sum_{\ell=0}^{m-1} \|A_*^{\frac{\beta}{2}} E_*^{(\ell)}(t_*)\|\|f^{(m-1-\ell)}(0)\|_{L^2(\Omega)}$$

$$+ c \sum_{m=0}^{k} t_*^m \int_0^{t_*} \|A_*^{\frac{\beta}{2}} E_*(s)\|\|f^{(m)}(t_* - s)\|_{L^2(\Omega)} ds$$

$$\le c \sum_{m=0}^{k} t_*^m \sum_{\ell=0}^{m-1} t_*^{(1-\frac{\beta}{2})\alpha-1-\ell} \|f^{(m-1-\ell)}(0)\|_{L^2(\Omega)}$$

$$+ c \sum_{m=0}^{k} t_*^m \int_0^{t_*} s^{(1-\frac{\beta}{2})\alpha-1} \|f^{(m)}(t_* - s)\|_{L^2(\Omega)} ds$$

$$\le c \sum_{m=0}^{k-1} t_*^{(1-\frac{\beta}{2})\alpha+m} \|f^{(m)}(0)\|_{L^2(\Omega)}$$

$$+ c \sum_{m=0}^{k} t_*^m \int_0^{t_*} s^{(1-\frac{\beta}{2})\alpha-1} \|f^{(m)}(t_* - s)\|_{L^2(\Omega)} ds.$$

Next, we simplify the second summation. The following identity: for $m < k$,

$$f^{(m)}(s) = \sum_{j=0}^{k-m-1} f^{(m+j)}(0) \frac{s^j}{j!} + \frac{1}{(k-m)!} \int_0^s (s - \xi)^{k-m-1} f^{(k)}(\xi) d\xi, \qquad (6.62)$$

and Theorem 2.1 imply

$$\int_0^{t_*} (t_* - s)^{(1-\frac{\beta}{2})\alpha-1} \|f^{(m)}(s)\|_{L^2(\Omega)} ds \le \int_0^{t_*} (t_* - s)^{(1-\frac{\beta}{2})\alpha-1}$$

$$\times \left(\sum_{j=0}^{k-m-1} \|f^{(m+j)}(0)\|_{L^2(\Omega)} \frac{s^j}{j!} + \frac{1}{(k-m)!} \int_0^s (s - \xi)^{k-m-1} \|f^{(k)}(\xi)\|_{L^2(\Omega)} d\xi \right) ds$$

$$\le c \sum_{j=0}^{k-m-1} t_*^{(1-\frac{\beta}{2})\alpha+j} \|f^{(m+j)}(0)\|_{L^2(\Omega)} + c t_*^{k-m} \int_0^{t_*} (t_* - s)^{(1-\frac{\beta}{2})\alpha-1} \|f^{(k)}(s)\|_{L^2(\Omega)} ds.$$

Combining these estimates gives the desired assertion for $\beta \in [0, 2)$. For $\beta = 2$, by Lemma 6.2 and integration by parts (and the identity $I - F_*(0) = 0$),

$$A_* \int_0^{t_*} E_*(s) f^{(m)}(t_* - s) ds = \int_0^{t_*} (I - F_*(s))' f^{(m)}(t_* - s) ds$$

$$= (I - F_*(t_*)) f^{(m)}(0) + \int_0^{t_*} (I - F_*(s)) f^{(m+1)}(t_* - s) ds,$$

and thus

$$\|A_* I(t_*)\|_{L^2(\Omega)} \le c \sum_{m=0}^{k} t_*^m \Big(\sum_{\ell=0}^{m-1} \|A_* E_*^{(\ell)}(t_*)\| \|f^{(m-1-\ell)}(0)\|_{L^2(\Omega)}$$

$$+ \|f^{(m)}(0)\| + \int_0^{t_*} \|f^{(m+1)}(s)\|_{L^2(\Omega)} ds \Big).$$

Then repeating the preceding argument completes the proof.   □

Now we can present the regularity result for the inhomogeneous problem.

Theorem 6.16 *If $a(x,t)$ satisfies (6.48)–(6.49), $u_0 \equiv 0$, then for all $t \in I$ and $k = 0, \dots, K$, the solution $u(t)$ to problem (6.47) satisfies for any $\beta \in [0, 2)$*

$$\|(t^k u(t))^{(k)}\|_{\dot{H}^\beta(\Omega)} \le c \sum_{j=0}^{k-1} t^{(1-\frac{\beta}{2})\alpha+j} \|f^{(j)}(0)\|_{L^2(\Omega)}$$

$$+ c t^k \int_0^t (t-s)^{(1-\frac{\beta}{2})\alpha-1} \|f^{(k)}(s)\|_{L^2(\Omega)} ds,$$

and similarly for $\beta = 2$,

$$\|(t^k u(t))^{(k)}\|_{\dot{H}^\beta(\Omega)} \le c \sum_{j=0}^{k} t^j \|f^{(j)}(0)\|_{L^2(\Omega)} + c t^k \int_0^t \|f^{(k+1)}(s)\|_{L^2(\Omega)} ds.$$

Proof Similar to Theorem 6.15, the proof is based on mathematical induction. Let $v_k(t) = t^k u(t)$ and $W_k(t) = t^k E_*(t)$. For $k = 0$, by the representation (6.58), we have

$$A_*^{\frac{\beta}{2}} u(t_*) = \int_0^{t_*} A_*^{\frac{\beta}{2}} E_*(t_* - s) f(s) ds + \int_0^{t_*} A_*^{\frac{\beta}{2}} E_*(t_* - s)(A_* - A(s)) u(s) ds.$$

Then for $\beta \in [0, 2)$, by Theorem 6.4(ii) and Lemma 6.5 there holds

$$\|A_*^{\frac{\beta}{2}} u(t_*)\|_{L^2(\Omega)} \le \int_0^{t_*} \|A_*^{\frac{\beta}{2}} E_*(t_* - s)\| \|f(s)\|_{L^2(\Omega)} ds$$

$$+ \int_0^{t_*} \|A_* E_*(t_* - s)\| \|A_*^{\frac{\beta}{2}} (I - A_*^{-1} A(s)) u(s)\|_{L^2(\Omega)} ds$$

$$\le c \int_0^{t_*} (t_* - s)^{(1-\frac{\beta}{2})\alpha-1} \|f(s)\|_{L^2(\Omega)} ds + c \int_0^{t_*} \|A_*^{\frac{\beta}{2}} u(s)\|_{L^2(\Omega)} ds.$$

The case $\beta = 2$ follows similarly from Lemma 6.2 and integration by parts:

$$\|A_* u(t_*)\|_{L^2(\Omega)} \le c \|f(0)\|_{L^2(\Omega)} + c \int_0^{t_*} \|f'(s)\|_{L^2(\Omega)} ds + c \int_0^{t_*} \|A_* u(s)\|_{L^2(\Omega)} ds.$$

Then Gronwall's inequality in Theorem 4.1 gives the assertion for the case $k = 0$. Now suppose it holds up to $k - 1 < K$, and we prove it for $k \le K$. Now note that

$$v_k^{(k)}(t) = \frac{d^k}{dt^k}\left(t^k \int_0^t E_*(t-s)f(s)ds\right)$$

$$+ \sum_{m=0}^{k} C(k,m)\frac{d^k}{dt^k}\int_0^t W_m(t-s)(A_* - A(s))v_{k-m}(s)ds.$$

This, Lemmas 6.7 and 6.8 and triangle inequality give for $\beta \in [0,2)$

$$\|A_*^{\frac{\beta}{2}}v_k^{(k)}(t)|_{t=t_*}\|_{L^2(\Omega)} \le c\int_0^{t_*}\|A_*^{\frac{\beta}{2}}v_k^{(k)}(s)\|_{L^2(\Omega)}ds$$

$$+ c\sum_{m=0}^{k-1}t_*^{(1-\frac{\beta}{2})\alpha+m}\|f^{(m)}(0)\|_{L^2(\Omega)} + ct_*^k\int_0^{t_*}(t_*-s)^{(1-\frac{\beta}{2})\alpha-1}\|f^{(k)}(s)\|_{L^2(\Omega)}ds,$$

and similarly, for $\beta = 2$,

$$\|A_*v_k^{(k)}(t)|_{t=t_*}\|_{L^2(\Omega)} \le c\int_0^{t_*}\|A_*v_k^{(k)}(s)\|_{L^2(\Omega)}ds + c\sum_{m=0}^{k}t_*^m\|f^{(m)}(0)\|_{L^2(\Omega)}$$

$$+ ct_*^k\int_0^{t_*}\|f^{(k+1)}(s)\|_{L^2(\Omega)}ds.$$

This and Gronwall's inequality from Theorem 4.1 complete the induction step. $\square$

The following result is needed in the proof of Theorem 6.16.

Lemma 6.8 *Under the conditions in Theorem 6.16, for any $\beta \in [0,2]$ and $m = 0,\ldots,k$, there holds for any $\beta \in [0,2)$,*

$$\left\|A_*^{\frac{\beta}{2}}\frac{d^k}{dt^k}\int_0^t(t-s)^{k-m}E_*(t-s)(A_*-A(s))s^m u(s)ds|_{t=t_*}\right\|_{L^2(\Omega)}$$

$$\le c\int_0^{t_*}\|A_*^{\frac{\beta}{2}}(s^k u(s))^{(k)}\|_{L^2(\Omega)}ds + c\sum_{m=0}^{k-1}t_*^{(1-\frac{\beta}{2})\alpha+m}\|f^{(m)}(0)\|_{L^2(\Omega)}$$

$$+ ct_*^k\int_0^{t_*}(t_*-s)^{(1-\frac{\beta}{2})\alpha-1}\|f^{(k)}(s)\|_{L^2(\Omega)}ds,$$

and for $\beta = 2$,

$$\left\|A_*\frac{d^k}{dt^k}\int_0^t(t-s)^{k-m}E_*(t-s)(A_*-A(s))s^m u(s)ds|_{t=t_*}\right\|_{L^2(\Omega)}$$

$$\le c\int_0^{t_*}\|A_*(s^k u(s))^{(k)}\|_{L^2(\Omega)}ds + c\sum_{m=0}^{k}t_*^m\|f^{(m)}(0)\|_{L^2(\Omega)}$$

$$+ ct_*^k\int_0^{t_*}\|f^{(k+1)}(s)\|_{L^2(\Omega)}ds.$$

Proof Let $v_k = t^k u(t)$ and $W_k(t) = t^k E_*(t)$. By the induction hypothesis and (6.62), for $\ell < m$, we have

$$\|A_* v_m^{(\ell)}(s)\|_{L^2(\Omega)} \le cs^{m-\ell}\Big(\sum_{j=0}^{\ell} s^j \|f^{(j)}(0)\|_{L^2(\Omega)} + s^\ell \int_0^s \|f^{(\ell+1)}(\xi)\|_{L^2(\Omega)}d\xi \Big)$$

$$\le cs^{m-\ell}\Big(\sum_{j=0}^{k-1} s^j \|f^{(j)}(0)\|_{L^2(\Omega)} + s^{k-1} \int_0^s \|f^{(k)}(\xi)\|_{L^2(\Omega)}d\xi \Big). \quad (6.63)$$

We denote the term in the bracket by $T(s; f, k)$. Now similar to the proof of Lemma 6.6, let $I_m(t)$ be the integral on the left-hand side. Then in view of the identity

$$I_m^{(k)}(t) = \sum_{\ell=0}^{k-m} C(k-m, \ell) \underbrace{\int_0^t W_m^{(m)}(A_* - A(t_* - s))^{(k-m-\ell)} v_{k-m}^{(\ell)}(s)ds}_{I_{m,\ell}(t)},$$

it suffices to bound the integrand $\widetilde{I}_{m,\ell}(s)$ of the integral $I_{m,\ell}(t_*)$, $\ell = 0, 1, \ldots, k - m$. Below we discuss the cases $\beta \in [0, 2)$ and $\beta = 2$ separately, due to the difference in singularity, as in the proof of Lemma 6.6.

Case (i): $\beta \in [0, 2)$. For the case $\ell < k$, Theorem 6.4(ii) and Lemma 6.5 lead to

$$\|A_*^{\frac{\beta}{2}} \widetilde{I}_{m,\ell}(s)\|_{L^2(\Omega)} \le \|A_*^{\frac{\beta}{2}} W_m^{(m)}\| \| (A_* - A(t_* - s))^{(k-m-\ell)} v_{k-m}^{(\ell)}(t_* - s)\|_{L^2(\Omega)}$$

$$\le \begin{cases} cs^{(1-\frac{\beta}{2})\alpha-1} s \|A_* v_{k-m}^{(k-m)}(t_* - s)\|_{L^2(\Omega)}, & \ell = k - m, \\ cs^{(1-\frac{\beta}{2})\alpha-1} \|A_* v_{k-m}^{(\ell)}(t_* - s)\|_{L^2(\Omega)}, & \ell < k - m, \end{cases}$$

$$\le \begin{cases} cs^{(1-\frac{\beta}{2})\alpha} T(t_* - s; f, k), & \ell = k - m, \\ cs^{(1-\frac{\beta}{2})\alpha-1}(t_* - s)^{k-m-\ell} T(t_* - s; f, k), & \ell < k - m, \end{cases}$$

where the last step is due to (6.63). Note that for $\ell < k$, the derivation requires $\beta \in [0, 2)$. Similarly, for the case $\ell = k$ (and thus $m = 0$),

$$\|A_*^{\frac{\beta}{2}} \widetilde{I}_{0,k}(s)\|_{L^2(\Omega)} \le c\|A_*^{\frac{\beta}{2}} v_k^{(k)}(t_* - s)\|_{L^2(\Omega)}. \quad (6.64)$$

Case (ii): $\beta = 2$. Note that for $\ell < k$, the derivation in case (i) requires $\beta \in [0, 2)$. When $\ell < k$ and $\beta = 2$, using the identity (6.61) and Theorem 6.4 and repeating the argument of Lemma 6.6, we obtain

$$\|A_* \widetilde{I}_{m,\ell}(s)\|_{L^2(\Omega)} \le \begin{cases} c(t_* - s)^{k-m-\ell-1} T(t_* - s; f, k), & \ell < k - 1, \\ cT(t_* - s; f, k) + c\|A_* v_k^{(k)}(t_* - s)\|_{L^2(\Omega)}, & \ell = k - 1, \end{cases}$$

Combining the preceding estimates, integrating from 0 to t_* in s and then applying Gronwall's inequality from Theorem 4.2 complete the proof. $\qquad\square$

Remark 6.6 The regularity results in Theorems 6.15 and 6.16 are identical with that for subdiffusion with a time-independent elliptic operator. All the constants in these theorems depend on k but are uniformly bounded as $\alpha \to 1^-$.

6.4 Nonlinear Subdiffusion

Now we discuss nonlinear subdiffusion, for which a general theory is still unavailable. Thus we present only elementary results for three model problems, i.e., semilinear problems with Lipschitz nonlinearity, Allen-Cahn equation and compressible Navier-Stokes system, to illustrate the main ideas. One early contribution to nonlinear problems is [LRYZ13] with Neumann boundary conditions; see also [KY17] for semilinear diffusion wave. Other nonlinear subdiffusion -type problems that have been studied include Navier-Stokes equations [dCNP15], Hamilton-Jacobi equation [GN17, CDMI19], and p-Laplace equation [LRdS18], Keller-Segel equations [LL18b], and fully nonlinear Cauchy problems [TY17]. The recent monograph [GW20] treats semilinear problems extensively.

6.4.1 Lipschitz nonlinearity

In this part, we analyze the following semilinear subdiffusion problem

$$\begin{cases} \partial_t^\alpha u + Au = f(u), & t > 0, \\ \quad u(0) = u_0, \end{cases} \tag{6.65}$$

where A is the negative Laplacian $-\Delta$ (equipped with a zero Dirichlet boundary condition). The solution of this problem depends very much on the behavior of $f(u)$. The solution may not exist, or blow up in finite time, just as in the case of fractional ODEs, cf. Proposition 4.7 in Chapter 4. If the nonlinearity f is globally Lipschitz, the situation is similar to the standard case, as shown in [JLZ18, Theorem 3.1]. The analysis employs a fixed point argument, equipped with Bielecki norm [Bie56]. Using the solution representation (6.24), the solution u satisfies

$$u(t) = F(t)u_0 + \int_0^t E(t-s)f(u(s))\,\mathrm{d}s. \tag{6.66}$$

Theorem 6.17 *Let $u_0 \in \dot{H}^2(\Omega)$, and let $f : \mathbb{R} \to \mathbb{R}$ be Lipschitz continuous. Then problem (6.65) has a unique solution u such that*

$$u \in C^\alpha(\bar{I}; L^2(\Omega)) \cap C(\bar{I}; \dot{H}^2(\Omega)), \quad \partial_t^\alpha u \in C(\bar{I}; L^2(\Omega)), \tag{6.67}$$

$$\partial_t u(t) \in L^2(\Omega) \text{ and } \|\partial_t u(t)\|_{L^2(\Omega)} \le ct^{\alpha-1} \quad \forall t \in I. \tag{6.68}$$

Proof The lengthy proof is divided into four steps.

Step 1: Existence and uniqueness. We denote by $C(\overline{I}; L^2(\Omega))_\lambda$ the space $C(\overline{I}; L^2(\Omega))$ equipped with the norm $\|v\|_\lambda := \max_{t \in \overline{I}} \|e^{-\lambda t} v(t)\|_{L^2(\Omega)}$, which is equivalent to the standard norm of $C(\overline{I}; L^2(\Omega))$ for any fixed $\lambda > 0$. Then we define a map $M : C(\overline{I}; L^2(\Omega))_\lambda \to C(\overline{I}; L^2(\Omega))_\lambda$ by

$$Mv(t) = F(t)u_0 + \int_0^t E(t-s)f(v(s))ds.$$

For any $\lambda > 0$, $u \in C(\overline{I}; L^2(\Omega))$ is a solution of (6.65) if and only if u is a fixed point of the map $M : C(\overline{I}; L^2(\Omega))_\lambda \to C(\overline{I}; L^2(\Omega))_\lambda$. It remains to prove that for some $\lambda > 0$, the map $M : C(\overline{I}; L^2(\Omega))_\lambda \to C(\overline{I}; L^2(\Omega))_\lambda$ has a unique fixed point. In fact, the definition of M and Theorem 6.4(ii) directly yield for any $v_1, v_2 \in C(\overline{I}; L^2(\Omega))_\lambda$,

$$\|e^{-\lambda t}(Mv_1(t) - Mv_2(t))\|_{L^2(\Omega)}$$
$$= \left\| e^{-\lambda t} \int_0^t E(t-s)(f(v_1(s)) - f(v_2(s)))ds \right\|_{L^2(\Omega)}$$
$$\le ce^{-\lambda t} \int_0^t (t-s)^{\alpha-1} \|v_1(s) - v_2(s)\|_{L^2(\Omega)}ds$$
$$\le c \int_0^t (t-s)^{\alpha-1} e^{-\lambda(t-s)} \max_{s \in [0,T]} \|e^{-\lambda s}(v_1(s) - v_2(s))\|_{L^2(\Omega)}ds$$
$$= c\lambda^{-\alpha} \left(\int_0^1 (1-\zeta)^{\alpha-1} (\lambda t)^\alpha e^{-\lambda t(1-\zeta)}d\zeta \right) \|v_1 - v_2\|_\lambda,$$

where the second identity follows by changing variables $s = t\zeta$. Now direct computation gives $\lambda^{-\alpha} \int_0^1 (1-\zeta)^{\alpha-1}(\lambda t)^\alpha e^{-\lambda t(1-\zeta)}d\zeta \le c\lambda^{-\frac{\alpha}{2}} T^{\frac{\alpha}{2}}$. Thus, we obtain

$$\|e^{-\lambda t}(Mv_1(t) - Mv_2(t))\|_{L^2(\Omega)} \le c\lambda^{-\frac{\alpha}{2}} T^{\frac{\alpha}{2}} \|v_1 - v_2\|_\lambda. \qquad (6.69)$$

Thus, by choosing a sufficiently large λ, the map M is contractive on $C(\overline{I}; L^2(\Omega))_\lambda$. Theorem A.10 implies that M has a unique fixed point, or a unique solution of (6.65).

Step 2: $C^\alpha(\overline{I}; L^2(\Omega))$ regularity. Consider the difference quotient for $\tau > 0$

$$\frac{u(t+\tau) - u(t)}{\tau^\alpha} = \frac{F(t+\tau) - F(t)}{\tau^\alpha}u_0 + \frac{1}{\tau^\alpha} \int_t^{t+\tau} E(s)f(u(t+\tau-s))ds$$
$$+ \int_0^t E(s)\frac{f(u(t+\tau-s)) - f(u(t-s))}{\tau^\alpha}ds =: \sum_{i=1}^3 I_i(t,\tau).$$
$$(6.70)$$

By Theorem 6.4(iii) is that $\tau^{-\alpha}\|F(t+\tau) - F(t)\| \le c$, which implies $\|I_1(t,\tau)\|_{L^2(\Omega)} \le c$. By appealing to Theorem 6.4(ii), we have

$$\|I_2(t,\tau)\|_{L^2(\Omega)} = \left\|\frac{1}{\tau^\alpha}\int_t^{t+\tau} E(s)f(u(t+\tau-s))ds\right\|_{L^2(\Omega)}$$

$$\leq c\frac{1}{\tau^\alpha}\int_t^{t+\tau} s^{\alpha-1}ds = \frac{c}{\alpha}\frac{(t+\tau)^\alpha - t^\alpha}{\tau^\alpha} \leq c.$$

By the Lipschitz continuity of f, we have

$$e^{-\lambda t}\|I_3(t,\tau)\|_{L^2(\Omega)} = \left\|e^{-\lambda t}\int_0^t E(t-s)\frac{f(u(s+\tau))-f(u(s))}{\tau^\alpha}ds\right\|_{L^2(\Omega)}$$

$$\leq c_1\int_0^t e^{-\lambda(t-s)}(t-s)^{\alpha-1}e^{-\lambda s}\left\|\frac{u(s+\tau)-u(s)}{\tau^\alpha}\right\|_{L^2(\Omega)}ds.$$

By substituting the estimates of $I_i(t,\tau)$, $i = 1, 2, 3$, into (6.70) and denoting $W_\tau(t) = e^{-\lambda t}\tau^{-\alpha}\|u(t+\tau)-u(t)\|_{L^2(\Omega)}$, we obtain

$$W_\tau(t) \leq c + c_1\int_0^t e^{-\lambda(t-s)}(t-s)^{\alpha-1}W_\tau(s)ds \leq c + c_1\left(\frac{T}{\lambda}\right)^{\frac{\alpha}{2}}\max_{s\in\bar{I}}W_\tau(s),$$

where the last inequality can be derived as (6.69). By choosing a sufficiently large λ and taking maximum of the left-hand side with respect to $t \in \bar{I}$, it implies $\max_{t\in\bar{I}}W_\tau(t) \leq c$, which further yields $\tau^{-\alpha}\|u(t+\tau)-u(t)\|_X \leq ce^{\lambda t} \leq c$, where c is independent of τ. Thus, we have proved $\|u\|_{C^\alpha(\bar{I};X)} \leq c$.

Step 3: $C(\bar{I}; D(A))$ *regularity.* By applying the operator A to both sides of (6.66) and using the identity $I - F(t) = \int_0^t AE(t-s)ds$ from Lemma 6.2, we obtain

$$Au(t) = AF(t)u_0 + \int_0^t AE(t-s)f(u(s))ds$$

$$= (AF(t)u_0 + (I-F(t))f(u(t))) \tag{6.71}$$

$$+ \int_0^t AE(t-s)(f(u(s))-f(u(t)))ds := I_4(t) + I_5(t).$$

By Theorem 6.4(ii) and the $C^\alpha(\bar{I}; L^2(\Omega))$ regularity from Step 2, we have

$$\|I_5(t)\|_{L^2(\Omega)} = \left\|\int_0^t AE(t-s)(f(u(s))-f(u(t)))ds\right\|_{L^2(\Omega)}$$

$$\leq \int_0^t \frac{c\|u(s)-u(t)\|_{L^2(\Omega)}}{t-s}ds \leq c\int_0^t |t-s|^{\alpha-1}ds \leq ct^\alpha, \quad \forall t \in I.$$

Theorem 6.4(i) implies that $I_5(t)$ is continuous for $t \in I$, and the last inequality implies that $I_5(t)$ is also continuous at $t = 0$. Hence, $I_5 \in C(\bar{I}; L^2(\Omega))$. Moreover, Theorem 6.4(i) gives $I_4 \in C(\bar{I}; L^2(\Omega))$ and

$$\|I_4(t)\|_{L^2(\Omega)} \leq c\|Au_0 + f(u(t))\|_{L^2(\Omega)} \leq c.$$

Substituting the estimates of $I_4(t)$ and $I_5(t)$ into (6.71) yields $\|Au\|_{C(\bar{I};L^2(\Omega))} \le c$, which further implies $\|u\|_{C(\bar{I};D(A))} \le c$. The regularity result $u \in C(\bar{I};D(A))$ yields $\partial_t^\alpha u = Au + f(u) \in C(\bar{I};L^2(\Omega))$.

Step 4: Estimate of $\|u'(t)\|_{L^2(\Omega)}$. By differentiating (6.66) with respect to t, we obtain

$$u'(t) = F'(t)u_0 + E(t)f(u_0) + \int_0^t E(s)f'(u(t-s))u'(t-s)ds$$

$$= E(t)(-Au_0 + f(u_0)) + \int_0^t E(t-s)f'(u(s))u'(s)ds.$$

By multiplying this equation by $t^{1-\alpha}$, we get

$$t^{1-\alpha}u'(t) = t^{1-\alpha}E(t)(-Au_0 + f(u_0))$$
$$+ \int_0^t t^{1-\alpha}s^{\alpha-1}E(t-s)f'(u(s))s^{1-\alpha}u'(s)ds,$$

which directly implies that

$$e^{-\lambda t}t^{1-\alpha}\|u'(t)\|_{L^2(\Omega)} \le e^{-\lambda t}t^{1-\alpha}\|E(t)\|\| - Au_0 + f(u_0)\|_{L^2(\Omega)}$$
$$+ \int_0^t e^{-\lambda(t-s)}t^{1-\alpha}s^{\alpha-1}(t-s)^{\alpha-1}\|f'(u(s))\|_{L^\infty(\Omega)}e^{-\lambda s}s^{1-\alpha}\|u'(s)\|_{L^2(\Omega)}ds$$
$$\le ce^{-\lambda t}\| - Au_0 + f(u_0)\|_{L^2(\Omega)} + c\lambda^{-\frac{\alpha}{2}}T^{\frac{\alpha}{2}}\max_{s\in\bar{I}}e^{-\lambda s}s^{1-\alpha}\|u'(s)\|_{L^2(\Omega)}.$$

where the last line follows similarly as (6.69). By choosing a sufficiently large λ and taking maximum of the left-hand side with respect to $t \in \bar{I}$, it implies $\max_{t\in\bar{I}}\|e^{-\lambda t}t^{1-\alpha}u'(t)\|_{L^2(\Omega)} \le c$, which further yields (6.68). □

Remark 6.7 For smooth u_0 and f, in the absence of extra compatibility conditions, the regularity results (6.67)–(6.68) are sharp with respect to the Hölder continuity in time. The regularity (6.67) is identical with Theorem 6.13. If f is smooth but not Lipschitz, and problem (6.65) has a unique bounded solution, then $f(u)$ and $f'(u)$ are still bounded. In this case, the estimates (6.67)–(6.68) are still valid.

One can prove regularity results under weaker assumption on initial data u_0 [AK19, Theorems 3.1 and 3.2]. The proof is similar to Theorem 6.17, and thus it is omitted. We refer interested readers also to [WZ20, Section 2] for further regularity estimates in $L^\infty(\Omega)$.

Theorem 6.18 *Let* $u_0 \in \dot{H}^\gamma(\Omega)$, $\gamma \in [0,2]$, *and* $f : \mathbb{R} \to \mathbb{R}$ *be Lipschitz continuous. Then problem (6.65) has a unique solution* u. *Further, the following statements hold.*

(i) *If* $\gamma \in (0,2]$, *then*

$$u \in C^{\frac{\gamma}{2}\alpha}(\bar{I};L^2(\Omega)) \cap C(I;\dot{H}^2(\Omega)), \quad \partial_t^\alpha u \in C(I;L^2(\Omega)),$$

$$\partial_t u \in L^2(\Omega), \quad \|\partial_t u(t)\|_{L^2(\Omega)} \le ct^{\frac{\gamma}{2}\alpha-1}, \quad t \in I.$$

(ii) *If* $\gamma = 0$, *then for* $p < \frac{1}{\alpha}$, $u \in C(\bar{I}; L^2(\Omega)) \cap L^p(I; \dot{H}^2(\Omega)) \cap C(I; \dot{H}^{2-\epsilon}(\Omega))$, $\partial_t^\alpha u \in L^p(I; L^2(\Omega))$, $\partial_t u(t) \in L^2(\Omega)$, $\|\partial_t u(t)\|_{L^2(\Omega)} \leq ct^{\frac{\gamma}{2}\alpha-1}$ *for any* $t \in I$ *and there hold*

$$\|u\|_{L^p(I;\dot{H}^2(\Omega))} + \|\partial_t^\alpha u\|_{L^p(I;L^2(\Omega))} \leq c(\|u_0\|_{L^2(\Omega)} + \|f(0)\|_{L^2(\Omega)}),$$

$$\|u(t)\|_{\dot{H}^{2-\epsilon}(\Omega)} \leq ct^{-\frac{2-\epsilon}{2}\alpha}\|u_0\|_{L^2(\Omega)} + c\epsilon^{-1}t^{\frac{\epsilon}{2}\alpha}\|f(0)\|_{L^2(\Omega)}, \quad \forall t \in I.$$

6.4.2 Allen-Cahn equation

The global Lipschitz continuity on the nonlinearity f is restrictive, and does not cover many important examples in practice. Thus, it is of interest to relax the condition to locally Lipschitz or Hölder continuity. Following [TYZ19, DYZ20] (see also [LS21] for gradient flow), we illustrate this with the following time-fractional Allen-Cahn equation

$$\begin{cases} \partial_t^\alpha u + Au = u - u^3, & \text{in } Q, \\ u(0) = u_0, & \text{in } \Omega. \end{cases} \tag{6.72}$$

When $\alpha = 1$, the model recovers the classical Allen-Cahn equation [AC72], which is often employed to describe the process of phase separation in multi-component alloy systems. The nonlinear term $f(u) = u - u^3$ can be rewritten as $f(u) = -F'(u)$, with $F(u) = \frac{1}{4}(1-u^2)^2$ being the standard double-well potential. Below, the operator A is taken to be $-\Delta$ with its domain $D(A) = H_0^1(\Omega) \cap H^2(\Omega)$. Problem (6.72) is the Euler-Lagrange equation of the following energy

$$E(u) = \int_\Omega \left(\frac{|\nabla u|^2}{2} + F(u) \right) dx.$$

Now we show the uniqueness and existence of a solution [DYZ20, Theorem 2.1].

Theorem 6.19 *For every* $u_0 \in H_0^1(\Omega)$, *there exists a unique weak solution to problem* (6.72) *such that* $u \in W^{\alpha,p}(I; L^2(\Omega)) \cap L^p(I; \dot{H}^2(\Omega))$, *for all* $p \in [2, \frac{2}{\alpha})$. *Further, if* $u_0 \in \dot{H}^2(\Omega)$, *it satisfies* $u \in L^\infty(Q)$.

Proof The proof applies the standard Galerkin procedure. We divide the lengthy proof into four steps.

Step 1. Galerkin approximation. Let $\{(\lambda_j, \varphi_j)\}_{j=1}^\infty$ be the eigenpairs of the operator A. For every $n \in \mathbb{N}$, let $X_n = \text{span}\{\varphi_j\}_{j=1}^n$. Let $u^n \in X_n$ such that

$$(\partial_t^\alpha u^n, v) + (\nabla u^n, \nabla v) = (f(u^n), v), \quad \forall v \in X_n \quad \text{and} \quad u^n(0) = P_n u_0, \tag{6.73}$$

where $P_n : L^2(\Omega) \to X_n$ is the $L^2(\Omega)$ orthogonal projection, defined by $(\phi, v) = (P_n\phi, v)$ for all $\phi \in L^2(\Omega)$ and $v \in X_n$. The existence and uniqueness of a local solution $u \in C(\bar{I}; \mathbb{R}^n) \cap C(I; \mathbb{R}^n)$ to (6.73) can be proved, since f is smooth and hence locally Lipschitz continuous; see Corollary 4.3 in Chapter 4. Letting $v = u^n$ in (6.73) and using the inequality

$$\tfrac{1}{2}\partial_t^\alpha \|u^n(t)\|_{L^2(\Omega)}^2 \le (\partial_t^\alpha u^n(t), u^n(t)),$$

cf. Lemma 2.11, we obtain

$$\partial_t^\alpha \|u^n(t)\|_{L^2(\Omega)}^2 + 2\|\nabla u^n(t)\|_{L^2(\Omega)}^2 + 2\|u^n(t)\|_{L^4(\Omega)}^4 \le 2\|u^n(t)\|_{L^2(\Omega)}^2.$$

This, Gronwall's inequality in Theorem 4.2 and $L^2(\Omega)$-stability of P_n yield

$$\|u^n(t)\|_{L^2(\Omega)} \le c_T \|u^n(0)\|_{L^2(\Omega)} \le c_T \|u_0\|_{L^2(\Omega)},$$

with c_T independent of n. Thus, $u^n \in C(\bar{I}; L^2(\Omega))$, and it is a global solution.

Step 2. Weak solution. Now we show that the sequence $\{u^n\}_{n=1}^\infty$ converges to a weak solution of problem (6.72) as $n \to \infty$ using an energy argument. To this end, taking $v = -\Delta u^n \in X_n$ in (6.73) and noting the identity

$$-\left(u^n - (u^n)^3, \Delta u^n\right) = \|\nabla u^n\|_{L^2(\Omega)}^2 - 3\|u^n|\nabla u^n|\|_{L^2(\Omega)}^2,$$

we obtain

$$\partial_t^\alpha \|\nabla u^n(t)\|_{L^2(\Omega)}^2 + 2\|\Delta u^n(t)\|^2 + 6\|u^n|\nabla u^n|(t)\|_{L^2(\Omega)}^2 \le 2\|\nabla u^n(t)\|_{L^2(\Omega)}^2.$$

This and Theorem 4.2 imply $\|\nabla u^n\|_{L^\infty(I;L^2(\Omega))} \le c_T \|\nabla u_0\|_{L^2(\Omega)}$, where c_T is independent of n. Then by Sobolev embedding in Theorem A.3, $u_n \in L^\infty(I; L^6(\Omega))$ for $\Omega \subset \mathbb{R}^d$ with $d \le 3$. Consequently, $\|f(u^n)\|_{L^\infty(I;L^2(\Omega))} \le c$. Since $u_0 \in H_0^1(\Omega)$, by the maximal L^p regularity in Theorems 6.8 and 6.11,

$$\|\partial_t^\alpha u^n\|_{L^p(I;L^2(\Omega))} + \|A u^n\|_{L^p(I;L^2(\Omega))} \le c_{p,\kappa}, \quad \forall p \in [2, \tfrac{2}{\alpha}).$$

where the constant $c_{p,\kappa}$ is independent of n. This implies

$$u^n \in W^{\alpha,p}(I; L^2(\Omega)) \cap L^p(I; \dot{H}^2(\Omega)),$$

which compactly embeds into $C(\bar{I}; L^2(\Omega))$ for $p \in (\tfrac{1}{\alpha}, \tfrac{2}{\alpha})$, by Sobolev embedding. Thus, there exists a function u and a subsequence, still denoted by $\{u^n\}$ such that as $n \to \infty$

$$u^n \longrightarrow u \quad \text{weak-}* \text{ in } \quad L^\infty(I; H_0^1(\Omega)),$$
$$\partial_t^\alpha u^n \longrightarrow \zeta \quad \text{weakly in } \quad L^p(I; L^2(\Omega)),$$
$$u^n \longrightarrow u \quad \text{weakly in } \quad L^p(I; \dot{H}^2(\Omega)),$$
$$u^n \longrightarrow u \quad \text{in } \quad C(\bar{I}; L^2(\Omega)).$$

Next, we claim $\zeta = \partial_t^\alpha u$. Indeed, for any $v \in H_0^1(\Omega)$ and $\phi \in C_0^\infty(I)$, by integration by parts, we have (with ${}_t\partial_T^\alpha \phi(t)$ being the right-sided Djrbashian-Caputo fractional derivative)

$$\int_I (\zeta(t), v)\phi(t)dt = \lim_{n\to\infty} \int_I (\partial_t^\alpha u^n(t), v)\phi(t)dt$$

$$= \lim_{n\to\infty} \int_I (u^n, v)_t\partial_T^\alpha \phi(t)dt + \phi(T)(({}_0 I_t^{1-\alpha} u^n)(T), v) - ({}_t I_T^{1-\alpha}\phi)(0)(u^n(0), v)$$

$$= \lim_{n\to\infty} \int_I (u^n, v)_t\partial_T^\alpha \phi(t)dt = \int_I (u, v)_t\partial_T^\alpha \phi(t)dt = \int_I (\partial_t^\alpha u(t), v)\phi(t)dt.$$

Thus, upon passing to the limit in (6.73), the function u satisfies

$$(\partial_t^\alpha u, v) + (\nabla u, \nabla v) = (f(u), v), \quad \forall v \in H_0^1(\Omega). \tag{6.74}$$

Moreover, by the convergence of u^n in $C(\bar{I}; L^2(\Omega))$, $u^n(0)$ converges to u_0 in $L^2(\Omega)$ and so $u(0) = u_0$. Thus, problem (6.72) admits a weak solution in $W^{\alpha,p}(I; L^2(\Omega)) \cap L^p(I; \dot{H}^2(\Omega))$ with $p \in [2, \frac{2}{\alpha})$.

Step 3. Uniqueness. Next, we prove the uniqueness. Let u_1 and u_2 be two weak solutions of problem (6.72). Then the difference $w = u_1 - u_2$ satisfies

$$(\partial_t^\alpha w(t), v) + (\nabla w(t), \nabla v) = (f(u_1) - f(u_2), v), \quad \forall v \in H_0^1(\Omega),$$

with $w(0) = 0$. Taking $v = w(t)$ and noting the inequalities

$$(f(u_1) - f(u_2))(u_1 - u_2) = (u_1 - u_2)^2(1 - u_1^2 - u_1 u_2 - u_2^2) \le (u_1 - u_2)^2,$$

and Lemma 2.11, we obtain $\partial_t^\alpha \|w(t)\|_{L^2(\Omega)}^2 \le 2\|w(t)\|_{L^2(\Omega)}^2$, with $w(0) = 0$. The Gronwall's inequality in Theorem 4.2 implies $w \equiv 0$, i.e., $u_1 = u_2$.

Step 4. L^∞ bound. Now we derive the $L^\infty(Q)$ bound. The preceding argument implies $f(u) \in L^\infty(I; L^2(\Omega))$. Since $u_0 \in \dot{H}^2(\Omega)$, we apply the maximal L^p regularity in Theorems 6.8 and 6.11, and obtain

$$\|\partial_t^\alpha u\|_{L^p(I;L^2(\Omega))} + \|Au\|_{L^p(I;L^2(\Omega))} \le c, \quad \forall p \in (1, \infty),$$

which implies $u \in W^{\alpha,p}(I; L^2(\Omega)) \cap L^p(I; \dot{H}^2(\Omega))$. Then, by means of real interpolation with a sufficiently large exponent p, we deduce $u \in L^\infty(Q)$. This completes the proof the theorem. $\qquad \square$

The next result gives the solution regularity for smooth initial data $u_0 \in \dot{H}^2(\Omega)$ [DYZ20, Theorem 2.2].

Theorem 6.20 *If $u_0 \in \dot{H}^2(\Omega)$, then the solution u to problem (6.72) satisfies that for any $\beta \in [0, 1)$ and $t \in I$*

$$u \in C^{\alpha}(\bar{I}; L^2(\Omega)) \cap C(\bar{I}; \dot{H}^2(\Omega)), \quad \partial_t^{\alpha} u \in C(\bar{I}; L^2(\Omega)); \tag{6.75}$$

$$Au \in C(I; H^{2\beta}(\Omega)), \quad \|A^{1+\beta} u(t)\|_{L^2(\Omega)} \le ct^{-\beta\alpha}; \tag{6.76}$$

$$\partial_t u(t) \in C((I; H^{2\beta}(\Omega)) \quad and \quad \|A^{\beta} \partial_t u(t)\|_{L^2(\Omega)} \le ct^{\alpha(1-\beta)-1}, \tag{6.77}$$

where the constant c depends on $\|Au_0\|_{L^2(\Omega)}$ *and T.*

Proof The regularity estimate (6.75) has already been shown in Theorem 6.17. It suffices to show (6.76)–(6.77). We employ the representation (6.66):

$$A^{1+\beta} u(t) = A^{\beta} F(t)(Au_0) + \int_0^t A^{\beta} E(t-s)[A(u-u^3)(s)]\mathrm{d}s.$$

By Theorem 6.4(ii)–(iii), we obtain

$$\|A^{1+\beta} u(t)\|_{L^2(\Omega)} \le ct^{-\beta\alpha}\|Au_0\|_{L^2(\Omega)} + \int_0^t (t-s)^{(1-\beta)\alpha-1}\|A(u-u^3)(s)\|_{L^2(\Omega)}\mathrm{d}s.$$

Next, we treat the term in the integral. In view of the identity $\Delta u^3 = 6u|\nabla u|^2 + 3u^2\Delta u$, the estimate $u \in C(\bar{I}; \dot{H}^2(\Omega))$ from (6.75) and $u \in L^{\infty}(Q)$ from Theorem 6.19, we deduce

$$\|Af(u)\|_{L^2(\Omega)} \le c\|\Delta u\|_{L^2(\Omega)} + c\||u|\nabla u|^2\|_{L^2(\Omega)} + c\|u^2\Delta u\|_{L^2(\Omega)} \le c_T,$$

with c_T independent of t. Hence, we obtain $\|A^{1+\beta} u(t)\|_{L^2(\Omega)} \le ct^{-\beta\alpha}$. Last, we prove (6.77). The case $\beta = 0$ is given in Theorem 6.17. For $\beta \in [0, 1)$,

$$A^{\beta} u'(t) = A^{\beta-1} F'(t)(Au_0) + \frac{\mathrm{d}}{\mathrm{d}t} \int_0^t A^{\beta} E(t-s)[(u-u^3)(s)]\mathrm{d}s$$

$$= A^{\beta-1} F'(t)(Au_0) + A^{\beta} E(t)[(u-u^3)(0)]$$

$$+ \int_0^t A^{\beta} E(t-s)[(u'-3u^2u')(s)]\mathrm{d}s = \sum_{i=1}^3 \mathrm{I}_i.$$

By Theorem 6.4, the terms I_1 and I_2 are respectively bounded by

$$\|\mathrm{I}_1\|_{L^2(\Omega)} \le ct^{(1-\beta)\alpha-1}\|\Delta u_0\|_{L^2(\Omega)},$$

$$\|\mathrm{I}_2\|_{L^2(\Omega)} \le ct^{(1-\beta)\alpha-1}\|u_0 - u_0^3\|_{L^2(\Omega)} \le ct^{(1-\beta)\alpha-1}\|Au_0\|_{L^2(\Omega)}.$$

Similarly, the third term I_3 follows by

$$\|\mathrm{I}_3\|_{L^2(\Omega)} \le c \int_0^t (t-s)^{\alpha-1}\|A^{\beta} u'(s)\|_{L^2(\Omega)}\mathrm{d}s.$$

The last three estimates and the Gronwall's inequality in Theorem 4.2 give the desired result. Last, the continuity follows directly from that of the solution operators $E(t)$ and $F(t)$, cf. Theorem 6.4. □

Last, we give two interesting qualitative results on problem (6.72), i.e., maximum principle and energy dissipation. The following maximum principle holds: if $|u_0| \leq 1$, then $\|u(t)\|_{L^\infty(\Omega)}$ is also bounded by 1. We refer to Section 6.5 for a more detailed account on maximum principle for the linear subdiffusion model.

Proposition 6.3 Let $u_0 \in \dot{H}^2(\Omega)$ with $|u_0(x)| \leq 1$. Then the solution u of problem (6.72) satisfies

$$|u(x,t)| \leq 1 \quad \forall(x,t) \in Q.$$

Proof Suppose that the minimum of u is smaller than -1 and achieved at $(x_0, t_0) \in Q$. Then by the regularity estimates (6.75) and (6.77) in Theorem 6.20, and Proposition 2.5(i), we deduce $\partial_t^\alpha u(x_0, t_0) \leq 0$. By Theorem 6.20 and Sobolev embedding theorem, Δu is continuous in $\overline{\Omega}$, and hence $\Delta u(x_0, t_0) \geq 0$. Consequently, we obtain

$$0 \geq \partial_t^\alpha u(x_0, t_0) - \Delta u(x_0, t_0) = u(x_0, t_0) - u(x_0, t_0)^3 > 0,$$

which leads to a contradiction. The upper bound can be proved similarly. □

The last result gives a (weak) energy dissipation law.

Proposition 6.4 If the solution u to problem (6.72) belongs to $W^{1, \frac{2}{2-\alpha}}(I; L^2(\Omega)) \cap L^2(I; \dot{H}^2(\Omega))$, then there holds $E(u(T)) \leq E(u_0)$.

Proof Taking $v = \partial_t u$ in (6.74) and integrating over $(0, T)$ lead to,

$$\int_0^T (\partial_t^\alpha u(t), \partial_t u(t)) + (\nabla u(t), \nabla \partial_t u(t)) \mathrm{d}t = \int_0^T (f(u(t)), \partial_t u(t)) \, \mathrm{d}t$$

Under the given regularity assumption on u, by Proposition 2.2,

$$\frac{1}{2} \int_0^T \frac{\mathrm{d}}{\mathrm{d}t} \int_\Omega |\nabla u(t)|^2 \, \mathrm{d}x \, \mathrm{d}t + \int_0^T \frac{\mathrm{d}}{\mathrm{d}t} \int_\Omega F(u(t)) \, \mathrm{d}x \, \mathrm{d}t \leq 0.$$

This immediately implies the desired bound. □

6.4.3 Compressible Navier-Stokes problem

This part is concerned with time-fractional compressible Navier-Stokes equations. The standard counterpart describes the dynamics of Newtonian fluids. In the incompressible case with constant density, the existence and uniqueness of a weak solution in 2D have been proved. However, in 3D case, global weak solutions may not be unique. The existence and uniqueness of global smooth solutions are still open. Let $\Omega \subset \mathbb{R}^d$, $d = 2, 3$ be an open bounded domain with a smooth boundary. The time-fractional compressible Navier-Stokes equations read

$$\begin{cases} \partial_t^\alpha u + u \cdot \nabla u + (\nabla u) \cdot u + (\nabla \cdot u)u = \Delta u, & \text{in } Q, \\ \qquad\qquad\qquad\qquad\qquad\qquad u = 0, & \text{on } \partial_L Q, \\ \qquad\qquad\qquad\qquad\qquad u(\cdot, 0) = u_0, & \text{in } \Omega. \end{cases}$$

Note that ∇u is a tensor, given by $(\nabla u)_{ij} = \partial_{x_i} u_j$. This system can also be formulated in the conservative form

$$\begin{cases} \partial_t^\alpha u + \nabla \cdot (u \otimes u) + \frac{1}{2}\nabla(|u|^2) = \Delta u, & \text{in } Q, \\ \qquad\qquad\qquad\qquad\qquad u = 0, & \text{on } \partial_L Q, \qquad\qquad (6.78) \\ \qquad\qquad\qquad\qquad u(\cdot, 0) = u_0, & \text{in } \Omega. \end{cases}$$

The notation $u \otimes u$ denotes the tensor product, i.e., $u \otimes u = [u_i u_j]$, $i, j = 1, \ldots, d$.

6.4.3.1 Compactness criteria

We employ some compactness criteria to show the existence of weak solutions of problem (6.78). For linear evolution equations, proving the existence of weak solutions is relatively easy. Indeed, one only needs weak compactness, which is guaranteed by boundedness in reflexive spaces; see Section 6.1. However, for non-linear evolution equations, strong compactness criteria, e.g., Aubin-Lions lemma, are often needed. Below we present two strong compactness criteria. The proof relies crucially on the following estimates on the time shift operator $\tau_h u(t) := u(t + h)$ [LL18b, Proposition 3.4].

Proposition 6.5 *Fix $T > 0$. Let E be a Banach space and $\alpha \in (0, 1)$. Suppose that $u \in L^1_{loc}(I; E)$ has a weak Djrbashian-Caputo fractional derivative $\partial_t^\alpha u \in L^p(I; E)$ associated with initial value $u_0 \in E$. Let $r_0 = \infty$ if $p\alpha \geq 1$ and $r_0 = \frac{p}{1-p\alpha}$ if $p\alpha < 1$. Then, there exists $c > 0$ independent of h and u such that*

$$\|\tau_h u - u\|_{L^r(0,T-h;E)} \leq \begin{cases} ch^{\alpha + \frac{1}{r} - \frac{1}{p}} \|\partial_t^\alpha u\|_{L^p(I;E)}, & r \in [p, r_0), \\ ch^\alpha \|\partial_t^\alpha u\|_{L^p(I;E)}, & r \in [1, p]. \end{cases}$$

Proof Let $f := \partial_t^\alpha u \in L^p(I; E)$. Then by Proposition 4.6, $u(t) = u(0) + {}_0I_t^\alpha f(t)$. Let $K_1(s, t; h) := (t + h - s)^{\alpha - 1}$ and $K_2(s, t; h) := (t - s)^{\alpha - 1} - (t + h - s)^{\alpha - 1}$. Then

$$\tau_h u(t) - u(t) = \frac{1}{\Gamma(\alpha)}\Big(\int_t^{t+h} K_1(s, t; h) f(s) \mathrm{d}s + \int_0^t K_2(s, t; h) f(s) \mathrm{d}s \Big),$$

so that

$$\int_0^{T-h} \|\tau_h u - u\|_E^r \mathrm{d}t \leq \frac{2^r}{(\Gamma(\alpha))^r}\Big(\int_0^{T-h} \Big(\int_t^{t+h} K_1 \|f\|_E(s)\mathrm{d}s \Big)^r \mathrm{d}t \\ + \int_0^{T-h} \Big(\int_0^t K_2 \|f\|_E(s)\mathrm{d}s \Big)^r \mathrm{d}t \Big).$$

First, we consider the case $r \geq p$ and $\frac{1}{r} > \frac{1}{p} - \alpha$. We denote $I_1 = (t, t + h)$ and $I_2 = (0, t)$. Let $\frac{1}{r} + 1 = \frac{1}{q} + \frac{1}{p}$. By Hölder inequality, we have for $i = 1, 2$,

$$\int_{I_i} K_i \|f\|_E(s) \mathrm{d}s \leq \left(\int_{I_i} K_i^q \|f\|_E^p \mathrm{d}s \right)^{\frac{1}{r}} \left(\int_{I_i} K_i^q \mathrm{d}s \right)^{\frac{r-q}{qr}} \left(\int_{I_i} \|f\|_E^p \mathrm{d}s \right)^{\frac{r-p}{pr}}.$$

Next, we bound the three terms on the right-hand side. Direct computation shows

$$\left(\int_{I_i} \|f\|_E^p \mathrm{d}s \right)^{\frac{r-p}{pr}} \leq \|f\|_{L^p(I;E)}^{1-\frac{p}{r}} \quad \text{and} \quad \int_{I_1} K_1^q \mathrm{d}s = \frac{h^{q(\alpha-1)+1}}{q(\alpha-1)+1}.$$

It follows from the inequality $(a + b)^q \geq a^q + b^q$ for $q \geq 1$, $a, b \geq 0$ that $K_2^q \leq (t - s)^{q(\alpha-1)} - (t + h - s)^{q(\alpha-1)}$. Since $q(\alpha - 1) + 1 > 0$, we find

$$\int_0^t K_2^q \mathrm{d}s \leq \frac{t^{q(\alpha-1)+1} - (t + h)^{q(\alpha-1)+1} + h^{q(\alpha-1)+1}}{q(\alpha - 1) + 1} \leq c h^{q(\alpha-1)+1}.$$

Combining the preceding estimates gives

$$\int_0^{T-h} \|\tau_h u - u\|_E^r \mathrm{d}t \leq c h^{(q(\alpha-1)+1)\frac{r-q}{q}} \left(\int_0^T \|f\|_E^p(s) \left(\int_{0 \wedge s - h}^s K_1^q \mathrm{d}t \right) \mathrm{d}s \right.$$
$$\left. + \int_0^{T-h} \|f\|_E^p(s) \left(\int_s^{T-h} K_2^q \mathrm{d}t \right) \mathrm{d}s \right).$$

Direct computation shows

$$\int_{0 \wedge s - h}^s K_1^q \mathrm{d}t \leq \frac{h^{q(\alpha-1)+1}}{q(\alpha - 1) + 1}$$

and

$$\int_s^{T-h} K_2^q \mathrm{d}t \leq \int_s^{T-h} (t - s)^{q(\alpha-1)} \mathrm{d}t - \int_s^{T-h} (t - s + h)^{q(\alpha-1)} \mathrm{d}t \leq \frac{h^{q(\alpha-1)+1}}{q(\alpha - 1) + 1}.$$

Consequently,

$$\|\tau_h u - u\|_{L^r(0,T-h;E)} \leq c h^{\alpha+\frac{1}{r}-\frac{1}{p}} \|\partial_t^\alpha u\|_{L^p(I;E)}.$$

This immediately implies the desired estimate. Next, we consider the case $r < p$. Note that for $r = p$, the first part implies

$$\|\tau_h u - u\|_{L^p(0,T-h;E)} \leq c h^\alpha \|\partial_t^\alpha u\|_{L^p(I;E)}.$$

Then, by Hölder inequality,

$$\|\tau_h u - u\|_{L^r(0,T-h;E)} \leq \|1\|_{L^{\frac{rp}{p-r}}(0,T-h;E)} \|\tau_h u - u\|_{L^p(0,T-h;E)}$$

$$\leq T^{\frac{1}{r}-\frac{1}{p}} \|\tau_h u - u\|_{L^p(0,T-h;E)}.$$

This completes the proof of the proposition.   □

Now we can state two compactness criteria of Aubin-Lions type [LL18b, Theorems 4.1 and 4.2], with "assignment" in the sense of Theorem 2.15.

Theorem 6.21 *Let $T > 0$, $\alpha \in (0, 1)$ and $p \in [1, \infty)$. Let E, E_0 and E_1 be three Banach spaces such that $E_1 \hookrightarrow E \hookrightarrow E_0$. Let the set $U \subset L^1_{loc}(I; E_1)$ satisfy*

(i) *There exists $c_1 > 0$ such that for any $u \in U$, $\sup_{t \in I} {_0}I_t^\alpha \|u\|_{E_1}^p (t) \leq c_1$.*
(ii) *There exist $r \in (\frac{p}{1+p\alpha}, \infty) \cap [1, \infty)$ and $c_2 > 0$ such that for any $u \in U$, there is an assignment of initial value u_0 for u and $\partial_t^\alpha u$ satisfies $\|\partial_t^\alpha u\|_{L^r(I;E_0)} \leq c_2$.*

Then, U is relatively compact in $L^p(I; E)$.

Proof Note that for any $\alpha \in (0, 1)$, there holds

$$\|u\|_{L^p(I;E_1)}^p \leq T^{1-\alpha} {_0}I_t^\alpha \|u\|_{E_1}^p (T). \tag{6.79}$$

Thus, (i) implies that U is bounded in $L^p(I; E)$. Next, let r_0 be the number defined in Proposition 6.5. If $r < \frac{1}{\alpha}$, then $r_0 = \frac{r}{1-r\alpha} > p$, since $r > \frac{p}{1+p\alpha}$. Otherwise, $r_0 = \infty > p$. Together with the condition $r \geq 1$, Proposition 6.5 and (ii) imply $\|\tau_h u - u\|_{L^p(I;E)} \to 0$ uniformly. Hence, the conditions in Theorem A.6 are fulfilled, and the relative compactness of U in $L^p(I; E)$ follows.   □

Theorem 6.22 *Let $T > 0$, $\alpha \in (0, 1)$ and $p \in [1, \infty)$. Let E, E_0 and E_1 be three Banach spaces such that $E_1 \hookrightarrow E \hookrightarrow E_0$. Let the set $U \subset L^1_{loc}(I; E_1)$ satisfy*

(i) *There exist $r_1 \in [1, \infty)$ and $c_1 > 0$ such that for any $u \in U$, $\sup_{t \in I} {_0}I_t^\alpha \|u\|_{E_1}^{r_1} \leq c_1$.*
(ii) *There exists $p_1 \in (p, \infty]$ such that U is bounded in $L^{p_1}(I; E)$.*
(iii) *There exist $r_2 \in [1, \infty)$ and $c_2 > 0$ such that for any $u \in U$, there is an assignment of initial value u_0 for u so that $\partial_t^\alpha u$ satisfies $\|\partial_t^\alpha u\|_{L^{r_2}(I;E_0)} \leq c_2$.*

Then, U is relatively compact in $L^p(I; E)$.

Proof (i) and (6.79) imply that U is bounded in $L^p(I; E)$. It follows from (iii) and Proposition 6.5 that $\|\tau_h u - u\|_{L^1(0,T-h;E)} \to 0$ uniformly. Hence, by Theorem A.6, U is relatively compact in $L^1(I; E)$. Since U is bounded in $L^{p_1}(I; E)$ with $p_1 > p$, cf. (ii), the relative compactness of U in $L^p(I; E)$ follows from Theorem A.7.   □

Next, we state an abstract weak convergence result for the Djrbashian-Caputo fractional derivative [LL18b, Proposition 3.5]. The notation E' denotes the dual space of the space E.

Proposition 6.6 *Let E be a reflexive Banach space, $\alpha \in (0, 1)$ and $T > 0$. Let $u_n \to u$ in $L^p(I; E)$, $p \geq 1$, and the Djrbashian-Caputo fractional derivatives $\partial_t^\alpha u_n$ with initial values $u_{0,n}$ are bounded in $L^r(I; E)$, $r \in [1, \infty)$.*

(i) *There is a subsequence of $u_{0,n}$ converging weakly in E to some $u_0 \in E$.*

(ii) *If $r > 1$, there exists a subsequence of $\partial_t^\alpha u_{n_k}$ converging weakly to f and u_{0,n_k} converging weakly to u_0, and f is the Djrbashian-Caputo fractional derivative of u with initial value u_0 so that $u(t) = u_0 + {_0}I_t^\alpha f(t)$. Further, if $r \geq \frac{1}{\alpha}$, then, $u(0^+) = u_0$ in E in the sense of Lebesgue point.*

Proof (i). Let $f_n = \partial_t^\alpha u_n$. By Theorem 2.2, $u_n(t) - u_{0,n} = {_0}I_t^\alpha f_n$ is bounded in $L^{r_1}(I; E)$ where $r_1 \in [1, \frac{r}{1-r\alpha} - \epsilon)$ if $r < \frac{1}{\alpha}$ or $r_1 \in [1, \infty)$ if $r > \frac{1}{\alpha}$. Then, $u_n(t) - u_{0,n}$ is bounded in $L^{p_1}(I; E)$, with $p_1 = \min(r_1, p)$. Since u_n converges in L^p and thus in L^{p_1}, then $u_{0,n}$ is bounded in $L^{p_1}(I; E)$. Hence, $u_{0,n}$ is bounded in E. Since E is reflexive, there is a subsequence u_{0,n_k} converging weakly to u_0 in E.

(ii). Take a subsequence such that u_{0,n_k} converges weakly to u_0 and $\partial_t^\alpha u_{n_k}$ to f weakly in $L^r(I; E)$ (since $r > 1$). For any $\varphi \in C_c^\infty[0, T)$ and $\phi \in E'$, we have $\int_0^T {_t}\partial_T^\alpha \varphi(u_{n_k}(t) - u_{0,n_k}) dt = \int_0^T \varphi f_{n_k} dt$. With $\langle \cdot, \cdot \rangle$ being the duality pairing between $L^{\frac{r}{r-1}}(I; E')$ and $L^r(I; E)$, the preceding identity and the relations $\phi\varphi, \phi_t\partial_T^\alpha \varphi \in L^{\frac{r}{r-1}}(I; E')$ imply $\langle \phi_t\partial_T^\alpha \varphi, u_{n_k}(t) - u_{0,n_k} \rangle = \langle \phi\varphi, f_{n_k} \rangle$ (with ${_t}\partial_T^\alpha$ being the right-sided Djrbashian-Caputo fractional derivative). Taking the limit $k \to \infty$ and using the weak convergence yield $\langle \phi_t\partial_T^\alpha \varphi, u - u_0 \rangle = \langle \phi\varphi, f \rangle$. Since ϕ is arbitrary and $f \in L^r(I; E)$, there holds $\int_0^T {_t}\partial_T^\alpha \varphi(u(t) - u_0) dt = \int_0^T \varphi f dt$. Hence, f is the Djrbashian-Caputo fractional derivative of u with initial value u_0 in the weak sense, and there holds $u = u_0 + {_0}I_t^\alpha f(t)$. The last claim follows from Theorem 2.15. □

6.4.3.2 Existence of weak solutions

Now we employ the compactness criteria to establish the existence of weak solutions. Motivated by Lemma 2.6, we define a weak solution as follows.

Definition 6.2 Let $u_0 \in L^2(\Omega)$ and $q_1 \min(2, \frac{4}{d})$. A function

$$u \in L^\infty(I; L^2(\Omega)) \cap L^2(I; H_0^1(\Omega)) \quad \text{with } \partial_t^\alpha u \in L^{q_1}(I; H^{-1}(\Omega))$$

is called a weak solution to problem (6.78), if

$$\int_0^T \int_\Omega (u(x, t) - u_0)_t\partial_T^\alpha v \, dxdt + \int_0^T \int_\Omega \left(-\nabla v : u \otimes u - \frac{1}{2}\nabla \cdot v|u|^2\right) dxdt$$

$$= \int_0^T \int_\Omega u \cdot \Delta v \, dxdt, \quad \forall v \in C_c^\infty([0, T) \times \Omega; \mathbb{R}^d).$$

If u is defined on $\mathbb{R}_+$ and its restriction on any interval $[0, T)$, for any $T > 0$, is a weak solution, u is called a global weak solution.

The following existence result holds for problem (6.78) [LL18b, Theorem 5.2].

Theorem 6.23 *For any $u_0 \in L^2(\Omega)$, there exists a global weak solution to problem (6.78) in the sense of Definition 6.2. Further, if $\max(\frac{1}{2}, \frac{d}{4}) \leq \alpha < 1$, there is a global weak solution continuous at $t = 0$ in the $H^{-1}(\Omega)$ norm.*

To prove the theorem, we employ a Galerkin procedure as in Section 6.1, and compactness criteria in Section 6.4.3.1. To this end, let $\{\varphi_j\}_{j=1}^{\infty}$ be an orthogonal basis of both $H_0^1(\Omega)$ and $L^2(\Omega)$ and orthonormal in $L^2(\Omega)$, and let P_n be the orthogonal projection onto $\operatorname{span}\{\varphi_j\}_{j=1}^n$ in $L^2(\Omega)$. Then it defines a bounded linear operator in both $L^2(\Omega)$ and $H_0^1(\Omega)$. Given $u_0 = \sum_{j=1}^{\infty} u_{0,j}\varphi_j(x) \in L^2(\Omega)$, we seek an approximate solution $u_n(x,t)$ of the form

$$u_n(t) = \sum_{j=1}^n c_{n,j}(t)\varphi_j$$

with $c_{n,j}(0) = u_{0,j}$, $j = 1,\ldots,n$, and $\mathbf{c}_n := (c_{n,1},\ldots,c_{n,n}) \in \mathbb{R}^{d\times n}$ is continuous in time t. The approximation u_n solves the Galerkin problem:

$$\begin{cases} (\varphi_j, \partial_t^\alpha u_n) + (\varphi_j, \nabla \cdot (u_n \otimes u_n)) + \frac{1}{2}(\varphi_j, \nabla|u_n|^2) = (\varphi_j, \Delta u_n), & j = 1,\ldots,n \\ u_n(0) = P_n u_0. \end{cases}$$
$$(6.80)$$

The following unique existence result and *a priori* bound hold.

Lemma 6.9 (i) *For any $n \geq 1$, there exists a unique solution u_n to problem* (6.80) *that is continuous on $\overline{\mathbb{R}}_+$, satisfying*

$$\|u_n\|_{L^\infty(\mathbb{R}_+;L^2(\Omega))} \leq \|u_0\|_{L^2(\Omega)} \quad \text{and} \quad \sup_{0 \leq t < \infty} {}_0I_t^\alpha\|\nabla u_n\|_{L^2(\Omega)}^2(t) \leq \frac{1}{2}\|u_0\|_{L^2(\Omega)}^2.$$

(ii) *There exists a function $u \in L^\infty(\mathbb{R}_+;L^2(\Omega)) \cap L_{loc}^2(\overline{\mathbb{R}}_+;H_0^1(\Omega))$ and a subsequence n_j such that $u_{n_j} \to u$ in $L_{loc}^2(\overline{\mathbb{R}}_+;L^2(\Omega))$. Further, $\partial_t^\alpha u \in L_{loc}^{q_1}(\overline{\mathbb{R}}_+;H^{-1}(\Omega))$ with $q_1 = \min(2, \frac{4}{d})$.*

Proof Note that problem (6.80) is equivalent to an autonomous fractional ODE

$$\begin{cases} \partial_t^\alpha \mathbf{c}_n(t) = \mathbf{F}_n(\mathbf{c}_n), & t > 0, \\ \mathbf{c}_n(0) = (u_{0,1},\ldots,u_{0,n}), \end{cases}$$
$$(6.81)$$

where $\mathbf{F}_n$ is a quadratic vector-valued function of $\mathbf{c}_n$, and hence smooth.
(i) By Proposition 4.7 in Chapter 4, the solution $\mathbf{c}_n(t)$ exists on $[0, t_*^n)$, with the blow-up time t_*^n either $t_*^n = \infty$ or $t_*^n < \infty$ and $\limsup_{t \to t_*^{n-}} |\mathbf{c}_n| = \infty$. Since $\mathbf{F}_n$ is quadratic, by Theorem 4.12, $\mathbf{c}_n \in C^1(0, t_*^n) \cap C[0, t_*^n)$ and hence, $u_n \in C^1((0, t_*^n); H_0^1(\Omega)) \cap C([0, t_*^n); H_0^1(\Omega))$. For $u_n = \sum_{j=1}^n c_{n,j}(t)\varphi_j$, by Lemma 2.12, using (6.80) and the identity

$$(u_n, \nabla \cdot (u_n \otimes u_n)) + \frac{1}{2}(u_n, \nabla|u_n|^2) = \frac{1}{2}\int_\Omega \nabla \cdot (|u|^2 u)\mathrm{d}x,$$

we have

$$\langle u_n, \partial_t^\alpha u_n \rangle + (u_n, \nabla \cdot (u_n \otimes u_n)) + \frac{1}{2}(u_n, \nabla|u_n|^2) = -\|\nabla u_n\|_{L^2(\Omega)}^2.$$

Hence,

$$\tfrac{1}{2}\partial_t^\alpha \|u_n\|_{L^2(\Omega)}^2(t) \le -\|\nabla u_n\|_{L^2(\Omega)}^2,$$

and it implies

$$\|u_n(t)\|_{L^2(\Omega)}^2 + 2\,{}_0I_t^\alpha \|\nabla u_n\|_{L^2(\Omega)}^2(t) \le \|u_0\|_{L^2(\Omega)}^2.$$

Thus, $t_*^n = \infty$, and the first assertion also follows.

(ii) Take a test function $v \in L^{p_1}(I; H_0^1(\Omega))$, with $p_1 = \max(2, \frac{4}{4-d})$ (its conjugate exponent $q_1 = \min(2, \frac{d}{4})$) and $\|v\|_{L^{p_1}(I;H_0^1(\Omega))} \le 1$. For $v_n := P_n v$, the stability of P_n in $H_0^1(\Omega)$ implies $\|v_n\|_{L^{p_1}(I;H_0^1(\Omega))} \le c(\Omega, T)$. Since $v_n \in \operatorname{span}\{\varphi_j\}_{j=1}^n$, there holds

$$(v, \partial_t^\alpha u_n) = (v_n, \partial_t^\alpha u_n) = -(v_n, \nabla \cdot (u_n \otimes u_n)) - \tfrac{1}{2}(v_n, \nabla|u_n|^2) + (v_n, \Delta u_n).$$

This in particular implies

$$|(v, \partial_t^\alpha u_n)| = |(v_n, -\nabla \cdot (u_n \otimes u_n) - \tfrac{1}{2}\nabla(|u_n|^2) + \Delta u_n)|$$
$$\le c \int_0^T \|\nabla v_n |u_n|^2\|_{L^1(\Omega)}\mathrm{d}t + \int_0^T \|\nabla v_n\|_{L^2(\Omega)}\|\nabla u_n\|_{L^2(\Omega)}\mathrm{d}t.$$

Using Gagliardo-Nirenberg inequality in Theorem A.4,

$$\|u\|_{L^4(\Omega)} \le c\|u\|_{L^2(\Omega)}^{1-\frac{d}{4}}\|\nabla u\|_{L^2(\Omega)}^{\frac{d}{4}},$$

the first term can be estimated as

$$\int_0^T \|\nabla v_n |u_n|^2\|_{L^1(\Omega)}\mathrm{d}t \le \int_0^T \|\nabla v_n\|_{L^2(\Omega)}\|u_n\|_{L^4(\Omega)}^2\mathrm{d}t$$
$$\le \left(\int_0^T \|\nabla v_n\|_{L^2(\Omega)}^{\frac{4}{4-d}}\mathrm{d}t\right)^{\frac{4-d}{4}} \left(\int_0^T \|\nabla u_n\|_{L^2(\Omega)}^2\mathrm{d}t\right)^{\frac{d}{4}}.$$

This and the bound $\|v_n\|_{L^{p_1}(I;H_0^1(\Omega))} \le c(\Omega, T)$ imply

$$\|\partial_t^\alpha u_n\|_{L^{q_1}(I;H^{-1}(\Omega))} \le c,$$

with $q_1 = \min(2, \frac{4}{d})$. In sum, we have obtained

$$\sup_{t \in I} {}_0I_t^\alpha \|\nabla u_n\|_{L^2(\Omega)}^2(t) \le c, \quad u_n \in L^\infty(I; L^2(\Omega)), \quad \|\partial_t^\alpha u_n\|_{L^{q_1}(I;H^{-1}(\Omega))} \le c.$$

By Theorem 6.22, there is a subsequence $\{u_{n_j}\}_{j=1}^\infty$ that converges in $L^p(I; L^2(\Omega))$ for any $p \in [1, \infty)$. By Proposition 6.6, u has a weak Djrbashian-Caputo fractional derivative with initial value u_0 such that $\partial_t^\alpha u \in L^{q_1}(I; H^{-1}(\Omega))$. By a standard diagonal argument, u is defined on $\mathbb{R}_+$ and $\partial_t^\alpha u \in L^{q_1}_{loc}(\overline{\mathbb{R}}_+; H^{-1}(\Omega))$ such that $u_{n_j} \to u$ in $L^2_{loc}(\overline{\mathbb{R}}_+; L^2(\Omega))$. Upon passing to a further subsequence, we may assume that u_{n_j} also converges a.e. to u in $\overline{\mathbb{R}}_+ \times \Omega$. Clearly, for any $t_1 < t_2$, $\int_{t_1}^{t_2} \|u_n\|^2_{L^2(\Omega)} dt \leq \|u_0\|^2_{L^2(\Omega)}(t_2 - t_1)$, and thus by Fatou's lemma, $\int_{t_1}^{t_2} \|u\|^2_{L^2(\Omega)} dt \leq \|u_0\|^2_{L^2(\Omega)}(t_2 - t_1)$, which implies $u \in L^\infty(\mathbb{R}_+; L^2(\Omega))$. Fix any $T > 0$, since u_{n_j} is bounded in $L^2(I; H^1_0(\Omega))$. Then, it has a further subsequence that converges weakly in $L^2(I; H^1_0(\Omega))$. By a standard diagonal argument, there is a subsequence that converges weakly in $L^2_{loc}(\overline{\mathbb{R}}_+; H^1_0(\Omega))$. The limit must be u by pairing with a smooth test function. Hence, $u \in L^2_{loc}(\overline{\mathbb{R}}_+; H^1_0(\Omega))$. □

Now, we can prove Theorem 6.23:

Proof By Lemma 6.9, there is a convergent subsequence in $L^p(I; L^2(\Omega))$ for any $p \in [1, \infty)$, with the limit denoted by u. For any $v \in C^\infty_c([0, T) \times \Omega; \mathbb{R}^d)$, let $v_n = P_n v$. Since v is smooth in t and vanishes at T, so is v_n, and

$$_t\partial^\alpha_T v_n \to {}_t\partial^\alpha_T v \quad \text{in } L^{p_1}(I; H^1_0(\Omega)).$$

Fix $n_0 \geq 1$, and for $n_j \geq n_0$, by integration by parts and noting the fact $_t\partial^\alpha_T v_{n_0} \in \text{span}\{\varphi_\ell\}_{\ell=1}^{n_j}$, we have

$$\int_0^T \int_\Omega (u_{n_j} - u_0)_t \partial^\alpha_T v_{n_0} dx dt = \int_0^T \int_\Omega v_{n_0} \partial^\alpha_t u_{n_j} dx dt$$

$$= \int_0^T \int_\Omega -v_{n_0} \nabla \cdot (u_{n_j} \otimes u_{n_j}) - \tfrac{1}{2} v_{n_0} \nabla |u_{n_j}|^2 + v_{n_0} \Delta u_{n_j} dx dt$$

$$= \int_0^T \int_\Omega \nabla (v_{n_0} : u_{n_j} \otimes u_{n_j} + \tfrac{1}{2} \nabla \cdot v_{n_0} |u_{n_j}|^2 - \nabla v_{n_0} : \nabla u_{n_j}) dx dt.$$

By Lemma 6.9, taking $j \to \infty$, we have

$$\int_0^T \int_\Omega (u - u_0)_t \partial^\alpha_T v_{n_0} dx dt = \int_0^T \int_\Omega (\nabla v_{n_0} : u \otimes u + \tfrac{1}{2} \nabla \cdot v_{n_0} |u|^2 - \nabla v_{n_0} : \nabla u) dx dt.$$

Then, taking $n_0 \to \infty$, by the convergence $v_n \to \varphi$ in $L^p(I; H^1_0(\Omega))$ for any $p \in (1, \infty)$, the weak formulation holds. Further, if $q_1 \geq \frac{1}{\alpha}$ or $\alpha \geq \max(\frac{1}{2}, \frac{d}{4})$, by Lemma 6.9 (ii) and Theorem 2.15, u is continuous at $t = 0$. □

6.5 Maximum Principles

For standard parabolic equations, there are many important qualitative properties, e.g., maximum principle, Hopf's lemma and unique continuation property. These properties are less understood for subdiffusion. The unique continuation principle states for any solution u supported on $t \geq 0$, if $u = 0$ in a subdomain $\omega \subset \Omega$ over $(0, T)$, then $u = 0$ in the parabolic cylinder Q, under certain circumstances [LN16, JLLY17], but the general case seems still open. There is also Hopf's lemma for 1D subdiffusion

[Ros16], generalizing that for standard parabolic problems [Fri58]. The proof mostly follows that for the classical parabolic problem as presented in [Can84, Section 15.4], but with $E_\alpha(-t)$ in place of e^{-t}. The maximum principle has been extensively studied [Luc09, LRY16, LY17]. We describe several versions of weak maximum principles below; see the review [LY19b] for further results, including more general subdiffusion -type models, e.g., multi-term and distributed order.

We discuss the maximum principle for the following model:

$$\begin{cases} \partial_t^\alpha u = \mathcal{L}u + f, & \text{in } Q, \\ u = g, & \text{on } \partial_L Q, \\ u(0) = u_0, & \text{in } \Omega, \end{cases} \tag{6.82}$$

where u_0 and g are given functions on $\overline{\Omega}$ and $\partial_L \overline{Q}$, respectively, and f is a given function on $\overline{Q}$. The elliptic operator $\mathcal{L} \equiv \mathcal{L}_{a,q}$ is defined by

$$\mathcal{L}u(x) = \sum_{i,j=1}^{d} \partial_{x_j}(a_{ij}(x)\partial_{x_i}u(x)) - q(x)u(x),$$

where the matrix $a = [a_{ij}]$ is symmetric and strongly elliptic, and $q \in C(\overline{\Omega})$ with $q \geq 0$ in $\overline{\Omega}$. Unless otherwise stated, we also assume $a_{ij} \in C^1(\overline{\Omega})$, and $g(x,t) = 0$.

The following weak maximum principle due to Luchko [Luc09, Theorem 2] holds for classical solutions. The condition $\partial_t^\alpha u, \partial_{x_i x_j}^2 u \in C(\overline{Q})$ is restrictive, in view of the limited smoothing properties of the solution operators.

Theorem 6.24 *Let $\alpha \in (0, 1)$, $q \geq 0$. Suppose that u is a continuous function on $\overline{Q}$, with $\partial_t^\alpha u, \partial_{x_i x_j}^2 u \in C(\overline{Q})$. Then the following statements hold.*

(i) *If $(\partial_t^\alpha + \mathcal{L})u(x,t) \leq 0$ for all $(x,t) \in \overline{Q}$, then*

$$u(x,t) \leq \max(0, m), \quad \text{with } m := \max\Big(\sup_{x \in \overline{\Omega}} u(x,0), \quad \sup_{(x,t) \in \partial_L \overline{Q}} u(x,t) \Big).$$

(ii) *If $(\partial_t^\alpha + \mathcal{L})u(x,t) \geq 0$ for all $(x,t) \in \overline{Q}$, then*

$$\min(m, 0) \leq u(x,t), \quad \text{with } m := \min\Big(\inf_{x \in \overline{\Omega}} u(x,0), \quad \inf_{(x,t) \in \partial_L \overline{Q}} u(x,t) \Big).$$

Proof It suffices to prove (i), and (ii) follows by considering $-u$. We present two slightly different proofs.

Proof 1. This proof is taken from [Luc09, Theorem 2]. Assume that the statement does not hold, i.e., there exists $(x_0, t_0) \in Q$ such that $u(x_0, t_0) > m > 0$. Let $\epsilon = u(x_0, t_0) - m > 0$, and let $w(x,t) := u(x,t) + \frac{\epsilon}{2}\frac{T-t}{T}$, for $(x,t) \in \overline{Q}$. Then w satisfies $w(x,t) \leq u(x,t) + \frac{\epsilon}{2}$, for $(x,t) \in \overline{Q}$, and for any $(x,t) \in \partial_p \overline{Q}$

$$w(x_0, t_0) \geq u(x_0, t_0) = \epsilon + m \geq \epsilon + u(x,t) \geq \epsilon + w(x,t) - \frac{\epsilon}{2} = w(x,t) + \frac{\epsilon}{2}.$$

Thus, w cannot attain its maximum on $\partial_p Q$. Let w attain its maximum over $\overline{Q}$ at $(x_1, t_1) \in Q$, then $w(x_1, t_1) \geq w(x_0, t_0) \geq \epsilon + m > \epsilon$. Now Proposition 2.5(i) and the necessary conditions for a maximum at (x_1, t_1) give

$$\partial_t^\alpha w(x_1, t_1) \geq 0, \quad \nabla w(x_1, t_1) = 0, \quad \Delta w(x_1, t_1) \leq 0.$$

By the definition of w, $\partial_t^\alpha u(x, t) = \partial_t^\alpha w(x, t) + (2T\Gamma(2 - \alpha))^{-1}\epsilon t^{1-\alpha}$. Consequently,

$$0 \geq \partial_t^\alpha u(x_1, t_1) - \mathcal{L}u(x_1, t_1)$$
$$= \partial_t^\alpha w(x_1, t_1) + (2T\Gamma(2 - \alpha))^{-1}\epsilon t_1^{1-\alpha} - \nabla \cdot (a\nabla w)(x_1, t_1) + q(w - (2T)^{-1}\epsilon(T - t_1))$$
$$\geq (2T\Gamma(2 - \alpha))^{-1}\epsilon t_1^{1-\alpha} + q\epsilon(1 - \tfrac{T-t_1}{2T}) > 0,$$

leading to a contradiction.

Proof 2. Since u is continuous on $\overline{Q}$, there exists a point $(x_0, t_0) \in \overline{Q}$ such that $u(x, t) \leq u(x_0, t_0)$, for all $(x, t) \in \overline{Q}$. If $u(x_0, t_0) \leq 0$, the desired inequality follows directly. Hence, it suffices to consider $u(x_0, t_0) > 0$, and we distinguish two cases. (a) If $(x_0, t_0) \in \partial_p Q$, then $m = u(x_0, t_0)$, and thus the assertion follows. (b) If $(x_0, t_0) \in Q$, then $\sum_{i,j=1}^d \partial_{x_j}(a_{ij}\partial_{x_i}u(x_0, t_0)) \leq 0$, $qu(x_0, t_0) \geq 0$, and the condition $(\partial_t^\alpha + \mathcal{L})u(x_0, t_0) \leq 0$ implies $\partial_t^\alpha u(x_0, t_0) \leq 0$, while Proposition 2.5(i) implies $\partial_t^\alpha u(x_0, t_0) \geq 0$. Hence, we obtain

$$\partial_t^\alpha u(x_0, t_0) = 0 \quad \text{and} \quad u(x_0, t) \leq u(x_0, t_0) \quad \text{for all } 0 \leq t \leq t_0.$$

Now Proposition 2.5(ii) yields $u(x_0, t_0) = u(x_0, 0)$, and hence the assertion follows. $\square$

The maximum principle in Theorem 6.24 allows us to derive *a priori* estimates and continuity results on classical solutions of problem (6.82) [Luc09, Theorem 4].

Corollary 6.3 *If u is a classical solution to problem (6.82), then there holds*

$$\|u\|_{C(\overline{Q})} \leq \max\left(\|u_0\|_{C(\overline{\Omega})}, \|g\|_{C(\partial_L \overline{Q})}\right) + \Gamma(\alpha + 1)^{-1}T^\alpha\|f\|_{C(\overline{Q})}.$$

Proof Let

$$w(x, t) = u(x, t) - c_f t^\alpha, \quad (x, t) \in \overline{Q},$$

with $c_f = \Gamma(\alpha + 1)^{-1}\|f\|_{C(\overline{Q})}$. Then w is a classical solution of problem (6.82) with a source $\overline{f}(x, t) = f(x, t) - \|f\|_{C(\overline{Q})} - qc_f t^\alpha$, and the boundary condition $\overline{g}(x, t) = g(x, t) - c_f t^\alpha$, in place of f and g, respectively. Clearly, $\overline{f}$ satisfies $\overline{f}(x, t) \leq 0$, for $(x, t) \in \overline{Q}$. Then the maximum principle in Theorem 6.24 applied to w leads to

$$w(x, t) \leq \max(\sup_\Omega u_0, \sup_{\partial_L Q} g), \quad (x, t) \in \overline{Q}.$$

Consequently, for u, we have

$$u(x, t) = w(x, t) + c_f t^\alpha \leq \max(\|u_0\|_{C(\overline{\Omega})}, \|g\|_{C(\partial_L \overline{Q})}) + c_f T^\alpha.$$

Similar, the minimum principle from Theorem 6.24(ii) applied to the auxiliary function $w(x, t) = u(x, t) + c_f t^\alpha$, $(x, t) \in \overline{Q}$ leads to

$$u(x,t) \geq -\max(\|u_0\|_{C(\overline{\Omega})}, \|g'\|_{C(\partial_L \overline{Q})}) - c_f T^\alpha.$$

Combining the last two estimates completes the proof of the corollary. $\qquad\square$

Next, we extend the weak maximum principle in Theorem 6.24 to weak solutions [LY17, Lemma 3.1]. Note that we discuss only the case of a zero boundary condition, i.e., $g \equiv 0$.

Lemma 6.10 *Let $u_0 \in L^2(\Omega)$ and $f \in L^2(I; L^2(\Omega))$, with $u_0 \geq 0$ a.e. in Ω and $f \geq 0$ a.e. in Q, and a_{ij} and q be smooth. Then the solution u to problem (6.82) satisfies $u \geq 0$ a.e. in Q.*

Proof First, we show the assertion for $u_0 \in C_0^\infty(\Omega)$ and $f \in C_0^\infty(Q)$. Then the solution u satisfies $u \in C^1(I; C^2(\overline{\Omega})) \cap C(\overline{Q})$ and $\partial_t u \in L^1(I; L^2(\Omega))$, i.e., $u(t)$ is a strong solution. Indeed, by the solution theory in Section 6.2, u is given by

$$u(t) = F(t)u_0 + \int_0^t E(t-s)f(s)\mathrm{d}s,$$

where the solution operators $F(t)$ and $E(t)$ associated with $\mathcal{L}_{a,q}$, cf. (6.21) and (6.22). Thus, for any $\gamma > 0$,

$$\|A^\gamma u(t)\|_{L^2(\Omega)} \leq \|F(t)A^\gamma u_0\|_{L^2(\Omega)} + \left\|\int_0^t E(t-s)A^\gamma f(s)\mathrm{d}s\right\|_{L^2(\Omega)}$$
$$\leq c\|A^\gamma u_0\|_{L^2(\Omega)} + c\int_0^t (t-s)^{\alpha-1}\|A^\gamma f(s)\|_{L^2(\Omega)}\mathrm{d}s,$$

and similarly, since $f \in C_0^\infty(Q)$,

$$\|A^\gamma u'(t)\|_{L^2(\Omega)} \leq \|F'(t)A^\gamma u_0\|_{L^2(\Omega)} + \left\|\int_0^t E(t-s)A^\gamma f'(s)\mathrm{d}s\right\|_{L^2(\Omega)}$$
$$\leq ct^{\alpha-1}\|A^\gamma u_0\|_{L^2(\Omega)} + c\int_0^t (t-s)^{\alpha-1}\|A^\gamma f'(s)\|_{L^2(\Omega)}\mathrm{d}s.$$

By Sobolev embedding, we deduce the desired claim by choosing a sufficiently large $\gamma > 0$. Now by the maximum principle in Theorem 6.24, the assertion holds if $u_0 \in C_0^\infty(\Omega)$ and $f \in C_0^\infty(Q)$. Now suppose $u_0 \in L^2(\Omega)$ and $f \in L^2(I; L^2(\Omega))$. Then there exist sequences $\{u_0^n\}_{n=1}^\infty \subset C_0^\infty(\Omega)$ and $\{f^n\}_{n=1}^\infty \subset C_0^\infty(Q)$ such that $u_0^n \to u_0$ in $L^2(\Omega)$ and $f^n \to f$ in $L^2(I; L^2(\Omega))$. Then the solution u^n corresponding to u_0^n and f^n satisfies $u^n \geq 0$ a.e. in Q, $n \in \mathbb{N}$. Moreover, since $u^n \to u$ in $L^2(I; H^2(\Omega))$, we obtain $u \geq 0$ a.e. in Q. $\qquad\square$

Next, we extend Lemma 6.10 to weak solutions, but removing the restriction on the sign of the coefficient $q \in C(\overline{\Omega})$ [LY17, Theorem 2.1].

Theorem 6.25 *Let $u_0 \in L^2(\Omega)$ and $f \in L^2(I; L^2(\Omega))$, with $u_0 \geq 0$ a.e. in Ω and $f \geq 0$ a.e. in Q, and the coefficients a_{ij} and q are smooth. Then the solution u to problem (6.82) satisfies $u \geq 0$ a.e. in Q.*

Proof Let $c_q = \|q\|_{C(\overline{\Omega})}$, and $\tilde{q}(x) = c_q + q(x)$. Then $\tilde{q}(x) \geq 0$ on $\overline{\Omega}$. Then problem (6.82) can be rewritten as

$$\begin{cases} \partial_t^\alpha u - \mathcal{L}_{a,\tilde{q}} u = f + c_q u, & \text{in } Q, \\ u = 0, & \text{on } \partial_L Q, \\ u(0) = u_0, & \text{in } \Omega. \end{cases}$$

Then the weak solution u satisfies

$$u(t) = F(t)u_0 + \int_0^t E(t-s)(f(s) + c_q u(s))ds, \quad \forall t > 0, \tag{6.83}$$

where the solution operators F and E are associated with the elliptic operator $\mathcal{L}_{a,\tilde{q}}$. Now we define a sequence $\{u^n\}_{n=0}^\infty$ by $u^0 = 0$ and

$$u^{n+1} = F(t)u_0 + \int_0^t E(t-s)(f(s) + c_q u^n(s))ds, \quad n = 0, 1, 2, \dots$$

The sequence $\{u^n\}_{n=0}^\infty$ converges in $L^2(I; L^2(\Omega))$. In fact, with $v^{n+1} = u^{n+1} - u^n$, $n = 0, 1, \dots$, we obtain $v^1 = u^1$ and

$$v^{n+1}(t) = c_q \int_0^t E(t-s)v^n(s)ds, \quad n = 1, 2, \dots.$$

Let $c_0 = \|u^1(t)\|_{C(\overline{I};L^2(\Omega))}$. By Theorem 6.4(iv) and 3.5,

$$\|v^{n+1}\|_{L^2(\Omega)} \leq \frac{c_q}{\Gamma(\alpha)} \int_0^t (t-s)^{\alpha-1} \|v^n(s)\|_{L^2(\Omega)} ds.$$

By means of mathematical induction, we obtain

$$\|v^n(t)\|_{L^2(\Omega)} \leq c_q^{n-1} c_0 \Gamma((n-1)\alpha + 1)^{-1} t^{(n-1)\alpha}, \quad 0 < t < T, \; n = 2, \dots.$$

Hence,

$$\|v^n\|_{C(\overline{I};L^2(\Omega))} \leq (c_q T^\alpha)^{n-1} c_0 \Gamma((n-1)\alpha + 1)^{-1}, \quad n = 2, 3, \dots.$$

Since $u^n = \sum_{k=1}^n v^k$, the series $\sum_{k=1}^\infty v^k$ converges: by Stirling's formula (2.7),

$$\lim_{k\to\infty} \frac{(c_q T^\alpha)^k c_0}{\Gamma(k\alpha + 1)} \left(\frac{(c_q T^\alpha)^{k-1} c_0}{\Gamma((k-1)\alpha + 1)} \right)^{-1} = \lim_{k\to\infty} c_q T^\alpha \frac{\Gamma((k-1)\alpha + 1)}{\Gamma(k\alpha + 1)} < 1,$$

Thus, the sequence u^n converges in $C(\overline{I}; L^2(\Omega))$, and by construction, to a fixed point of (6.83), i.e., to the weak solution u of problem (6.82). Last, we show the nonnegativity of u. By the definition of u^n, since $u_0 \geq 0$ and $f \geq 0$, Lemma 6.10 yields $u^1 \geq 0$ in Q. Then $f + c_q u^1 \geq 0$ in Q and we can apply Lemma 6.10 again to conclude $u^2 \geq 0$ in Q. Repeating the arguments yields $u^n \geq 0$ in Q for all $n \in \mathbb{N}$. Since $u^n \to u$ in $C(\overline{I}; L^2(\Omega))$, $u \geq 0$ holds a.e. in Q. □

Theorem 6.25 directly yields the following comparison property.

Corollary 6.4 *Let* $u_{0,1}, u_{0,2} \in L^2(\Omega)$ *and* $f_1, f_2 \in L^2(I; L^2(\Omega))$ *satisfy* $u_{0,1} \geq u_{0,2}$ *in* Ω *and* $f_1 \geq f_2$ *in* Q, *respectively, and the coefficients* a_{ij} *and* q *be smooth. Let* u_i *be the solution to problem* (6.82) *with data* $u_{0,i}$ *and source* f_i, $i = 1, 2$. *Then there holds* $u_1(x, t) \geq u_2(x, t)$ *a.e. in* Q.

Now we fix $f \geq 0$ and $u_0 \geq 0$, and denote by $u(q)$ the weak solution to problem (6.82) with the potential q. Then the following comparison principle holds.

Corollary 6.5 *Let* $f \in L^2(I; L^2(\Omega))$ *and* $u_0 \in L^2(\Omega)$, *with* $f \geq 0$ *in* Q *and* $u_0 \geq 0$ *a.e. in* Ω, *and the coefficients* a_{ij} *be smooth. Let* $q_1, q_2 \in C(\overline{\Omega})$ *satisfy* $q_1 \geq q_2$ *in* Ω, *and* $u(q_i)$ *be the solution to problem* (6.82) *corresponding to* q_i. *Then* $u(q_1) \leq u(q_2)$ *in* Q.

Proof Let $w = u(q_1) - u(q_2)$. Then w satisfies

$$\begin{cases} \partial_t^\alpha w - \mathcal{L}_{a,q_1} w = -(q_1 - q_2)u(q_2), & \text{in } Q, \\ w = 0, & \text{on } \partial_L Q, \\ w(0) = 0, & \text{in } \Omega. \end{cases}$$

Since $f \geq 0$ in Q and $u_0 \geq 0$ in Ω, Theorem 6.25 implies $u(q_2) \geq 0$ in Q. Hence, $(q_1 - q_2)u(q_2) \geq 0$ in Q. Now invoking Theorem 6.25 leads to $w \leq 0$ in Q. □

One can strengthen the weak maximum principle to a nearly strong one for the homogeneous problem, but the version as is known for the standard diffusion equation remains elusive. First, we state an auxiliary lemma [LRY16, Lemma 3.1].

Lemma 6.11 *Let* $G(x, y, t)$ *be the Green function, defined by*

$$G(x, y, t) = \sum_{n=1}^{\infty} E_{\alpha,1}(-\lambda_n t^\alpha)\varphi_n(x)\varphi_n(y).$$

Then for any fixed $x \in \Omega$ *and* $t > 0$, *there hold*

$$G(x, \cdot, t) \in L^2(\Omega) \quad \text{and} \quad G(x, \cdot, t) \geq 0 \text{ a.e. in } \Omega.$$

Proof Using Green's function $G(x, y, t)$, the solution u is given by

$$u(x, t) = \int_\Omega G(x, y, t)u_0(y)\mathrm{d}y. \tag{6.84}$$

We claim that for any fixed $x \in \Omega$ and $t > 0$, $G(x, \cdot, t) \geq 0$ a.e. in Ω. Actually, assume that there exist $x_1 \in \Omega$ and $t_1 > 0$ such that the Lebesgue measure of the set $\omega := \{G(x_1, \cdot, t_1) < 0\} \subset \Omega$ is positive. Then

$$v_{\chi_\omega}(x_1, t_1) = \int_\Omega G(x_1, y, t_1) \chi_\omega(y) dy < 0,$$

where χ_ω is the characteristic function of ω satisfying $\chi_\omega \in L^2(\Omega)$ and $\chi_\omega \geq 0$. Meanwhile, Theorem 6.24 implies $v_{\chi_\omega}(x_1, t_1) \geq 0$, leading to a contradiction. □

Then the following version of strong maximum principle holds [LRY16, Theorem 1.1]. Note that for the standard diffusion equation, we have $\mathcal{E}_x = \emptyset$. It remains elusive to prove this for subdiffusion.

Theorem 6.26 *Let u be the solution to problem (6.82) with $u_0 \in L^2(\Omega)$, $u_0 \geq 0$ and $u_0 \neq 0$, $f \equiv 0$. Then for any $x \in \Omega$, the set $\mathcal{E}_x := \{t > 0 : u(x, t) \leq 0\}$ is at most finite.*

Proof By Theorem 6.7 and the continuous embedding $H^2(\Omega) \hookrightarrow C(\overline{\Omega})$ for $d \leq 3$, cf. Theorem A.3, we have $u \in C(\mathbb{R}_+; C(\overline{\Omega}))$, for $u_0 \in L^2(\Omega)$. By the weak maximum principle in Theorem 6.25, the solution $u(x, t) \geq 0$ a.e. in Q. Thus,

$$\mathcal{E}_x := \{t > 0 : u(x, t) \leq 0\} = \{t > 0 : u(x, t) = 0\}$$

coincides with the zero point set of $u(x, t)$ as a function of $t > 0$. Assume that there exists some $x_0 \in \Omega$ such that the set $\mathcal{E}_{x_0}$ is not finite. Then $\mathcal{E}_{x_0}$ contains at least one accumulation point $t_* \in [0, \infty]$. We treat the three cases: (i) $t_* = \infty$, (ii) $t_* \in \mathbb{R}_+$ and (iii) $t_* = 0$ separately.

(i). If $t_* = \infty$, then by definition, there exists $\{t_j\}_{j=1}^\infty \subset \mathcal{E}_{x_0}$ such that $t_j \to \infty$ as $j \to \infty$ and $u(x_0, t_j) = 0$. Since

$$u(x_0, t) = \sum_{n=1}^\infty E_{\alpha,1}(-\lambda_n t^\alpha)(u_0, \varphi_n)\varphi_n(x_0),$$

the asymptotic of $E_{\alpha,1}(z)$ in Theorem 3.2 implies

$$u(x_0, t) = \frac{1}{\Gamma(1-\alpha)t^\alpha} \sum_{n=1}^\infty \lambda_n^{-1}(u_0, \varphi_n)\varphi_n(x_0) + O(t^{-2\alpha}) \sum_{n=1}^\infty \lambda_n^{-2}(u_0, \varphi_n)\varphi_n(x_0),$$

as $t \to \infty$. Setting $t = t_i$ with large i in the identity, multiplying both sides by t_i^α and passing i to ∞, we obtain $v(x_0) = 0$ and $v := \sum_{n=1}^\infty \lambda_n^{-1}(u_0, \varphi_n)\varphi_n$. Then the function v satisfies $\mathcal{L}v = u_0$ in Ω with $v = 0$ on $\partial\Omega$. The classical weak maximum principle for elliptic PDEs indicates $v \geq 0$ in $\overline{\Omega}$. Moreover, since v attains its minimum at $x_0 \in \Omega$, the strong maximum principle implies $v \equiv 0$ and thus $u_0 = \mathcal{L}v = 0$, contradicting the assumption $u_0 \neq 0$. Hence, ∞ cannot be an accumulation point of $\mathcal{E}_{x_0}$.

(ii). Now suppose that the set of zeros $\mathcal{E}_{x_0}$ admits an accumulation point $t_* \in \mathbb{R}_+$. By the analyticity of $u : \mathbb{R}_+ \to \dot{H}^2(\Omega) \subset C(\overline{\Omega})$ from Proposition 6.1, $u(x_0, t)$ is analytic with respect to $t > 0$. Hence, $u(x_0, t)$ vanishes identically if its zeros accumulate at some finite and nonzero point t_*. This case reduces to case (i) and eventually leads to a contradiction.

(iii). Since $u(x_0, t)$ is not analytic at $t = 0$, we have to treat the case $t_* = 0$ differently. Then there exists $\{t_i\}_{i=1}^{\infty} \subset \mathcal{E}_{x_0}$ such that $t_i \to 0$ as $i \to \infty$ and further,

$$u(x_0, t_i) = \int_{\Omega} G(x_0, y, t_i)u_0(y)dy = 0, \quad i = 1, 2, \ldots.$$

By Lemma 6.11, $G(x_0, \cdot, t) \geq 0$ and $u_0 \geq 0$, we deduce

$$G(x_0, y, t_i)u_0(y) = 0, \quad \forall i = 1, 2, \ldots, \quad \text{a.e. } y \in \Omega.$$

Since $u_0 \not\equiv 0$, $G(x_0, \cdot, t_i)$ vanish in the set $\omega := \{u_0 > 0\}$ whose Lebesgue measure is positive. By (6.84), this indicates

$$\sum_{n=1}^{\infty} E_{\alpha,1}(-\lambda_n t_i^{\alpha})\varphi_n(x_0)\varphi_n(x) = 0 \quad \text{a.e. in } \omega, \, i = 1, 2 \ldots.$$

Now we choose $\psi \in C_0^{\infty}(\omega)$ arbitrarily as the initial data of problem (6.82) (and $f \equiv 0$) and study

$$u_{\psi}(x_0, t) = \int_{\Omega} G(x_0, y, t)\psi(y)dy = \sum_{n=1}^{\infty} E_{\alpha,1}(-\lambda_n t^{\alpha})(\psi, \varphi_n)\varphi_n(x_0). \quad (6.85)$$

Let $\psi_n := (\psi, \varphi_n)\varphi_n(x_0)$. The series (6.85) converges in $C[0, \infty)$. Indeed, by Sobolev embedding $H^2(\Omega) \hookrightarrow C(\overline{\Omega})$ for $d \leq 3$ in Theorem A.3,

$$|\varphi_n(x_0)| \leq c\|\varphi_n\|_{H^2(\Omega)} \leq c\|A\varphi_n\|_{L^2(\Omega)} = c\lambda_n.$$

Since $\psi \in C_0^{\infty}(\omega) \subset D(A^3)$, we have

$$|(\psi, \varphi_n)| = \lambda_n^{-3}(A^3\psi, \varphi_n) \leq \lambda_n^{-3}\|A^3\psi\|_{L^2(\Omega)}\|\varphi_n\|_{L^2(\Omega)} \leq c\lambda_n^{-3}\|\psi\|_{C^6(\overline{\omega})}.$$

Since $\lambda_n = O(n^{-\frac{2}{d}})$ as $n \to \infty$ (Weyl's law), and $0 \leq E_{\alpha,1}(-x) \leq 1$ on $\mathbb{R}_+$ cf. Theorem 3.6, we deduce

$$|E_{\alpha,1}(-\lambda_n t^{\alpha})\psi_n| \leq c\|\psi\|_{C^6(\overline{\omega})}\lambda_n^{-2} \leq cn^{-\frac{4}{d}}.$$

Thus, for $d \leq 3$,

$$\sum_{n=1}^{\infty} |E_{\alpha,1}(-\lambda_n t^{\alpha})\psi_n| < \infty, \quad \forall t > 0, \psi \in C_0^{\infty}(\omega).$$

Thus, $u_{\psi}(x_0, t) \in C[0, \infty)$. Similarly, for $\ell = 0, 1, 2, \ldots$,

$$|(\psi, \varphi_n)| = \lambda_n^{-(\ell+3)}|(A^{\ell+3}\psi, \varphi_n)| \leq c\lambda_n^{-(\ell+3)}\|\psi\|_{C^{2(\ell+3)}(\overline{\omega})},$$

implying

$$\sum_{n=1}^{\infty} |\lambda_n^{\ell}\psi_n| \leq c\|\psi\|_{C^{2(\ell+3)}(\overline{\omega})} \sum_{n=1}^{\infty} \lambda_n^{-2} < \infty,$$

for all $\ell = 0, 1, 2, \ldots$ and $\psi \in C_0^{\infty}(\omega)$. Moreover, since $E_{\alpha,\beta}(-t)$ is uniformly bounded for all $t \geq 0$ (for any $\alpha \in (0, 1)$ and $\beta \in \mathbb{R}$; see Theorem 3.2), we have

$$\sum_{n=1}^{\infty} |\lambda_n^{\ell} E_{\alpha,\beta}(-\lambda_n t^{\alpha})\psi_n| < \infty, \quad \forall \ell = 0, 1, \ldots,$$

for any $\beta > 0$, $t \geq 0$ and $\psi \in C_0^{\infty}(\omega)$. Utilizing the identity

$$E_{\alpha,1+\ell\alpha}(z) = \Gamma(1 + \ell\alpha)^{-1} + zE_{\alpha,1+(\ell+1)\alpha}(z), \quad \ell = 0, 1, 2, \ldots. \tag{6.86}$$

with $\ell = 0$, we split $u_{\psi}(x_0, t)$ as

$$u_{\psi}(x_0, t) = \sum_{n=1}^{\infty} \psi_n - t^{\alpha} \sum_{n=1}^{\infty} \lambda_n E_{\alpha,\alpha+1}(-\lambda_n t^{\alpha})\psi_n,$$

where the boundedness of the summations have been shown. Taking $t = t_i$ and letting $i \to \infty$, we obtain $\sum_{n=1}^{\infty} \psi_n = 0$, i.e.,

$$u_{\psi}(x_0, t) = \sum_{n=1}^{\infty} (-\lambda_n t^{\alpha})E_{\alpha,\alpha+1}(-\lambda_n t^{\alpha})\psi_n.$$

For $t > 0$, we divide both sides by $-t^{\alpha}$ and take $\ell = 1$ in (6.86) to deduce

$$-t^{-\alpha}u_{\psi}(x_0, t) = \frac{1}{\Gamma(1 + \alpha)} \sum_{n=1}^{\infty} \lambda_n\psi_n - t^{\alpha} \sum_{n=1}^{\infty} E_{\alpha,2\alpha+1}(-\lambda_n t^{\alpha})\psi_n.$$

Again taking $t = t_i$ and then passing $i \to \infty$ yields $\sum_{n=1}^{\infty} \lambda_n\psi_n = 0$, and thus

$$u_{\psi}(x_0, t) = \sum_{n=1}^{\infty} (-\lambda_n t^{\alpha})^2 E_{\alpha,2\alpha+1}(-\lambda_n t^{\alpha})\psi_n.$$

Repeating the process and by mathematical induction, we obtain

$$u_{\psi}(x_0, t) = \sum_{n=1}^{\infty} (-\lambda_n t^{\alpha})^{\ell} E_{\alpha,\ell\alpha+1}(-\lambda_n t^{\alpha})\psi_n, \quad \forall \ell = 0, 1, \ldots.$$

Now it suffices to prove

$$\lim_{\ell \to \infty} \sum_{n=0}^{\infty} (-\lambda_n t^{\alpha})^{\ell} E_{\alpha,\ell\alpha+1}(-\lambda_n t^{\alpha})\psi_n = 0, \quad \forall t > 0. \tag{6.87}$$

In fact, with $\eta = \lambda_n t^{\alpha}$,

$$(-\lambda_n t^\alpha)^\ell E_{\alpha,\ell\alpha+1}(-\lambda_n t^\alpha) = (-\eta)^\ell \sum_{k=0}^{\infty} \frac{(-\eta)^k}{\Gamma(k\alpha + \ell\alpha + 1)} = \sum_{k=\ell}^{\infty} \frac{(-\eta)^k}{\Gamma(k\alpha + 1)},$$

coincides with the summation after the ℓth term in the defining series of $E_{\alpha,1}(-\eta)$. Since the series is uniformly convergent with respect to $\eta \geq 0$, we obtain

$$\lim_{\ell\to\infty} (-\lambda_n t^\alpha)^\ell E_{\alpha,\ell\alpha+1}(-\lambda_n t^\alpha) = 0, \quad \forall n = 1, 2, \ldots, \forall t \geq 0,$$

which together with the boundedness of $\sum_{n=1}^{\infty} |\psi_n|$, yields (6.87). Since $u_\psi(x_0, t)$ is independent of ℓ, we deduce

$$u_\psi(x_0, t) = \sum_{n=1}^{\infty} E_{\alpha,1}(-\lambda_n t^\alpha)\psi_n = 0, \quad t \geq 0, \forall \psi \in C_0^\infty(\omega),$$

Since by Lemma 3.2, $\int_0^\infty e^{-zt} E_{\alpha,1}(-\lambda t^\alpha)dt = \frac{z^{\alpha-1}}{z^\alpha + \lambda}$ (which is analytically extended to $\mathfrak{R}(z) > 0$), and the above series for u_ψ converges in $C[0,\infty)$, we derive

$$z^{\alpha-1} \sum_{n=1}^{\infty} \frac{\psi_n}{z^\alpha + \lambda_n} = 0, \quad \forall \mathfrak{R}(z) > 0, \psi \in C_0^\infty(\omega),$$

i.e.,

$$\sum_{n=1}^{\infty} \frac{\psi_n}{\zeta + \lambda_n} = 0, \quad \mathfrak{R}(\zeta) > 0, \forall \psi \in C_0^\infty(\omega). \tag{6.88}$$

By a similar argument for the convergence of (6.85), we deduce that the series in (6.88) is convergent on any compact set in $\mathbb{C} \setminus \{-\lambda_n\}_{n=1}^\infty$, and the analytic continuation in ζ yields that (6.88) holds for $\zeta \in \mathbb{C} \setminus \{-\lambda_n\}_{n=1}^\infty$. Especially, since the first eigenvalue λ_1 is simple, we can choose a small circle around $-\lambda_1$ which does not contain $-\lambda_n$ ($n \geq 2$). Integrating (6.88) on this circle yields

$$\psi_1 = (\varphi_1, \psi)\varphi_1(x_0) = 0, \quad \psi \in C_0^\infty(\omega).$$

Since $\psi \in C_0^\infty(\omega)$ is arbitrary, $\varphi_1(x_0)\varphi_1 = 0$ a.e. in ω, contradicting the strict positivity of the first eigenfunction φ_1. Therefore, $t_* = 0$ cannot be an accumulation point of $\mathcal{E}_{x_0}$.

In summary, for any $x \in \Omega$, $\mathcal{E}_x$ cannot possess any accumulation point, i.e., the set $\mathcal{E}_x$ is at most finite. This completes the proof. $\qquad\square$

Corollary 6.6 *Let $u_0 \in L^2(\Omega)$ satisfy $u_0 > 0$ a.e. in Ω, $f \equiv 0$, and u be the solution to problem (6.82). Then $u > 0$ in Q.*

Proof Recall that the solution u allows a pointwise definition if $u_0 \in L^2(\Omega)$, and $f \geq 0$ in $\Omega \times (0,\infty)$. Assume that there exists $x_0 \in \Omega$ and $t_0 > 0$ such that $u(x_0, t_0) = 0$. Employing the representation (6.84), $\int_\Omega G(x_0, y, t_0)u_0(y)dy = 0$. Since $G(x_0, \cdot, t_0) \geq 0$ by Lemma 6.11 and $u_0 > 0$, there holds $G(x_0, \cdot, t_0) = 0$, that is

$$\sum_{n=1}^{\infty} E_{\alpha,1}(-\lambda_n t_0^\alpha)\varphi_n(x_0)\varphi_n = 0 \quad \text{in } \Omega.$$

Since $\{\varphi_n\}_{n=1}^{\infty}$ is an orthonormal basis in $L^2(\Omega)$, we obtain

$$E_{\alpha,1}(-\lambda_n t_0^\alpha)\varphi_n(x_0) = 0, \quad n = 1, 2, \ldots,$$

especially $E_{\alpha,1}(-\lambda_1 t_0^\alpha)\varphi_1(x_0) = 0$. However, this leads to a contradiction, since $E_{\alpha,1}(-\lambda_1 t_0^\alpha) > 0$, cf. Corollary 3.2, and $\varphi_1(x_0) > 0$. Therefore, such a pair (x_0, t_0) cannot exist. □

The last corollary combines Theorem 6.26 with Theorem 6.25 [LY17, Corollary 2.2].

Corollary 6.7 For $\Omega \subset \mathbb{R}^d$, $d = 1, 2, 3$, let $u_0 \geq 0$ a.e. in Ω, $u_0 \not\equiv 0$, and $f \equiv 0$. Then the weak solution u to problem (6.82) satisfies $u \in C(I; C(\overline{\Omega}))$ and for each $x \in \Omega$ the set $\{t : t > 0 \cap u(x, t) \leq 0\}$ is at most finite.

Proof If $q \geq 0$, the statement is already proved in Theorem 6.26. It suffices to show the statement without the condition $q \geq 0$. By Theorem 6.7, under the given conditions, the weak solution u to problem (6.82) belongs to $C(I; H^2(\Omega))$. For $d \leq 3$, Sobolev embedding implies $u \in C(I; C(\overline{\Omega}))$. Now let v be the weak solution to problem (6.82) with the coefficient $q + \|q\|_{C(\overline{\Omega})}$ and $f \equiv 0$. Since $u_0 \geq 0$ and $q + \|q\|_{C(\overline{\Omega})} \geq 0$ hold, by Theorem 6.26, there holds that for an arbitrary but fixed $x \in \Omega$, there exists an at most finite set $\mathcal{E}_x$ such that $v(x, t) \leq 0$, $t \in \mathcal{E}_x$. Since $q \leq q + \|q\|_{C(\overline{\Omega})}$ in Ω, Corollary 6.5 leads to the inequality $u(x, t) \geq v(x, t)$, which implies the desired assertion. □

For the standard parabolic problem, i.e., $\alpha = 1$, there holds $\mathcal{E}_x = \emptyset$. This holds also for subdiffusion with $q \equiv 0$ [LY19b, Theorem 9].

Theorem 6.27 Let $f \equiv 0$, $u_0 \in D(A^\gamma)$, with $\gamma > \frac{d}{4}$, and $u_0 \geq 0$ and $u_0 \not\equiv 0$. Then the solution u is strictly positive, i.e., $u(x, t) > 0$ for any $(x, t) \in Q$.

The regularity condition on u_0 implies $u \in C(\overline{I}; C(\overline{\Omega}))$. The proof employs a weak Harnack inequality [Zac13b, Theorem 1.1]. For arbitrary fixed $\delta \in (0, 1)$, $t_0 \geq 0$, $r > 0$, $\tau > 0$ and $x_0 \in \Omega$, let $B(x_0, r) := \{x \in \mathbb{R}^d : |x - x_0| < r\}$ and

$$Q_-(x_0, t_0, r) := B(x_0, \delta r) \times (t_0, t_0 + \delta \tau r^{\frac{2}{\alpha}}),$$

$$Q_+(x_0, t_0, r) := B(x_0, \delta r) \times (t_0 + (2 - \delta)\tau r^{\frac{2}{\alpha}}, t_0 + 2\tau r^{\frac{2}{\alpha}}).$$

By $|Q_-(x_0, t_0, r)|$ denotes the Lebesgue measure of the set $Q_-(x_0, t_0, r)$ in $\mathbb{R}^d \times \mathbb{R}$. The proof of the lemma is lengthy: it relies on a Yosida regularization of the singular kernel (cf. Lemma 6.1), and uses Moser iteration technique and a lemma of E. Bombieri and E. Giusti. Thus, we refer interested readers to the original paper for the complete proof.

Lemma 6.12 *Let $0 < \delta < 1$, $\tau > 0$ be fixed. Then for any $t_0 \geq 0$, $0 < p < \frac{2+d\alpha}{2+(d-2)\alpha}$ and $r > 0$ with $t_0 + 2\tau r^{\frac{2}{\alpha}} \leq T$ and $B(x_0, 2r) \subset \Omega$. Then there holds*

$$\left(|Q_-(x_0, t_0, r)|^{-1} \int_{Q_-(x_0,t_0,r)} u(x,t)^p \, dx dt \right)^{\frac{1}{p}} \leq c \inf_{(x,t)\in Q_+(x_0,t_0,r)} u(x,t).$$

Now we can give the proof of Theorem 6.27.

Proof The proof proceeds by contradiction. Assume that there exists $(x_0, t_0) \in Q$ such that $u(x_0, t_0) = 0$. Choose $r > 0$ sufficiently small so that $B(x_0, 2r) \subset \Omega$. Now set $\tilde{t}_0 = t_0 - s > 0$, for a sufficiently small $s > 0$, and $\tau = sr^{-\frac{2}{\alpha}}(2 - \frac{\delta}{2})^{-1}$. Then $\tau r^{\frac{2}{\alpha}} = s(2 - \frac{\delta}{2})^{-1}$ is sufficiently small, so that $\tilde{t}_0 + 2\tau r^{\frac{2}{\alpha}} \leq T$. Since

$$(2 - \delta)\tau r^{\frac{2}{\alpha}} = (2 - \delta)s(2 - \tfrac{\delta}{2})^{-1} < s \quad \text{and} \quad 2\tau r^{\frac{2}{\alpha}} = 2s(2 - \tfrac{\delta}{2})^{-1} > s,$$

we have

$$t_0 \in (\tilde{t}_0 + (2 - \delta)\tau r^{\frac{2}{\alpha}}, \tilde{t}_0 + 2\tau r^{\frac{2}{\alpha}}).$$

Hence,

$$(x_0, t_0) \in Q_+(x_0, \tilde{t}_0, r).$$

By Theorem 6.25, for $u_0 \geq 0$, the inequality $u(x, t) \geq 0$ holds on $\overline{Q}$. Therefore,

$$\inf_{(x,t)\in Q_+(x_0,\tilde{t}_0,r)} u(x, t) = 0.$$

Lemma 6.12 yields that there exists $t_1 > 0$ such that

$$u(x, t) = 0, \quad (x, t) \in B(x_0, \delta r) \times (\tilde{t}_0, \tilde{t}_0 + t_1).$$

By the uniqueness of u from Theorem 6.9, we obtain $u(x, t) = 0$ for $x \in \Omega$ and $0 \leq t \leq T$. This contradicts the condition $u_0 \neq 0$, and completes the proof. $\square$

6.6 Inverse Problems

In this section, we discuss several inverse problems for subdiffusion: given partial information about the solution to the subdiffusion model (6.17), we seek to recover one or several unknown problem data, e.g., boundary condition, initial condition, fractional order, or unknown coefficient(s) in the model. Such problems arise in many practical applications, and represent a very important class of applied problems. Mathematically they exhibit dramatically different features than direct problems analyzed so far in the sense that inverse problems are often ill-posed in the sense of Hadamard: the solution may not exist and may be nonunique, and if it does exist, it does not depend continuously on the given problem data. Consequently, they require different solution techniques. The goal of this section is to give a flavor of inverse problems, by illustrating the following model problems: backward subdiffusion, inverse source problem, order determination and inverse potential problem. The literature on inverse problems for subdiffusion is vast; see the introductory survey [JR15], and reviews [LLY19b] (inverse source problems), [LY19a] (fractional orders) and [LLY19a] (inverse coefficient problem) for further pointers. The standard reference for inverse problems for PDES is [Isa06].

The discussions below use extensively various properties of Mittag-Leffler function $E_{\alpha,\beta}(z)$, cf. (3.1) in Chapter 3, especially the complete monotonicity in Theorem 3.5. Thus, the discussion is largely restricted to the case of time-independent coefficients, and the extension to more general case is largely open.

6.6.1 Backward subdiffusion

Backward subdiffusion is one classical example of inverse problems for subdiffusion. Consider the following subdiffusion problem (with $\alpha \in (0,1)$)

$$\begin{cases} \partial_t^\alpha u = \Delta u, & \text{in } Q, \\ u = 0, & \text{on } \partial_L Q, \\ u(0) = u_0, & \text{in } \Omega. \end{cases} \tag{6.89}$$

This problem has a unique solution u that depends continuously on u_0, e.g.,

$$\|u(t)\|_{L^2(\Omega)} \le \|u_0\|_{L^2(\Omega)},$$

in view of the smoothing property of the operator $F(t)$ in Theorem 6.4(iv). Backward subdiffusion reads: Given the terminal data $g = u(T)$, with $T > 0$ fixed, can one recover the initial data u_0? With the eigenpairs $\{(\lambda_n, \varphi_n)\}_{n=1}^\infty$ of the Dirichlet $-\Delta$, the solution u to problem (6.89) is given by

$$u(x,t) = \sum_{n=1}^\infty (u_0, \varphi_n) E_{\alpha,1}(-\lambda_n t^\alpha) \varphi_n(x). \tag{6.90}$$

Thus, the exact terminal data $g = u(T)$ is given by

$$g = \sum_{n=1}^\infty (u_0, \varphi_n) E_{\alpha,1}(-\lambda_n T^\alpha) \varphi_n,$$

and u_0 and g are related by

$$(g, \varphi_n) = E_{\alpha,1}(-\lambda_n T^\alpha)(u_0, \varphi_n), \quad n = 1, 2, \dots. \tag{6.91}$$

Formally, u_0 is given by

$$u_0 = \sum_{n=1}^{\infty} \frac{(g, \varphi_n)}{E_{\alpha,1}(-\lambda_n T^{\alpha})} \varphi_n,$$

if the series does converge in a suitable norm (note that $E_{\alpha,1}(-\lambda_n T^{\alpha})$ does not vanish, cf. Corollary 3.3). The formula (6.91) is very telling. For $\alpha = 1$, it reduces to

$$(u_0, \varphi_n) = e^{\lambda_n T}(g, \varphi_n).$$

It shows clearly the severely ill-posed nature of backward diffusion: the perturbation in (g, φ_n) is amplified by $e^{\lambda_n T}$ in the recovered expansion coefficient (u_0, φ_n). Even for a small index n, (u_0, φ_n) can be huge, if T is not exceedingly small. This implies a very bad stability, and one can only expect a logarithmic-type result. For $\alpha \in (0, 1)$, by Theorem 3.2, $E_{\alpha,1}(-t)$ decays linearly on $\mathbb{R}_+$, and thus $E_{\alpha,1}(-\lambda_n T^{\alpha})^{-1}$ grows only linearly in λ_n, i.e., $E_{\alpha,1}(-\lambda_n T^{\alpha})^{-1} \sim T^{\alpha} \lambda_n$, cf. Theorem 3.6, which is very mild compared to $e^{\lambda_n T}$ for $\alpha = 1$, indicating a much better behavior. Thus, $|(u_0, \varphi_n)| \leq c|(g, \lambda_n \varphi_n)|$ or $|(u_0, \varphi_n)| \leq c|(g, \Delta \varphi_n)|$, and by integration by part twice, if $g \in \dot{H}^2(\Omega)$,

$$|(u_0, \varphi_n)| \leq c|(\Delta g, \varphi_n)|.$$

Roughly, backward subdiffusion amounts to two spatial derivative loss. More precisely, we have the following stability estimate [SY11, Theorem 4.1].

Theorem 6.28 *Let $T > 0$ be fixed, and $\alpha \in (0, 1)$. For any $g \in \dot{H}^2(\Omega)$, there exists a unique $u_0 \in L^2(\Omega)$ and a weak solution $u \in C(\bar{I}; L^2(\Omega)) \cap C(I; \dot{H}^2(\Omega))$ to problem (6.89) such that $u(T) = g$, and*

$$c_1 \|u_0\|_{L^2(\Omega)} \leq \|u(T)\|_{H^2(\Omega)} \leq c_2 \|u_0\|_{L^2(\Omega)}.$$

Proof The second inequality is already proved in Theorem 6.7. For $g \in \dot{H}^2(\Omega)$,

$$c_1' \|g\|_{H^2(\Omega)} \leq \sum_{n=1}^{\infty} \lambda_n^2 (g, \varphi_n)^2 \leq c_2' \|g\|_{H^2(\Omega)}^2.$$

By Corollary 3.3, $E_{\alpha,1}(-\lambda_n t^{\alpha})$ does not vanish on $\mathbb{R}_+$. Thus, we may let

$$c_n = \frac{(g, \varphi_n)}{E_{\alpha,1}(-\lambda_n T^{\alpha})} \quad \text{and} \quad u_0 = \sum_{n=1}^{\infty} c_n \varphi_n.$$

Clearly, the solution $u(x, t)$ to problem (6.89) with this choice satisfies $g = u(\cdot, T)$. Further, by Theorem 3.6,

$$\sum_{n=1}^{\infty} c_n^2 \leq \sum_{n=1}^{\infty} (g, \varphi_n)^2 (1 + \Gamma(1 - \alpha)\lambda_n T^{\alpha})^2 \leq c_{T,\alpha} \sum_{n=1}^{\infty} (1 + \lambda_n^2)(g, \varphi_n)^2.$$

Thus, there holds, $\|u_0\|_{L^2(\Omega)} \leq c\|u(\cdot, T)\|_{H^2(\Omega)}$, showing the first inequality. $\square$

Thus, backward subdiffusion is well-posed for $g \in \dot{H}^2(\Omega)$ and $u_0 \in L^2(\Omega)$. In practice, however, g is measured and corrupted by rough noise, and it should be studied for $g \in L^2(\Omega)$, and thus is ill-posed, i.e., the solution may not exist, and even if it does exist, it is unstable with respect to perturbations in the data g. Indeed, the forward map from the initial data u_0 to the terminal data $u(T)$ is compact in $L^2(\Omega)$: for $u_0 \in L^2(\Omega)$, Theorem 6.7 implies $u(T) \in \dot{H}^2(\Omega)$, and the space $\dot{H}^2(\Omega)$ is compactly embedded into $L^2(\Omega)$, cf. Theorem A.3. Ill-posedness of this type is characteristic of many inverse problems, and specialized solution techniques are required to obtain reasonable approximations. It is instructive to compare the problem for subdiffusion and normal diffusion. Intuitively, for the backward problem, the history mechanism of subdiffusion retains the complete dynamics of the physical process all the way to time $t = 0$, including u_0. The parabolic case has no such memory effect and hence, the coupling between the current and previous states is weak. The big difference between the cases $0 < \alpha < 1$ and $\alpha = 1$ might lead to a belief that "Inverse problems for FDES are always less ill-conditioned than their classical counterparts." or "Models based on the fractional derivative paradigm can escape the curse of being strongly ill-posed," However, this statement turns out to be quite false.

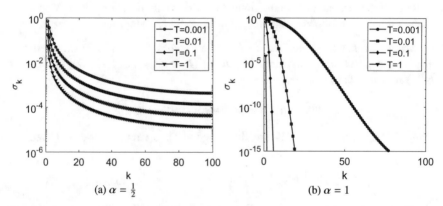

Fig. 6.2 The singular value spectrum of the forward map F from the initial data to the final time data for backward (sub)diffusion, for (a) $\alpha = \frac{1}{2}$ and (b) $\alpha = 1$, at four different time instances.

A relatively complete picture for an inverse problem can be gained by the singular value spectrum of the forward map. Fig. 6.2 shows the singular value spectra for $\alpha = 1$ and $\alpha = \frac{1}{2}$. When $\alpha = 1$, the singular values σ_k decay exponentially to zero, which agrees with the factor $e^{\lambda_n T}$, but there is considerable difference with the change of time scale. For $T = 1$ there is likely at most one usable singular value, and still only a handful at $T = 0.1$. In the fractional case, due to the very different decay rate, there are a considerable number of singular values available even for larger times. However, there is still the initially fast decay, especially as α tends to zero. It is instructive to compare the singular value spectrum at a fixed time T. At $T = 0.01$, the singular values with smaller indices (those less than about 25) are larger for $\alpha = 1$ than $\alpha = 0.5$, and thus at this value for T, normal diffusion should allow superior reconstructions (under suitable conditions on the noise in g, of course). The situation reverses when singular values of larger indices are considered and quickly shows the infeasibility of recovering high-frequency information for backward diffusion, whereas in the fractional case many more Fourier modes might be attainable (again depending on the noise in the data).

6.6.2 Inverse source problems

Now we focus on the basic subdiffusion model (with $\alpha \in (0, 1)$)

$$\begin{cases} \partial_t^\alpha u - \Delta u = f, & \text{in } Q, \\ \quad\quad\quad u = 0, & \text{on } \partial_L Q, \\ \quad\quad\quad u(0) = u_0, & \text{in } \Omega, \end{cases} \tag{6.92}$$

with suitable boundary condition. The inverse problem of interest is to recover the source term f from additional data. We shall not attempt a general f, since we are looking at only final time data or lateral boundary data (or internal data). In order to have a unique determination, we look for only a space- or time-dependent component of f. By the linearity of problem (6.92), we may assume $u_0 = 0$. There is a "folklore" theorem for the standard diffusion equation: inverse problems where the data is not aligned in the same direction as the unknown are almost surely severely ill-posed, but usually only mildly so when these directions do align. There is a huge body of literature on inverse source problems, with different combinations of the data and unknowns; Inverse source problems for the classical diffusion equation have been extensively studied; see e.g., [Can68, Can84, IY98, CD98].

First, we consider the case of recovering the component $p(t)$ in $f(x, t) = q(x)p(t)$ from the function u at a fixed point $x_0 \in \Omega$ (or the flux at the boundary). The exact data $g(t)$ is readily available:

$$g(t) = \sum_{n=1}^{\infty} \int_0^t (t - s)^{\alpha-1} E_{\alpha,\alpha}(-\lambda_n (t - s)^\alpha) p(s) \, ds \, (q, \varphi_n) \varphi_n(x_0).$$

In particular, if $q = \varphi_n$ for some $n \in \mathbb{N}$ then the problem becomes

$$g(t) = \int_0^t R(t - s) p(s) \, ds,$$

with $R(t) = t^{\alpha-1} E_{\alpha,\alpha}(-\lambda_n t^\alpha)$. This is a Volterra integral equation of the first kind for $p(t)$, and the behavior of the solution hinges on the behavior of $t^{\alpha-1} E_{\alpha,\alpha}(-\lambda_n t^\alpha)$ near $t = 0$. Since $E_{\alpha,\alpha}(-\lambda_n t^\alpha) \to \Gamma(\alpha)^{-1}$, as $t \to 0^+$, it is weakly singular with the leading singularity $t^{\alpha-1}$. In particular, applying the operator ∂_t^α to both sides and then using the differentiation formula

$$\partial_t^\alpha \int_0^t s^{\alpha-1} E_{\alpha,\alpha}(-\lambda s^\alpha) g(t-s) ds = g(t) - \lambda \int_0^t s^{\alpha-1} E_{\alpha,\alpha}(-\lambda s^\alpha) g(t-s) ds, \quad (6.93)$$

lead to

$$\tilde{g}(t) = (\partial_t^\alpha g)(t) = p(t) - \lambda_n \int_0^t R(t-s) p(s) \, ds,$$

which is a Volterra integral equation of the second kind, where the kernel R is integrable. Thus, following the method of successive approximations (see, e.g., the proof of Proposition 4.4), we can prove that it has a unique solution $p(t)$ that depends continuously on R and $\tilde{g}$, and further, a bound can be derived using Gronwall's inequality in Theorem 4.2 in Section 4.1. Hence, the inverse problem is mildly ill-posed requiring only a fractional derivative of order α loss on the data g and $\|p\|_{L^\infty(I)} \le c \|\partial_t^\alpha g\|_{L^\infty(I)}$. The assertion holds in the general case. Consider the following initial boundary value problem

$$\begin{cases} \partial_t^\alpha u = \Delta u + q(x) p(t), & \text{in } Q, \\ u = 0, & \text{on } \partial_L Q, \\ u(\cdot, 0) = 0, & \text{in } \Omega. \end{cases} \quad (6.94)$$

Then the inverse source problem reads: given $x_0 \in \Omega$, determine $p(t)$, $t \in \bar{I}$ from $u(x_0, t)$, $t \in \bar{I}$. The following result gives a two-sided stability estimate [SY11, Theorem 4.4]. The regularity requirement in Theorem 6.29 is slightly lower than that stated in [SY11, Theorem 4.4], which was obtained in [LLY19b, Theorem 1].

Theorem 6.29 *Let $q \in \dot{H}^\gamma(\Omega)$, with $\gamma > \frac{d}{2}$, $q(x_0) \neq 0$. Then the solution u to problem (6.94) with $p \in C(\bar{I})$ satisfies*

$$c_0 \|\partial_t^\alpha u(x_0, \cdot)\|_{L^\infty(I)} \le \|p\|_{C(\bar{I})} \le c_1 \|\partial_t^\alpha u(x_0, \cdot)\|_{L^\infty(I)}.$$

Proof The first estimate is direct from Section 6.2. Since $p \in C(\bar{I})$ and $q \in \dot{H}^\gamma(\Omega)$, we obtain

$$u = \sum_{n=1}^\infty \int_0^t p(s)(q, \varphi_n)(t-s)^{\alpha-1} E_{\alpha,\alpha}(-\lambda_n(t-s)^\alpha) ds \varphi_n \quad \text{in } L^2(I; H^2(\Omega)).$$

By the differentiation formula (6.93), we deduce

$$\partial_t^\alpha u = pq + \sum_{n=1}^\infty (-\lambda_n) \int_0^t p(s)(q, \varphi_n)(t-s)^{\alpha-1} E_{\alpha,\alpha}(-\lambda_n(t-s)^\alpha) ds \varphi_n, \quad (6.95)$$

in $L^2(Q)$. Consequently,

$$p(t) = \frac{1}{q(x_0)} \left\{ \partial_t^\alpha u(x_0, t) + \int_0^t K(x_0, s) p(t-s) ds \right\},$$

with $K(x, t) = t^{\alpha-1} \sum_{n=1}^{\infty} \lambda_n E_{\alpha,\alpha}(-\lambda_n t^{\alpha})(q, \varphi_n) \varphi_n(x)$. Let $\epsilon := \frac{\gamma}{2} - \frac{d}{4} > 0$. Then by the asymptotics of $E_{\alpha,\alpha}(z)$ in Theorem 3.2, we have

$$\|K(\cdot, t)\|^2_{\dot{H}^{\frac{d}{2}+\epsilon}(\Omega)} = t^{2(\alpha-1)} \sum_{n=1}^{\infty} |\lambda_n^{1-\frac{\epsilon}{2}} E_{\alpha,\alpha}(-\lambda_n t^{\alpha})|^2 |\lambda_n^{\epsilon+\frac{d}{4}}(q, \varphi_n)|^2$$

$$\leq c^2 t^{2(\frac{\epsilon}{2}\alpha-1)} \sum_{n=1}^{\infty} \left(\frac{(\lambda_n t^{\alpha})^{1-\frac{\epsilon}{2}}}{1+\lambda_n t^{\alpha}}\right)^2 |\lambda_n^{\frac{\gamma}{2}}(q, \varphi_n)|^2 \leq (c\|q\|_{\dot{H}^{\gamma}(\Omega)} t^{\frac{\epsilon}{2}\alpha-1})^2.$$

Using the Sobolev embedding $\dot{H}^{\frac{d}{2}+\epsilon}(\Omega) \hookrightarrow C(\overline{\Omega})$, we obtain

$$|K(x_0, t)| \leq \|K(\cdot, t)\|_{C(\overline{\Omega})} \leq c\|K(\cdot, t)\|_{\dot{H}^{\frac{d}{2}+\epsilon}(\Omega)} \leq c\|q\|_{\dot{H}^{\gamma}(\Omega)} t^{\frac{\epsilon}{2}\alpha-1}.$$

Consequently,

$$|p(t)| \leq c\|\partial_t^{\alpha} u(x_0, t)\|_{L^{\infty}(I)} + c\|q\|_{\dot{H}^{\gamma}(\Omega)} \int_0^t s^{\frac{\epsilon}{2}\alpha-1} |p(t-s)| d.$$

Applying the Gronwall inequality in Theorem 4.1 yields the second identity. $\qquad\square$

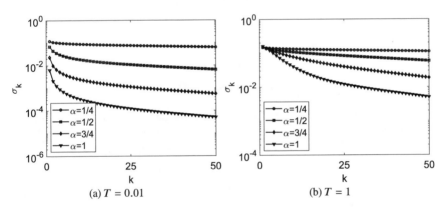

Fig. 6.3 Singular value spectrum for the inverse source problem of recovering $p(t)$ from $\partial_x u(0)$.

Figure 6.3 shows the singular value spectrum for the forward map. There is a slight increase in ill-posedness as α increases, which agrees with the stability estimate: the recovery amounts to taking the αth-order fractional derivative. Thus, fractional diffusion can mitigate the degree of ill-posedness of the concerned inverse problem. For α close to zero, it effectively behaves as if it were well-posed.

Next, we look at the case of recovering a space-dependent (and time-independent) source $f(x)$ from the time trace of the solution u at a fixed point $x_0 \in \Omega$. The solution u to the direct problem is given by

$$u(x, t) = \sum_{n=1}^{\infty} \lambda_n^{-1}\left(1 - E_{\alpha,1}(-\lambda_n t^{\alpha})\right)(f, \varphi_n)\varphi_n(x),$$

and hence,

$$g(t) = \sum_{n=1}^{\infty} \lambda_n^{-1}\varphi_n(x_0)\left(1 - E_{\alpha,1}(-\lambda_n t^{\alpha})\right)(f, \varphi_n).$$

This formula is very informative. First, the choice of the point x_0 has to be strategic in the sense that it should satisfy the condition $\varphi_n(x_0) \neq 0$ for all $n \in \mathbb{N}$. If $\varphi_n(x_0) = 0$, then the nth mode (f, φ_n) is not recoverable. The condition $\varphi_n(x_0) \neq 0$ is almost impossible to arrange in practice: for $\Omega = (0, 1)$, the Dirichlet eigenfunctions are $\varphi_n(x) = \sin(n\pi x)$ and all points of the form $q\pi$ for q rational must be excluded for x_0 in order to satisfy the condition. Second, upon simple algebraic operations, the inverse problem amounts to representing $g(t)$ in terms of $\{E_{\alpha,1}(-\lambda_n t^{\alpha})\}_{n=1}^{\infty}$. These functions all decay to zero polynomially in t, and thus almost linearly dependent, indicating the ill-posed nature of the inverse problem. This shows a stark contrast in the degree of ill-posedness between recovering a time and a space-dependent source component from the same time-dependent trace data. It is also very much in line with the "folklore" theorem.

To gain further insights, consider $\Omega = (0, 1)$ and

$$\begin{cases} \partial_t^{\alpha} u = \partial_{xx} u + f, & \text{in } Q, \\ \partial_x u(0, \cdot) = 0, & \text{in } I, \\ -\partial_x u(1, \cdot) = 0, & \text{in } I, \\ u(\cdot, 0) = u_0, & \text{in } \Omega. \end{cases} \tag{6.96}$$

For $u_0 \in L^2(\Omega)$ and $f \in L^2(\Omega)$, there exists a unique solution $u \in C(\bar{I}; L^2(\Omega)) \cap C(I; \dot{H}^2(\Omega))$. Indeed, let $\{(\lambda_n, \varphi_n)\}_{n=0}^{\infty}$ be the eigenpairs of the Neumann Laplacian on the domain Ω. Since $\lambda_0 = 0$ (with $\varphi_0(x) \equiv 1$), the solution u is given by

$$u(x, t) = \sum_{n=0}^{\infty}(u_0, \varphi_n)E_{\alpha,1}(-\lambda_n t^{\alpha})\varphi_n(x)$$

$$+ \frac{t^{\alpha}}{\Gamma(\alpha + 1)}(f, \varphi_0) + \sum_{n=1}^{\infty}\lambda_n^{-1}(f, \varphi_n)(1 - E_{\alpha,1}(-\lambda_n t^{\alpha}))\varphi_n(x).$$

The inverse problem is to determine $f(x)$ from the lateral trace data $g(t) = u(0, t)$ for $t \in \bar{I}$. The next result gives an affirmative answer to the question [ZX11, Theorem 1]. The proof indicates that the data to the unknown relation in the inverse source problem amounts to analytic continuation, which is known to be severely ill-posed.

Theorem 6.30 *The data $g(t)$ uniquely determines the source term $f(x)$.*

Proof The desired uniqueness follows directly from Lemma 6.13 below. □

Lemma 6.13 *Let $w, \widetilde{w} \in C(\overline{I}; L^2(\Omega)) \cap C(I; \dot{H}^2(\Omega))$ be solutions of (6.96) for with $f \equiv 0$ and the initial conditions u_0 and $\widetilde{u}_0$, respectively. Then $w(0, t) = \widetilde{w}(0, t)$ for $t \in \overline{I}$ implies $u_0(x) = \widetilde{u}_0(x)$ in $L^2(\Omega)$.*

Proof Since the eigenfunctions $\{\varphi_n\}_{n=0}^{\infty}$ of the Neumann Laplacian form a complete orthogonal basis in $L^2(\Omega)$, the initial conditions can be represented by $u_0(x) = \sum_{n=0}^{\infty} a_n \varphi_n(x)$ and $\widetilde{u}_0(x) = \sum_{n=0}^{\infty} \widetilde{a}_n \varphi_n(x)$. It suffices to show $a_n = \widetilde{a}_n$, $n = 0, 1, \ldots$, given the data $w(0, t) = \widetilde{w}(0, t)$. Since $w(0, t) = \widetilde{w}(0, t)$ for $t \in \overline{I}$, we have

$$\sum_{n=0}^{\infty} a_n E_{\alpha,1}(-\lambda_n t^\alpha) = \sum_{n=0}^{\infty} \widetilde{a}_n E_{\alpha,1}(-\lambda_n t^\alpha), \quad t \in \overline{I}.$$

Moreover, by the Riemann-Lebesgue lemma and analyticity of the Mittag-Leffler function, both sides are analytic in $t > 0$. By the unique continuation property of real analytic functions, we have

$$\sum_{n=0}^{\infty} a_n E_{\alpha,1}(-\lambda_n t^\alpha) = \sum_{n=0}^{\infty} \widetilde{a}_n E_{\alpha,1}(-\lambda_n t^\alpha), \quad \forall t \geq 0.$$

Since $a_n \to 0$, by Theorem 3.6,

$$|e^{-t\mathcal{R}(z)} \sum_{n=0}^{\infty} a_n E_{\alpha,1}(-\lambda_n t^\alpha)| \leq e^{-t\mathcal{R}(z)} \left(|a_0| + \sum_{n=1}^{\infty} \frac{\Gamma(1+\alpha)|a_n|}{\lambda_n t^\alpha} \right)$$

$$\leq c e^{-t\mathcal{R}(z)} t^{-\alpha} \sum_{n=1}^{\infty} \lambda_n^{-1} \leq c t^{-\alpha} e^{-t\mathcal{R}(z)},$$

and the function $e^{-t\mathcal{R}(z)} t^{-\alpha}$ is integrable for $t \in \mathbb{R}_+$ for fixed z with $\mathcal{R}(z) > 0$. By Lebesgue's dominated convergence theorem and Lemma 3.2, we deduce

$$\int_0^{\infty} e^{-zt} \sum_{n=0}^{\infty} a_n E_{\alpha,1}(-\lambda_n t^\alpha) dt = \sum_{n=0}^{\infty} a_n \frac{z^{\alpha-1}}{z^\alpha + \lambda_n},$$

which implies

$$\sum_{n=0}^{\infty} \frac{b_n}{\eta + \lambda_n} = 0, \quad \mathcal{R}(\eta) > 0, \tag{6.97}$$

where $\eta = z^\alpha$ and $b_n = a_n - \widetilde{a}_n$. Since $\lim_{n\to\infty}(a_n - \widetilde{a}_n) \to 0$, we can analytically continue in η, so that the identity (6.97) holds for $\eta \in \mathbb{C} \setminus \{-\lambda_n\}_{n\geq 0}$. Last, we deduce $b_0 = 0$ from the identity by taking a suitable disk which includes 0 and does not include $\{-\lambda_n\}_{n=1}^{\infty}$. By the Cauchy integral formula, integrating the identity along the disk gives $2\pi i\, b_0 = 0$, i.e., $a_0 = \widetilde{a}_0$. Upon repeating the argument, we obtain $a_n = \widetilde{a}_n$, $n = 1, 2, \ldots$, which completes the proof of the lemma. $\square$

The proof of the theorem indicates that the inverse source problem amounts analytic continuation, which is known to be severely ill-posed. In particular, it does not show any beneficial effect of subdiffusion over normal diffusion.

6.6.3 Determining fractional order

This is perhaps the most obvious inverse problem for FDEs! In theory, the fractional-order α in the subdiffusion model can be determined by the tail decay rate of the waiting time-distribution in microscopic descriptions, as the derivation in Section 1.2 indicates. However, in practice, it often cannot be determined directly, and has to be inferred from experimental data, which leads to a nonlinear inverse problem. In many cases the asymptotic behavior of the solution u can be used to determine α. To show the feasibility of the recovery, assume that in problem (6.89), u_0 is taken to be φ_n, and the additional data is the flux at $x_0 \in \partial\Omega$, i.e., $\frac{\partial u}{\partial \nu}(x_0, t) = h(t)$, for $0 < t \le T$. Then $h(t)$ uniquely determines α if x_0 satisfies $\frac{\partial \varphi_n}{\partial \nu}(x_0) \ne 0$. Indeed, $h(t)$ is given by

$$h(t) = E_{\alpha,1}(-\lambda_n t^\alpha)(u_0, \varphi_n)\frac{\partial \varphi_n}{\partial \nu}(x_0) = E_{\alpha,1}(-\lambda_n t^\alpha)\frac{\partial \varphi_n}{\partial \nu}(x_0),$$

It follows from the series expansion of $E_{\alpha,1}(-\lambda t^\alpha)$ that

$$E_{\alpha,1}(-\lambda t^\alpha) = 1 - \lambda\Gamma(1+\alpha)^{-1}t^\alpha + O(\lambda^2 t^{2\alpha}),$$

Thus, one can recover α by looking at the small time behavior of $h(t)$.

Now we make this reasoning more precise. In problem (6.89), suppose that we can measure $g(t) = u(x_0, t)$ close to $t = 0$. The following result shows that the asymptotic of $g(t)$ at $t = 0$ uniquely determines the fractional-order α [HNWY13, Theorem 1]; see [LHY20, Theorem 2] for the stability issue and [Kia20, Theorem 2.1] and [JK21, Theorem 1.2] for determining the fractional order without knowing the initial condition or medium. The formula involves taking one derivative with respect to t, which is ill-posed, even though only mildly so. It is worth noting that besides the smoothness condition, the result actually needs only the value $u_0(x_0)$.

Theorem 6.31 *In problem* (6.89), *let* $u_0 \in C_0^\infty(\Omega)$ *with* $\Delta u_0(x_0) \ne 0$. *Then*

$$\alpha = \lim_{t \to 0^+} \frac{t\frac{\partial u}{\partial t}(x_0, t)}{u(x_0, t) - u_0(x_0)}.$$

Proof Since $u_0 \in C_0^\infty(\Omega)$, we deduce by integration by parts repeatedly that for any $\ell \in \mathbb{N}$, there holds

$$|(u_0, \varphi_n)| = \lambda_n^{-1}|(\Delta u_0, \varphi_n)| = \lambda_n^{-\ell}|(\Delta^\ell u_0, \varphi_n)| \le c(\ell)\lambda_n^{-\ell}. \tag{6.98}$$

Meanwhile, by Sobolev embedding in Theorem A.3, we have for any $\kappa > \frac{d}{2}$,

$$\|\varphi_n\|_{L^\infty(\Omega)} \leq c\|\varphi_n\|_{H^\kappa(\Omega)} \leq c\|(-\Delta)^{\frac{\kappa}{2}}\varphi_n\|_{L^2(\Omega)} \leq c\lambda_n^{\frac{\kappa}{2}}. \tag{6.99}$$

For $t \in I$,

$$u(x_0, t) = \sum_{n=1}^{\infty} (u_0, \varphi_n)\varphi_n(x_0)E_{\alpha,1}(-\lambda_n t^\alpha),$$

which converges in $C(\overline{I})$. Therefore, for $t \in I$

$$\frac{\partial u}{\partial t}(x_0, t) = \sum_{n=1}^{\infty} -\lambda_n(u_0, \varphi_n)\varphi_n(x_0)t^{\alpha-1}E_{\alpha,\alpha}(-\lambda_n t^\alpha).$$

Meanwhile, by the definition of $E_{\alpha,\alpha}(z)$, $E_{\alpha,\alpha}(-\lambda_n t^\alpha) = \Gamma(\alpha)^{-1} + t^\alpha r_n(t)$, where $r_n(t)$ is defined by $r_n(t) = \frac{E_{\alpha,\alpha}(-\lambda_n t^\alpha)-\Gamma(\alpha)^{-1}}{t^\alpha}$. The function $r_n(t)$ is continuous at $t = 0$, the limit $\lim_{t\to 0^+} r_n(t)$ exists, and by the complete monotonicity of $E_{\alpha,2\alpha}(-t)$, cf. Corollary 3.2,

$$|r_n(t)| = |\lambda_n||E_{\alpha,2\alpha}(-\lambda_n t^\alpha)| \leq \Gamma(2\alpha)^{-1}|\lambda_n|, \quad t \geq 0, n \in \mathbb{N}. \tag{6.100}$$

Consequently,

$$\lim_{t\to 0^+} t^{1-\alpha}\frac{\partial u}{\partial t}(x_0, t) = \frac{1}{\Gamma(\alpha)}\sum_{n=1}^{\infty} -\lambda_n(u_0, \varphi_n)\varphi_n(x_0)$$

$$+ \lim_{t\to 0^+} t^\alpha \sum_{n=1}^{\infty} -\lambda_n(u_0, \varphi_n)\varphi_n(x_0)r_n(t).$$

The second summation can be bounded using (6.98)–(6.100) and Sobolev embedding theorem by

$$\left|\sum_{n=1}^{\infty} -\lambda_n(u_0, \varphi_n)\varphi_n(x_0)r_n(t)\right| \leq \sum_{n=1}^{\infty} \frac{|\lambda_n|^2}{\Gamma(2\alpha)}|(u_0, \varphi_n)\varphi_n(x_0)|$$

$$\leq \sum_{n=1}^{\infty} \frac{|\lambda_n|^2}{\Gamma(2\alpha)}\frac{c(\ell)}{|\lambda_n|^\ell}c|\lambda_n|^{\frac{\kappa}{2}}.$$

For any sufficiently large $\ell \in \mathbb{N}$, we have

$$\sup_{t\in\overline{I}}\left|\sum_{n=1}^{\infty} -\lambda_n(u_0, \varphi_n)\varphi_n(x_0)r_n(t)\right| < \infty.$$

Hence,

$$\lim_{t\to 0^+} t^{1-\alpha}\frac{\partial u}{\partial t}(x_0, t) = \Gamma(\alpha)^{-1}\Delta u_0(x_0). \tag{6.101}$$

Meanwhile, since

$$E_{\alpha,1}(-\lambda_n t^\alpha) = 1 - \Gamma(\alpha+1)^{-1}\lambda_n t^\alpha + t^{2\alpha}\lambda_n^2 E_{\alpha,2\alpha+1}(-\lambda_n t^\alpha),$$

we have

$$u(x_0, t) = \sum_{n=1}^{\infty}(u_0, \varphi_n)\varphi_n(x_0) + t^\alpha \sum_{n=1}^{\infty}\frac{-\lambda_n(u_0, \varphi_n)\varphi_n(x_0)}{\Gamma(\alpha+1)}$$

$$+ t^{2\alpha}\sum_{n=1}^{\infty}\lambda_n^2 E_{\alpha,2\alpha+1}(-\lambda_n t^\alpha)(u_0, \varphi_n)\varphi_n(x_0)$$

$$= u_0(x_0) + \Gamma(\alpha+1)^{-1}\Delta u_0(x_0)t^\alpha + t^{2\alpha}\widetilde{r}(t),$$

where $\sup_{\overline{T}}|\widetilde{r}(t)| < \infty$. Consequently,

$$\lim_{t\to 0^+} t^{-\alpha}(u(x_0, t) - u_0(x_0)) = \Gamma(\alpha+1)^{-1}\Delta u_0(x_0). \qquad (6.102)$$

Combining the identities (6.101)–(6.102), the assumption $\Delta u_0(x_0) \neq 0$, and the recursion $\Gamma(\alpha+1) = \alpha\Gamma(\alpha)$ complete the proof of the theorem. $\qquad\square$

6.6.4 Inverse potential problem

Parameter identifications for PDES encompass a broad range of inverse problems. The earliest contribution for subdiffusion is [CNYY09], which proves the uniqueness of simultaneously recovering the diffusion coefficient and fractional order α from the lateral data at one single spatial point, in the one-dimensional case. The proof relies on relevant results on the inverse Sturm-Liouville problem. This piece of pioneering work has inspired much further research on inverse problems for subdiffusion. See also [KOSY18] for uniqueness of the diffusion coefficient in multi-dimension. We illustrate with an inverse potential problem from the terminal data proved in [JZ21, Theorem 1.1]; see [CY97, Theorem 1] for standard parabolic problems and the works [ZZ17, KR19] for other results on the inverse problem. Specifically, let $u \equiv u(q)$ be the solution to

$$\begin{cases} \partial_t^\alpha u = \Delta u + qu, & \text{in } Q, \\ u = 0, & \text{on } \partial_L Q, \\ u(\cdot, 0) = u_0, & \text{in } \Omega. \end{cases}$$

The inverse problem reads: given the terminal data g, recover $q \in L^2(\Omega)$ such that

$$u(q)(\cdot, T) = g \quad \text{in } \Omega. \qquad (6.103)$$

The direct problem for $q \in L^2(\Omega)$ is not covered in Section 6.2, since so far we have assumed that the coefficients a and q are smooth. Following the work [JZ21], we illustrate how the assumption on the potential q can be relaxed. Throughout this part, $A = -\Delta$ with its domain $D(A) = \dot{H}^2(\Omega)$ and and the graph norm denoted by $\|\cdot\|_{D(A)}$. Then given $q \in L^2(\Omega)$, consider the following abstract Cauchy problem

$$\begin{cases} \partial_t^\alpha u(t) + Au(t) = qu(t), & \text{in } I, \\ \qquad\qquad u(0) = u_0. \end{cases} \tag{6.104}$$

We prove that for suitably smooth u_0 and any $q \in L^2(\Omega)$, problem (6.104) has a unique classical solution $u = u(q) \in C^\alpha(\overline{I}; L^2(\Omega)) \cap C(\overline{I}; D(A))$. The analysis is based on a "perturbation" argument. Specifically, we view $qu(t)$ as the inhomogeneous term and obtain that the solution $u(t)$ satisfies

$$u(t) = U(t) + \int_0^t E(t - s)qu(s)ds,$$

where $U(t) = F(t)u_0$, and the solution operators $F(t)$ and $E(t)$ are associated with the operator A. Now we can specify the function analytic setting. Let $\frac{3}{4} < \gamma < 1$ and $0 < \beta < (1 - \gamma)\alpha$ be fixed and set

$$X = C^\beta(\overline{I}; D(A^\gamma)) \cap C(\overline{I}; D(A)), \tag{6.105}$$

with the norm given by $\|v\|_X = \|v\|_{C^\beta(\overline{I};D(A^\gamma))} + \|v\|_{C(\overline{I};D(A))}$. Then for every $q \in L^2(\Omega)$, we define an associated operator $L(q)$ by

$$[L(q)f](t) = \int_0^t E(t - s)qf(s)ds, \quad \forall f \in C^\beta(\overline{I}; D(A^\gamma)).$$

The next result gives the mapping property of the operator $L(q)$.

Lemma 6.14 *For any $q \in L^2(\Omega)$, $L(q)$ maps $C^\beta(\overline{I}; D(A^\gamma))$, with $\frac{3}{4} < \gamma < 1$, into X.*

Proof Let $f \in C^\beta(\overline{I}; D(A^\gamma))$, $q \in L^2(\Omega)$ and let $g(t) = qf(t), 0 \leq t \leq T$. We split the function $L(q)f$ into two terms $L(q)f = v_1 + v_2$, with

$$v_1(t) = \int_0^t E(t - s)(g(s) - g(t))ds \quad \text{and} \quad v_2(t) = \int_0^t E(t - s)g(t)ds.$$

Since $f \in C^\beta(\overline{I}; D(A^\gamma))$, by Sobolev embedding theorem, $g \in C^\beta(\overline{I}; L^2(\Omega))$, and thus by Lemma 6.4, $v_1 \in C^\beta(\overline{I}; D(A)) \subset X$. Next, for $t \in \overline{I}, \tau > 0$ such that $t + \tau \leq T$,

$$v_2(t + \tau) - v_2(t) = \int_t^{t+\tau} E(s)g(t + \tau)ds + \int_0^t E(s)[g(t + \tau) - g(t)]ds.$$

Thus, by Theorem 6.4, we deduce

$$\|A^\gamma(v_2(t + \tau) - v_2(t))\|_{L^2(\Omega)} \leq c_{\alpha,\gamma}\|g\|_{C(\overline{I};L^2(\Omega))} \int_t^{t+\tau} s^{(1-\gamma)\alpha-1}ds$$

$$+ c_{\alpha,\gamma}\tau^\beta\|g\|_{C^\beta(\overline{I};L^2(\Omega))} \int_0^t s^{(1-\gamma)\alpha-1}ds.$$

Since $\int_t^{t+\tau} s^{(1-\gamma)\alpha-1}ds \leq ((1 - \gamma)\alpha)^{-1}\tau^{(1-\gamma)\alpha}$, we obtain

$$\tau^{-\beta}\||A^{\gamma}[v_2(t+\tau)-v_2(t)]\||_{L^2(\Omega)}$$

$$\leq((1-\gamma)\alpha)^{-1}c_{\alpha,\gamma}(\tau^{(1-\gamma)\alpha-\beta}+T^{(1-\gamma)\alpha})\|g\|_{C^{\beta}(\bar{I};L^2(\Omega))}.$$

Since $\beta < (1-\gamma)\alpha$, $v_2 \in C^{\beta}(\bar{I};L^2(\Omega))$. It remains to show $Av_2 \in C(\bar{I};L^2(\Omega))$. This follows from the identity $-Av_2(t) = -\int_0^t AE(t-s)g(t)ds = (F(t)-I)g(t)$, in view of Lemma 6.2. Since $F(t)-I$ is continuous on $L^2(\Omega)$, cf. Lemma 6.3, the desired assertion follows. This completes the proof of the lemma. $\qquad\square$

Lemma 6.15 *If $q \in L^2(\Omega)$, then $I - L(q)$ is boundedly invertible in X.*

Proof The proof proceeds by the argument of equivalent norm family, following [Bie56]: we equip X with an equivalent family of norms $\| \cdot \|_{\lambda}$, $\lambda \geq 0$, defined by

$$\|f\|_{\lambda} = \sup_{t \in \bar{I}} e^{-\lambda t}[\|f(t)\|_{L^2(\Omega)} + \|A^{\gamma}f(t)\|_{L^2(\Omega)}]$$

$$+ \sup_{0 \leq s < t \leq T} e^{-\lambda(t+1)}|t-s|^{-\beta}\|f(s)-f(t)\|_{D(A^{\gamma})} + \sup_{t \in \bar{I}} e^{-\lambda(t+2)}\|Af(t)\|_{L^2(\Omega)},$$

which is equivalent to the norm on X, and then prove the invertibility by choosing λ suitably. For $f \in X$, let $v = L(q)f$. Then by Sobolev embedding in Theorem A.3 and Theorem 6.4,

$$e^{-\lambda t}\|v(t)\|_{L^2(\Omega)} = e^{-\lambda t}\|\int_0^t E(t-s)qf(s)ds\|_{L^2(\Omega)}$$

$$\leq c\int_0^t e^{-\lambda(t-s)}\|E(t-s)\|\|q\|_{L^2(\Omega)}e^{-\lambda s}\|f(s)\|_{C(\bar{I};D(A^{\gamma}))}ds$$

$$\leq c\int_0^t e^{-\lambda s}s^{\alpha-1}ds\|q\|_{L^2(\Omega)}\|f\|_{\lambda} \leq c\lambda^{-\alpha}\|q\|_{L^2(\Omega)}\|f\|_{\lambda}.$$

Similarly,

$$e^{-\lambda t}\|A^{\gamma}v(t)\|_{L^2(\Omega)} \leq c\int_0^t e^{-\lambda(t-s)}\|A^{\gamma}E(t-s)\|\|q\|_{L^2(\Omega)}e^{-\lambda s}\|f(s)\|_{C(\bar{I};D(A^{\gamma}))}ds$$

$$\leq c\int_0^t e^{-\lambda s}s^{(1-\gamma)\alpha-1}ds\|q\|_{L^2(\Omega)}\|f\|_{\lambda} \leq c\lambda^{-(1-\gamma)\alpha}\|q\|_{L^2(\Omega)}\|f\|_{\lambda}.$$

Meanwhile, for $t \in [0,T)$ and $\tau > 0$ with $t + \tau \leq T$, we have

$$v(t+\tau) - v(t) = \int_0^t E(s)q[f(t+\tau-s) - f(t-s)]ds$$

$$+ \int_t^{t+\tau} E(s)qf(t+\tau-s)ds,$$

which directly implies

$$e^{-\lambda(t+\tau+1)} \|A^\gamma(v(t+\tau) - v(t))\|_{L^2(\Omega)}$$

$$\leq c\tau^\beta \|q\|_{L^2(\Omega)} \|f\|_\lambda \left(\int_0^t e^{-\lambda s} s^{(1-\gamma)\alpha-1} ds + \tau^{-\beta} \int_t^{t+\tau} e^{-\lambda(s+1)} s^{(1-\gamma)\alpha-1} ds \right).$$

This and the choice $\beta < (1-\gamma)\alpha$ give

$$e^{-\lambda(t+\tau+1)} \tau^{-\beta} \|A^\gamma(v(t+\tau) - v(t))\|_{L^2(\Omega)} \leq c\|q\|_{L^2(\Omega)} \|f\|_\lambda (\lambda^{-(1-\gamma)\alpha} + e^{-\lambda}).$$

In the same way, we deduce

$$\tau^{-\beta} e^{-\lambda(t+\tau+1)} \|v(t+\tau) - v(t)\|_{L^2(\Omega)} \leq c_T \|q\|_{L^2(\Omega)} \|f\|_\lambda (\lambda^{-\alpha} + e^{-\lambda}).$$

Combining the preceding two estimates gives

$$\sup_{0 \leq s < t \leq T} e^{-\lambda(t+1)} |t-s|^{-\beta} \|v(t) - v(s)\|_{D(A^\gamma)} \leq c_T \|q\|_{L^2(\Omega)} \|f\|_\lambda (\lambda^{-(1-\gamma)\alpha} + e^{-\lambda}).$$

Next, in view of Lemmas 6.2 and 6.3, we deduce

$$-Av(t) = \int_0^t -AE(t-s)q[f(s) - f(t)] + F(t)qf(t) - qf(t).$$

Then the preceding argument and Theorem 6.4 lead to

$$e^{-\lambda(t+2)} \|Av(t)\|_{L^2(\Omega)} \leq ce^{-\lambda(t+2)} \|q\|_{L^2(\Omega)} \|f(t)\|_{D(A^\gamma)}$$

$$+ c \int_0^t e^{-\lambda} \|AE(t-s)\| \|q\|_{L^2(\Omega)} e^{-\lambda(t+1)} \|f(s) - f(t)\|_{D(A^\gamma)} ds$$

$$\leq c\|q\|_{L^2(\Omega)} \|f\|_\lambda e^{-\lambda} \left(\int_0^t (t-s)^{\beta-1} ds + 1 \right) \leq c_T \|q\|_{L^2(\Omega)} \|f\|_\lambda e^{-\lambda}.$$

Combining the preceding estimates implies

$$\|L(q)f\|_\lambda \leq c_T (e^{-\lambda} + \lambda^{-(1-\gamma)\alpha}) \|q\|_{L^2(\Omega)} \|f\|_\lambda.$$

It follows directly from this estimate that $\|L(q)f\|_\lambda$ tends to zero as $\lambda \to \infty$, and thus the operator norm $\|L(q)\|_\lambda < 1$ if λ is large enough, which shows the lemma. $\square$

Now we can state the unique solvability of the Cauchy problem (6.104).

Proposition 6.7 *If $q \in L^2(\Omega)$ and $u_0 \in D(A^{1+\gamma})$. Then problem (6.104) has a unique classical solution $u(q) = (I - L(q))^{-1}U$.*

Proof Since $u_0 \in D(A^{1+\gamma})$, $U \in X$. Thus, by Lemma 6.15, problem (6.104) has a unique solution $u = (I - L(q))^{-1}U \in X$. Further, u is the classical solution to problem (6.104), since $qu \in C^\beta(\overline{I}; L^2(\Omega))$ and $u_0 \in D(A)$. $\square$

We denote by λ_1 the smallest eigenvalue of A, and $\bar{\varphi}_1$ the corresponding nonnegative eigenfunction, normalized by $\|\bar{\varphi}_1\|_{L^\infty(\Omega)} = 1$. Further, let

$$c_\alpha = \sup_{t \geq 0} t E_{\alpha,\alpha}(-t), \qquad (6.106)$$

Proposition 6.8 gives an upper bound on c_α, which implies $c_\alpha < \alpha$. Then the following stability estimate holds: for small time T, the inverse problem is locally Lipschitz stable. The proof of the theorem employs the implicit function theorem, and certain estimates on the solution operators with sharp constants.

Theorem 6.32 *Let* $\frac{3}{4} < \gamma < 1, 0 < \epsilon < 1 - \frac{c_\alpha}{\alpha}, \mu_0, \mu_1 > 0$ *such that* $1 \leq \frac{\mu_1}{\mu_0} < \frac{(1-\epsilon)\alpha}{c_\alpha}$. *Let* $u_0 \in D(A^{1+\gamma})$, *with*

$$\mu_0 \lambda_1 \bar{\varphi}_1 \leq -\Delta u_0 \leq \mu_1 \lambda_1 \bar{\varphi}_1 \qquad (6.107)$$

and set $\omega = \{x \in \Omega : \bar{\varphi}_1(x) \geq 1 - \epsilon\}$. *Then there exists a constant* $\theta > 0$ *depending only on* $\frac{\mu_1}{(1-\epsilon)\mu_0}$ *and* α *such that if* $\lambda_1 T^\alpha < \theta$, *then there is* V, *a neighborhood of* 0 *in* $L^2(\omega)$ *and a constant* c *such that*

$$\|q_1 - q_2\|_{L^2(\omega)} \leq c\|u(q_1)(T) - u(q_2)(T)\|_{D(A)}, \quad \forall q_1, q_2 \in V.$$

Remark 6.8 The regularity condition $u_0 \in D(A^{1+\gamma})$ is to ensure the well-posedness of the direct problem with $q \in L^2(\Omega)$. The condition (6.107) is to ensure pointwise lower and upper bounds on the solution $u(0)(T)$, and the set of u_0 satisfying (6.107) is a convex subset of $D(A^{1+\gamma})$. The condition $\lambda_1 T^\alpha < \theta$ dictates that either T or λ_1 should be sufficiently small, the latter of which holds if the domain Ω is large, since λ_1 tends to zero as the volume of Ω tends to infinity.

Below we assume that u_0 satisfies the condition of Theorem 6.32. In view of the Sobolev embedding $D(A^{1+\gamma}) \hookrightarrow C^{2,\delta}(\overline{\Omega})$ for some $\delta > 0$, the function $U(t) \in C^{2+\delta, \frac{2+\delta}{2}\alpha}(\overline{Q})$ (see Theorem 7.9 in Chapter 7), and satisfies

$$\begin{cases} \partial_t^\alpha U = \Delta U, & \text{in } Q, \\ U(0) = u_0, & \text{in } \Omega, \\ U = 0, & \text{on } \partial_L Q. \end{cases}$$

The next result collects several properties of the function $U(t)$.

Lemma 6.16 *The following properties hold on the function* $U(t)$.

(i) $\mu_0 \bar{\varphi}_1(x) \leq u_0(x) \leq \mu_1 \bar{\varphi}_1(x)$, $x \in \overline{\Omega}$.
(ii) $\mu_0 E_{\alpha,1}(-\lambda_1 t^\alpha)\bar{\varphi}_1(x) \leq U(x,t) \leq \mu_1 E_{\alpha,1}(-\lambda_1 t^\alpha)\bar{\varphi}_1(x)$, $(x,t) \in \overline{\Omega} \times [0,T]$.
(iii) $0 \leq -\partial_t^\alpha U(x,t) \leq \|\Delta u_0\|_{L^\infty(\Omega)} \leq \mu_1 \lambda_1$, $(x,t) \in \Omega \times [0,T]$.

Proof Part (i). Since $-\Delta \bar{\varphi}_1 = \lambda_1 \bar{\varphi}_1$, we obtain $\Delta(\mu_1 \bar{\varphi}_1 - u_0) \leq 0$. But $\mu_1 \varphi_1 - u_0$ is zero on the boundary $\partial \Omega$. Then $\mu_1 \bar{\varphi}_1 - u_0$ is nonnegative, by the elliptic maximum principle. The first inequality of (i) can be obtained similarly. Part (ii) follows from the maximum principle in Section 6.5. We only prove (iii). Let $w(x,t) = \partial_t^\alpha U(x,t)$. Then w satisfies

$$\begin{cases} \partial_t^\alpha w = \Delta w, & \text{in } \Omega \times (0,T], \\ w(0) = \Delta u_0, & \text{in } \Omega, \\ w = 0, & \text{on } \partial \Omega \times [0,T]. \end{cases}$$

By assumption, $\Delta u_0 \leq 0$ in Ω, and thus by the maximum principle, $0 \leq -w(x,t) \leq \|\Delta u_0\|_{L^\infty(\Omega)}$. This implies assertion (iii). $\square$

Let ω be defined as in Theorem 6.32. Lemma 6.16(ii) implies that $\frac{1}{U(T)|_\omega}$ extended by zero outside ω belongs to $L^\infty(\Omega)$. We define an operator $P_T : L^2(\omega) \to L^2(\omega)$ by

$$q \mapsto \int_0^T -AE(T-s)U(T)^{-1}|_\omega [U(s) - U(T)]q\, ds + F(T)q.$$

This operator arises from the linearization of the forward map.

The next result gives an upper bound on the constant c_α defined in (6.106). In particular, it indicates that $c_\alpha < \alpha < 1$, which is crucial for proving Theorem 6.32.

Proposition 6.8 *For any* $\alpha \in (0,1)$, $c_\alpha := \sup_{t \geq 0} t E_{\alpha,\alpha}(-t) \leq \frac{\alpha^2 \pi}{\sin(\alpha\pi) + \alpha\pi}$.

Proof Let $f(t) = t E_{\alpha,\alpha}(-t)$. By the asymptotics and complete monotonicity of $E_{\alpha,\alpha}(-t)$, $f(t)$ is nonnegative on $\mathbb{R}_+$, and tends to zero as $t \to \infty$. Thus, there exists a maximum. Let $w(t) = t^\alpha E_{\alpha,\alpha}(-t^\alpha)$. By the Cristoffel-Darboux-type formula in Exercise 3.12 and a limiting argument,

$$\int_0^t s^{\alpha-1} E_{\alpha,\alpha}(-s^\alpha) E_{\alpha,1}(-(t-s)^\alpha) ds = \frac{d}{d\lambda} \lambda t^\alpha E_{\alpha,\alpha+1}(\lambda t^\alpha)|_{\lambda=-1},$$

which upon simplification gives directly the formula

$$w(t) = \int_0^t (t-s)^{\alpha-1} E_{\alpha,\alpha}(-(t-s)^\alpha) \alpha E_{\alpha,1}(-s^\alpha) ds. \tag{6.108}$$

Since $0 \leq E_{\alpha,1}(-t) \leq 1$ and $E_{\alpha,\alpha}(-t) \geq 0$, cf. Theorem 3.5, we deduce

$$w(t) \leq \alpha \int_0^t (t-s)^{\alpha-1} E_{\alpha,\alpha}(-(t-s)^\alpha) ds = \alpha(1 - E_{\alpha,1}(-t^\alpha)) < \alpha.$$

Meanwhile, by Theorem 3.6, there holds

$$w(t) = \int_0^t (t-s)^{\alpha-1} E_{\alpha,\alpha}(-(t-s)^\alpha) \alpha E_{\alpha,1}(-s^\alpha) ds$$

$$\leq \alpha\Gamma(\alpha+1) \int_0^t (t-s)^{\alpha-1} E_{\alpha,\alpha}(-(t-s)^\alpha) s^{-\alpha} ds$$

$$= \alpha\Gamma(\alpha+1) \int_0^t \sum_{k=0}^\infty \frac{(-1)^k}{\Gamma(k\alpha+\alpha)} (t-s)^{k\alpha+\alpha-1} s^{-\alpha} ds.$$

Using the identity $\int_0^t (t-s)^{a-1} s^{b-1} ds = B(a,b) t^{a+b-1}$ for $a, b > 0$, we deduce

$$w(t) \leq \alpha\Gamma(\alpha + 1) \sum_{k=0}^{\infty} \frac{(-1)^k}{\Gamma(k\alpha + \alpha)} \frac{\Gamma(k\alpha + \alpha)\Gamma(1 - \alpha)}{\Gamma(k\alpha + 1)} t^{k\alpha}$$

$$\leq \alpha\Gamma(\alpha + 1)\Gamma(1 - \alpha) \sum_{k=0}^{\infty} \frac{(-1)^k}{\Gamma(k\alpha + 1)} t^{k\alpha} = \alpha\Gamma(1 - \alpha)\Gamma(1 + \alpha)E_{\alpha,1}(-t^{\alpha}).$$

Now by the recursion identity (2.2) and reflection identity (2.3) for $\Gamma(z)$, $\Gamma(1 - \alpha)\Gamma(1 + \alpha) = \alpha\Gamma(1 - \alpha)\Gamma(\alpha) = \frac{\alpha\pi}{\sin(\alpha\pi)}$. Combining these estimates leads to

$$\sup_{t\geq 0} t E_{\alpha,\alpha}(-t) \leq \alpha \max_{t\geq 0} \min\left(\frac{\alpha\pi}{\sin(\alpha\pi)} E_{\alpha,1}(-t^{\alpha}), 1 - E_{\alpha,1}(-t^{\alpha}) \right). \tag{6.109}$$

By the complete monotonicity of $E_{\alpha,1}(-t)$, $\frac{\alpha\pi}{\sin(\alpha\pi)} E_{\alpha,1}(-t^{\alpha})$ is monotonically decreasing, whereas $1 - E_{\alpha,1}(-t^{\alpha})$ is monotonically increasing. Thus, one simple upper bound is obtained by equating these two terms, which directly gives $E_{\alpha,1}(-t_*^{\alpha}) = \frac{\sin(\alpha\pi)}{\sin(\alpha\pi)+\alpha\pi}$. Upon substituting it back to (6.109) and noting the complete monotonicity of $E_{\alpha,1}(-t)$, we complete the proof.   □

Remark 6.9 Note that the identities

$$\lim_{\alpha\to 0^+} \frac{\alpha\pi}{\alpha\pi + \sin\alpha\pi} = \frac{1}{2} \quad \text{and} \quad \lim_{\alpha\to 1^-} \frac{\alpha\pi}{\alpha\pi + \sin\alpha\pi} = 1,$$

and the function $f(\alpha) = \frac{\alpha\pi}{\alpha\pi+\sin\alpha\pi}$ is strictly increasing in α over the interval $(0, 1)$. Thus, the factor is strictly less than 1 for any $\alpha \in (0, 1)$. Note also that for the limiting case $\alpha = 1$, the constant $c_1 = \sup_{t\geq 0} te^{-t} = e^{-1}$, which is much sharper than the preceding bound. Since the function $E_{\alpha,\alpha}(-t)$ is actually continuous in α, one may refine the bound on c_α slightly for α close to unit.

Remark 6.10 Proposition 6.8 provides an upper bound on c_α. In Fig. 6.4(a), we plot the function $\alpha^{-1} t E_{\alpha,\alpha}(-t)$ versus t. Clearly, for any fixed α, $t E_{\alpha,\alpha}(-t)$ first increases with t and then decreases, and there is only one global maximum. The maximum is always achieved at some $t^* \in [0.8, 1]$, a fact that remains to be proved, and the maximum value decreases with α. The ratio $\frac{c_\alpha}{\alpha}$ versus the upper bound $\frac{\alpha\pi}{\sin\alpha\pi+\alpha\pi}$ is shown in Fig. 6.4(b). Note that $\frac{c_\alpha}{\alpha}$ is strictly increasing with respect to α, and the upper bound in Proposition 6.8 is about three times larger than the optimal one $\frac{c_\alpha}{\alpha}$, since the derivation employs upper bounds of $E_{\alpha,1}(-t)$ that are valid on $\mathbb{R}_+$, instead of sharper ones on a finite interval, e.g., $[0, 1]$.

The next result gives the invertibility of the operator $I - P_T$ on $L^2(\omega)$.

Lemma 6.17 *Under the assumptions of Theorem 6.32, there exists a $\theta > 0$ depending only on α and ϵ such that if $\lambda_1 T^{\alpha} < \theta$, then the operator $I - P_T$ has a bounded inverse in $B(L^2(\omega))$.*

Proof First, we bound $\|tAE(t)\|$. Using the eigenpairs $\{(\lambda_j, \varphi_j)\}_{j=1}^{\infty}$ of the operator A, we deduce for $v \in L^2(\Omega)$, $E(t)v = \sum_{j=1}^{\infty} t^{\alpha-1} E_{\alpha,\alpha}(-\lambda_j t^{\alpha})(\varphi_j, v)\varphi_j$. Thus,

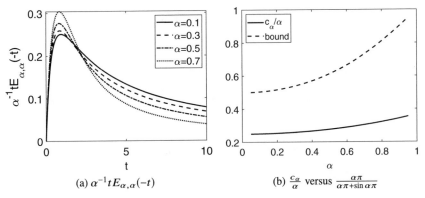

(a) $\alpha^{-1}tE_{\alpha,\alpha}(-t)$ (b) $\frac{c_\alpha}{\alpha}$ versus $\frac{\alpha\pi}{\alpha\pi+\sin\alpha\pi}$

Fig. 6.4 The function $\alpha^{-1}tE_{\alpha,\alpha}(-t)$ and its maximum $\frac{c_\alpha}{\alpha}$ versus the upper bound $\frac{\alpha\pi}{\alpha\pi+\sin\alpha\pi}$ in Proposition 6.8.

$$\|tAE(t)v\|^2 = \sum_{j=1}^{\infty}(\lambda_j t^\alpha E_{\alpha,\alpha}(-\lambda_j t^\alpha))^2(v,\varphi_j)^2.$$

By Proposition 6.8, $\|AE(t)\| \le c_\alpha t^{-1}$. Meanwhile, using the governing equation for $U(t)$, we have $U(t) - U(0) = {}_0I_t^\alpha \Delta U(t)$. This and the fact $\Delta U(x,t) \le 0$ imply

$$U(t) - U(T) = ({}_0I_t^\alpha \Delta U)(t) - ({}_0I_t^\alpha \Delta U)(T)$$

$$= \frac{1}{\Gamma(\alpha)} \int_0^t [(T-s)^{\alpha-1} - (t-s)^{\alpha-1}](-\Delta U(s))ds$$

$$+ \frac{1}{\Gamma(\alpha)} \int_t^T (T-s)^{\alpha-1}(-\Delta U(s))ds.$$

Since $(T-s)^{\alpha-1} - (t-s)^{\alpha-1} \le 0$ and $-\Delta U(x,t) \ge 0$ in $\Omega\times[0,T]$, by Lemma 6.16(iii)

$$U(t) - U(T) \le \frac{1}{\Gamma(\alpha)} \int_t^T (T-s)^{\alpha-1}(-\Delta U(s))ds \le \frac{(T-t)^\alpha}{\Gamma(\alpha+1)}\mu_1\lambda_1.$$

Similarly,

$$U(T) - U(t) \le \frac{1}{\Gamma(\alpha)} \int_0^t [(t-s)^{\alpha-1} - (T-s)^{\alpha-1}](-\Delta U(s))ds$$

$$\le \mu_1\lambda_1\Gamma(\alpha+1)^{-1}(t^\alpha + (T-t)^\alpha - T^\alpha) \le (T-t)^\alpha \Gamma(\alpha+1)^{-1}\mu_1\lambda_1.$$

Consequently, there holds

$$\|U(s) - U(T)\|_{L^\infty(\Omega)} \le \Gamma(\alpha+1)^{-1}\mu_1\lambda_1(T-s)^\alpha.$$

Lemma 6.16(ii) implies

$$\|U(T)^{-1}|_\omega\|_{L^\infty(\Omega)} \le (\mu_0(1-\epsilon)E_{\alpha,1}(-\lambda_1 T^\alpha))^{-1}.$$

The preceding two estimates and Theorem 6.4(iv) imply

$$\|P_T\|_{B(L^2(\omega))}$$

$$\le \int_0^T \|AE(T-s)\|\|U(T)^{-1}|_\omega\|_{L^\infty(\Omega)}\|U(s)-U(T)\|_{L^\infty(\Omega)}ds + \|F(T)\|$$

$$\le \int_0^T c_\alpha(T-s)^{-1}\frac{\lambda_1\mu_1}{\Gamma(\alpha+1)}(T-s)^\alpha\frac{1}{\mu_0(1-\epsilon)E_{\alpha,1}(-\lambda_1 T^\alpha)}ds + E_{\alpha,1}(-\lambda_1 T^\alpha)$$

$$= \frac{c_\alpha\mu_1}{\mu_0(1-\epsilon)\alpha\Gamma(\alpha+1)E_{\alpha,1}(-\lambda_1 T^\alpha)}\lambda_1 T^\alpha + E_{\alpha,1}(-\lambda_1 T^\alpha).$$

Let $m(x)$ be defined by

$$m(x) = \frac{c_\alpha\mu_1}{\mu_0(1-\epsilon)\alpha\Gamma(\alpha+1)}\frac{x}{E_{\alpha,1}(-x)} + E_{\alpha,1}(-x).$$

Straightforward computation shows

$$m'(x) = \frac{c_\alpha\mu_1}{\mu_0(1-\epsilon)\alpha\Gamma(\alpha+1)}\frac{E_{\alpha,1}(-x) - xE'_{\alpha,1}(-x)}{E_{\alpha,1}(-x)^2} + E'_{\alpha,1}(-x).$$

Thus, $m(0) = 1$ and by Proposition 6.8,

$$m'(0) = \frac{c_\alpha\mu_1}{\mu_0(1-\epsilon)\alpha\Gamma(\alpha+1)} - \frac{1}{\Gamma(\alpha+1)} = \left[\frac{c_\alpha\mu_1}{\mu_0(1-\epsilon)\alpha} - 1\right]\frac{1}{\Gamma(\alpha+1)} < 0,$$

under the given conditions on ϵ, μ_0 and μ_1 in Theorem 6.32. Thus, there exists a $\theta > 0$ such that whenever $x < \theta$, $m(x) < 1$, and accordingly, for $\lambda_1 T^\alpha$ sufficiently close to zero, P_T is a contraction on $L^2(\omega)$. Then by Neumann series expansion, $I - P_T$ is invertible and $(I - P_T)^{-1}$ is bounded. $\square$

Now we define a trace operator tr : $X \to D(A)$, $v \mapsto v(T)$. Then tr $\in B(X, D(A))$ and $\|\text{tr}\|_{B(X,D(A))} \le 1$. Finally, we can present the proof of Theorem 6.32.

Proof We define the mapping:

$$K : L^2(\omega) \to L^2(\omega), \quad q \mapsto [-Au(q)(T)]|_\omega = [-A\text{tr}(I - L(q))^{-1}U]|_\omega.$$

Clearly, K is continuously Fréchet differentiable, cf. Lemma 6.15, and its derivative K' at $q \in L^2(\omega)$ in the direction p is given by

$$K'(q)[p] = [-A\text{tr}(I - L(q))^{-1}L(p)(I - L(q))^{-1}U]|_\omega.$$

Let $Q_T = K'(0) = [-A\text{tr}L(\cdot)U]|_\omega$. Then

$$Q_T(p) = \left[\int_0^T -AE(T-s)p[U(s)-U(T)]ds + (F(T) - I)pU(T)\right]\Big|_\omega.$$

We define a multiplication operator $M : L^2(\omega) \rightarrow L^2(\omega)$, $p \mapsto U(T)p$. Then M is invertible, and its inverse is exactly the multiplication operator by $\frac{1}{U(T)}|_\omega$. Consequently, $Q_T M^{-1} = P_T - I$. By Lemma 6.17, $(P_T - I)^{-1}$ belongs to $B(L^2(\omega))$. Therefore, Q_T has a bounded inverse and $Q_T^{-1} = M^{-1}(P_T - I)^{-1}$. By the implicit function theorem, K is locally a C^1-diffeomorphism from a neighborhood of 0 onto a neighborhood of $K(0)$. In particular, K^{-1} is Lipschitz continuous in a neighborhood of $K(0)$. Then Theorem 6.32 follows by noting the following inequality

$$\|Au(q_1)(T)|_\omega - Au(q_2)(T)|_\omega\|_{L^2(\omega)} \leq \|u(q_1)(T) - u(q_2)(T)\|_{D(A)}, \forall q_1, q_2 \in L^2(\omega).$$

6.7 Numerical Methods

Now we describe some numerical methods for subdiffusion models. One outstanding challenge lies in the accurate and efficient discretization of the Djrbashian-Caputo fractional derivative $\partial_t^\alpha u$. Roughly speaking, there are two predominant classes of numerical methods for time stepping, i.e., convolution quadrature (CQ) and finite difference-type methods, e.g., L1 scheme and L1-2 scheme. The former relies on approximating the (Riemann-Liouville) fractional derivative in the Laplace domain, whereas the latter approximates $\partial_t^\alpha u$ directly by piecewise polynomials. These two approaches have their pros and cons: CQ is often easier to analyze, since by construction, it inherits excellent numerical stability of the underlying schemes for ODEs, but it is restricted to uniform grids. Finite difference-type methods are very flexible in construction and implementation and generalize to nonuniform grids, but often challenging to analyze. Generally, these schemes are only low-order, unless restrictive compatibility conditions are fulfilled. One promising idea is to employ suitable corrections to restore the desired high-order convergence.

In this section, we review these two popular classes of time-stepping schemes on uniform grids, following [JLZ19a], where many further references and technical details can be found. Specifically, let $\{t_n = n\tau\}_{n=0}^N$ be a uniform partition of the time interval $[0, T]$, with a time step size $\tau = \frac{T}{N}$.

6.7.1 Convolution quadrature

Convolution quadrature (CQ) was systematically developed by Christian Lubich in a series of pioneering works [Lub86, Lub88, LST96, CLP06] first for fractional integrals, and then for parabolic equations with memory and fractional diffusion wave equations. It has been widely applied in discretizing the Riemann-Liouville fractional derivative. It requires only that Laplace transform of the kernel be known. Specifically, CQ approximates the Riemann-Liouville fractional derivative $R\partial_t^\alpha \varphi(t) = \frac{d}{dt} \frac{1}{\Gamma(1-\alpha)} \int_0^t (t-s)^{-\alpha} \varphi(s) ds$ (with $\varphi(0) = 0$) by a discrete convolution (with the shorthand notation $\varphi^n = \varphi(t_n)$)

$$\bar{\partial}_\tau^\alpha \varphi^n := \frac{1}{\tau^\alpha} \sum_{j=0}^{n} b_j \varphi^{n-j}. \tag{6.110}$$

The weights $\{b_j\}_{j=0}^{\infty}$ are the coefficients in the power series expansion

$$\delta_\tau(\zeta)^\alpha = \frac{1}{\tau^\alpha} \sum_{j=0}^{\infty} b_j \zeta^j, \tag{6.111}$$

where $\delta_\tau(\zeta)$ is the characteristic polynomial of a linear multistep method for ODEs. There are several possible choices of the characteristic polynomial, e.g., backward differentiation formula (BDF), trapezoidal rule, Newton-Gregory method and Runge-Kutta methods. The most popular one is the backward differentiation formula of order k (BDFk), $k = 1, \ldots, 6$, for which $\delta_\tau(\zeta)$ is given by

$$\delta_\tau(\zeta) := \frac{1}{\tau} \sum_{j=1}^{k} \frac{1}{j}(1 - \zeta)^j, \quad j = 1, 2, \ldots.$$

The case $k = 1$, i.e., backward Euler CQ, is also known as Grünwald-Letnikov approximation, cf. Section 2.3.3. Then the weights b_j are given explicitly by

$$b_0 = 1 \quad \text{and} \quad b_j = -j^{-1}(\alpha - j + 1)b_{j-1}, \quad j \geq 1.$$

The CQ discretization of the subdiffusion model first reformulates it using $^R\partial_t^\alpha \varphi$, using the defining relation for the (regularized) Djrbashian-Caputo fractional derivative $\partial_t^\alpha \varphi(t) = {}^R\partial_t^\alpha(\varphi - \varphi(0))$ in (2.37), into the form

$$^R\partial_t^\alpha(u - u_0) - \Delta u = f.$$

Then the time stepping scheme based on CQ is to find approximations U^n to the exact solution $u(t_n)$ by

$$\bar{\partial}_\tau^\alpha(U - u_0)^n - \Delta U^n = f(t_n), \quad n = 1, \ldots, N, \tag{6.112}$$

with $U^0 = u_0$. When combined with spatially semidiscrete schemes, e.g., Galerkin finite element methods, we arrive at fully discrete schemes. We discuss only the temporal error for time-stepping schemes, and omit the spatial errors. The backward Euler CQ has the first-order accuracy [JLZ16c, Theorems 3.5 and 3.6].

Theorem 6.33 Let $u_0 \in \dot{H}^\gamma(\Omega)$, $\gamma \in [0, 2]$, and $f \in C(\bar{I}; L^2(\Omega))$ with $\int_0^t (t - s)^{\alpha-1} \|f'(s)\|_{L^2(\Omega)} ds < \infty$ for any $t \in \bar{I}$. Let U^n be the solutions of the scheme (6.112). Then there holds

$$\|u(t_n) - U^n\|_{L^2(\Omega)} \leq c\tau\left(t^{\frac{\gamma}{2}\alpha-1}\|u_0\|_{\dot{H}^\gamma(\Omega)} + t_n^{\alpha-1}\|f(0)\|_{L^2(\Omega)}\right.$$
$$\left. + \int_0^{t_n}(t_n - s)^{\alpha-1}\|f'(s)\|_{L^2(\Omega)}ds\right).$$

If the exact solution u is smooth and has a sufficient number of vanishing derivatives at $t = 0$, then the approximation U^n converges at a rate of $O(\tau^k)$ uniformly in time t for BDFk convolution quadrature [Lub88, Theorem 3.1]. However, in practice, it generally only exhibits a first-order accuracy when solving subdiffusion problems even for smooth u_0 and f [CLP06, JLZ16c], since the requisite compatibility condition is not satisfied. This loss of accuracy is one distinct feature for most time-stepping schemes since they are usually derived under the assumption that u is smooth in time, which, according to the regularity theory in Section 6.2, holds only if the problem data satisfy certain compatibility conditions. In summary, they tend to lack robustness with respect to the regularity of problem data.

One promising idea is initial correction. To restore the second-order accuracy for BDF2 CQ, one may correct the first step of the scheme and obtain

$$\begin{cases} \bar{\partial}_\tau^\alpha(U - u_0)^1 - \Delta U^1 = \frac{1}{2}(\Delta u_0 + f(0)) + f(t_1), \\ \bar{\partial}_\tau^\alpha(U - u_0)^n - \Delta U^n = f(t_n), \quad 2 \leq n \leq N. \end{cases} \tag{6.113}$$

When compared with the vanilla CQ scheme (6.112), the additional terms $\frac{1}{2}(\Delta u_0 + f(0))$ at the first step are constructed so as to improve the overall accuracy of the scheme to $O(\tau^2)$ for an initial data $v \in D(\Delta)$ and a possibly incompatible right-hand side f. The difference between the corrected scheme (6.113) and the standard scheme (6.112) lies in the first step, and hence it is very easy to implement. The scheme (6.113) satisfies the following error estimate [JLZ16c, Theorems 3.8 and 3.9].

Theorem 6.34 Let $u_0 \in \dot{H}^\gamma(\Omega)$, $\gamma \in [0,2]$, $f \in C^1(\bar{I}; L^2(\Omega))$ and $\int_0^t(t - s)^{\alpha-1}\|f^{(2)}(s)\|_{L^2(\Omega)}ds < \infty$ for any $t \in \bar{I}$. Then for the solutions U^n to the scheme (6.113), there holds for any $t_n > 0$.

$$\|U^n - u(t_n)\|_{L^2(\Omega)} \leq c\tau^2\left(t_n^{\frac{\gamma}{2}\alpha-2}\|u_0\|_{\dot{H}^\gamma(\Omega)} + t_n^{\alpha-2}\|f(0)\|_{L^2(\Omega)} + t_n^{\alpha-1}\|f'(0)\|_{L^2(\Omega)}\right.$$
$$\left. + \int_0^{t_n}(t_n - s)^{\alpha-1}\|f^{(2)}(s)\|_{L^2(\Omega)}ds\right).$$

6.7.2 Piecewise polynomial interpolation

Now we describe time stepping schemes based on piecewise polynomial approximation, and are essentially of finite difference nature. The idea has been very popular [SW06, LX07, Ali15], and the most prominent one is the L1 scheme due to Lin and Xu [LX07]. It employs piecewise linear interpolation, and hence the name L1 scheme. First, we derive the approach from Taylor expansion [LX07, Section 3]. Recall the following Taylor expansion formula with an integral remainder:

$$f(t) = f(s) + f'(s)(t - s) + \int_s^t f''(\xi)(t - \xi)d\xi \quad \forall t, s \in I.$$

Applying the identity to the function $u(t)$ at $t = t_j$ and $t = t_{j+1}$, respectively, gives

$$u(t_j) = u(s) + u'(s)(t_j - s) + \int_s^{t_j} u''(\xi)(t_j - \xi)d\xi,$$

$$u(t_{j+1}) = u(s) + u'(s)(t_{j+1} - s) + \int_s^{t_{j+1}} u''(\xi)(t_{j+1} - \xi)d\xi.$$

Subtracting the second identity from the first and dividing both sides by τ yields

$$u'(s) = \frac{u(t_{j+1}) - u(t_j)}{\tau} - \tau^{-1}\int_s^{t_{j+1}} u''(\xi)(t_{j+1} - \xi)d\xi + \tau^{-1}\int_s^{t_j} u''(\xi)(t_j - \xi)d\xi.$$

Thus, for all $0 \le n \le N - 1$, we have

$$\partial_t^\alpha u(t_n) = \frac{1}{\Gamma(1 - \alpha)} \sum_{j=0}^{n-1} \int_{t_j}^{t_{j+1}} u'(s)(t_n - s)^{-\alpha}ds$$

$$= \frac{1}{\Gamma(1 - \alpha)} \sum_{j=0}^{n-1} \frac{u(t_{j+1}) - u(t_j)}{\tau} \int_{t_j}^{t_{j+1}} (t_n - s)^{-\alpha}ds + r_\tau^n$$

$$= \sum_{j=0}^{n-1} b_j \frac{u(t_{n-j}) - u(t_{n-1-j})}{\tau^\alpha} + r_\tau^n$$

$$= \underbrace{\tau^{-\alpha}\left[b_0 u(t_n) - b_{n-1}u(t_0) + \sum_{j=1}^{n-1}(b_j - b_{j-1})u(t_{n-j})\right]}_{=:L_1^n(u)} + r_\tau^n,$$

where the weights b_j are give by $b_j = \Gamma(2-\alpha)^{-1}((j+1)^{1-\alpha} - j^{1-\alpha})$, $j = 0, 1, \ldots, N-1$ and r_τ^n is the local truncation error defined by

$$r_\tau^n = \frac{1}{\tau\Gamma(1 - \alpha)} \sum_{j=0}^{n-1} \int_{t_j}^{t_{j+1}} \left[-\int_s^{t_{j+1}} u''(\xi)\frac{t_{j+1} - \xi}{(t_n - s)^\alpha}d\xi + \int_s^{t_j} u''(\xi)\frac{t_j - \xi}{(t_n - s)^\alpha}d\xi\right]ds.$$

In essence, the scheme approximates the function u by a continuous piecewise linear interpolation, in a manner similar to the backward Euler method. It can be viewed as a fractional analogue of the latter. It was shown in [LX07, (3.3)] that the local truncation error of the approximation is of order $O(\tau^{2-\alpha})$; see also [JLLZ15].

Lemma 6.18 *There exists some constant c independent of τ such that*

$$|r_\tau^n| \le c \max_{0 \le t \le t_n} |u''(t)| \tau^{2-\alpha}.$$

Proof Changing the integration order gives

$$r_\tau^n = \frac{1}{\tau \Gamma(1-\alpha)} \sum_{j=0}^{n-1} \int_{t_j}^{t_{j+1}} u''(\xi) \Bigg[-(t_{j+1} - \xi) \int_{t_j}^{\xi} \frac{ds}{(t_n - s)^\alpha}$$

$$+ (t_j - \xi) \int_{\xi}^{t_{j+1}} \frac{ds}{(t_n - s)^\alpha} \Bigg] d\xi = \frac{1}{\tau \Gamma(2-\alpha)} \sum_{j=0}^{n-1} \int_{t_j}^{t_{j+1}} u''(\xi) R_j^n(\xi) d\xi,$$

where the auxiliary function $R_j^n(\xi)$ is defined by

$$R_j^n(\xi) = (t_n - \xi)^{1-\alpha} \tau - (t_{j+1} - \xi)(t_n - t_j)^{1-\alpha} + (t_j - \xi)(t_n - t_{j+1})^{1-\alpha}.$$

We claim $R_j^n(\xi)$ is nonnegative for all $\xi \in [t_j, t_{j+1}]$. Indeed, we have $R_j^n(t_j) = R_j^n(t_{j+1}) = 0$, and $\frac{d^2}{d\xi^2} R_j^n(\xi) = (1-\alpha)(-\alpha)(t_n - \xi)^{-1-\alpha} \tau \le 0$, for $0 < \alpha < 1$. That is, $R_j^n(\xi)$ is a concave function. The concavity implies the desired nonnegativity $R_j^n(\xi) \ge 0$ for all $r \in [t_j, t_{j+1}]$. Thus, with $c_u = \max_{0 \le \xi \le t_n} |u''(\xi)|$, there holds

$$r_\tau^n \le \frac{c_u}{\tau \Gamma(2-\alpha)} \sum_{j=0}^{n-1} \int_{t_j}^{t_{j+1}} R_j^n(\xi) d\xi.$$

It suffices to bound the integrals on the right-hand side. Since

$$\int_{t_j}^{t_{j+1}} R_j^n(\xi) d\xi = \frac{\tau^{3-\alpha}}{2(2-\alpha)} \left[2(n-j)^{2-\alpha} - 2(n-1-j)^{2-\alpha} \right.$$

$$\left. -(2-\alpha)[(n-j)^{1-\alpha} + (n-1-j)^{1-\alpha}] \right],$$

by a simple change of variables

$$\sum_{j=0}^{n-1} \int_{t_j}^{t_{j+1}} R_j^n(\xi) d\xi = \frac{\tau^{3-\alpha}}{2} \sum_{j=0}^{n-1} s_j,$$

with

$$s_j = \frac{2}{2-\alpha} \left[(j+1)^{2-\alpha} - j^{2-\alpha} \right] - \left[(j+1)^{1-\alpha} + j^{1-\alpha} \right]$$

$$= j^{1-\alpha} \left(\frac{2j}{2-\alpha} ((1 + j^{-1})^{2-\alpha} - 1) - (1 + j^{-1})^{1-\alpha} - 1 \right).$$

Next, we claim that the sum $\sum_{j=0}^{n-1} s_j$ is uniformly bounded independent of n, which completes the proof. Indeed, by binomial expansion, the following expansions hold

$$(1 + j^{-1})^{2-\alpha} = 1 + (2 - \alpha)j^{-1} + \frac{(2 - \alpha)(1 - \alpha)}{2!}j^{-2} + \frac{(2 - \alpha)(1 - \alpha)(-\alpha)}{3!}j^{-3}$$
$$+ \frac{(2 - \alpha)(1 - \alpha)(-\alpha)(-\alpha - 1)}{4!}j^{-4} + \ldots,$$
$$(1 + j^{-1})^{1-\alpha} = 1 + (1 - \alpha)j^{-1} + \frac{(1 - \alpha)(-\alpha)}{2!}j^{-2} + \frac{(1 - \alpha)(-\alpha)(-\alpha - 1)}{3!}j^{-3} + \ldots$$

Consequently, we deduce

$$|s_j| = j^{1-\alpha} \left| \left(\frac{1}{2!} - \frac{2}{3!} \right)(1 - \alpha)\alpha j^{-2} + \left(\frac{1}{3!} - \frac{2}{4!} \right)(1 - \alpha)\alpha(-\alpha - 1)j^{-3} + \ldots \right|$$
$$\leq \frac{1}{3!}(1 - \alpha)\alpha j^{-1-\alpha} \left(1 + j^{-1} + j^{-2} + \ldots \right) \leq \frac{2}{3!}(1 - \alpha)\alpha j^{-1-\alpha} \leq j^{-1-\alpha}.$$

Therefore, the series $\sum_{j=0}^{\infty} s_j$ converges for all $\alpha > 0$. Meanwhile, $s_j = 0$ for $\alpha = 0, 1$. Hence, there exists $c > 0$, independent of α and k such that

$$\sum_{j=0}^{k} \left\{ \frac{2}{2 - \alpha}[(j + 1)^{2-\alpha} - j^{2-\alpha}] - [(j + 1)^{1-\alpha} + j^{1-\alpha}] \right\} \leq c.$$

Combining these estimates yields the desired claim. See Fig. 6.5 for an illustration. □

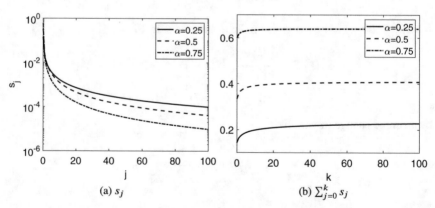

Fig. 6.5 The coefficient s_j and the partial sum $\sum_{j=0}^{k} s_j$.

It is noteworthy that the local truncation error in Lemma 6.18 requires that the solution u be twice continuously differentiable in time, which generally does not hold for solutions to subdiffusion (or fractional ODEs), according to the regularity theory in Section 6.2. Since its first appearance, the L1 scheme has been widely used in practice, and currently it is one of the most popular and successful numerical methods for solving subdiffusion models. With the L1 scheme in time, we arrive at the following time stepping scheme: Given $U^0 = v$, find $U^n \in \dot{H}^1(\Omega)$ for $n = 1, 2, \ldots, N$

$$L_1^n(U) - \Delta U^n = f(t_n). \tag{6.114}$$

We have the following error estimate for the scheme (6.114) [JLZ16a, Theorems 3.10 and 3.13] (for the homogeneous problem). It was derived by means of discrete Laplace transform, and the proof is technical, since the discrete Laplace transform of the weights b_j involves the fairly wieldy polylogarithmic function. Formally, the error estimate is nearly identical to that for the backward Euler CQ.

Theorem 6.35 *Let* $u_0 \in \dot{H}^\gamma(\Omega)$, $\gamma \in [0, 2]$, *and* $f \in C(\bar{I}; L^2(\Omega))$ *with* $\int_0^t (t - s)^{\alpha-1} \|f'(s)\|_{L^2(\Omega)} ds < \infty$ *for any* $t \in \bar{I}$. *Let* U^n *be the solutions of the scheme* (6.114). *Then there holds*

$$\|u(t_n) - U^n\|_{L^2(\Omega)} \le c\tau \left(t^{\frac{\gamma}{2}\alpha-1} \|u_0\|_{\dot{H}^\gamma(\Omega)} + t_n^{\alpha-1} \|f(0)\|_{L^2(\Omega)} \right.$$
$$\left. + \int_0^{t_n} (t_n - s)^{\alpha-1} \|f'(s)\|_{L^2(\Omega)} ds \right).$$

Thus, in contrast to the $O(\tau^{2-\alpha})$ rate expected from the local truncation error in Lemma 6.18 for smooth solutions, the L1 scheme is generally only first-order accurate, just as the backward Euler CQ, even for smooth initial data or source term. Similar to the BDF2 CQ scheme, Yan et al [YKF18] developed the following correction scheme, and proved that it can achieve an accuracy $O(\tau^{2-\alpha})$ for general problem data. Interestingly, this correction is identical with that in (6.113)

$$\begin{cases} L_1^1(U) - \Delta U^1 = \frac{1}{2}(\Delta u_0 + f(0)) + f(t_1), \\ L_1^n(U) - \Delta U^n = f(t_n), \quad 2 \le n \le N. \end{cases}$$

There have been several works in extending the L1 scheme to high-order schemes by using high-order polynomials and superconvergent points. For example, the so-called L1-2 scheme applies a piecewise linear approximation on the first subinterval, and a quadratic approximation on the remaining subintervals to improve the numerical accuracy; see, e.g., Exercise 6.18 for related discussions.

Exercises

Exercise 6.1 This exercise is to provide additional details for Lemma 6.1.

(i) Let $h_n \in L_{loc}^1(\mathbb{R}_+)$ be the resolvent kernel associated with $nk_{1-\alpha}$, i.e.,

$$h_n + n(h_n * k_{1-\alpha})(t) = nk_{1-\alpha}(t), \quad t > 0, n \in \mathbb{N}.$$

Prove that the function h_n is uniquely defined.
(ii) Prove that $k_n = h_n * k_\alpha \to k_\alpha$ in $L^1(I)$.
(iii) Prove that for any $f \in L^p(I)$, $k_n * f \to k_\alpha * f$.

Exercise 6.2 Derive (6.21)–(6.22) from their Laplace transforms (6.25)–(6.26).

Exercise 6.3 Prove the other estimate in Corollary 6.2 using properties of $E_{\alpha,\alpha}(-t)$.

Exercise 6.4 Prove the existence and uniqueness for problem (6.17) with $u_0 \equiv 0$ and a nonzero source f, in a manner similar to Theorem 6.6.

Exercise 6.5 Consider problem (6.17) with $u_0 = 0$, and $f \in L^\infty(I; \dot{H}^\gamma(\Omega))$, $-1 \leq \gamma \leq 1$. Show that there exists a unique weak solution $u \in L^2(I; \dot{H}^{2+\gamma}(\Omega))$ with $\partial_t^\alpha u \in L^2(I; \dot{H}^\gamma(\Omega))$, and

$$\|u\|_{L^2(I;\dot{H}^{2+\gamma}(\Omega))} + \|\partial_t^\alpha u\|_{L^2(I;\dot{H}^\gamma(\Omega))} \leq c\|f\|_{L^2(I;\dot{H}^\gamma(\Omega))}. \tag{6.115}$$

This result can be proved in several steps.

(i) Use Laplace transform to prove $\partial_t^\alpha \int_0^t E(t-s)f(s)ds = f - \int_0^t AE(t-s)f(s)ds$.
(ii) Using the property of $E_{\alpha,1}(-t)$ to prove $\int_0^t |s^{\alpha-1}E_{\alpha,\alpha}(-\lambda_j s^\alpha)|ds \leq \lambda_j^{-1}$.
(iii) Use (ii) and Young's inequality to prove

$$\left\|\partial_t^\alpha \int_0^t (f(s), \varphi_j)(t-s)^{\alpha-1}E_{\alpha,\alpha}(-\lambda_j(t-s)^\alpha)\,ds\right\|_{L^2(I)}^2 \leq c\int_0^T |(f(t), \varphi_j)|^2 dt.$$

(iv) Prove the estimate (6.115) using (iii).
(v) Show that for any $\gamma \leq \beta < \gamma + 2$, $\lim_{t \to 0^+} \|u(t)\|_{\dot{H}^\beta(\Omega)} = 0$.

Exercise 6.6 Prove the uniqueness and existence for the following Neumann subdiffusion problem

$$\begin{cases} \partial_t^\alpha u = \mathcal{L}u + f, & \text{in } Q, \\ \partial_n u = 0, & \text{on } \partial_L Q, \\ u(\cdot, 0) = u_0, & \text{in } \Omega, \end{cases}$$

where $\partial_n u$ denotes the unit outward normal derivative of u on the boundary $\partial_L Q$.

Exercise 6.7 Prove Theorem 6.18.

Exercise 6.8 Consider the following diffusion wave problem (with $\alpha \in (1, 2)$)

$$\begin{cases} \partial_t^\alpha u = \Delta u + f, & \text{in } Q, \\ u = 0, & \text{on } \partial_L Q, \\ u(\cdot, 0) = u_0, & \text{in } \Omega, \\ u'(\cdot, 0) = u_1, & \text{in } \Omega. \end{cases}$$

(i) Derive the solution representation using the separation of variables technique.
(ii) Discuss the Sobolev regularity of the solution.
(iii) Is there any maximum principle for the problem?

Exercise 6.9 Let $0 < \alpha_0 < \alpha_1 < 1$. Consider the following two-term subdiffusion:

$$\begin{cases} \partial_t^{\alpha_1} u + \partial_t^{\alpha_0} u = \Delta u + f & \text{in } Q, \\ u = 0 & \text{on } \partial_L Q, \\ u(\cdot, 0) = u_0 & \text{in } \Omega. \end{cases}$$

(i) Derive the solution representation using the separation of variables technique.
(ii) Discuss the existence and uniqueness of the solution.
(iii) Discuss the asymptotic of the solution for the homogeneous problem.

Exercise 6.10 Consider the following lateral Cauchy problem with $\alpha \in (0, 1)$,

$$\begin{cases} \partial_t^{\alpha} u = \partial_{xx}^2 u, & 0 < x < r_1, 0 < t < r_2, \\ u(0, t) = f(t), & 0 < t < r_2, \\ -u_x(0, t) = g(t), & 0 < t < r_2, \end{cases}$$

where r_1, r_2 are positive constants, and f and g are known functions. A simple idea is to look for a solution of the form of a power series $u(x, t) = \sum_{j=0}^{\infty} a_j(t) x^j$, where the coefficients a_j are to be determined.

(i) Determine the recursion relation for the coefficients a_j, and derive closed-form coefficients $a_j(t)$.
(ii) Discuss the conditions on the data f and g that are sufficient to ensure the convergence of the formal series in (i).
(iii) Discuss the impact of the fractional-order α on the behavior of the solution (assuming existence).

Exercise 6.11 [Ros16] Consider the following one-dimensional Stefan problem for subdiffusion with $\alpha \in (0, 1)$, and $a, b \in \mathbb{R}$:

$$\begin{cases} \partial_t^{\alpha} u = \partial_{xx}^2, & 0 < x < t^{\frac{\alpha}{2}}, 0 < t \le T, \\ u(0, t) = a, & 0 < t \le T, \\ u(t^{\frac{\alpha}{2}}, t) = b, & 0 < t \le T. \end{cases}$$

Show that the function $u(x, t) = a + (1 - W_{-\frac{\alpha}{2}, 1}(-1))^{-1}(b - a)(1 - W_{-\frac{\alpha}{2}, 1}(-xt^{-\frac{\alpha}{2}}))$ is a solution.

Exercise 6.12 Under the setting of Theorem 6.31, derive an alternative inversion formula when $\Delta u(x_0) = 0$.

Exercise 6.13 [HNWY13] This problem is concerned with determining the fractional-order α, as Theorem 6.31. Let $u_0 \in C_0^{\infty}(\Omega)$, $u_0 \ge 0$ or ≤ 0, $\neq 0$ on $\overline{\Omega}$. Then there holds $\alpha = -\lim_{t \to \infty} t u(x_0, t)^{-1} \frac{\partial u}{\partial t}(x_0, t)$.

Exercise 6.14 Consider the following diffusion wave problem with $1 < \alpha < 2$:

$$\begin{cases} \partial_t^{\alpha} u = \Delta u, & \text{in } Q, \\ u = 0, & \text{on } \partial_L Q, \\ u(\cdot, 0) = u_0, & \text{in } \Omega, \\ u'(\cdot, 0) = u_1, & \text{in } \Omega. \end{cases}$$

Discuss the possibility of simultaneous recovering u_0 and u_1 from $g_1 = u(x, T_1)$ and $g_2 = u(x, T_2)$ for some $T_2 > T_1 > 0$.

Exercise 6.15 What about recovering u_0 in (6.89) from the time trace $g(t) = u_x(0, t)$?

Exercise 6.16 Prove the following bound for backward Euler CQ: $|b_j^{(\alpha)}| \le c j^{-\alpha-1}$.

Exercise 6.17 One can rewrite the L1 approximation $L_1^n(U)$ into the form

$$L_1^n(U) = \sum_{j=0}^{n} b_{n-j}^{(n)} U^j.$$

Then the following statements hold.

(i) For any $n \ge 1$, $\sum_{j=0}^{n} b_j^{(n)} = 0$.
(ii) For any $n \ge 1$, $\sum_{j=0}^{n} b_{n-j}^{(n)} t_n = \Gamma(2 - \alpha)^{-1} \tau^\alpha$.
(iii) For any $n \ge 1$, $b_j^{(n)} < 0$, $j = 1, \ldots, n$.

Exercise 6.18 One natural idea to extend the L1 scheme for the Djrbashian-Caputo fractional derivative is to employ high-order polynomial interpolation instead of linear interpolation. This exercise is to explore the idea of using piecewise quadratic interpolants on a uniform grid $t_i = i\tau$, $i = 0, 1, \ldots, N$, with $\tau = \frac{T}{N}$, which leads to the so-called L1-2 scheme [GSZ14].

(i) Find the piecewise linear interpolant $\Pi_{1,1} u$ of u on $[0, \tau]$, and piecewise quadratic interpolant $\Pi_{2,i} u$ on $[t_{i-2}, t_i]$, $i \ge 2$.
(ii) Develop an approximation scheme to $\partial_t^\alpha u$ using $\Pi_{1,1} u$ on $[0, \tau]$ and $\Pi_{2,i} u$ on $[t_{i-1}, t_i]$, $i \ge 2$.
(iii) Derive the local truncation error of the scheme under the assumption $u \in C^3(\bar{I})$.

Chapter 7
Subdiffusion: Hölder Space Theory

In this chapter, we discuss Hölder regularity of the solutions to subdiffusion models with variable coefficients, using fundamental solutions corresponding to constant coefficients. The issue of Hölder regularity has not been extensively discussed. Our description follows largely the works [KV13, Kra16]. The one- and multi-dimensional cases will be discussed separately due to the significant differences in fundamental solutions. The overall analysis strategy follows closely that for standard parabolic equations [LSU68, Chapter IV].

7.1 Fundamental Solutions

First, we derive fundamental solutions, and discuss the associated fractional θ functions. Fundamental solutions for subdiffusion have been extensively studied [SW89, Koc90, Mai96, MLP01, EK04, Psk09]. In the work [Koc90], Kochubei derived the expression for the fundamental solutions $G_\alpha(x,t)$ and $\overline{G}_\alpha(x,t)$ in terms of Fox's H-functions, see also [SW89]. The work [EK04] gives several estimates. More recently, Pskhu [Psk09, Section 3] derived alternative representations of $G_\alpha(x,t)$ and $\overline{G}_\alpha(x,t)$ using the Wright function $W_{\rho,\mu}(z)$, cf. (3.24) in Chapter 3, and also gave several estimates. See the works [KKL17, DK19, DK20, HKP20, DK21] for further estimates in L^p spaces. The survey [Don20] provides an overview on the estimates on fundamental solutions. We present only the representations via Wright function.

7.1.1 Fundamental solutions

Consider subdiffusion in $\mathbb{R}^d$ ($d = 1, 2, 3$)

$$\begin{cases} \partial_t^\alpha u - \Delta u = f, & \text{in } \mathbb{R}^d \times \mathbb{R}_+, \\ u(0) = u_0, & \text{in } \mathbb{R}^d. \end{cases} \tag{7.1}$$

© The Author(s), under exclusive license to Springer Nature Switzerland AG 2021
B. Jin, *Fractional Differential Equations*, Applied Mathematical Sciences 206,
https://doi.org/10.1007/978-3-030-76043-4_7

Throughout the initial data u_0 and the source f are assumed to be smooth and have compact supports, i.e., there exists an $R > 0$ such that

$$u_0(x), f(x, t) = 0, \quad \forall |x| \geq R > 0. \tag{7.2}$$

The next result gives a solution representation to problem (7.1) using the Wright function $W_{\rho,\mu}(z)$, cf. (3.24) in Chapter 3, see Sections A.3.1 and A.3.2 for Laplace and Fourier transforms, respectively.

Proposition 7.1 *Under Assumption (7.2), the solution u of problem (7.1) is given by*

$$u(x, t) = \int_{\mathbb{R}^d} G_\alpha(x - y, t)u_0(y)dy + \int_0^t \int_{\mathbb{R}^d} \overline{G}_\alpha(x - y, t - s)f(y, s)dyds, \tag{7.3}$$

with $G_\alpha(x, t)$ and $\overline{G}_\alpha(x, t)$ given, respectively, by

$$G_\alpha(x, t) = \begin{cases} 2^{-1}t^{-\frac{\alpha}{2}}W_{-\frac{\alpha}{2}, 1-\frac{\alpha}{2}}(-|x|t^{-\frac{\alpha}{2}}), & d = 1, \\ (4\pi)^{-\frac{d}{2}} \int_0^\infty \lambda^{-\frac{d}{2}} e^{-\frac{|x|^2}{4\lambda}} t^{-\alpha} W_{-\alpha, 1-\alpha}(-\lambda t^{-\alpha})d\lambda, & d = 2, 3, \end{cases} \tag{7.4}$$

$$\overline{G}_\alpha(x, t) = \begin{cases} 2^{-1}t^{\frac{\alpha}{2}-1}W_{-\frac{\alpha}{2}, \frac{\alpha}{2}}(-|x|t^{-\frac{\alpha}{2}}), & d = 1, \\ (4\pi)^{-\frac{d}{2}} \int_0^\infty \lambda^{-\frac{d}{2}} e^{-\frac{|x|^2}{4\lambda}} t^{-1} W_{-\alpha, 0}(-\lambda t^{-\alpha})d\lambda, & d = 2, 3. \end{cases} \tag{7.5}$$

Proof Applying Fourier transform and Laplace transform to (7.1) with respect to x and t, denoted by $\tilde{}$ and $\hat{}$, respectively, and using Lemma 2.9 give

$$z^\alpha \widehat{\tilde{u}}(\xi, z) + \xi^2 \widehat{\tilde{u}}(\xi, z) = z^{\alpha-1}\tilde{u}_0(\xi) + \widehat{\tilde{f}}(\xi, z),$$

i.e.,

$$\widehat{\tilde{u}}(\xi, z) = (z^\alpha + \xi^2)^{-1} z^{\alpha-1}\tilde{u}_0(\xi) + (z^\alpha + \xi^2)^{-1}\widehat{\tilde{f}}(\xi, z).$$

Next we discuss the cases $d = 1$ and $d = 2, 3$ separately. For $d = 1$, by Lemma 3.2, the solution $\tilde{u}(\xi, t)$ is given by

$$\tilde{u}(\xi, t) = E_{\alpha, 1}(-\xi^2 t^\alpha)\tilde{u}_0(\xi) + \int_0^t (t - s)^{\alpha-1} E_{\alpha, \alpha}(-\xi^2(t - s)^\alpha)\tilde{f}(\xi, s)ds.$$

The desired assertion follows by applying Fourier transform relation of $W_{\rho,\mu}(-|x|)$ in Proposition 3.4. For $d = 2, 3$, we rewrite $\widehat{\tilde{u}}$ into

$$\widehat{\tilde{u}}(\xi, z) = \int_0^\infty z^{\alpha-1} e^{-(z^\alpha+|\xi|^2)\lambda} d\lambda \tilde{u}_0(\xi) + \int_0^\infty e^{-(z^\alpha+|\xi|^2)\lambda} \widehat{\tilde{f}}(\xi, z)d\lambda.$$

In view of the identities $\mathcal{F}^{-1}[e^{-|\xi|^2\lambda}](x) = (4\pi\lambda)^{-\frac{d}{2}} e^{-\frac{|x|^2}{4\lambda}}$ and $\mathcal{L}^{-1}[z^{-\mu} e^{-z^\alpha\lambda}](t) = t^{\mu-1}W_{-\alpha, \mu}(-\lambda t^{-\alpha})$, cf. Proposition 3.3, the convolution formulas for Laplace and Fourier transforms yield the desired representation. $\qquad\square$

Remark 7.1 Note that for subdiffusion, there are two fundamental solutions, corresponding to u_0 and f, respectively. They are identical in the limit case $\alpha = 1$, i.e., normal diffusion, which is given by $G_1(x,t) = (4\pi t)^{-\frac{d}{2}} e^{-\frac{|x|^2}{4t}}$. When $d = 2, 3$, $G_\alpha(x,t)$ and $\overline{G}_\alpha(x,t)$ can be represented by

$$G_\alpha(x,t) = \int_0^\infty G_1(x,\lambda) t^{-\alpha} W_{-\alpha,1-\alpha}(-\lambda t^{-\alpha}) d\lambda,$$

$$\overline{G}_\alpha(x,t) = \int_0^\infty G_1(x,\lambda) t^{-1} W_{-\alpha,0}(-\lambda t^{-\alpha}) d\lambda.$$

That is, $G_\alpha(x,t)$ and $\overline{G}_\alpha(x,t)$ are both mixtures of $G_1(x,t)$, but with different mixing density. Further, $G_\alpha(x,t)$ and $\overline{G}_\alpha(x,t)$ are related by $G_\alpha(x,t) = {_0I_t^{1-\alpha}}\overline{G}_\alpha(x,t)$, in view of the identity (3.28) in Chapter 3. See also Lemma 7.1(iii).

Remark 7.2 When $d = 1$, we can rewrite $G_\alpha(x,t)$ using Mainardi's M function $M_\mu(z)$ (cf. Exercise 3.19) as $G_\alpha(x,t) = 2^{-1} t^{-\frac{\alpha}{2}} M_{\frac{\alpha}{2}}(|x|t^{-\frac{\alpha}{2}})$. For $\alpha \in (0,1)$, for every $t > 0$, $x \mapsto G_\alpha(x,t)$ is not differentiable at $x = 0$, cf. Fig. 3.7, which contrasts sharply with $G_1(x,t)$, which is C^∞ in space for every $t > 0$.

The next result collects elementary properties of $G_\alpha(x,t)$ and $\overline{G}_\alpha(x,t)$. For a vector $x \in \mathbb{R}^d$, $d > 1$, the notation x' denotes the subvector $(x_1, \ldots, x_{d-1}) \in \mathbb{R}^{d-1}$, and $^R\partial_t^\alpha$ denotes the Riemann-Liouville fractional derivative, cf. Definition 2.2.

Lemma 7.1 *The following statements hold:*

(i) *The functions $G_\alpha(x,t)$ and $\overline{G}_\alpha(x,t)$ are positive, and so is $-\partial_{x_d}\overline{G}_\alpha(x,t)$ for $x_d > 0$.*

(ii) *The following identities hold:*

$$\int_{\mathbb{R}^d} G_\alpha(x,t) dx = 1, \quad \forall t > 0, \tag{7.6}$$

$$\int_{\mathbb{R}^d} \overline{G}_\alpha(x,t) dx = \Gamma(\alpha)^{-1} t^{\alpha-1}, \quad \forall t > 0, \tag{7.7}$$

$$\begin{cases} \int_0^\infty \partial_x \overline{G}_\alpha(x,t) dt = -\frac{1}{2}, & d = 1, \\ \int_{\mathbb{R}^{d-1}} \int_0^\infty \partial_{x_d} \overline{G}_\alpha(x,t) dt\, dx' = -\frac{1}{2}, & d = 2, 3, \end{cases} \quad \forall x_d > 0. \tag{7.8}$$

(iii) $\Delta\overline{G}_\alpha(x,t) = \partial_t G_\alpha(x,t)$, *if $x \neq 0$, i.e., $\overline{G}_\alpha(x,t) = {^R\partial_t^{1-\alpha}} G_\alpha(x,t)$.*

Proof (i) is direct from the integral representations of $G_\alpha(x,t)$ and $\overline{G}_\alpha(x,t)$, and Lemma 3.5 and Theorem 3.8. (iii) follows from direct computation. For (ii), we analyze separately $d = 1$ and $d = 2, 3$. When $d = 1$, by the identity (3.26) and changing variables $r = xt^{-\frac{\alpha}{2}}$, we deduce

$$\int_{-\infty}^\infty G_\alpha(x,t) dx = \frac{1}{2} \int_{-\infty}^\infty W_{-\frac{\alpha}{2},1-\frac{\alpha}{2}}(-|r|) dr = \int_{-\infty}^0 \frac{d}{dr} W_{-\frac{\alpha}{2},1}(r) dr = W_{-\frac{\alpha}{2},1}(0) = 1.$$

This shows the identity (7.6). The equality (7.7) follows analogously:

$$\int_{-\infty}^{\infty} \overline{G}_\alpha(x,t)\mathrm{d}x = \frac{1}{2t^{1-\alpha}} \int_{-\infty}^{\infty} W_{-\frac{\alpha}{2},\frac{\alpha}{2}}(-|r|)\mathrm{d}r$$

$$= t^{\alpha-1} \int_{-\infty}^{0} \frac{\mathrm{d}}{\mathrm{d}r} W_{-\frac{\alpha}{2},\alpha}(r)\mathrm{d}r = t^{\alpha-1} W_{-\frac{\alpha}{2},\alpha}(0) = \frac{t^{\alpha-1}}{\Gamma(\alpha)}.$$

Likewise, the identity (7.8) follows from the formula (3.27) as

$$\int_0^{\infty} \partial_x \overline{G}_\alpha(x,t)\mathrm{d}t = \int_0^{\infty} -\frac{1}{2t} W_{-\frac{\alpha}{2},0}(-|x|t^{-\frac{\alpha}{2}})\mathrm{d}t$$

$$= -\frac{1}{2}\int_0^{\infty} \frac{\mathrm{d}}{\mathrm{d}t} W_{-\frac{\alpha}{2},1}(-|x|t^{-\frac{\alpha}{2}})\mathrm{d}t = -\frac{1}{2}W_{-\frac{\alpha}{2},1}(0) = -\frac{1}{2}.$$

Next, when $d = 2, 3$, by the definition of $G_\alpha(x,t)$, we have

$$\int_{\mathbb{R}^d} G_\alpha(x,t)\mathrm{d}x = \int_0^{\infty} \left(\int_{\mathbb{R}^d} (4\pi)^{-\frac{d}{2}} \lambda^{-\frac{d}{2}} e^{-\frac{|x|^2}{4\lambda}} \mathrm{d}x \right) t^{-\alpha} W_{-\alpha,1-\alpha}(-\lambda t^{-\alpha})\mathrm{d}\lambda$$

$$= \int_0^{\infty} t^{-\alpha} W_{-\alpha,1-\alpha}(-\lambda t^{-\alpha})\mathrm{d}\lambda = \int_0^{\infty} W_{-\alpha,1-\alpha}(-\eta)\mathrm{d}\eta = 1,$$

where the last line follows from the formula (3.26). This shows (7.6). The identity (7.7) can be shown analogously. Last, by the definition of $\overline{G}_\alpha$,

$$-2\partial_{x_d}\overline{G}_\alpha(x,t) = \frac{1}{(4\pi)^{\frac{d}{2}}} \int_0^{\infty} \frac{x_d}{\lambda} \lambda^{-\frac{d}{2}} e^{-\frac{|x|^2}{4\lambda}} t^{-1} W_{-\alpha,0}(-\lambda t^{-\alpha})\mathrm{d}\lambda.$$

Thus, the recursion formula $W_{-\alpha,0}(-z) = \alpha z W_{-\alpha,1-\alpha}(-z)$ from (3.33), and changing variables $y' = x'\lambda^{-\frac{1}{2}}$, $\eta = \lambda t^{-\alpha}$ and $\mu = x_d \lambda^{-\frac{1}{2}}$ imply

$$-2\int_{\mathbb{R}^{d-1}} \int_0^{\infty} \partial_{x_d}\overline{G}_\alpha(x,t)\mathrm{d}t\mathrm{d}x'$$

$$= 2(4\pi)^{-\frac{d}{2}} \int_0^{\infty} e^{-\frac{\mu^2}{4}} \mathrm{d}\mu \int_0^{\infty} W_{-\alpha,1-\alpha}(-\eta)\mathrm{d}\eta \int_{\mathbb{R}^{d-1}} e^{-\frac{|y'|^2}{4}} \mathrm{d}y'.$$

This shows the identity (7.8) and completes the proof of the lemma.   □

Remark 7.3 Lemma 7.1 indicates that the fundamental solution $G_\alpha(x,t)$ is actually a PDF for any fixed $t > 0$, i.e., $G_\alpha(x,t) \geq 0$ and $\int_{\mathbb{R}^d} G_\alpha(x,t)\,\mathrm{d}x = 1$. Further, we can compute all the moments for $d = 1$: by the integral identity (3.32), the moments (of even order) of $G_\alpha(x,t)$ are given by

$$\mu_{2n}(t) := \int_{-\infty}^{\infty} x^{2n} G_\alpha(x,t)\,\mathrm{d}x = \frac{\Gamma(2n+1)}{\Gamma(\alpha n+1)} t^{\alpha n}, \quad n = 0, 1, \ldots, t \geq 0.$$

In particular, the variance $\sigma^2(t) := \mu_2(t) = 2\Gamma(\alpha + 1)^{-1}t^\alpha$ is consistent with anomalously slow diffusion expected for subdiffusion processes. In the limit of $\alpha \to 1^-$, it does recover the linear growth for normal diffusion.

The representation (7.3) motivates the following two integral operators:

$$\mathcal{I}u_0(x, t) = \int_{\mathbb{R}^d} G_\alpha(x - y, t)u_0(y)\mathrm{d}y, \qquad (7.9)$$

$$\mathcal{V}f(x, t) = \int_0^t \int_{\mathbb{R}^d} \overline{G}_\alpha(x - y, t - s)f(y, s)\mathrm{d}y\mathrm{d}s. \qquad (7.10)$$

Their properties will be central to the Hölder regularity analysis. We first give two preliminary results. The first result gives the continuity of the potential $\mathcal{I}u_0$ at $t = 0$.

Lemma 7.2 *For any uniformly bounded $u_0 \in C(\mathbb{R}^d)$,*

$$\lim_{t \to 0^+} \mathcal{I}u_0(x, t) = u_0(x)$$

uniformly for all x in a compact subset of $\mathbb{R}^d$. The limit holds uniformly for all x in compact subsets of a piecewise continuous function u_0, and also for every point x of continuity of any integrable u_0.

Proof We prove only the case $d = 1$. By (7.7), for any fixed x,

$$\int_{\mathbb{R}} G_\alpha(x - y, t)u_0(y)\mathrm{d}y - u_0(x) = \int_{\mathbb{R}} G_\alpha(x - y, t)(u_0(y) - u_0(x))\mathrm{d}y := \sum_{i=1}^3 I_i(t),$$

with the three terms I_1, I_2 and I_3 denoting the integral over $(-\infty, x - \delta)$, $(x - \delta, x + \delta)$ and $(x + \delta, \infty)$, respectively, where $\delta > 0$ is to be chosen. Since $u_0 \in C(\mathbb{R})$, it is uniformly continuous on any compact subset S of $\mathbb{R}$. For such a subset S and for each $\epsilon > 0$, there exists a $\delta_\epsilon > 0$ such that $|u_0(x) - u_0(y)| < \frac{\epsilon}{3}$ holds for every $x, y \in S$ with $|x - y| < \delta_\epsilon$. Further, we may assume $(x - \delta_\epsilon, x + \delta_\epsilon) \subset S$. Thus, for $\delta = \delta_\epsilon$, by Lemma 7.1(i) and (7.6), we have

$$|I_2(t)| \leq \int_{x-\delta_\epsilon}^{x+\delta_\epsilon} G_\alpha(x - y, t)|u_0(x) - u_0(y)|\mathrm{d}y \leq \frac{\epsilon}{3}.$$

For the term I_3, by changing variables $\xi = |x - y|t^{-\frac{\alpha}{2}}$, we have

$$|I_3(t)| \leq 2\|u_0\|_{C(\mathbb{R})} \int_{x+\delta_\epsilon}^\infty G_\alpha(x - y, t)\mathrm{d}y \leq 2\|u_0\|_{C(\mathbb{R})} \int_{-\infty}^{-\delta_\epsilon t^{-\frac{\alpha}{2}}} W_{-\frac{\alpha}{2}, 1 - \frac{\alpha}{2}}(\xi)\mathrm{d}\xi.$$

Now the convergence of the integral $\int_{-\infty}^\infty W_{-\frac{\alpha}{2}, 1 - \frac{\alpha}{2}}(-|\xi|)\mathrm{d}\xi$ implies that there exists $t_\epsilon > 0$, such that for all $t < t_\epsilon$, $|I_3| < \frac{\epsilon}{3}$. The term I_1 can be bounded similarly. Combining these estimates gives the first claim. The rest follows by a density argument. $\square$

The next result gives the continuity of the volume potential $\mathcal{V}f$. For an index $\ell \in \mathbb{N}_0^d$, $D_x^\ell u$ denotes the mix derivative $D_x^\ell u(x) = \frac{\partial^{|\ell|}}{\partial x_1^{\ell_1} \ldots \partial x_d^{\ell_d}} u(x)$, with $|\ell| = \sum_{i=1}^d \ell_i$.

Lemma 7.3 *For any bounded $f \in C(\mathbb{R}^d \times \mathbb{R}_+)$, which is uniformly Hölder continuous with an exponent $\delta \in (0,1)$ with respect to x, the function $u = \mathcal{V}f$ satisfies the following properties:*

(i) u, $\partial_t^\alpha u$, and $D_x^\ell u$, with $|\ell| = 1,2$, are continuous.

(ii) *For any $x \in \mathbb{R}^d$ and $t > 0$, there holds $\partial_t^\alpha u(x,t) = \Delta u(x,t) + f(x,t)$.*

Proof We give the proof only for the case $d = 1$. For any $0 < \tau < \frac{t}{2}$, let

$$u_\tau(x,t) = \int_0^{t-\tau} \int_\mathbb{R} \overline{G}_\alpha(x-y, t-s) f(y,s) \mathrm{d}y \mathrm{d}s.$$

Since the singularity of the kernel $\overline{G}_\alpha(x-y, t-s)$ occurs at $x = y$, $t = s$ and $t - s \geq \tau > 0$, u_τ is continuously differentiable and by Leibniz's rule, the functions

$$\partial_x u_\tau(x,t) = \int_0^{t-\tau} \int_\mathbb{R} \partial_x \overline{G}_\alpha(x-y, t-s) f(y,s) \mathrm{d}y \mathrm{d}s$$

$$\partial_{xx}^2 u_\tau(x,t) = \int_0^{t-\tau} \int_\mathbb{R} \partial_{xx}^2 \overline{G}_\alpha(x-y, t-s) f(y,s) \mathrm{d}y \mathrm{d}s$$

are continuous. By Lemma 7.1(i) and (7.7), we can bound $|u(x,t) - u_\tau(x,t)|$ by

$$|u(x,t) - u_\tau(x,t)| \leq \int_{t-\tau}^t \int_\mathbb{R} \overline{G}_\alpha(x-y, t-s)|f(y,s)|\mathrm{d}y\mathrm{d}s$$

$$\leq \Gamma(\alpha)^{-1} \|f\|_{C(\mathbb{R} \times \overline{I})} \int_{t-\tau}^t (t-s)^{\alpha-1} \mathrm{d}s = \Gamma(\alpha+1)^{-1} \|f\|_{C(\mathbb{R} \times \overline{I})} \tau^\alpha.$$

Thus, as $\tau \to 0^+$, u is the uniform limit of continuous functions u_τ, and hence u is also continuous. The proof shows also $|u(x,t)| \leq ct^\alpha \|f\|_{C(\mathbb{R} \times \overline{I})}$, and thus u is continuous at $t = 0$ and is equal to zero. Let

$$g(x,t) = \int_0^t \int_\mathbb{R} \partial_x \overline{G}_\alpha(x-y, t-s) f(y,s) \mathrm{d}y \mathrm{d}s.$$

By Lemma 7.1(i), $\partial_x \overline{G}_\alpha(x,t)$ does not change sign for $x > 0$, and by the differentiation formula (3.26), $\partial_x \overline{G}_\alpha(x,t) = -(2t)^{-1} W_{-\frac{\alpha}{2},0}(-|x|t^{-\frac{\alpha}{2}})$. Then with the change of variable $r = xt^{-\frac{\alpha}{2}}$ and the differentiation formula (3.26), we deduce

$$\int_{-\infty}^\infty |\partial_y \overline{G}_\alpha(y,t)|\mathrm{d}y = \frac{1}{t} \int_0^\infty W_{-\frac{\alpha}{2},0}(-xt^{-\frac{\alpha}{2}}) \mathrm{d}x$$

$$= t^{\frac{\alpha}{2}-1} \int_0^\infty W_{-\frac{\alpha}{2},0}(-r)\mathrm{d}r = -t^{\frac{\alpha}{2}-1} \int_0^\infty \frac{\mathrm{d}}{\mathrm{d}r} W_{-\frac{\alpha}{2},\frac{\alpha}{2}}(-r)\mathrm{d}r = \frac{t^{\frac{\alpha}{2}-1}}{\Gamma(\frac{\alpha}{2})}.$$

This identity implies

$$|g(x,t) - \partial_x u_\tau(x,t)| \leq \left| \int_{t-\tau}^t \int_{\mathbb{R}} \partial_x \overline{G}_\alpha(x - y, t - s) f(y,s) dy ds \right|$$

$$\leq \|f\|_{C(\mathbb{R} \times \overline{I})} \int_{t-\tau}^t \int_{\mathbb{R}} |\partial_x \overline{G}_\alpha(x - y, t - s)| dy ds$$

$$= \frac{\|f\|_{C(\mathbb{R} \times \overline{I})}}{\Gamma(\frac{\alpha}{2})} \int_{t-\tau}^t (t - s)^{\frac{\alpha}{2} - 1} ds = \frac{\|f\|_{C(\mathbb{R} \times \overline{I})} \tau^{\frac{\alpha}{2}}}{\Gamma(\frac{\alpha}{2} + 1)}.$$

Thus, g is the uniform limit of the continuous functions $\partial_x u_\tau$, and it is continuous. Further, $g(x,t)$ is continuous at $t = 0$ and is equal to zero. It follows from the identity $u_\tau(x,t) = \int_{x_0}^x \partial_x u_\tau(y,t) dy + u_\tau(x_0,t)$ and the uniform convergence of u_τ to u and $\partial_x u_\tau$ to g on compact subsets of $\mathbb{R}$, $t > 0$ that $u(x,t) = \int_{x_0}^x g(y,t) dy + u(x_0,t)$, whence the identity

$$\partial_x u(x,t) = g(x,t) = \int_0^t \int_{\mathbb{R}} \partial_x \overline{G}_\alpha(x - y, t - s) f(y,s) dy ds$$

exists and continuous. Next, by the identity (7.7), we have

$$\int_{\mathbb{R}} \partial_t^\alpha \overline{G}_\alpha(x - y, t - s) dy = \int_{\mathbb{R}} \partial_{xx}^2 \overline{G}_\alpha(x - y, t - s) dy$$

$$= \partial_{xx}^2 \int_{\mathbb{R}} \overline{G}_\alpha(x - y, t - s) dy = 0.$$

The preceding two identities allow rewriting $\partial_{xx}^2 u_\tau(x,t)$ as

$$\partial_{xx}^2 u_\tau(x,t) = \int_0^{t-\tau} \int_{\mathbb{R}} \partial_{xx} \overline{G}_\alpha(x - y, t - s)(f(y,s) - f(x,s)) dy ds.$$

Let

$$h(x,t) = \int_0^t \int_{\mathbb{R}} \partial_{xx}^2 \overline{G}_\alpha(x - y, t - s)(f(y,s) - f(x,s)) dy ds.$$

Then by the Hölder continuity of f,

$$|h(x,t) - \partial_{xx}^2 u_\tau(x,t)| \leq \|f\|_{C(\overline{I};C^\alpha(\mathbb{R}))} \int_{t-\tau}^t \int_{\mathbb{R}} |\partial_{xx}^2 \overline{G}_\alpha(x, t - s)| |x|^\delta dx ds.$$

By Lemmas 7.1(iii) and 7.6(ii) below, for any $t > 0$, with $r = |x| t^{-\frac{\alpha}{2}}$, there holds

$$\int_{\mathbb{R}} |\partial_{xx}^2 \overline{G}_\alpha(x,t)| |x|^\delta dx \leq c \int_{\mathbb{R}} t^{-1-\frac{\alpha}{2}} (1 + r^{1+\frac{2}{\alpha}})^{-1} |x|^\delta dx \leq c t^{\frac{\delta}{2} \alpha - 1}.$$

Thus, we obtain $|h(x,t) - \partial_{xx}^2 u_\tau(x,t)| \leq c\|f\|_{C([0,T];C^\delta(\mathbb{R}))} \tau^{\frac{\delta}{2} \alpha}$. Thus, $h(x,t)$ is the uniform limit of the continuous functions $\partial_{xx}^2 u_\tau(x,t)$, and it is also continuous. Since $\partial_x u_\tau(x,t) = \int_{x_0}^x \partial_{xx}^2 u_\tau(y,t) dy + \partial_x u_\tau(x_0,t)$, there holds

$$\partial_x u(x,t) = \int_{x_0}^x h(y,t) \mathrm{d}y + \partial_x u(x_0,t),$$

whence $\partial_{xx}^2 u(x,t) = h(x,t)$ exists and is continuous. Last, consider the Djrbashian-Caputo fractional derivative $\partial_t^\alpha u(x,t)$. Since $|u(x,t)| \to 0$ as $t \to 0^+$, the Riemann-Liouville and Djrbashian-Caputo fractional derivatives are identical. Thus,

$$
\begin{aligned}
\partial_t^\alpha u(x,t) &= \frac{\mathrm{d}}{\mathrm{d}t} \frac{1}{\Gamma(1-\alpha)} \int_0^t (t-\eta)^{-\alpha} \int_0^\eta \int_{\mathbb{R}} f(y,s) \overline{G}_\alpha(x-y,\eta-s) \mathrm{d}y \mathrm{d}s \mathrm{d}\eta \\
&= \frac{\mathrm{d}}{\mathrm{d}t} \frac{1}{\Gamma(1-\alpha)} \int_0^t \int_s^t (t-\eta)^{-\alpha} \int_{\mathbb{R}} f(y,s) \overline{G}_\alpha(x-y,\eta-s) \mathrm{d}y \mathrm{d}\eta \mathrm{d}s \\
&= \frac{\mathrm{d}}{\mathrm{d}t} \int_0^t {}_s I_t^{1-\alpha} \int_{\mathbb{R}} f(y,s) \overline{G}_\alpha(x-y,t-s) \mathrm{d}y \mathrm{d}s.
\end{aligned}
$$

Now by (7.7),

$$\int_{\mathbb{R}} |f(y,s) {}_s I_t^{1-\alpha} \overline{G}_\alpha(x-y,t-s)| \mathrm{d}y \le c \|f\|_{C(\mathbb{R} \times [0,t])}.$$

Hence, $\partial_t^\alpha u(x,t)$ is continuous. Using this identity gives

$$
\begin{aligned}
\partial_t^\alpha u(x,t) &= \frac{\mathrm{d}}{\mathrm{d}t} \int_0^t \int_{\mathbb{R}} (f(y,s) - f(x,s))_s I_t^{1-\alpha} \overline{G}_\alpha(x-y,t-s) \mathrm{d}y \mathrm{d}s \\
&\quad + \frac{\mathrm{d}}{\mathrm{d}t} \int_0^t f(x,s) \int_{\mathbb{R}} {}_s I_t^{1-\alpha} \overline{G}_\alpha(x-y,t-s) \mathrm{d}y \mathrm{d}s.
\end{aligned}
$$

In view of (7.7), $\int_{\mathbb{R}} {}_s I_t^{1-\alpha} \overline{G}_\alpha(x-y,t-s) \mathrm{d}y = 1$ and thus we have

$$\partial_t^\alpha u(x,t) = \int_0^t \int_{\mathbb{R}} (f(y,s) - f(x,s))_s^R \partial_t^\alpha \overline{G}_\alpha(x-y,t-s) \mathrm{d}y \mathrm{d}s + f(x,t).$$

This and the expression for $\partial_{xx}^2 u(x,t)$ above show the desired identity in (ii). $\qquad \square$

7.1.2 Fractional θ-functions

For normal diffusion, the fundamental solution $G_1(x,t)$ can be used to solve problems in domains with special geometry, using the so-called θ-function [Can84, Chapter 6], which holds also for $G_\alpha(x,t)$ and $\overline{G}_\alpha(x,t)$. We illustrate the idea with the unit interval $\Omega = (0,1)$, using fractional analogues of the θ-function. Consider

$$\begin{cases} \partial_t^\alpha u = \partial_{xx}^2 u + f, & \text{in } \Omega \times I, \\ u(0) = u_0, & \text{in } \Omega, \\ u(0, \cdot) = g_1, & \text{on } I, \\ u(1, \cdot) = g_2, & \text{on } I. \end{cases} \tag{7.11}$$

To see the extension, we assume $g_1 \equiv g_2 \equiv 0$ and $f \equiv 0$. First, we make an odd extension of u_0 from Ω to $(-1, 0)$, and then periodically to $\mathbb{R}$ with a period 2, denoted by $\tilde{u}_0$. Then the solution, still denoted by u, of the extended problem is given by

$$u(x, t) = \int_{-\infty}^{\infty} G_\alpha(x - y, t)\tilde{u}_0(y)dy$$

$$= \int_0^1 \Big[\sum_{m=-\infty}^{\infty} G_\alpha(x + 2m - y, t) - \sum_{m=-\infty}^{\infty} G_\alpha(x + 2m + y, t) \Big] u_0(y)dy.$$

Thus, we define two (fractional) functions $\theta_\alpha(x, t)$ and $\bar{\theta}_\alpha(x, t)$ by

$$\theta_\alpha(x, t) = \sum_{m=-\infty}^{\infty} G_\alpha(x + 2m, t) \quad \text{and} \quad \bar{\theta}_\alpha(x, t) = \sum_{m=-\infty}^{\infty} \overline{G}_\alpha(x + 2m, t).$$

Both functions coincide with the classical θ function as $\alpha \to 1^-$. They were studied in [LZ14], from which the discussions below are adapted. The following lemmas collect useful properties of θ_α and $\bar{\theta}_\alpha$. Due to the presence of the kink at $x = 0$ for $G_\alpha(x, t)$ and $\overline{G}_\alpha(x, t)$, the functions θ_α and $\bar{\theta}_\alpha$ are not differentiable at $x = 0, \pm 2, \pm 4, \ldots$ and not globally smooth for $x \in \mathbb{R}$, which differs from the classical θ function.

Lemma 7.4 *The functions θ_α and $\bar{\theta}_\alpha$ are even in x and C^∞ for $x \in (0, 2)$ and $t > 0$.*

Proof We give the proof only for θ_α, and $\bar{\theta}_\alpha$ can be analyzed similarly. Let $r_m = |x + 2m|t^{-\frac{\alpha}{2}}$, $m \in \mathbb{Z}$. Then $\theta_\alpha(x, t)$ can be rewritten as

$$\theta_\alpha(x, t) = \sum_{m=-\infty}^{\infty} \tfrac{1}{2} t^{-\frac{\alpha}{2}} W_{-\frac{\alpha}{2}, 1-\frac{\alpha}{2}}(-r_m).$$

Recall the asymptotics of $W_{\rho, \nu}(z)$ in Theorem 3.7 (see also Exercise 3.19(ii))

$$W_{-\mu, 1-\mu}(-r) \sim A r^a e^{-br^c}, \quad r \to \infty,$$

with $A = (2\pi(1 - \mu)\mu^{\frac{1-2\mu}{1-\mu}})^{-\frac{1}{2}}$, $a = (2\mu - 1)(2 - 2\mu)^{-1}$, $b = (1 - \mu)\mu^{\frac{\mu}{1-\mu}}$ and $c = (1 - \mu)^{-1}$. Then with $\mu = \frac{\alpha}{2}$, there are constants $A > 0$, $a < 0$, $b > 0$ and $c \in (1, 2)$ such that

$$|\theta_\alpha(x, t)| < c_1 \sum_{m=-\infty}^{\infty} t^{-\frac{\alpha}{2}} A r_m^a e^{-br_m^c}.$$

Now for a fixed $t > 0$, there exists an $M \in \mathbb{N}$ such that $r_m > 1$ for all $m > M$, and thus $e^{-br_m^c} < e^{-br_m}$. It suffices to consider the terms with $m > M$. Let $u_m(x, t) =$

$At^{-\frac{\alpha}{2}} r_m^a e^{-br_m}$. Then for any $t_0 > 0$, $t \geq t_0 > 0$, there is an integer M such that $|r_{m+1}| - |r_m| = 2t^{-\frac{\alpha}{2}}$ for all $m > M$. Upon restricting m to this range, we obtain

$$\lim_{m \to \infty} \frac{u_{m+1}}{u_m} = e^{-2bt^{-\frac{\alpha}{2}}} < 1,$$

since $b > 0$ and $t \geq t_0 > 0$. Thus, the series $\sum_{m=0}^{\infty} u_m(x,t)$ converges uniformly for $x \in \mathbb{R}$ and $t \geq t_0 > 0$. The series $\sum_{m=-\infty}^{0} u_m(x,t)$ can be analyzed similarly. Thus, the series for $\theta_\alpha(x,t)$ is uniformly convergent for $x \in \mathbb{R}$ and $t \geq t_0 > 0$. Similarly, we can deduce that all partial derivatives of θ_α are uniformly convergent for $x \in (0,2)$ and $t > 0$, and thus θ_α belongs to $C^\infty(\mathbb{R}_+)$ in t and in $C^\infty(0,2)$ in x. $\square$

The following asymptotic expansion at $x = 0$ is immediate:

$$\theta_\alpha(0,t) = c_\alpha t^{-\frac{\alpha}{2}} + 2 \sum_{m=1}^{\infty} G_\alpha(2m,t) := c_\alpha t^{-\frac{\alpha}{2}} + H(t), \qquad (7.12)$$

where $(4\pi)^{-\frac{1}{2}} < c_\alpha = \frac{1}{2\Gamma(1-\frac{\alpha}{2})} < \frac{1}{2}$, and $H(t) \in C^\infty[0,\infty)$ with $H^{(m)}(0) = 0$, $m \in \mathbb{N}_0$.

Lemma 7.5 *The following relations hold:*

$$\lim_{x \to 0^\pm} \mathcal{L}\left[\partial_x \theta_\alpha(x,t)\right](z) = \mp \tfrac{1}{2} z^{\alpha-1}, \qquad (7.13)$$

$$\lim_{x \to 0^\pm} \mathcal{L}\left[\partial_x \bar{\theta}_\alpha(x,t)\right](z) = \mp \tfrac{1}{2}, \qquad (7.14)$$

$$\lim_{t \to 0^+} \theta_\alpha(x,t) = 0, \quad \forall x \in (0,2), \qquad (7.15)$$

$$\lim_{x \to 1} \partial_x \theta_\alpha(x,t) = 0, \quad \forall t > 0. \qquad (7.16)$$

Proof By (3.26) and the uniform convergence in Lemma 7.4, we have

$$\lim_{x \to 0^\pm} \partial_x \theta_\alpha(x,t) = \lim_{x \to 0^\pm} \partial_x G_\alpha(x,t) = \lim_{x \to 0^\pm} \mp \tfrac{1}{2} t^{-\alpha} W_{-\frac{\alpha}{2}, 1-\alpha}(-|x|t^{-\frac{\alpha}{2}}).$$

This and Proposition 3.3 give the identity (7.13). For any $\varphi(t) \in C_0^\infty(\mathbb{R}_+)$, ${}^R\partial_t^{1-\alpha} \varphi(t)$ exists and is continuous and further ${}_0 I_t^{1-\alpha} \varphi(0) = 0$, and thus by Lemma 2.7, $\mathcal{L}[{}^R\partial_t^{1-\alpha}\varphi](z) = z^{1-\alpha} \mathcal{L}[\varphi](z)$. By Lemma 2.6, we obtain

$$\lim_{x \to 0^\pm} \mathcal{L}\left[\int_0^t \partial_x \bar{\theta}_\alpha(x,t-s)\varphi(s)\,ds\right](z) = \lim_{x \to 0^\pm} \mathcal{L}\left[\int_0^t \partial_x \theta_\alpha(x,t-s) {}^R\partial_s^{1-\alpha}\varphi(s)\,ds\right](z)$$

$$= \lim_{x \to 0^\pm} \mathcal{L}\left[\partial_x \theta_\alpha(x,t)\right](z)\mathcal{L}\left[{}^R\partial_t^{1-\alpha}\varphi\right](z) = \mp \tfrac{1}{2} z^{\alpha-1} z^{1-\alpha} \mathcal{L}[\varphi](z) = \mathcal{L}[\mp \tfrac{1}{2}\varphi](z),$$

from which (7.14) follows. The identity (7.15) is direct from Theorem 3.7: for any $x \in (0,2)$,

$$\lim_{t \to 0^+} |\theta_\alpha(x,t)| \leq \lim_{t \to 0^+} ct^{-\frac{\alpha}{2}} e^{-\sigma t^{-\frac{\alpha}{2-\alpha}} |x|^{\frac{2}{2-\alpha}}} = 0.$$

Last, to show the identity (7.16), for $t > 0$:

$$\partial_x \theta_\alpha(x,t)|_{x=1} = -\frac{1}{2t^\alpha} \sum_{m=1}^{\infty} \sum_{k=0}^{\infty} \frac{(-|1 + 2m|t^{-\frac{\alpha}{2}})^k}{k!\Gamma(-k\frac{\alpha}{2} + (1-\alpha))}$$

$$+ \frac{1}{2t^\alpha} \sum_{m=1}^{\infty} \sum_{k=0}^{\infty} \frac{(-|1 - 2m|t^{-\frac{\alpha}{2}})^k}{k!\Gamma(-k\frac{\alpha}{2} + (1-\alpha))} - \frac{1}{2t^\alpha} \sum_{k=0}^{\infty} \frac{(-t^{-\frac{\alpha}{2}})^k}{k!\Gamma(-k\frac{\alpha}{2} + (1-\alpha))} = 0.$$

Last, the continuity of $\partial_x \theta_\alpha(x,t)$ in x at $x = 1$ from Lemma 7.4 gives (7.16). □

Now we can state a solution representation for problem (7.11). The uniqueness of the solution has to be shown independently.

Theorem 7.1 *For piecewise continuous functions* f, u_0, g_1 *and* g_2, *the following representation gives a solution* u *to problem* (7.11)

$$u(x,t) = \sum_{i=1}^{4} u_i,$$

where the functions u_i *are defined by*

$$u_1(x,t) = \int_0^1 (\theta_\alpha(x-y,t) - \theta_\alpha(x+y,t))u_0(y)\,dy,$$

$$u_2(x,t) = -2 \int_0^t \partial_x \bar{\theta}_\alpha(x,t-s)g_1(s)\,ds,$$

$$u_3(x,t) = 2 \int_0^t \partial_x \bar{\theta}_\alpha(x-1,t-s)g_2(s)\,ds,$$

$$u_4(x,t) = \int_0^t \int_0^1 [\bar{\theta}_\alpha(x-y,t-s) - \bar{\theta}_\alpha(x+y,t-s)]f(y,s)\,dy\,ds.$$

Proof By the definition of the function θ_α, we have

$$\partial_t^\alpha u_1(x,t) = \int_0^1 (\partial_t^\alpha \theta_\alpha(x+y,t) - \partial_t^\alpha \theta_\alpha(x-y,t))u_0(y)dy$$

$$= \int_0^1 \partial_{xx}^2 [\theta_\alpha(x+y,t) - \theta_\alpha(x-y,t)]u_0(y)dy = \partial_{xx}^2 u_1(x,t).$$

Next, we consider the term u_2. By Lemma 2.6,

$$\int_0^t {}^R_s\partial_t^{1-\alpha} \phi(t-s)\varphi(s)\,ds = \int_0^t \phi(t-s)\,{}^R\partial_s^{1-\alpha}\varphi(s)\,ds$$

for any $\phi, \varphi \in AC(\bar{I})$ with $\phi(0) = \varphi(0) = 0$. Then for $g_1 \in AC(\bar{I})$ with $g_1(0) = 0$, Lemma 7.1(iii) gives

$$\partial_t^\alpha u_2(x,t) = -2\partial_t^\alpha \int_0^t \partial_x \bar\theta_\alpha(x,t-s)g_1(s)\,ds$$

$$= -2\partial_t^\alpha \int_0^t \partial_x \theta_\alpha(x,t-s)^R\partial_s^{1-\alpha}g_1(s)\,ds,$$

since by (7.13), $\lim_{t\to 0^+}\partial_x\theta_\alpha(x,t) = 0$. Thus,

$$\partial_t^\alpha u_2(x,t) = \frac{-2}{\Gamma(1-\alpha)}\int_0^t (t-s)^{-\alpha}\partial_x\theta_\alpha(x,0)^R\partial_s^{1-\alpha}g_1(s)ds$$

$$+ \frac{-2}{\Gamma(1-\alpha)}\int_0^t (t-s)^{-\alpha}\int_0^s \partial_x\partial_s\theta_\alpha(x,s-\xi)^R\partial_\xi^{1-\alpha}g_1(\xi)\,d\xi ds.$$

Since $\lim_{t\to 0^+}\partial_x\theta_\alpha(x,t) = 0$, by Theorem 2.1, we deduce

$$\partial_t^\alpha u_2(x,t) = -2\int_0^t \partial_{xs}^R\partial_t^\alpha\theta_\alpha(x,t-s)^R\partial_t^{1-\alpha}g_1(s)\,ds$$

$$= -2\int_0^t \partial_{xxx}^3\theta_\alpha(x,t-s)^R\partial_t^{1-\alpha}g_1(s)\,ds$$

$$= -2\int_0^t \partial_{xx}^2(\partial_x\bar\theta_\alpha)(x,t-s)g_1(s)\,ds = \partial_{xx}^2 u_2(x,t),$$

where changing the order of derivatives $^R\partial_t^{1-\alpha}$ and ∂_x is justified by the uniform convergence of the series for θ_α and its fractional derivatives, see Lemma 7.4. The assertion holds also for piecewise continuous g_1 by a density argument. The other terms u_3 and u_4 can be verified similarly. Thus, the representation satisfies the governing equation in (7.11). The initial condition follows from the construction of θ_α and Lemma 7.2 by

$$\lim_{t\to 0^+} u_1(x,t) = \lim_{t\to 0^+}\int_0^1 [\theta_\alpha(x-y,t) - \theta_\alpha(x+y,t)]u_0(y)dy$$

$$= \lim_{t\to 0^+}\int_{-\infty}^\infty G_\alpha(x-y,t)\tilde u_0(y)dy = u_0(x),$$

for every point $x \in \overline\Omega$ of continuity of u_0, where $\tilde u_0$ is the periodic extension of u_0 to $\mathbb{R}$ described above. The boundary conditions follow from Lemma 7.5, and we analyze only the term u_2. At $x = 0$, (7.14) in Lemma 7.5 implies

$$\lim_{x\to 0^+} u_2(x,t) = -2\lim_{x\to 0^+}\int_0^t \partial_x\bar\theta_\alpha(x,t-s)g_1(s)ds = g_1(t).$$

At $x = 1$, (7.16) and Lebesgue's dominated convergence theorem imply

$$\lim_{x\to 1^-} u_2(x,t) = -2\lim_{x\to 1^-}\int_0^t \partial_x\bar\theta_\alpha(x,t-s)g_1(s)ds = 0.$$

The initial and boundary conditions for the other terms can be treated similarly. □

A similar argument gives a Neumann analogue of Theorem 7.1.

Theorem 7.2 *Let* $\Omega = (0, 1)$. *Then for piecewise smooth functions* u_0, f, g_1 *and* g_2, *the solution* u *to*

$$
\begin{cases}
\partial_t^\alpha u - \partial_{xx}^2 u = f, & in \ \Omega \times I, \\
-\partial_x u(0, \cdot) = g_1, & on \ I, \\
\partial_x u(1, \cdot) = g_2, & on \ I, \\
u(0) = u_0, & in \ \Omega
\end{cases}
$$

is represented by

$$
u(x, t) = \int_0^1 \left[\theta_\alpha(x - y, t) + \theta_\alpha(x + y, t) \right] u_0(y) \, dy
$$

$$
- 2 \int_0^1 \bar{\theta}_\alpha(x, t - s) g_1(s) \, ds - 2 \int_0^1 \bar{\theta}_\alpha(x - 1, t - s) g_2(s) \, ds
$$

$$
+ \int_0^t \int_0^1 \left[\bar{\theta}_\alpha(x - y, t - s) + \bar{\theta}_\alpha(x + y, t - s) \right] f(y, s) \, dy \, ds.
$$

7.2 Hölder Regularity in One Dimension

Now we turn to Hölder regularity of solutions of time-fractional diffusion, which have different regularity pickups in the space and time variables. Thus, anisotropic Hölder spaces will be used extensively. Let $\Omega \subset \mathbb{R}^d$ be an open bounded domain and $T > 0$ be the final time. We denote by $Q = \Omega \times (0, T]$ the usual parabolic cylinder. For any $\theta \in (0, 1)$ and $\alpha \in (0, 1)$, the notation $C^{\theta, \frac{\theta}{2}\alpha}(\overline{Q})$ denotes the set of functions v defined in $\overline{Q}$ and having a finite norm

$$
\|v\|_{C^{\theta, \frac{\theta}{2}\alpha}(\overline{Q})} = \|v\|_{C(\overline{Q})} + [v]_{C^{\theta, \frac{\theta}{2}\alpha}(\overline{Q})},
$$

with

$$
\|v\|_{C(\overline{Q})} = \sup_{(x,t)\in\overline{Q}} |v(x,t)|, \quad [v]_{C^{\theta, \frac{\theta}{2}\alpha}(\overline{Q})} = [v]_{x,\overline{Q}}^{(\theta)} + [v]_{t,\overline{Q}}^{(\frac{\theta}{2}\alpha)},
$$

$$
[v]_{x,\overline{Q}}^{(\theta)} = \sup_{(x,t),(y,t)\in\overline{Q}, x\neq y} \frac{|v(x,t) - v(y,t)|}{|x - y|^\theta},
$$

$$
[v]_{t,\overline{Q}}^{(\frac{\theta}{2}\alpha)} = \sup_{(x,t),(x,t')\in\overline{Q}, t\neq t'} \frac{|v(x,t) - v(x,t')|}{|t - t'|^{\frac{\theta}{2}\alpha}}.
$$

The notations $[\cdot]_{x,\overline{Q}}^{(\theta)}$ and $[\cdot]_{t,\overline{Q}}^{(\theta)}$ denote Hölder seminorm in space and time, respectively. We use extensively the space $C^{2+\theta, \frac{2+\theta}{2}\alpha}(\overline{Q})$ in this chapter.

Definition 7.1 A function $v : \overline{Q} \to \mathbb{R}$ is said to be in the space $C^{2+\theta, \frac{2+\theta}{2}\alpha}(\overline{Q})$ if and only if the function v and its spatial derivatives $D^\ell v(x,t)$, $|\ell| = 0, 1, 2$, (left-sided) Djrbashian-Caputo fractional derivatives in time t based at $t = 0$ $\partial_t^\alpha v$ are continuous and the following norms are finite (with $m = 0, 1$):

$$\|v\|_{C^{2+\theta, \frac{2+\theta}{2}\alpha}(\overline{Q})} = \sum_{|\ell|+2m \leq 2} \|\partial_t^{m\alpha} D_x^\ell v\|_{C(\overline{Q})} + [v]_{C^{2+\theta, \frac{2+\theta}{2}\alpha}(\overline{Q})},$$

$$[v]_{C^{2+\theta, \frac{2+\theta}{2}\alpha}(\overline{Q})} = \sum_{|\ell|+2m=2} [\partial_t^{m\alpha} D_x^\ell v]_{C^{\theta, \frac{\theta}{2}\alpha}(\overline{Q})} + \sum_{|\ell|=1} [D_x^\ell v]_{t,\overline{Q}}^{(\frac{1+\theta}{2}\alpha)}.$$

Further, the notation $C_0^{k+\theta, \frac{k+\theta}{2}\alpha}(\overline{Q})$ denotes the subspace of $C^{k+\theta, \frac{k+\theta}{2}\alpha}(\overline{Q})$, $k = 0, 1, 2$, consisting of functions $v(x,t)$ such that $\partial_t^{m\alpha} v|_{t=0} = 0$, where $2m \leq k$.

It is worth noting that with the help of local coordinates and partition of unity, these spaces can also be defined on manifolds. In the case $\alpha = 1$, the spaces $C^{k+\theta, \frac{k+\theta}{2}\alpha}(\overline{Q})$ coincide with the classical Hölder spaces $C^{k+\theta, \frac{k+\theta}{2}}(\overline{Q})$ for standard parabolic problems [LSU68, Chapter I].

In this section, we derive $C^{2+\theta, \frac{2+\theta}{2}\alpha}(\overline{Q})$ regularity of the solutions u to one-dimensional subdiffusion problems, and leave the more complex multi-dimensional case to Section 7.3. The overall analysis strategy parallels closely that of the standard parabolic case [LSU68]. This is achieved using mapping properties of the operators $\mathcal{I}$ and $\mathcal{V}$ in Hölder spaces. The analysis below begins with the real line $\mathbb{R}$, and then half real line $\mathbb{R}_+ = (0, \infty)$ and last a bounded interval with variable coefficients. In the first two cases, the analysis relies on sharp mapping properties of suitable integral operators in Hölder spaces, which in turn boils down certain estimates on the Wright function $W_{\rho,\mu}(z)$ (cf. (3.24) in Chapter 3) appearing in the fundamental solutions, cf. Lemma 7.6. In the last case, the analysis employs freezing coefficient and reduction to the whole space and half space.

7.2.1 Subdiffusion in $\mathbb{R}$

First, we consider the case $\Omega = \mathbb{R}$. The analysis uses extensively the following pointwise bounds on $G_\alpha(x,t)$ and $\overline{G}_\alpha(x,t)$ and their mixed derivatives.

Lemma 7.6 *The following statements hold with $r = |x|t^{-\frac{\alpha}{2}}$, and $\sigma(\alpha) > 0$.*

(i) $t^{\frac{\alpha}{2}} G_\alpha(x,t) + t^{1-\frac{\alpha}{2}} \overline{G}_\alpha(x,t) + t|\partial_x \overline{G}_\alpha(x,t)| \leq c e^{-\sigma(\alpha) r^{\frac{1}{1-\frac{\alpha}{2}}}}$.

(ii) *For any $x \neq 0$ and $t > 0$,*

$$|\partial_t G_\alpha(x,t)| \leq ct^{-1-\frac{\alpha}{2}}(1+r^{1+\frac{2}{\alpha}})^{-1}, \quad |\partial_{xxx}^3 \overline{G}_\alpha(x,t)| \leq ct^{-1-\alpha}(1+r^{2+\frac{2}{\alpha}})^{-1},$$

$$|\partial_{xxt}^3 \overline{G}_\alpha(x,t)| \leq ct^{-2-\frac{\alpha}{2}}(1+r^{1+\frac{2}{\alpha}})^{-1}.$$

Proof In view of (7.4) and (3.26), we have

$$\partial_x \overline{G}_\alpha(x,t) = -(2t)^{-1} W_{-\frac{\alpha}{2},0}(-|x|t^{-\frac{\alpha}{2}})\text{sgn}(x).$$

Thus, the assertions in (i) follow directly from Theorem 3.8(i) and (iv) for the first two and the third, respectively. Next, the formulas (3.26) and (3.27) imply

$$\partial_t G_\alpha(x,t) = t^{-\frac{\alpha}{2}-1} W_{-\frac{\alpha}{2},-\frac{\alpha}{2}}(-|x|t^{-\frac{\alpha}{2}}),$$

$$\partial^3_{xxx} \overline{G}_\alpha(x,t) = -2^{-1} t^{-1-\alpha} W_{-\frac{\alpha}{2},-\alpha}(-|x|t^{-\frac{\alpha}{2}})\text{sgn}(x),$$

$$\partial^3_{xxt} \overline{G}_\alpha(x,t) = 2^{-1} t^{-2-\frac{\alpha}{2}} W_{-\frac{\alpha}{2},-\frac{\alpha}{2}-1}(-|x|t^{-\frac{\alpha}{2}}).$$

Then the first two estimates in (ii) follow from Theorem 3.8(iii). Now the identity (3.33) and Theorem 3.8(iii) lead to

$$|W_{-\frac{\alpha}{2},-\frac{\alpha}{2}-1}(-r)| \le cr(1+r^{2+\frac{2}{\alpha}})^{-1} + c(1+r^{1+\frac{2}{\alpha}})^{-1} \le c(1+r^{1+\frac{2}{\alpha}})^{-1}.$$

This shows directly the third estimate, and completes the proof of the lemma.   □

We begin with several integral bounds on $G_\alpha(x,t)$ and $\overline{G}_\alpha(x,t)$.

Lemma 7.7 *For any $\delta > 0$ and $\theta \in (0,1)$, the following estimates hold:*

(i) $\int_0^t \int_{-\infty}^\infty |\partial_s G_\alpha(x,s)||x|^\theta dx ds \le \frac{c}{\theta\alpha} t^{\frac{\theta\alpha}{2}}$;

(ii) $\int_0^t \left| \int_{|x|>\delta} \partial^2_{xx} \overline{G}_\alpha(x,s)dx \right| ds \le c$;

(iii) $\int_0^t \int_{|x|<\delta} |\partial^2_{xx} \overline{G}_\alpha(x,s)||x|^\theta dx ds \le \frac{c}{\theta}\delta^\theta$;

(iv) $\int_0^t \int_{|x|>\delta} |\partial^3_{xxx} \overline{G}_\alpha(x,s)||x|^\theta dx ds \le \frac{c}{1-\theta}\delta^{\theta-1}$.

Proof First, by Lemma 7.6(ii) and changing variables $r = xs^{-\frac{\alpha}{2}}$, we have

$$\int_0^t \int_{-\infty}^\infty |\partial_s G_\alpha(x,s)||x|^\theta dx ds \le c \int_0^t \int_{-\infty}^\infty \frac{s^{-1-\frac{\alpha}{2}}}{1+(|x|s^{-\frac{\alpha}{2}})^{1+\frac{2}{\alpha}}} |x|^\theta dx ds$$

$$= c \int_0^t s^{-1+\frac{\theta\alpha}{2}} ds \int_0^\infty \frac{r^\theta}{1+r^{1+\frac{2}{\alpha}}} dr,$$

which, upon integration, gives (i). Second, by (3.26) and Theorem 3.8(i)

$$\left| \int_{|x|>\delta} \partial^2_{xx} \overline{G}_\alpha(x,t)dx \right| \le c| \lim_{x\to\infty} \partial_x \overline{G}_\alpha(x,t) - \partial_x \overline{G}_\alpha(\delta,t)| = ct^{-1} W_{-\frac{\alpha}{2},0}(-\delta t^{-\frac{\alpha}{2}}).$$

The differentiation relation (3.27) with $n = 1$ gives

$$\int_0^t \left| \int_{|x|>\delta} \partial^2_{xx} \overline{G}_\alpha(x,s)dx \right| ds \le c \int_0^t \frac{d}{ds} W_{-\frac{\alpha}{2},1}(-\delta s^{-\frac{\alpha}{2}}) ds.$$

This and Theorem 3.8(i) show (ii). Third, (iii) follows from Lemmas 7.1(iii) and 7.6(ii) as

$$\int_0^t \int_{|x|<\delta} |\partial^2_{xx} \overline{G}_\alpha(x,s)||x|^\theta dx ds \le c \int_0^\delta x^{\theta-1} dx \int_0^\infty \frac{dr}{1+r^{1+\frac{2}{\alpha}}} \le \frac{c}{\theta}\delta^\theta.$$

Last, by Lemma 7.6(ii) and changing variables $r = xs^{-\frac{\alpha}{2}}$, we obtain

$$\int_0^t \int_{|x|>\delta} |\partial_{xxx}^3 \overline{G}_\alpha(x,s)||x|^\theta dx ds \leq c \int_\delta^\infty x^\theta \int_0^t \frac{s^{-\alpha-1}}{1+(xs^{-\frac{\alpha}{2}})^{2+\frac{2}{\alpha}}} ds dx$$

$$\leq c \int_\delta^\infty x^{\theta-2} dx \int_0^\infty \frac{r}{1+r^{2+\frac{2}{\alpha}}} dr \leq \frac{c}{1-\theta}\delta^{\theta-1}.$$

This completes the proof of the lemma. $\square$

The next result bounds the volume potential $\mathcal{V}f$ in Hölder spaces.

Lemma 7.8 *Let $\theta \in (0,1)$, $f \in C^{\theta, \frac{\theta\alpha}{2}}(\overline{Q})$ and $f(x,t) \equiv 0$ for $|x| \geq R$, for some $R > 0$. Then the function $u(x,t) = \mathcal{V}f(x,t)$ satisfies*

$$\|u\|_{C(\overline{Q})} + \|\partial_{xx}^2 u\|_{C(\overline{Q})} \leq c(\|f\|_{C(\overline{Q})} + [f]_{x,\overline{Q}}^{(\theta)}),$$

$$[\partial_{xx}^2 u]_{x,\overline{Q}}^{(\theta)} + [\partial_{xx}^2 u]_{t,\overline{Q}}^{(\frac{\theta\alpha}{2})} \leq c[f]_{x,\overline{Q}}^{(\theta)}.$$

Proof The bound on $\|u\|_{C(\overline{Q})}$ is direct from Lemma 7.1(i) and the identity (7.7). Next, by applying (7.7) again and Lemma 7.1(iii) (see also the proof of Lemma 7.3), we can rewrite $\partial_{xx}^2 u$ as

$$\partial_{xx}^2 u(x,t) = \int_0^t \int_{\mathbb{R}} \partial_{xx}^2 \overline{G}_\alpha(x-y,t-s)[f(y,s)-f(x,s)]dy ds. \qquad (7.17)$$

Now Lemmas 7.1(iii) and 7.7(i) imply

$$\|\partial_{xx}^2 u\|_{C(\overline{Q})} \leq [f]_{x,\overline{Q}}^{(\theta)} \int_0^t \int_{\mathbb{R}} |\partial_{xx}\overline{G}_\alpha(x,s)||x|^\theta dx ds \leq \frac{c}{\theta\alpha}t^{\frac{\theta\alpha}{2}}[f]_{x,\overline{Q}}^{(\theta)}.$$

To bound the seminorm $[\partial_{xx}^2 u]_{x,\overline{Q}}^{(\theta)}$, we take any $x_1, x_2 \in \mathbb{R}$ with $x_2 > x_1$, and let $h := x_2 - x_1$. Using the identity (7.17), the difference $\partial_{xx}^2 u(x_2,t) - \partial_{xx}^2 u(x_1,t)$ can be split into (with $K = \{y \in \mathbb{R} : |x_1 - y| < 2h\}$)

$$\partial_{xx}^2 u(x_2,t) - \partial_{xx}^2 u(x_1,t) =: \int_0^t \int_K \partial_{xx}^2 \overline{G}_\alpha(x_1-y,t-s)[f(x_1,s)-f(y,s)]dy ds$$

$$+ \int_0^t \int_K \partial_{xx}^2 \overline{G}_\alpha(x_2-y,t-s)[f(y,s)-f(x_2,s)]dy ds$$

$$+ \int_0^t \int_{\mathbb{R}\backslash K} \partial_{xx}^2 \overline{G}_\alpha(x_2-y,t-s)[f(x_1,s)-f(x_2,s)]dy ds$$

$$+ \int_0^t \int_{\mathbb{R}\backslash K} [\partial_{xx}^2 \overline{G}_\alpha(x_1-y,t-s) - \partial_{xx}^2 \overline{G}_\alpha(x_2-y,t-s)][f(x_1,s)-f(y,s)]dy ds.$$

The four terms are denoted by I_i, $i = 1, \ldots, 4$. For the terms I_1 and I_2, the Hölder continuity of f and Lemma 7.7(iii) with $\delta = 2h$ and $\delta = 3h$, respectively, imply

$$|I_1| + |I_2| \le ch^\theta [f]_{x,\overline{Q}}^{(\theta)}.$$

By Lemma 7.7(ii) with $\delta = h$, we estimate I_3 by

$$|I_3| \le ch^\theta [f]_{x,\overline{Q}}^{(\theta)} \int_0^t \left| \int_{|x|>h} \partial_{xx}^2 \overline{G}_\alpha(x,s) dx \right| ds \le ch^\theta [f]_{x,\overline{Q}}^{(\theta)}.$$

Last, the mean value theorem and Lemma 7.7(iv) with $\delta = h$ give

$$|I_4| \le ch [f]_{x,\overline{Q}}^{(\theta)} \int_0^t \int_{|x|>h} |x|^\theta |\partial_{xxx}^3 \overline{G}_\alpha(\zeta,s)| d\zeta ds \le ch^\theta [f]_{x,\overline{Q}}^{(\theta)}.$$

These bounds show $[\partial_{xx}^2 u]_{x,\overline{Q}}^{(\theta)} \le c[f]_{x,\overline{Q}}^{(\theta)}$. To bound $[\partial_{xx}^2 u]_{t,\overline{Q}}^{(\frac{\theta\alpha}{2})}$, for any $t_1, t_2 \in \overline{I}$, $t_2 > t_1$, let $\tau = t_2 - t_1$, and we discuss the cases $t_1 > 2\tau$ and $t_1 \le 2\tau$ separately. If $t_1 > 2\tau$, by (7.17), the difference $\partial_{xx}^2 u(x,t_2) - \partial_{xx}^2 u(x,t_1)$ can be split into

$$\partial_{xx}^2 u(x,t_2) - \partial_{xx}^2 u(x,t_1) = \int_{\mathbb{R}} \int_{t_1-2\tau}^{t_2} \partial_{xx}^2 \overline{G}_\alpha(x-y,t_2-s)[f(y,s) - f(x,s)] ds dy$$

$$+ \int_{\mathbb{R}} \int_{t_1-2\tau}^{t_1} \partial_{xx}^2 \overline{G}_\alpha(x-y,t_1-s)(f(x,s) - f(y,s)) ds dy$$

$$+ \int_0^{t_1-2\tau} \int_{\mathbb{R}} [\partial_{xx}^2 \overline{G}_\alpha(x-y,t_2-s) - \partial_{xx}^2 \overline{G}_\alpha(x-y,t_1-s)][f(y,s) - f(x,s)] dy ds.$$

Next, we bound the three terms, denoted by II_i, $i = 1, 2, 3$. For II_1, changing variables $\xi = t_2 - s$ and then applying Lemmas 7.1(iii) and 7.7(i) with $t = \tau$ lead to

$$|II_1| \le c\tau^{\frac{\theta\alpha}{2}} [f]_{x,\overline{Q}}^{(\theta)}.$$

II_2 can be bounded similarly. For II_3, the mean value theorem and Lemma 7.6(ii) imply (with $r = |x - y| s^{-\frac{\alpha}{2}}$)

$$|II_3| \le c\tau [f]_{x,\overline{Q}}^{(\theta)} \int_\tau^t s^{\frac{\alpha\theta}{2}-2} \int_0^\infty \frac{r^\theta}{1 + r^{1+\frac{2}{\alpha}}} dr ds \le c[f]_{x,\overline{Q}}^{(\theta)} \tau^{\frac{\theta\alpha}{2}},$$

since $t > t_1 > 2\tau$. In sum, for $t_1 > 2\tau$, we obtain

$$[\partial_{xx}^2 u]_{t,\overline{Q}}^{(\frac{\alpha\theta}{2})} \le c[f]_{x,\overline{Q}}^{(\theta)} \tau^{\frac{\theta\alpha}{2}}.$$

Next, we show the estimate for the case $t_1 \le 2\tau$. Indeed,

$$\partial_{xx}^2 u(x,t_2) - \partial_{xx}^2 u(x,t_1) = \int_{-\infty}^\infty \int_0^{t_1} \partial_{xx}^2 \overline{G}_\alpha(x-y,t_1-s)[f(x,s) - f(y,s)] dy ds$$

$$+ \int_{-\infty}^\infty \int_0^{t_2} [f(y,s) - f(x,s)] \partial_{xx}^2 \overline{G}_\alpha(x-y,t_2-s) ds dy.$$

The bounds on the two terms, denoted by III_1 and III_2, follow directly from Lemmas 7.1(iii) and 7.7(i) by

$$|\mathrm{III}_1| + |\mathrm{III}_2| \le c[f]^{(\theta)}_{x,\overline{Q}} t_2^{\frac{\theta\alpha}{2}} \le c[f]^{(\theta)}_{x,\overline{Q}} \tau^{\frac{\theta\alpha}{2}}.$$

This gives the assertion for $t_1 \le 2\tau$, and completes the proof of the lemma. $\qquad\square$

The next lemma treats the potential $\mathcal{I}u_0$.

Lemma 7.9 *Let* $u_0 \in C^{2+\theta}(\mathbb{R})$ *and* $u_0(x) \equiv 0$ *for* $|x| \ge R$, *for some* $R > 0$. *Then the function* $u = \mathcal{I}u_0$ *satisfies*

$$\|u\|_{C(\overline{Q})} + \|\partial^2_{xx}u\|_{C(\overline{Q})} \le c\|u_0\|_{C^2(\mathbb{R})},$$

$$[\partial^2_{xx}u]^{(\theta)}_{x,\overline{Q}} + [\partial^2_{xx}u]^{(\frac{\alpha\theta}{2})}_{t,\overline{Q}} \le c[\partial^2_{xx}u_0]^{(\theta)}_{x,\mathbb{R}}.$$

Proof The proof is similar to Lemma 7.8. We only prove the second estimate. From the identity $\partial^2_{xx}u(x,t) = \int_{-\infty}^{\infty} G_\alpha(y,t)\partial^2_{xx}u_0(x-y)dy$, we deduce

$$\partial^2_{xx}u(x_1,t) - \partial^2_{xx}u(x_2,t) = \int_{-\infty}^{\infty} G_\alpha(y,t)(\partial^2_{xx}u_0(x_1-y) - \partial^2_{xx}u_0(x_2-y))dy.$$

Thus, Lemma 7.1(i) and (7.6) imply that for any $x_2 > x_1$,

$$|\partial^2_{xx}u(x_1,t) - \partial^2_{xx}u(x_2,t)| \le [\partial^2_{xx}u_0]^{(\theta)}_{x,\mathbb{R}}|x_1 - x_2|^\theta.$$

Next, we show Hölder regularity in time. Since G_α is even in x, by (7.6),

$$u(x,t) = \int_{-\infty}^{\infty} G_\alpha(y,t)u_0(x-y)dy = \int_{-\infty}^{\infty} G_\alpha(y,t)u_0(x+y)dy$$

$$= \frac{1}{2}\int_{-\infty}^{\infty} G_\alpha(y,t)(u_0(x-y) + u_0(x+y))dy$$

$$= u_0(x) + \frac{1}{2}\int_{-\infty}^{\infty} G_\alpha(y,t)(u_0(x-y) - 2u_0(x) + u_0(x+y))dy.$$

Now for any $t_2 > t_1$, let $\tau = t_2 - t_1$. Suppose first $\tau > t_1$. Then by the mean value theorem, there exists some $\xi \in (t_1,t_2)$ such that

$$|\partial^2_{xx}u(x,t_1) - \partial^2_{xx}u(x,t_2)|$$

$$= \frac{1}{2}\left|\int_{-\infty}^{\infty} (G_\alpha(y,t_1) - G_\alpha(y,t_2))\partial^2_{xx}(u_0(x-y) - 2u_0(x) + u_0(x+y))dy\right|$$

$$\le c\tau[\partial^2_{xx}u_0]^{(\theta)}_{x,\mathbb{R}}\int_{-\infty}^{\infty} |\partial_t G_\alpha(y,\xi)||y|^\theta dy,$$

which together with Lemma 7.6(ii) gives

$$|\partial_{xx}^2 u(x,t_1) - \partial_{xx}^2 u(x,t_2)| \le c\tau [\partial_{xx}^2 u_0]_{x,\mathbb{R}}^{(\theta)} \int_{-\infty}^{\infty} \frac{\xi^{-1-\frac{\alpha}{2}}}{1+(|y|\xi^{-\frac{\alpha}{2}})^{1+\frac{2}{\alpha}}} |y|^\theta dy$$

$$\le c\tau \xi^{\frac{\theta}{2}\alpha - 1}[\partial_{xx}^2 u_0]_{x,\mathbb{R}}^{(\theta)} \int_0^\infty \frac{r^\theta}{1+r^{1+\frac{2}{\alpha}}} dr \le c\tau^{\frac{\theta}{2}\alpha}[\partial_{xx}^2 u_0]_{x,\mathbb{R}}^{(\theta)},$$

where the last step follows from $\xi \in (t_1, t_2)$. The case $t_1 \le \tau$ follows similarly. $\square$

Now we can state the unique solvability of problem (7.1) in $\mathbb{R}$.

Theorem 7.3 *Let $\theta \in (0,1)$, and $f \in C^{\theta,\frac{\theta\alpha}{2}}(\overline{Q})$, $u_0 \in C^{2+\theta}(\mathbb{R})$, and the condition (7.2) be fulfilled. Then there exists a unique solution $u \in C^{2+\theta,\frac{2+\theta}{2}\alpha}(\overline{Q})$ of problem (7.1), given in (7.3), and it satisfies*

$$\|u\|_{C^{2+\theta,\frac{2+\theta}{2}\alpha}(\overline{Q})} \le c\left(\|f\|_{C^{\theta,\frac{\theta}{2}\alpha}(\overline{Q})} + \|u_0\|_{C^{2+\theta}(\mathbb{R})}\right).$$

Proof The regularity estimate is direct from Lemmas 7.8 and 7.9 and interpolation inequalities. It suffices to show the uniqueness of the solution $u \in C^{2+\theta,\frac{2+\theta}{2}\alpha}(\overline{Q})$ or, equivalently, the homogeneous problem has only the zero solution:

$$\begin{cases} \partial_t^\alpha u - \partial_{xx}^2 u = 0, & \text{in } Q, \\ u(0) = 0, & \text{in } \mathbb{R}, \end{cases}$$

following the strategy of [Sol76], Clearly, there exists a solution $u \in C^{2+\theta,\frac{2+\theta}{2}\alpha}(\overline{Q})$. Let $\chi_R \in C^\infty(\mathbb{R})$ be a smooth cut-off function satisfying $\chi_R(x) = 1$ for $|x| \le R$ and $\chi_R(x) = 0$ for $|x| \ge 2R$ and

$$\left|\frac{d^k}{dx^k}\chi_R(x)\right| \le cR^{-k}, \quad x \in \mathbb{R}, k = 1, 2, \ldots.$$

Let $u_R(x,t) = u(x,t)\chi_R(x) \in C^{2+\theta,\frac{2+\theta}{2}\alpha}(\overline{Q})$. It has a finite support and satisfies

$$\begin{cases} \partial_t^\alpha u_R - \partial_{xx}^2 u_R = f_R, & \text{in } \mathbb{R} \times (0,\infty), \\ u_R(x,0) = 0, & \text{in } \mathbb{R}, \end{cases}$$

with $f_R = 2\partial_x u \partial_x \chi_R + u\partial_{xx}^2 \chi_R$. It can be verified directly that

$$\|f_R\|_{C^{\theta,\frac{\theta}{2}\alpha}(\overline{Q})} \le cR^{-\theta}\|u\|_{C^{2+\theta,\frac{2+\theta}{2}\alpha}(\overline{Q})}.$$

By the properties of the functions u_R and χ_R, using Laplace transform, we obtain

$$u_R(x,t) = \int_0^t \int_{\mathbb{R}} \overline{G}_\alpha(x-y, t-s) f_R(y,s) dy ds.$$

Since f_R satisfies the conditions of Lemma 7.8, $\|u_R\|_{C(\overline{Q})} \le cR^{-\theta}\|u\|_{C^{2+\theta,\frac{2+\theta}{2}\alpha}(\overline{Q})}$. Letting R to infinity gives $u_R \equiv 0$, which completes the proof of the theorem. $\square$

7.2.2 Subdiffusion in $\mathbb{R}_+$

Consider the following initial boundary value problems on the positive real semi-axis $\Omega = \mathbb{R}_+$ and $Q = \Omega \times I$:

$$\begin{cases} \partial_t^\alpha u - \partial_{xx}^2 u = f, & \text{in } Q, \\ u(x,0) = u_0(x), & \text{in } \Omega, \\ u(0,t) = g(t), & \text{on } I, \\ u(x,t) \to 0 & \text{as } x \to \infty, \forall t \in \overline{I}, \end{cases} \tag{7.18}$$

and

$$\begin{cases} \partial_t^\alpha u - \partial_{xx}^2 u = f, & \text{in } Q, \\ u(x,0) = u_0(x), & \text{in } \Omega, \\ \partial_x u(0,t) = g(t), & \text{on } I, \\ u(x,t) \to 0 & \text{as } x \to \infty, \forall t \in \overline{I}. \end{cases} \tag{7.19}$$

We shall assume the following compatibility condition:

$$\begin{cases} u_0(0) = g(0), & \text{for } (7.18), \\ \partial_x u_0(0) = g(0), & \text{for } (7.19), \end{cases} \tag{7.20}$$

and there exists some $R > 0$ such that

$$u_0(x), \ f(x,t) \equiv 0, \quad \text{if } x > R, \forall t \in \overline{I}. \tag{7.21}$$

First, we discuss the compatibility condition (7.20) for problem (7.18). Without loss of generality, it suffices to study $g \equiv 0$. By taking the odd extension of u_0 to $\mathbb{R}$, i.e., $u_0(-x) = -u_0(x)$ for $x > 0$ and similarly for $f(x,t)$, then the solution u to problem (7.18) is given by

$$\begin{aligned} u(x,t) &= \int_{-\infty}^{\infty} G_\alpha(x-y,t) u_0(y)\, dy + \int_0^t \int_{-\infty}^{\infty} \overline{G}_\alpha(x-y,t-s) f(y,s) dy ds \\ &= \int_0^{\infty} (G_\alpha(x-y,t) - G_\alpha(x+y,t)) u_0(y)\, dy \\ &\quad + \int_0^t \int_0^{\infty} (\overline{G}_\alpha(x-y,t) - \overline{G}_\alpha(x+y,s)) f(y,s)\, dy ds. \end{aligned}$$

Clearly, for any integrable and bounded u_0 and f, u is continuous for any $t > 0$. It is natural to ask what happens as $t \to 0^+$. It turns out that this depends very much on the compatibility condition (7.20). If u_0 is continuous and compatible, i.e., $u_0(0) = 0$, then u is continuous up to $t = 0$. Now suppose that $u_0(0) \neq 0$, when $f \equiv 0$. Then u is discontinuous at $(x,t) = (0,0)$. To determine the nature of the discontinuity, we rewrite it as

$$u(x,t) = \int_{-\infty}^{x} G_\alpha(y,t) u_0(x-y)\, dy - \int_{x}^{\infty} G_\alpha(y,t) u_0(y-x)\, dy.$$

Let $\psi(x,t)$ be the solution corresponding to the case $u_0(x) = \chi_{\mathbb{R}_+}(x)$, the characteristic function of the set $\mathbb{R}_+$, i.e., $\psi(x,t) = \int_{-\infty}^{x} G_\alpha(y,t)\, dy - \int_{x}^{\infty} G_\alpha(y,t)\, dy$. By the identity (7.6), we can simplify the expression to

$$\psi(x,t) = 1 - 2\int_{x}^{\infty} G_\alpha(y,t)\, dy = 1 - t^{-\frac{\alpha}{2}} \int_{x}^{\infty} W_{-\frac{\alpha}{2}, 1-\frac{\alpha}{2}}(-|y| t^{-\frac{\alpha}{2}})\, dy,$$

and obtain ψ in the form of a similarity solution (for $x > 0$)

$$\psi(x,t) = \Psi(xt^{-\frac{\alpha}{2}}) \quad \text{with } \Psi(x) = 1 - \int_{x}^{\infty} W_{-\frac{\alpha}{2}, 1-\frac{\alpha}{2}}(-y)\, dy.$$

Note that the identities

$$\lim_{x \to 0^+} \psi(x,t) = \Psi(0) = 0, \quad \forall t > 0,$$

$$\lim_{t \to 0^+} \psi(x,t) = \Psi(\infty) = 1, \quad \forall x > 0.$$

Thus, $\psi(x,t)$ is not continuous at $(0,0)$, and it characterizes precisely the singular behavior with incompatible problem data. The solution u can be written as $u(x,t) = u_0(0)\psi(x,t) + w(x,t)$, where w is the solution to problem (7.18) with initial data $u_0(x) - u_0(0)$, and thus continuous at $(0,0)$. The preceding discussion applies also to problem (7.19). These discussions motivate the compatibility condition (7.20).

Now we derive Hölder regularity for problems (7.18) and (7.19), with $u_0 \equiv 0$ and $f \equiv 0$, since the general case can be analyzed using the linearity of the problems. Problems (7.18) and (7.19) can be analyzed similarly, and thus we discuss only the latter, i.e., $u \in C^{2+\theta, \frac{2+\theta}{2}\alpha}(\overline{Q})$ for $g(t) \in C^{\frac{1+\theta}{2}}(\overline{I})$. Specifically, applying Laplace transform (cf. Section A.3.1 for the definition) to problem (7.19) and Lemma 2.9, we obtain the following ODE for $\hat{u}(x,z)$:

$$\begin{cases} z^\alpha \hat{u}(x,z) - \partial_{xx}^2 \hat{u}(x,z) = 0, & x \in \Omega, \\ \hat{u}(x,z) = 0, & \text{if } x \to \infty, \\ \partial_x \hat{u}(0,z) = \hat{g}(z). \end{cases}$$

The unique solution $\hat{u}(x,z)$ is given by $\hat{u}(x,z) = -z^{-\frac{\alpha}{2}} e^{-xz^{\frac{\alpha}{2}}} \hat{g}(z)$, which together with inverse Laplace transform and Proposition 3.3 yields

$$u(x,t) = \int_{0}^{t} \widetilde{G}_\alpha(x, t-s) g(s)\, ds, \tag{7.22}$$

with the corresponding fundamental solution $\widetilde{G}_\alpha(x,t)$ given by

$$\widetilde{G}_\alpha(x,t) = -t^{\frac{\alpha}{2}-1} W_{-\frac{\alpha}{2}, \frac{\alpha}{2}}(-xt^{-\frac{\alpha}{2}}) = -2\overline{G}_\alpha(x,t).$$

Similarly, the solution $u(x, t)$ to problem (7.18) is given by

$$u(x, t) = \int_0^t \check{G}_\alpha(x, t - s)g(s)ds,$$

with $\check{G}_\alpha(x, t) = \mathcal{L}^{-1}[-e^{-xz^{\frac{\alpha}{2}}}] = -W_{-\frac{\alpha}{2}, 1}(-xt^{-\frac{\alpha}{2}}) = -2\partial_x \overline{G}_\alpha(x, t)$.

First, we give several integral estimates on $\widetilde{G}_\alpha$.

Lemma 7.10 *Let $\theta \in (0, 1)$, $x \in \Omega$, $t \in I$. Then the following estimates hold:*

(i) $\int_0^\delta |\partial_x \widetilde{G}_\alpha(x, s)|s^{\frac{1+\theta}{2}\alpha}ds \le c\delta^{\frac{1+\theta}{2}\alpha}$,

(ii) $\int_\delta^\infty |\partial_{xxs}^3 \widetilde{G}_\alpha(x, s)|s^{\frac{1+\theta}{2}\alpha}ds \le c\delta^{-1+\frac{\theta}{2}\alpha}$,

(iii) $\int_0^\delta |\partial_{xx}^2 \widetilde{G}_\alpha(x, s)|s^{\frac{1+\theta}{2}\alpha}ds \le c\delta^{\frac{\theta}{2}\alpha}$,

(iv) $\int_\delta^\infty |\partial_{xxx}^3 \widetilde{G}_\alpha(x, s)|s^{\frac{1+\theta}{2}\alpha}ds \le c\delta^{\frac{\theta-1}{2}\alpha}$ *and*

(v) $\int_0^t (t - s)^{-\alpha} \frac{\partial^m}{\partial x^m} \overline{G}_\alpha(x, s)ds \le ct^{-\frac{m+1}{2}\alpha}e^{-\kappa(xt^{-\frac{\alpha}{2}})^{\frac{1}{1-\frac{\alpha}{2}}}}$, $m = 0, 1$.

Proof The proof uses Lemma 7.6. Let $r = xs^{-\frac{\alpha}{2}}$. By Lemma 7.6(i),

$$\int_0^\delta |\partial_x \widetilde{G}_\alpha(x, s)|s^{\frac{1+\theta}{2}\alpha}ds \le c \int_0^\delta s^{-1}s^{\frac{1+\theta}{2}\alpha}ds \le c\delta^{\frac{1+\theta}{2}\alpha}.$$

Similarly, by Lemma 7.6(ii),

$$\int_\delta^\infty |\partial_{xxs}^3 \widetilde{G}_\alpha(x, s)|s^{\frac{1+\theta}{2}\alpha}ds \le c \int_\delta^\infty s^{-2-\frac{\alpha}{2}}s^{\frac{1+\theta}{2}\alpha}ds \le c\delta^{-1+\frac{\theta}{2}\alpha}.$$

Similarly, (iii) and (iv) follow from the identity $\partial_{xx}^2 \widetilde{G}_\alpha(x, t) = -2\partial_{xx}^2 \overline{G}_\alpha(x, t) = -2\partial_t G_\alpha(x, t)$, cf. Lemma 7.1(iii), and Lemma 7.6(ii) as

$$\int_0^\delta |\partial_{xx}^2 \overset{*}{\widetilde{G}}_\alpha(x, s)|s^{\frac{1+\theta}{2}\alpha}ds \le c \int_0^\delta s^{-1-\frac{\alpha}{2}}s^{\frac{1+\theta}{2}\alpha}ds \le c\delta^{\frac{\theta}{2}\alpha},$$

$$\int_\delta^\infty |\partial_{xxx}^3 \widetilde{G}_\alpha(x, s)|s^{\frac{1+\theta}{2}\alpha}ds \le c \int_\delta^\infty s^{-1-\alpha}s^{\frac{1+\theta}{2}\alpha}ds \le c\delta^{\frac{\theta-1}{2}\alpha}.$$

The last estimate is direct from Lemma 7.6(i) and the identity (3.28).   □

The next result gives several estimates on the surface potential $\int_0^t \widetilde{G}_\alpha(x, s)g(s)ds$.

Lemma 7.11 *Let $\theta \in (0, 1)$, $g \in C^{\frac{1+\theta}{2}\alpha}(\overline{I})$ with $g(0) = 0$, then the function $u(x, t) = \int_0^t \widetilde{G}_\alpha(x, s)g(s)ds$ satisfies*

$$\|\partial_{xx}^2 u\|_{C(\overline{Q})} \le c(\|g\|_{C(\overline{I})} + [g]_{t,\overline{I}}^{(\frac{1+\theta}{2}\alpha)}),$$

$$[\partial_{xx}^2 u]_{x,\overline{Q}}^{(\theta)} + [\partial_{xx}^2 u]_{t,\overline{Q}}^{(\frac{\theta}{2}\alpha)} \le c\|g\|_{C^{\frac{1+\theta}{2}\alpha}(\overline{I})}.$$

Proof Since $g(0) = 0$, we may extend $g(t)$ by zero for $t < 0$. Then by (7.8),

$$\partial_{xx}^2 u(x,t) = \int_{-\infty}^{t} \partial_{xx} \widetilde{G}_\alpha(x, t - s)(g(s) - g(t))ds + g(t) \int_{-\infty}^{t} \partial_{xx} \widetilde{G}_\alpha(x, t - s)ds$$

$$= \int_{-\infty}^{t} \partial_{xx} \widetilde{G}_\alpha(x, t - s)(g(s) - g(t))ds.$$

Then the bound on $\|\partial_{xx}^2 u\|_{C(\overline{Q})}$ follows by (with $t_0 > 0$ fixed)

$$|\partial_{xx}^2 u(x,t)| \leq \int_0^{t_0} |\partial_{xx} \widetilde{G}_\alpha(x,s)|s^{\frac{1+\theta}{2}\alpha}ds[g]_{t,\overline{I}}^{(\frac{1+\theta}{2}\alpha)} + \int_{t_0}^{\infty} |\partial_{xx}^2 \widetilde{G}_\alpha(x,s)|ds\|g\|_{C(\overline{I})}.$$

The first term can be bounded using Lemma 7.10(iii), and the second by Lemmas 7.1(iii) and 7.6(ii). Next, for any $x_1, x_2 \in \Omega$, with $h = |x_2 - x_1|^{\frac{2}{\alpha}}$, we have

$$\partial_{xx}^2 u(x_2,t) - \partial_{xx}^2 u(x_1,t) = \int_0^h \partial_{xx}^2 \widetilde{G}_\alpha(x_2,s)(g(t-s)-g(t))ds$$

$$+ \int_0^h \partial_{xx}^2 \widetilde{G}_\alpha(x_1,s)(g(t)-g(t-s))ds$$

$$+ \int_h^{\infty} [\partial_{xx}^2 \widetilde{G}_\alpha(x_2,s) - \partial_{xx}^2 \widetilde{G}_\alpha(x_1,s)](g(t-s)-g(t))ds := \sum_{i=1}^{3} I_i.$$

The term I_1 can be bounded by Lemma 7.10(iii) as

$$|I_1| \leq c[g]_{t,\overline{I}}^{(\frac{1+\theta}{2}\alpha)} \int_0^h |\partial_{xx}^2 \widetilde{G}_\alpha(x_2,s)|s^{\frac{1+\theta}{2}\alpha}ds \leq c[g]_{t,\overline{I}}^{(\frac{1+\theta}{2}\alpha)}|x_2-x_1|^\theta.$$

The term I_2 can be bounded similarly. By the mean value theorem and Lemma 7.10(iv), we have $|I_3| \leq c[g]_{t,\overline{I}}^{(\frac{1+\theta}{2}\alpha)}|x_2-x_1|^\theta$. This gives $[\partial_{xx}^2 u]_{x,Q}^{(\theta)} \leq c[g]_{t,\overline{I}}^{(\frac{1+\theta}{2}\alpha)}$. Last we bound $[\partial_{xx}^2 u]_{t,Q}^{(\frac{\theta}{2}\alpha)}$. For $t_1, t_2 \in \overline{I}$ and $t_2 > t_1$, let $\tau = t_2 - t_1$. If $t_1 > 2\tau$, then

$$\partial_{xx}^2 u(x,t_2) - \partial_{xx}^2 u(x,t_1) = \int_{2t_1-t_2}^{t_2} \partial_{xx}^2 \widetilde{G}_\alpha(x,t_2-s)(g(s)-g(t_2))ds$$

$$+ \int_{2t_1-t_2}^{t_1} \partial_{xx}^2 \widetilde{G}_\alpha(x,t_1-s)(g(t_1)-g(s))ds$$

$$+ \int_{-\infty}^{2t_1-t_2} \partial_{xx}^2 \widetilde{G}_\alpha(x,t_1-s)ds(g(t_1)-g(t_2))$$

$$+ \int_{-\infty}^{2t_1-t_2} (\partial_{xx}^2 \widetilde{G}_\alpha(x,t_2-s) - \partial_{xx}^2 \widetilde{G}_\alpha(x,t_1-s))(g(s)-g(t_2))ds := \sum_{i=1}^{4} II_i.$$

By Lemma 7.10(iii) with $\delta = \tau$,

$$|\text{II}_1| + |\text{II}_2| \leq c\tau^{\frac{\theta}{2}\alpha}[g]_{t,\overline{I}}^{(\frac{1+\theta}{2}\alpha)}.$$

To bound II_3, we apply Lemma 7.6(ii) and deduce

$$|\text{II}_3| \leq c\tau^{\frac{1+\theta}{2}\alpha}[g]_{t,\overline{I}}^{(\frac{1+\theta}{2}\alpha)} \int_\tau^\infty |\partial_{xx}^2 \widetilde{G}_\alpha(x,s)|ds$$

$$\leq c\tau^{\frac{1+\theta}{2}\alpha}[g]_{t,\overline{I}}^{(\frac{1+\theta}{2}\alpha)} \int_\tau^\infty s^{-1-\frac{\alpha}{2}}ds \leq c\tau^{\frac{\theta}{2}\alpha}[g]_{t,\overline{I}}^{(\frac{1+\theta}{2}\alpha)}.$$

Finally, the mean value theorem and Lemma 7.10(ii) with $\delta = \tau$ lead to $|\text{II}_4| \leq c\tau^{\frac{\theta}{2}\alpha}[g]_{t,\overline{I}}^{(\frac{1+\theta}{2}\alpha)}$. This shows the bound on $[\partial_{xx}^2 u]_{t,\overline{Q}}^{(\frac{\theta}{2}\alpha)}$. The case $t_1 \leq 2\tau$ can be analyzed similarly. Combining the preceding estimates completes the proof.   □

Now we can state the following existence, uniqueness and Hölder regularity estimate for problem (7.19).

Theorem 7.4 *Let $\theta \in (0,1)$, and $g \in C^{\frac{1+\theta}{2}\alpha}(\overline{I})$ with $g(0) = 0$. Then there exists a unique solution $u(x,t) \in C^{2+\theta,\frac{2+\theta}{2}\alpha}(\overline{Q})$ of problem (7.19) with $u_0 \equiv 0, f \equiv 0$, given by (7.22) and*

$$\|u\|_{C^{2+\theta,\frac{2+\theta}{2}\alpha}(\overline{Q})} \leq c\|g\|_{C^{\frac{1+\theta}{2}\alpha}(\overline{I})}.$$

Proof First, we prove the *a priori* estimate. By Lemma 7.11 and (7.19),

$$\|\partial_t^\alpha u\|_{C^{\theta,\frac{\theta}{2}\alpha}(\overline{Q})} \leq c\|\partial_{xx}^2 u\|_{C^{\theta,\frac{\theta}{2}\alpha}(\overline{Q})} \leq c\|g\|_{C^{\frac{1+\theta}{2}\alpha}(\overline{I})}.$$

Since $u_0 = 0$, $\|u\|_{C(\overline{Q})} \leq c\|\partial_t^\alpha u\|_{C(\overline{Q})}$. Then by interpolation, we obtain the desired estimate from Lemma 7.11. Next, we verify that the representation u in (7.22) solves problem (7.19). Recall the following identity:

$${}^R\partial_t^\alpha \int_0^t \phi(t-s)\varphi(s)ds = \int_0^t \varphi(t-s)\,{}^R\partial_s^\alpha \phi(s)ds + \varphi(t) \lim_{s\to 0^+} {}_0I_s^{1-\alpha}\phi(s)$$

and $\partial_t^\alpha \phi(t) = {}^R\partial_t^\alpha \phi(t)$, if $\phi(0) = 0$. Then Lemma 7.10(v) leads to

$$\partial_t^\alpha u(x,t) = \int_0^t g(t-s)\,{}^R\partial_s^\alpha \widetilde{G}_\alpha(x,s)ds + g(t) \lim_{t\to 0^+} {}_0I_t^{1-\alpha}\widetilde{G}_\alpha(x,t)$$

$$= \int_0^t g(t-s)\,{}^R\partial_s^\alpha \widetilde{G}_\alpha(x,s)ds.$$

This directly implies $\partial_t^\alpha u(x,t) = \partial_{xx}^2 u(x,t)$. By Lemma 7.6(i), $\lim_{t\to 0^+} u(x,t) = 0$ and $u(x,t) \to 0$ if $x \to \infty$. Last we verify the boundary condition at $x = 0$. Note that

$$|\partial_x u(x,t) - g(t)|$$

$$\leq | \int_x^\infty \partial_x \widetilde{G}_\alpha(x,s)(g(t-s) - g(t))ds | + | \int_0^x \partial_x \widetilde{G}_\alpha(x,s)(g(t-s) - g(t))ds |$$

$$\leq 2\|g\|_{C(\bar{I})} \int_x^\infty |\partial_x \widetilde{G}_\alpha(x,s)|ds + c[g]_{t,\bar{I}}^{(\frac{1+\theta}{2}\alpha)} \int_0^x |\partial_x \widetilde{G}_\alpha(x,s)|s^{\frac{1+\theta}{2}\alpha}ds.$$

Then Lemma 7.6(i) and Lemma 7.10(i) with $\delta = x$ give

$$|\partial_x u(x,t) - g(t)| \leq c\|g\|_{C^{\frac{1+\theta}{2}\alpha}(\bar{I})}(x^{1-\frac{\alpha}{2}} + x^{\frac{1+\theta}{2}\alpha}).$$

Thus, the function u in (7.22) satisfies $\partial_x u(x,t)|_{x=0} = g(t)$, and it solves problem (7.19). The uniqueness can be proved as Theorem 7.3. $\qquad\square$

Now we state the result in the general case, i.e., $f, u_0 \not\equiv 0$.

Theorem 7.5 *Let $\theta \in (0,1)$, conditions (7.20), and (7.21) hold, $f \in C^{\theta,\frac{\theta}{2}\alpha}(\bar{Q})$, $u_0 \in C^{2+\theta}(\bar{\Omega})$ and $g \in C^{\frac{1+\theta}{2}\alpha}(\bar{I})$. Then for any $T > 0$, there exists a unique solution u to problem (7.19) and*

$$\|u\|_{C^{2+\theta,\frac{2+\theta}{2}\alpha}(\bar{Q})} \leq c(\|g\|_{C^{\frac{1+\theta}{2}\alpha}(\bar{I})} + \|f\|_{C^{\theta,\frac{\theta}{2}\alpha}(\bar{Q})} + \|u_0\|_{C^{2+\theta}(\bar{\Omega})}).$$

Proof Let $\bar{f}$ and $\bar{u}_0$ be extensions of f and u_0, respectively, to $x < 0$ such that $\bar{f}$ and $\bar{u}_0$ are finitely supported and

$$\|\bar{f}\|_{C^{\theta,\frac{\theta}{2}\alpha}(\mathbb{R}\times\bar{I})} \leq c\|f\|_{C^{\theta,\frac{\theta}{2}\alpha}(\bar{Q})} \quad \text{and} \quad \|\bar{u}_0\|_{C^{2+\theta}(\mathbb{R})} \leq c\|u_0\|_{C^{2+\theta}(\bar{\Omega})}.$$

Now consider the following Cauchy problem:

$$\begin{cases} \partial_t^\alpha \bar{u} - \partial_{xx}^2 \bar{u} = \bar{f}, & \text{in } \mathbb{R} \times I, \\ \bar{u}(0) = \bar{u}_0, & \text{in } \mathbb{R}, \\ \bar{u}(x,t) \to 0, & \text{if } |x| \to \infty, \forall t \in \bar{I}. \end{cases}$$

By Theorem 7.3, there is a unique solution $\bar{u} \in C^{2+\theta,\frac{2+\theta}{2}\alpha}(\mathbb{R}\times\bar{I})$ and

$$\|\bar{u}\|_{C^{2+\theta,\frac{2+\theta}{2}\alpha}(\mathbb{R}\times\bar{I})} \leq c(\|\bar{f}\|_{C^{\theta,\frac{\theta}{2}\alpha}(\mathbb{R}\times\bar{I})} + \|\bar{u}_0\|_{C^{2+\theta}(\mathbb{R})})$$
$$\leq c(\|f\|_{C^{\theta,\frac{\theta}{2}\alpha}(\bar{Q})} + \|u_0\|_{C^{2+\theta}(\bar{\Omega})}).$$

The solution $u(x,t)$ to problem (7.19) can be split into $u(x,t) = \bar{u}(x,t) + \bar{\bar{u}}(x,t)$, where $\bar{\bar{u}}$ solves problem (7.19) with $f, u_0 \equiv 0$, and satisfies (7.20). Hence, we can bound $\bar{\bar{u}}$ by Theorem 7.4. This completes the proof. $\qquad\square$

An analogous result holds for problem (7.18).

Theorem 7.6 *Let $\theta \in (0,1)$, conditions (7.20), and (7.21) hold, and $f \in C^{\theta,\frac{\theta}{2}\alpha}(\bar{Q})$, $u_0 \in C^{2+\theta}(\bar{\Omega})$, $g \in C^{\frac{2+\theta}{2}\alpha}(\bar{I})$. Then for any $T > 0$, there exists a unique solution u to* *problem* (7.18) *and*

$$\|u\|_{C^{2+\theta, \frac{2+\theta}{2}\alpha}(\overline{Q})} \leq c\left(\|g\|_{C^{\frac{2+\theta}{2}\alpha}(\overline{I})} + \|f\|_{C^{\theta, \frac{\theta}{2}\alpha}(\overline{Q})} + \|u_0\|_{C^{2+\theta}(\overline{\Omega})}\right).$$

7.2.3 Subdiffusion on bounded intervals

Now we study subdiffusion on bounded intervals, with variable coefficients and low-order terms. We study the following subdiffusion model on the interval $\Omega = (0, 1)$:

$$\begin{cases} \partial_t^\alpha u - \mathcal{L}(t)u = f, & \text{in } Q, \\ -\partial_x u(0, t) = g_1(t), & \text{on } I, \\ \partial_x u(1, t) = g_2(t), & \text{on } I, \\ u(x, 0) = u_0(x), & \text{in } \Omega, \end{cases} \tag{7.23}$$

where the (possibly time dependent) elliptic operator $\mathcal{L}(t)$ is defined by

$$\mathcal{L}(t)u(x, t) = a_0(x, t)\partial_{xx}^2 u(x, t) - a_1(x, t)\partial_x u(x, t) - a_2(x, t)u(x, t).$$

The next result gives the Hölder regularity of the solution u.

Theorem 7.7 *Let* $\theta \in (0, 1)$, *and* $f \in C^{\theta, \frac{\theta}{2}\alpha}(\overline{Q})$, $u_0 \in C^{2+\theta}(\overline{\Omega})$, $g_i(t) \in C^{\frac{1+\theta}{2}\alpha}(\overline{I})$, $i = 1, 2$ *and* $a_0(x, t) \geq \delta_1 > 0$ *in* $\overline{Q}$, $a_i(x, t) \in C^{\theta, \frac{\theta}{2}\alpha}(\overline{Q})$, $i = 0, 1, 2$, *and the compatibility conditions* $g_1(0) = -\partial_x u_0(0)$ *and* $g_2(0) = \partial_x u_0(1)$. *Then there exists a unique solution* $u(x, t) \in C^{2+\theta, \frac{2+\theta}{2}\alpha}(\overline{Q})$ *of problem* (7.23) *and*

$$\|u\|_{C^{2+\theta, \frac{2+\theta}{2}\alpha}(\overline{Q})} \leq c\left(\|f\|_{C^{\theta, \frac{\theta}{2}\alpha}(\overline{Q})} + \|u_0\|_{C^{2+\theta}(\overline{\Omega})} + \sum_{i=1}^{2} \|g_i\|_{C^{\frac{1+\theta}{2}\alpha}(\overline{I})}\right), \tag{7.24}$$

where the constant c depends only on Ω and $\|a_i\|_{C^{\theta, \frac{\theta}{2}\alpha}(\overline{Q})}$, $i = 0, 1, 2$.

We prove the theorem by mathematical induction. First prove the result for the case $t \in [0, \tau]$ for small $\tau < \tau_0$, with τ_0 to be determined, and then extend the solution u from $[0, \tau]$ to $[0, T]$ by repeating the procedure. Below let $Q_\tau = \Omega \times (0, \tau]$.

Lemma 7.12 *Let the conditions of Theorem 7.7 hold. Then there exists a unique solution* $u(x, t) \in C^{2+\theta, \frac{2+\theta}{2}\alpha}(\overline{Q}_\tau)$ *of problem* (7.23) *for any* $\tau \in (0, \tau_0]$ *and*

$$\|u\|_{C^{2+\theta, \frac{2+\theta}{2}\alpha}(\overline{Q}_\tau)} \leq c\left(\|f\|_{C^{\theta, \frac{\theta}{2}\alpha}(\overline{Q}_\tau)} + \|u_0\|_{C^{2+\theta}(\overline{\Omega})} + \sum_{i=1}^{2} \|g_i\|_{C^{\frac{1+\theta}{2}\alpha}[0, \tau]}\right), \tag{7.25}$$

where the constant c depends only on Ω and $\|a_i\|_{C^{\theta, \frac{\theta}{2}\alpha}(\overline{Q}_\tau)}$, $i = 0, 1, 2$.

Proof If the following conditions hold

$$f(x,t) \in C_0^{\theta, \frac{\theta}{2}\alpha}(\overline{Q_\tau}), \quad g_i \in C_0^{\frac{1+\theta}{2}\alpha}[0, \tau], \ i = 1, 2, \quad u_0 \equiv 0, \quad (7.26)$$

then the desired assertion follows directly from Theorems 7.3 and 7.5, using the arguments of [LSU68, Chapter IV, Sections 4–7]. This is achieved by freezing the coefficients and then applying the contraction mapping theorem (after lengthy computations). The detailed argument can also be found in Lemma 7.21 below. To remove the conditions in (7.26), we follow the argument of [LSU68, Chapter IV, Section 4] using Hestenes-Whitney extension, and extend the functions u_0, f and a_i from Ω to $\mathbb{R}$ with compact supports, denoted by $\bar{u}_0$, $\bar{f}$ and $\bar{a}_i$, respectively, such that

$$\|\bar{u}_0\|_{C^{2+\theta}(\mathbb{R})} \leq c\|u_0\|_{C^{2+\theta}(\overline{\Omega})}, \quad \|\bar{f}\|_{C^{\theta, \frac{\theta}{2}\alpha}(\mathbb{R}\times[0,\tau])} \leq c\|f\|_{C^{\theta, \frac{\theta}{2}\alpha}(\overline{Q_\tau})},$$

$$\|\bar{a}_i\|_{C^{\theta, \frac{\theta}{2}\alpha}(\mathbb{R}\times[0,\tau])} \leq c\|a_i\|_{C^{\theta, \frac{\theta}{2}\alpha}(\overline{Q_\tau})}, \quad i = 0, 1, 2.$$

Let $\tilde{f} := \bar{f} + \bar{a}_0 \partial_{xx}^2 \bar{u}_0 - \bar{a}_1 \partial_x \bar{u}_0 - \bar{a}_2 \bar{u}_0$ for $(x,t) \in \mathbb{R} \times [0, \tau]$. Then the function $\tilde{f} \in C^{\theta, \frac{\theta}{2}\alpha}(\mathbb{R} \times [0, \tau])$ has a finite support and satisfies

$$\|\tilde{f}\|_{C^{\theta, \frac{\theta}{2}\alpha}(\mathbb{R}\times[0,\tau])} \leq c\left(\|u_0\|_{C^{2+\theta}(\overline{\Omega})} + \|f\|_{C^{\theta, \frac{\theta}{2}\alpha}(\overline{Q_\tau})}\right),$$

where c depends on $\|a_i\|_{C^{\theta, \frac{\theta}{2}\alpha}(\overline{Q_\tau})}$, $i = 0, 1, 2$. Let w be the solution to

$$\begin{cases} \partial_t^\alpha w - \partial_{xx}^2 w = \tilde{f} - \partial_{xx}^2 \bar{u}_0, & \text{in } \mathbb{R} \times [0, \tau], \\ w(0) = \bar{u}_0, & \text{in } \mathbb{R}. \end{cases}$$

By Theorem 7.3, there exists a unique solution $w \in C^{2+\theta, \frac{2+\theta}{2}\alpha}(\mathbb{R} \times [0, \tau])$ and

$$\|w\|_{C^{2+\theta, \frac{2+\theta}{2}\alpha}(\mathbb{R}\times[0,\tau])} \leq c\left(\|u_0\|_{C^{2+\theta}(\overline{\Omega})} + \|f\|_{C^{\theta, \frac{\theta}{2}\alpha}(\overline{Q_\tau})}\right).$$

Now we look for the solution u of problem (7.23) in the form $u = w + v$, where the function v solves

$$\begin{cases} \partial_t^\alpha v - a_0 \partial_{xx}^2 v + a_1 \partial_x v + a_2 v = f^*, & \text{in } Q_\tau, \\ -\partial_x v(0, t) = g_1^*, & \text{on } [0, \tau], \\ \partial_x v(1, t) = g_2^*, & \text{on } [0, \tau], \\ v(\cdot, 0) = 0, & \text{in } \Omega, \end{cases}$$

with $f^* = f + a_0 \partial_{xx}^2 w - a_1 \partial_x w - a_2 w - \tilde{f} + \partial_{xx}^2 \bar{u}_0 - \partial_{xx}^2 w$, $g_1^* = g_1 + \partial_x w(0, t)$ and $g_2^* = g_2 - \partial_x w(1, t)$. By the very construction of these functions, we have

$$\|f^*\|_{C^{\theta,\frac{\theta}{2}\alpha}(\overline{Q}_\tau)} + \sum_{i=1}^{2} \|g_i^*\|_{C^{\frac{1+\theta}{2}\alpha}[0,\tau]}$$

$$\leq c\left(\|f\|_{C^{\theta,\frac{\theta}{2}\alpha}(\overline{Q}_\tau)} + \|u_0\|_{C^{2+\theta}(\overline{\Omega})} + \sum_{i=1}^{2} \|g_i\|_{C^{\frac{1+\theta}{2}\alpha}[0,\tau]}\right),$$

where the constant c depends on $\|a_i\|_{C^{\theta,\frac{\theta}{2}\alpha}(\overline{Q}_\tau)}$, $i = 0, 1, 2$. Further, the conditions $g_i^*(0) = f^*(x, 0) = 0$ hold. Thus, the problem data for v satisfy the compatibility conditions in (7.26), and the preceding discussions indicate that there exists a unique solution $v \in C^{2+\theta,\frac{2+\theta}{2}\alpha}(\overline{Q}_\tau)$ with the desired estimate. Now the lemma follows from the properties of v and w.   □

Now we can present the proof of Theorem 7.7.

Proof By Lemma 7.12, it suffices to prove that problem (7.23) has a unique solution u for $t \in [\tau, T]$. First, we construct a function $\bar{u} \in C^{2+\theta,\frac{2+\theta}{2}\alpha}(\overline{Q})$ by

$$\begin{cases} \partial_t^\alpha \bar{u} - a_0 \partial_{xx}^2 \bar{u} + a_1 \partial_x \bar{u} + a_2 \bar{u} = f, & \text{in } Q, \\ -\partial_x \bar{u}(0, t) = g_1, & \text{on } I, \\ \partial_x \bar{u}(1, t) = g_2, & \text{on } I, \\ \bar{u} = u, & \text{in } Q_\tau, \end{cases}$$

where $u \in C^{2+\theta,\frac{2+\theta}{2}\alpha}(\overline{Q}_\tau)$ is the unique solution to problem (7.23) for $t \in [0, \tau]$, $\tau < \tau_0$, given in Lemma 7.12. Set $T = 2\tau$, and let $\tilde{u}$ and $\tilde{a}_i$ be the extensions of u and a_i from $\overline{\Omega}$ to $\mathbb{R}$ with finite supports and

$$\|\tilde{u}\|_{C^{2+\theta,\frac{2+\theta}{2}\alpha}(\mathbb{R}\times[0,\tau])} \leq c\|u\|_{C^{2+\theta,\frac{2+\theta}{2}\alpha}(\overline{Q}_\tau)},$$

$$\|\tilde{a}_i\|_{C^{\theta,\frac{\theta}{2}\alpha}(\mathbb{R}\times[0,\tau])} \leq c\|a_i\|_{C^{\theta,\frac{\theta}{2}\alpha}(\overline{Q}_\tau)}, \quad i = 0, 1, 2.$$

Let $g(x, t)$ be defined by

$$g(x, t) = \begin{cases} \partial_t^\alpha \tilde{u} - \partial_{xx}^2 \tilde{u}, & \text{if } (x, t) \in \mathbb{R} \times [0, \tau], \\ g(x, \tau), & \text{if } (x, t) \in \mathbb{R} \times [\tau, 2\tau]. \end{cases}$$

Then $g \in C^{\theta,\frac{\theta}{2}\alpha}(\overline{Q}_{2\tau})$. Indeed, by the regularity $\tilde{u} \in C^{2+\theta,\frac{2+\theta}{2}\alpha}(\overline{Q}_\tau)$, $\partial_t^\alpha \tilde{u}, \partial_{xx}^2 \tilde{u} \in C^{\theta,\frac{\theta}{2}\alpha}(\overline{Q}_\tau)$. Thus, $g \in C^{\theta,\frac{\theta}{2}\alpha}(\overline{Q}_\tau)$, and the construction of g ensures $g \in C^{\theta,\frac{\theta}{2}\alpha}(\overline{Q}_{2\tau})$: $[g]_{x,\overline{Q}_{2\tau}}^{(\theta)} \leq [g]_{x,\overline{Q}_\tau}^{(\theta)}$ and $[g]_{t,\overline{Q}_{2\tau}}^{(\frac{\theta}{2}\alpha)} \leq [g]_{t,\overline{Q}_\tau}^{(\frac{\theta}{2}\alpha)}$. Let w solve

$$\begin{cases} \partial_t^\alpha w - \partial_{xx}^2 w = g, & \text{in } \mathbb{R} \times (0, 2\tau), \\ w(0) = \tilde{u}(0), & \text{in } \mathbb{R}. \end{cases}$$

By the properties of the functions $\tilde{u}$ and $g(x, t)$, Theorem 7.3 implies that there exists a unique solution $w \in C^{2+\theta,\frac{2+\theta}{2}\alpha}(\mathbb{R} \times [0, 2\tau])$, with $\|w\|_{C^{2+\theta,\frac{2+\theta}{2}\alpha}(\mathbb{R}\times[0,2\tau])} \leq$

$c\|u\|_{C^{2+\theta,\frac{2+\theta}{2}\alpha}(\mathbb{R}\times[0,\tau])}$, and $w = \tilde{u}$ in $\mathbb{R} \times [0, \tau]$. Now we look for the solution $\bar{u}$ of problem (7.23) of the form $\bar{u} = w + v$, where v satisfies

$$\begin{cases} \partial_t^\alpha v - a_0 \partial_{xx}^2 v + a_1 \partial_x v + a_2 v = f^*, & \text{in } Q_{2\tau}, \\ -\partial_x v(0, t) = g_1^*(t), & \text{on } (0, 2\tau], \\ \partial_x v(1, t) = g_2^*(t), & \text{on } (0, 2\tau], \\ v = 0, & \text{in } Q_\tau, \end{cases}$$

where the last line is due to the construction of w, and f^* and g_i^* are defined by $f^* = f - \partial_t^\alpha w + a_0 \partial_{xx}^2 w - a_1 \partial_x w - a_2 w - \partial_{xx}^2 u_0 - \partial_{xx}^2 w$, $g_1^* = g_1 + \partial_x w(0, t)$ and $g_2^* = g_2 - \partial_x w(1, t)$. By the construction of w, $f^*(x, t) = g_i^*(t) = 0$, for any $t \in [0, \tau]$. Next, let $\bar{t} = t - \tau$, and shifted quantities: $\bar{v}(x, \bar{t}) = v(x, \bar{t} + \tau)$, for all $\bar{t} \in [-\tau, \tau]$, and $\bar{a}_i(x, \bar{t})$, $\bar{f}^*(x, \bar{t})$ and $\bar{g}_i^*(\bar{t})$, etc. Then $\bar{a}_i$, $\bar{f}^*$, and $\bar{g}_i^*$ satisfy the conditions of Theorem 7.7. Since $\bar{f}^*(x, \bar{t}) = \bar{g}_i^*(\bar{t}) = 0$ for $\bar{t} \in [-\tau, 0]$, we deduce $\bar{v}(x, \bar{t}) = 0$, if $\bar{t} \in [-\tau, 0]$. Further, we have $\partial_t^\alpha v = \partial_{\bar{t}}^\alpha \bar{v}$, for any $\bar{t} \in (0, \tau]$. Consequently,

$$\begin{cases} \partial_{\bar{t}}^\alpha \bar{v} - \bar{a}_0 \partial_{xx}^2 \bar{v} + \bar{a}_1 \partial_x \bar{v} + \bar{a}_2 \bar{v} = \bar{f}^*, & \text{in } Q_\tau, \\ -\partial_x \bar{v}(0, \bar{t}) = \bar{g}_1^*(\bar{t}), & \bar{t} \in (0, \tau), \\ \partial_x \bar{v}(1, \bar{t}) = \bar{g}_2^*(\bar{t}), & \bar{t} \in (0, \tau), \\ \bar{v}(0) = 0, & \text{in } \Omega. \end{cases}$$

Lemma 7.12 gives the unique solvability for $\bar{v}(x, \bar{t}) \in C^{2+\theta, \frac{2+\theta}{2}\alpha}(\overline{Q}_\tau)$, with a bound similar to (7.25). By changing back $t = \bar{t} + \tau$, we obtain the unique solvability for $v \in C^{2+\theta, \frac{2+\theta}{2}\alpha}(\overline{Q}_{2\tau})$. Thus, the properties of w and v allow constructing a unique solution $u \in C^{2+\theta, \frac{2+\theta}{2}\alpha}(\overline{Q}_{2\tau})$ of problem (7.23) for $t \in [\tau, 2\tau]$. Repeating this procedure, we prove the existence of the solution $u \in C^{2+\theta, \frac{2+\theta}{2}\alpha}(\overline{Q})$, and obtain the estimate (7.24). The uniqueness follows from the estimate (7.24).   □

Last we state a Hölder regularity result in the Dirichlet case:

$$\begin{cases} \partial_t^\alpha u - \mathcal{L}(t)u = f, & \text{in } Q, \\ u(0, t) = g_1(t), & \text{on } I, \\ u(1, t) = g_2(t), & \text{on } I, \\ u(x, 0) = u_0(x), & \text{in } \Omega. \end{cases} \tag{7.27}$$

The proof is analogous to Theorem 7.7, and thus omitted.

Theorem 7.8 *Let* $\theta, \alpha \in (0, 1)$ *and* $f \in C^{\theta, \frac{\theta}{2}\alpha}(\overline{Q})$, $u_0 \in C^{2+\theta}(\overline{\Omega})$, $g_i(t) \in C^{\frac{2+\theta}{2}\alpha}(\overline{I})$, $i = 1, 2$ *and* $a_0(x, t) \geq \delta_1 > 0$ *in* $\overline{Q}$, $a_i(x, t) \in C^{\theta, \frac{\theta}{2}\alpha}(\overline{Q})$, $i = 0, 1, 2$, *and the compatibility conditions* $g_1(0) = u_0(0)$, $g_2(0) = u_0(1)$, $\partial_t^\alpha g_1(0) = \mathcal{L}(0)u_0(0)$ *and* $\partial_t^\alpha g_2(0) = \mathcal{L}(0)u_0(1)$. *Then there exists a unique solution* $u(x, t) \in C^{2+\theta, \frac{2+\theta}{2}\alpha}(\overline{Q})$ *of problem* (7.27) *and*

$$\|u\|_{C^{2+\theta,\frac{2+\theta}{2}\alpha}(\overline{Q})} \le c\Big(\|f\|_{C^{\theta,\frac{\theta}{2}\alpha}(\overline{Q})} + \|u_0\|_{C^{2+\theta}(\overline{\Omega})} + \sum_{i=1}^{2} \|g_i\|_{C^{\frac{2+\theta}{2}\alpha}(\overline{I})}\Big),$$

where the constant c depends on Ω and $\|a_i\|_{C^{\theta,\frac{\theta}{2}\alpha}(\overline{Q})}$, $i = 0, 1, 2$.

7.3 Hölder Regularity in Multi-Dimension

Now we extend the analysis in Section 7.2 and derive Hölder regularity results for multi-dimensional subdiffusion. The derivation is more complex, since the fundamental solutions are more involved, especially the auxiliary problem with the oblique boundary conditions, cf. problem (7.36). Nonetheless, the overall analysis strategy is similar to Section 7.2, by first analyzing the whole space $\mathbb{R}^d$, then the half space $\mathbb{R}^d_+$ and last a smooth open bounded domain with variable coefficients.

7.3.1 Subdiffusion in $\mathbb{R}^d$

We shall need various technical estimates on the fundamental solutions $G_\alpha(x, t)$ and $\overline{G}_\alpha(x, t)$ that are needed in bounding various potentials, an essential step for deriving Hölder regularity of the solution. We begin with several elementary inequalities.

Lemma 7.13 *The following inequalities hold.*

(i) *For $B > 0$, $\zeta > 0$*

$$\int_0^\infty \eta^{1-\frac{m}{2}} e^{-B(\zeta\eta^{-1}+\eta^{\frac{1}{1-\alpha}})}d\eta \le c \begin{cases} 1, & m = 1, 2, 3, \\ |\ln\zeta| + 1, & m = 4, \\ \zeta^{2-\frac{m}{2}}, & m \ge 5. \end{cases}$$

(ii) *For $\zeta \ge 0$ and $\alpha \in (0, 1)$, there holds*

$$\frac{\zeta}{\eta} + \eta^{\frac{1}{1-\alpha}} \ge 2\zeta^{\frac{1}{2-\alpha}}, \quad \forall \eta \in (0, \infty). \tag{7.28}$$

(iii) *For $\beta \in (0, \frac{1}{2})$, and $c > 0$,*

$$(|\ln\zeta| + 1)e^{-c|\zeta|^{\frac{1}{1-\beta}}} \le c'|\zeta|^{-\frac{1}{4}}e^{-\frac{c}{2}|\zeta|^{\frac{1}{1-\beta}}}, \quad \forall \zeta > 0. \tag{7.29}$$

Proof Denote the integral in (i) by I. If $m = 1, 2, 3$, the estimate is direct:

$$I(\zeta) \le c \int_0^\infty \eta^{1-\frac{m}{2}} e^{-B\eta^{\frac{1}{1-\alpha}}} d\eta \le c.$$

If $m = 4$, we split the integral into two terms and obtain

$$\int_0^\zeta e^{-B\frac{\zeta}{\eta}}\eta^{-1}\,d\eta = \int_1^\infty e^{-Bs}s^{-1}\,ds \le c,$$

$$\int_\zeta^\infty e^{-B\eta^{\frac{1}{1-\alpha}}}\eta^{-1}\,d\eta = \ln\eta\, e^{-B\eta^{\frac{1}{1-\alpha}}}\big|_\zeta^\infty + \frac{B}{1-\alpha}\int_\zeta^\infty \ln\eta\,\eta^{\frac{\alpha}{1-\alpha}}e^{-B\eta^{\frac{1}{1-\alpha}}}\,d\eta$$

$$\le c\left(|\ln\zeta| + \int_0^\infty \ln\eta\,\eta^{\frac{\alpha}{1-\alpha}}e^{-B\eta^{\frac{1}{1-\alpha}}}\,d\eta\right) \le c(|\ln\zeta| + 1).$$

This shows the assertion for $m = 4$. Last, if $m \ge 5$, letting $\xi = \frac{\eta}{\zeta}$ gives

$$\mathrm{I}(\zeta) = \int_0^\infty (\xi\zeta)^{1-\frac{m}{2}}e^{-B(\xi^{-1}+(\zeta\xi)^{\frac{1}{1-\alpha}})}\zeta\,d\xi \le c\zeta^{2-\frac{m}{2}}\int_0^\infty \xi^{1-\frac{m}{2}}e^{-B\xi^{-1}}\,d\xi \le c\zeta^{2-\frac{m}{2}}.$$

This shows (i). (ii) follows since $\frac{\zeta}{\eta} + \eta^{\frac{1}{1-\alpha}}$ attains its global minimum at $\eta^* = \zeta^{\frac{1-\alpha}{2-\alpha}}$, and the minimum value is $2\zeta^{\frac{1}{2-\alpha}}$. (iii) holds, since by direct computation,

$$\zeta^{\frac{1}{4}}(|\ln\zeta| + 1)e^{-\frac{c}{2}\zeta^{\frac{1}{1-\beta}}} \le c'.$$

Indeed, for $\zeta \in (0, 1]$, there holds

$$\zeta^{\frac{1}{4}}(|\ln\zeta| + 1)e^{-\frac{c}{2}\zeta^{\frac{1}{1-\beta}}} \le \zeta^{\frac{1}{4}}|\ln\zeta| + \zeta^{\frac{1}{4}} \le (4 + e)e^{-1}.$$

Likewise, by the trivial inequality $1 + \zeta \le e^\zeta$ for $\zeta > 0$, we have $\ln\zeta \le \zeta$. This and the inequality $\zeta^a e^{-b\zeta} \le (\frac{a}{b})^a e^{-a}$, for $a, b > 0$ imply

$$\zeta^{\frac{1}{4}}(|\ln\zeta| + 1)e^{-\frac{c}{2}\zeta^{\frac{1}{1-\beta}}} \le \zeta^{\frac{1}{4}}(\zeta + 1)e^{-\frac{c}{2}\zeta} \le (\frac{1}{2c})^{\frac{1}{4}}e^{-\frac{1}{4}} + (\frac{5}{2c})^{\frac{5}{4}}e^{-\frac{5}{4}}.$$

This shows (iii) and completes the proof of the lemma. □

Next, we derive a pointwise bound on the mixed derivative $^R\partial_t^\nu D_x^\ell \overline{G}_\alpha(x, t)$ ($\ell \in \mathbb{N}_0^d$ is a multi-index), for any $\nu \in \mathbb{R}$, $|\ell| := \sum_{i=1}^d \ell_i \ge 0$, similar to Lemma 7.6. For $\nu < 0$, the Riemann-Liouville fractional derivative $^R\partial_t^\nu$ is identified with the Riemann-Liouville integral operator $_0I_t^{-\nu}$ of order $-\nu > 0$. The analysis relies on the following bound for the heat kernel $G_1(x, t)$:

$$|D_x^\ell G_1(x, t)| \le c_0 t^{-\frac{d+|\ell|}{2}}e^{-c_1\frac{|x|^2}{t}}. \tag{7.30}$$

Lemma 7.14 *Let $\nu \in \mathbb{R}$, $|\ell| \ge 0$. Then the function $\overline{G}_\alpha(x, t)$ defined in (7.4) satisfies*

$$|^R\partial_t^\nu D_x^\ell \overline{G}_\alpha(x, t)| \le ct^{(2-d-|\ell|)\frac{\alpha}{2}-\nu-1}\gamma_{p(\nu, |\ell|)}(r)e^{-c'r^{\frac{1}{1-\frac{\alpha}{2}}}}, \tag{7.31}$$

with $r = |x|t^{-\frac{\alpha}{2}}$, and

$$p(\nu, |\ell|) = \begin{cases} d + |\ell|, & \nu \in \mathbb{N}_0, \\ d + |\ell| + 2, & \nu \notin \mathbb{N}_0, \end{cases} \qquad \gamma_m(\zeta) = \begin{cases} 1, & m \le 3, \\ |\ln \zeta| + 1, & m = 4, \\ \zeta^{4-d}, & m \ge 5. \end{cases}$$

Proof It follows from (7.4) and the differentiation formula (3.28) for $W_{\rho,\mu}(z)$ that

$$^R\partial_t^\nu D_x^\ell \overline{G}_\alpha(x,t) = \int_0^\infty D_x^\ell G_1(x,\lambda) t^{-\nu-1} W_{-\alpha,-\nu}(-\lambda t^{-\alpha}) d\lambda.$$

If $\nu \in \mathbb{N}_0$, inequality (7.30) and Theorem 3.8(iv) give

$$|^R\partial_t^\nu D_x^\ell \overline{G}_\alpha(x,t)| \le c_0 \int_0^\infty \lambda^{-\frac{d+|\ell|}{2}} e^{-c_1 \frac{|x|^2}{\lambda}} \lambda t^{-\alpha-\nu-1} e^{-\sigma(\lambda t^{-\alpha})^{\frac{1}{1-\alpha}}} d\lambda,$$

for some $\sigma = \sigma(\alpha) > 0$. By changing variables $\eta = \lambda t^{-\alpha}$, we obtain

$$|^R\partial_t^\nu D_x^\ell \overline{G}_\alpha(x,t)| \le c_0 t^{(2-d-|\ell|)\frac{\alpha}{2}-\nu-1} \int_0^\infty \eta^{1-\frac{d+|\ell|}{2}} e^{-c_2(\frac{|x|^2}{t^\alpha}\frac{1}{\eta}+\eta^{\frac{1}{1-\alpha}})} d\eta,$$

with $c_2 = \min(c_1, \sigma)$. Now by the inequality (7.28), with $r = |x| t^{-\frac{\alpha}{2}}$,

$$|^R\partial_t^\nu D_x^\ell \overline{G}_\alpha(x,t)| \le c t^{(2-d-|\ell|)\frac{\alpha}{2}-\nu-1} e^{-\frac{c_2}{2}r^{\frac{1}{1-\frac{\alpha}{2}}}} \int_0^\infty \eta^{1-\frac{d+|\ell|}{2}} e^{-\frac{c_2}{2}(\frac{|x|^2}{t^\alpha}\frac{1}{\eta}+\eta^{\frac{1}{1-\alpha}})} d\eta.$$

The desired assertion follows from Lemma 7.13. Similarly, if $\nu \notin \mathbb{N}_0$, inequality (7.30) and Theorem 3.8(iv) give

$$|^R\partial_t^\nu D_x^\ell \overline{G}_\alpha(x,t)| \le c_0 \int_0^\infty \lambda^{-\frac{d+|\ell|}{2}} e^{-c_1 \frac{|x|^2}{\lambda}} t^{-\nu-1} e^{-\sigma(\lambda t^{-\alpha})^{\frac{1}{1-\alpha}}} d\lambda.$$

By changing variables $\eta = \lambda t^{-\alpha}$, we obtain

$$|^R\partial_t^\nu D_x^\ell \overline{G}_\alpha(x,t)| \le c_0 t^{(2-d-|\ell|)\frac{\alpha}{2}-\nu-1} \int_0^\infty \eta^{-\frac{d+|\ell|}{2}} e^{-c_2(\frac{|x|^2}{t^\alpha}\frac{1}{\eta}+\eta^{\frac{1}{1-\alpha}})} d\eta,$$

with $c_2 = \min(c_1, \sigma)$. Now by the inequality (7.28), with $r = |x| t^{-\frac{\alpha}{2}}$,

$$|^R\partial_t^\nu D_x^\ell \overline{G}_\alpha(x,t)| \le c t^{(2-d-|\ell|)\frac{\alpha}{2}-\nu-1} e^{-\frac{c_2}{2}r^{\frac{1}{1-\frac{\alpha}{2}}}} \int_0^\infty \eta^{-\frac{d+|\ell|}{2}} e^{-\frac{c_2}{2}(\frac{|x|^2}{t^\alpha}\frac{1}{\eta}+\eta^{\frac{1}{1-\alpha}})} d\eta.$$

The assertion for $\nu \notin \mathbb{N}_0$ follows from Lemma 7.13.   □

Now we give several estimates on the potentials associated with G_α and $\overline{G}_\alpha$.

Lemma 7.15 *For the functions $v = \mathcal{V} f$ and $w = \mathcal{I} u_0$, there hold*

$$\|v\|_{C^{2,0}(\overline{Q})} \le c\|f\|_{C^{\theta,0}(\overline{Q})},$$

$$[D_x^\ell v]_{x,Q}^{(\theta)} + [D_x^\ell v]_{t,Q}^{(\frac{\theta}{2}\alpha)} \le c[f]_{x,Q}^{(\theta)}, \quad |\ell| = 2,$$

$$\|w\|_{C^{2,0}(\overline{Q})} \le c\|u_0\|_{C^2(\overline{\mathbb{R}^d_+})},$$

$$[D_x^\ell w]_{x,Q}^{(2+\theta)} + [D_x^\ell w]_{t,Q}^{(\frac{\theta}{2}\alpha)} \le c\|u_0\|_{C^{2+\theta}(\overline{\mathbb{R}^d_+})}, \quad |\ell| = 2.$$

Proof The proof of the lemma is nearly identical with Lemmas 7.8 and 7.9, using Lemmas 7.1 and 7.14. We sketch only the estimates for v. First, we derive an auxiliary estimate. Lemma 7.14 with $v = 0$ and then changing variables $r = |x|t^{-\frac{\alpha}{2}}$ lead to

$$\int_0^\infty |D_x^\ell \overline{G}_\alpha(x,t)|dt \le c|x|^{2-d-|\ell|} \int_0^\infty r^{d+|\ell|-3}\gamma_{d+|\ell|}(r)e^{-c'r^{\frac{1}{1-\frac{\alpha}{2}}}} dr$$

$$\le c|x|^{2-d-|\ell|}, \quad \text{if } d + |\ell| \ge 3. \tag{7.32}$$

The bound on $\|v\|_{C^{2,0}(\overline{Q})}$ is direct to derive. By Lemma 7.1(ii), for any $|\ell| \ge 1$,

$$D_x^\ell v(x,t) = \int_0^t \int_{\mathbb{R}^d} D_x^\ell \overline{G}_\alpha(x-y,t-s)(f(y,s) - f(x,s))dyds.$$

To bound $[D_x^\ell v]_{x,Q}^{(\theta)}$, with $|\ell| = 2$, for any distinct $x, \bar{x} \in \mathbb{R}^d$, let $h := |x - \bar{x}|$ and $K = \{y \in \mathbb{R}^d : |\bar{x} - y| \le 2h\}$. Then we split $D_x^\ell u(x,t) - D_x^\ell u(\bar{x},t)$ into

$$D_x^\ell v(x,t) - D_x^\ell v(\bar{x},t) = \sum_{i=1}^4 I_i := \int_0^t \int_K D_x^\ell \overline{G}_\alpha(\bar{x}-y,t-s)[f(\bar{x},s) - f(y,s)]dyds$$

$$+ \int_0^t \int_K D_x^\ell \overline{G}_\alpha(x-y,t-s)[f(y,s) - f(x,s)]dyds$$

$$+ \int_0^t \int_{\mathbb{R}^d\setminus K} D_x^\ell \overline{G}_\alpha(x-y,t-s)[f(\bar{x},s) - f(x,s)]dyds$$

$$+ \int_0^t \int_{\mathbb{R}^d\setminus K} [D_x^\ell \overline{G}_\alpha(\bar{x}-y,t-s) - D_x^\ell \overline{G}_\alpha(x-y,t-s)][f(\bar{x},s) - f(y,s)]dyds.$$

For I_1, by (7.32), we have

$$I_1 \le [f]_{x,Q}^{(\theta)} \int_K |\bar{x} - y|^{\theta-d}dy \le c[f]_{x,Q}^{(\theta)}h^\theta.$$

The term I_2 can be bounded similarly. Next, similar to the 1D case, cf. Lemma 7.7(ii), for $|\ell| = 2$, we have

$$\int_0^t \left| \int_{\mathbb{R}^d\setminus K} D_x^\ell \overline{G}_\alpha(x-y,t-s)dy \right| ds \le c, \tag{7.33}$$

with which, we can bound I_3 by

$$|I_3| \le ch^\theta [f]_{x,\overline{Q}}^{(\theta)} \int_0^t \left| \int_{|x|>h} D_x^\ell \overline{G}_\alpha(x,s) dx \right| ds \le ch^\theta [f]_{x,\overline{Q}}^{(\theta)}.$$

Last, for I_4, the mean value theorem and (7.32) give for some $\zeta \in [\bar{x}-y, x-y]$,

$$|I_4| \le ch[f]_{x,\overline{Q}}^{(\theta)} \int_0^t \int_{|x|>h} |x|^\theta |\nabla D_x^\ell \overline{G}_\alpha(\zeta,s)| d\zeta ds$$

$$\le ch[f]_{x,\overline{Q}}^{(\theta)} \int_{|x|>h} |x|^{\theta-d-1} dx \le ch^\theta [f]_{x,\overline{Q}}^{(\theta)}.$$

These bounds on I_i show that for $|\ell| = 2$, $[D_x^\ell v]_{x,\overline{Q}}^{(\theta)} \le c[f]_{x,\overline{Q}}^{(\theta)}$. Next, we bound $[D_x^\ell v]_{t,\overline{Q}}^{\left(\frac{\theta\alpha}{2}\right)}$, with $|\ell| = 2$. For any $t_1, t_2 \in \overline{I}$, $t_2 > t_1$, let $\tau = t_2 - t_1$, and we discuss the two cases, i.e., $t_1 > 2\tau$ and $t_1 \le 2\tau$, separately. If $t_1 > 2\tau$, we split the difference $D_x^\ell v(x,t_2) - D_x^\ell v(x,t_1)$ into

$$D_x^\ell v(x,t_2) - D_x^\ell v(x,t_1) = \int_{\mathbb{R}^d} \int_{t_1-2\tau}^{t_2} D_x^\ell \overline{G}_\alpha(x-y,t_2-s)[f(y,s)-f(x,s)] ds dy$$

$$+ \int_{\mathbb{R}^d} \int_{t_1-2\tau}^{t_1} D_x^\ell \overline{G}_\alpha(x-y,t_1-s)(f(x,s)-f(y,s)) ds dy$$

$$+ \int_0^{t_1-2\tau} \int_{\mathbb{R}^d} [D_x^\ell \overline{G}_\alpha(x-y,t_2-s) - D_x^\ell \overline{G}_\alpha(x-y,t_1-s)][f(y,s)-f(x,s)] dy ds.$$

Next we bound the three terms, denoted by II_i, $i = 1,2,3$. For the term II_1, using Lemma 7.14 with $\nu = 0$ and $r = |x|s^{-\frac{\alpha}{2}}$,

$$|II_1| \le [f]_{x,\overline{Q}}^{(\theta)} \int_0^\tau \int_{\mathbb{R}^d} |x|^\theta |D_x^\ell G(x,s)| dx ds$$

$$\le c[f]_{x,\overline{Q}}^{(\theta)} \int_0^\tau \int_{\mathbb{R}^d} |x|^\theta s^{-d\frac{\alpha}{2}-1} \gamma_{2+d}(r) e^{-c'r^{\frac{1}{1-\frac{\alpha}{2}}}} dx ds$$

$$= c[f]_{x,\overline{Q}}^{(\theta)} \int_0^\tau s^{\frac{\theta}{2}\alpha-1} ds \int_{\mathbb{R}^d} |x|^\theta \gamma_{2+d}(|x|) e^{-c'|x|^{\frac{1}{1-\frac{\alpha}{2}}}} dx \le c\tau^{\frac{\theta}{2}\alpha} [f]_{x,\overline{Q}}^{(\theta)}.$$

The term II_2 can be bounded similarly. For the term II_3, the mean value theorem and Lemma 7.14 imply for some $t \in (t_1, t_2)$,

$$|II_3| \le c\tau[f]_{x,\overline{Q}}^{(\theta)} \int_\tau^t \int_{\mathbb{R}^d} |x|^\theta s^{-\frac{d}{2}\alpha-2} \gamma_{2+d}(r) e^{-c'r^{\frac{1}{1-\frac{\alpha}{2}}}} dx ds$$

$$= c\tau[f]_{x,\overline{Q}}^{(\theta)} \int_\tau^t s^{\frac{\theta}{2}\alpha-2} ds \int_{\mathbb{R}^d} |x|^\theta \gamma_{2+d}(|x|) e^{-c'|x|^{\frac{1}{1-\frac{\alpha}{2}}}} dx \le c\tau^{\frac{\theta}{2}\alpha} [f]_{x,\overline{Q}}^{(\theta)}.$$

In sum, for $t_1 > 2\tau$, we obtain $[D_x^\ell v]_{t,Q}^{(\frac{\alpha\theta}{2})} \le c[f]_{x,Q}^{(\theta)}\tau^{\frac{\theta\alpha}{2}}$. The case $t_1 \le 2\tau$ can be analyzed similarly, and the proof is omitted. $\qquad\qquad\square$

Using Lemma 7.15 and repeating the argument of Theorem 7.3, we can show the existence result of a solution $u \in C^{2+\theta,\frac{2+\theta}{2}\alpha}(\overline{Q})$ to problem (7.1).

Proposition 7.2 *Let* $u_0 \in C^{2+\theta}(\overline{\Omega})$ *and* $f \in C^{\theta,\frac{\theta}{2}\alpha}(\overline{Q})$ *satisfy (7.2). Then problem (7.1) has a unique solution* $u \in C^{2+\theta,\frac{2+\theta}{2}\alpha}(\overline{Q})$, *and*

$$\|u\|_{C^{2+\theta,\frac{2+\theta}{2}\alpha}(\overline{Q})} \le c\left(\|f\|_{C^{\theta,\frac{\theta}{2}\alpha}(\overline{Q})} + \|u_0\|_{C^{2+\theta}(\overline{\Omega})}\right).$$

7.3.2 Subdiffusion in $\mathbb{R}_+^d$

Let the half space $\Omega \equiv \mathbb{R}_+^d = \{x \in \mathbb{R}^d : x_d > 0\}$, with the lateral boundary $\partial_L Q = \{(x',0,t) : x' = (x_1,\ldots,x_{d-1}) \in \mathbb{R}^{d-1}, t \in (0,T)\}$. Consider the following two initial boundary value problems:

$$\begin{cases} \partial_t^\alpha v - \Delta v = 0, & \text{in } Q, \\ v(x,0) = 0, & \text{in } \Omega, \\ v \to 0 & \text{as } |x| \to \infty, \\ v(x',0,t) = g(x',t), & \text{on } \partial_L Q, \end{cases} \qquad (7.34)$$

and

$$\begin{cases} \partial_t^\alpha w - \Delta w = 0, & \text{in } Q, \\ w(x,0) = 0, & \text{in } \Omega, \\ w \to 0 & \text{as } |x| \to \infty, \\ \sum_{i=1}^d h_i \partial_{x_i} w(x',0,t) = g(x',t), & \text{on } \partial_L Q, \end{cases} \qquad (7.35)$$

where $h = (h_1,\ldots,h_d)$ is a constant vector. We shall assume

$$g \in \begin{cases} C_0^{2+\theta,\frac{2+\theta}{2}\alpha}(Q), & \text{if (7.34)} \\ C_0^{1+\theta,\frac{1+\theta}{2}\alpha}(Q), & \text{if (7.35)} \end{cases} \quad \text{and} \quad g(x',t) = 0, \quad \text{for } |x| > R > 0. \qquad (7.36)$$

The first part of the condition implies $g(x',t) = 0$ for $t \le 0$, and may be extended by 0 for $t < 0$, again denoted by g below. Further, the constant vector h in the oblique derivative $\sum_{i=1}^d h_i \partial_{x_i} w(x',0,t)$ satisfies the condition for some $m_0 \ge 0$ (with the shorthand $h' = (h_1,\ldots,h_{d-1})$, the first $d-1$ components of $h \in \mathbb{R}^d$)

$$h_d \le -\delta_0 < 0, \quad |h'| \le m_0. \qquad (7.37)$$

Clearly, in the case $|h'| = 0$, problem (7.35) recovers the standard Neumann problem. Now we derive the solution representations for problems (7.34) and (7.35). The fundamental solution $\overline{G}_\alpha(x, t)$ takes a complex form, cf. (7.38), which complicates the analysis of the associated potentials.

Proposition 7.3 *Let Assumptions* (7.36) *and* (7.37) *hold. Then the solutions of problems* (7.34) *and* (7.35) *can be, respectively, represented by*

$$v(x, t) = -2 \int_{-\infty}^{t} \int_{\mathbb{R}^{d-1}} \partial_{x_d} \overline{G}_\alpha(x' - y', x_d, t - s) g(y', s) dy' ds,$$

$$w(x, t) = \int_{-\infty}^{t} \int_{\mathbb{R}^{d-1}} \widetilde{G}_\alpha(x' - y', x_d, t - s) g(y', s) dy' ds,$$

where the fundamental solution $\widetilde{G}_\alpha(x, t)$ *is given by*

$$\widetilde{G}_\alpha(x, t) = -2 \int_0^\infty \partial_{x_d} \overline{G}_\alpha(x - h\lambda, t) d\lambda. \tag{7.38}$$

Proof Let $\mathcal{F}'$ be the Fourier transform in the tangent space variable x', i.e.,

$$\mathcal{F}'[v] = \int_{\mathbb{R}^{d-1}} v(x, t) e^{-ix' \cdot \xi} dx', \quad \text{with } \xi = (\xi_1, \dots, \xi_{d-1}) \in \mathbb{R}^{d-1}.$$

Taking tangent Fourier transform $\mathcal{F}'$ in x' and Laplace transform $\mathcal{L}$ in t, problem (7.34) reduces to the following ODE (with $\widehat{\widetilde{v}}(\xi, x_d, z) = \mathcal{L}[\mathcal{F}'[v]](\xi, x_d, z)$)

$$\begin{cases} z^\alpha \widehat{\widetilde{v}}(\xi, x_d, z) + |\xi|^2 \widehat{\widetilde{v}} - \widehat{\widetilde{v}}_{x_d x_d}(\xi, x_d, z) = 0, & x_d > 0 \\ \widehat{\widetilde{v}}(\xi, 0, z) = \widehat{\widetilde{g}}(\xi, z), \\ \widehat{\widetilde{v}}(\xi, x_d, 0) \to 0, & \text{as } x_d \to \infty. \end{cases}$$

The solution $\widehat{\widetilde{v}}$ is given by $\widehat{\widetilde{v}}(\xi, x_d, z) = e^{-\sqrt{z^\alpha + |\xi|^2} x_d} \widehat{\widetilde{g}}(\xi, z)$. It suffices to prove

$$\mathcal{L}[\mathcal{F}'[-2\partial_{x_d} \overline{G}_\alpha]](\xi, x_d, z) = e^{-\sqrt{z^\alpha + \xi|^2} x_d}. \tag{7.39}$$

Then the desired representation follows by the convolution formula. Indeed, in view of the identities $\mathcal{F}'\left[e^{-\frac{|x'|^2}{4\lambda}}\right](\xi) = (4\pi\lambda)^{\frac{d-1}{2}} e^{-\lambda|\xi|^2}$ and $\mathcal{L}\left[t^{-1} W_{-\alpha,0}(-\lambda t^\alpha)\right](z) = e^{-\lambda z^\alpha}$, cf. Proposition 3.3, we obtain

$$\mathcal{L}[\mathcal{F}'[-2\partial_{x_d} \overline{G}_\alpha(x, t)]](\xi, x_d, z) = \frac{1}{2\sqrt{\pi}} \int_0^\infty x_d \lambda^{-\frac{3}{2}} e^{-\frac{x_d^2}{4\lambda} - (z^\alpha + |\xi|^2)\lambda} d\lambda.$$

Changing variables $\zeta = \lambda^{-\frac{1}{2}} x_d$ leads to

$$\mathcal{L}[\mathcal{F}'[-2\partial_{x_d} \overline{G}_\alpha]](\xi, x_d, z) = \frac{1}{\sqrt{\pi}} \int_0^\infty e^{-\frac{\zeta^2}{4} - \frac{(z^\alpha + |\xi|^2) x_d^2}{\zeta^2}} d\zeta = e^{-\sqrt{z^\alpha + |\xi|^2} x_d},$$

using also the following identity (cf. [GR15, formula (3.325), p. 339])

$$\int_0^\infty e^{-ax^2 - \frac{b}{x^2}} \, dx = \frac{1}{2}\sqrt{\frac{\pi}{a}} e^{-2\sqrt{ab}}, \quad \forall a, b > 0$$

with $a = \frac{1}{4}$ and $b = (z^\alpha + |\xi|^2)x_d^2$. This shows the representation for v. Similarly, by applying Laplace and (tangent) Fourier transforms in t and x', respectively, problem (7.35) reduces to the following ODE:

$$\begin{cases} z^\alpha \widehat{\widetilde{w}}(\xi, x_d, z) + |\xi|^2 \widehat{\widetilde{w}}(\xi, x_d, z) - \widehat{\widetilde{w}}_{x_d x_d}(\xi, x_d, z) = 0, \quad x_d > 0 \\ \qquad (h_d \partial_{x_d} \widehat{\widetilde{w}}(\xi, 0, z) + ih' \cdot \xi \widehat{\widetilde{w}})|_{x_d=0} = \widehat{\widetilde{g}}(\xi, z), \\ \qquad \qquad \widehat{\widetilde{w}}(\xi, x_d, 0) \to 0, \quad \text{as } x_d \to \infty, \end{cases}$$

Under Assumption (7.37), the solution $\widehat{\widetilde{w}}$ is given by

$$\widehat{\widetilde{w}}(\xi, x_d, z) = \frac{e^{-\sqrt{z^\alpha + |\xi|^2}x_d} \widehat{\widetilde{g}}(\xi, z)}{-h_d\sqrt{z^\alpha + |\xi|^2} + ih' \cdot \xi} = \int_0^\infty e^{-\sqrt{z^\alpha + |\xi|^2}(x_d - h_d\lambda) - ih' \cdot \xi\lambda} \, d\lambda \widehat{\widetilde{g}}(\xi, z).$$

Now the identity (7.39) implies

$$\mathcal{L}^{-1}[\mathcal{F}'^{-1}[\widetilde{G}_\alpha]](x, t) = -2 \int_0^\infty \partial_{x_d} \overline{G}_\alpha(x - h\lambda, t) d\lambda.$$

It remains to prove that the representations satisfy the corresponding boundary conditions. For problem (7.34), the identity (7.8) implies the desired boundary condition. Meanwhile, for problem (7.35),

$$\sum_{i=1}^d h_i \partial_{x_i} \widetilde{G}_\alpha(x, t) = -2 \int_0^\infty \sum_{i=1}^d \partial^2_{x_d x_i} \overline{G}_\alpha(x - h\lambda, t) h_i d\lambda$$

$$= 2 \int_0^\infty \frac{d}{d\lambda} \partial_{x_d} \overline{G}_\alpha(x - h\lambda, t) d\lambda = -2\partial_{x_d} \overline{G}_\alpha(x, t).$$

This fact implies the desired oblique boundary condition:

$$\sum_{i=1}^d h_i \partial_{x_i} \int_0^t \int_{\mathbb{R}^{d-1}} \widetilde{G}_\alpha(x' - y', x_d, t - s)g(y', s)dy'ds|_{x_d=0} = g(x', t).$$

This completes the proof of the proposition.   □

Next, we give a simple lower bound on $|x - h\lambda|^2$ under Assumption (7.37).

Lemma 7.16 *Under Assumption* (7.37), *the following lower bound holds:*

$$\sum_{i=1}^d (x_i - h_i\lambda)^2 \geq c_0^2\left(\sum_{i=1}^{d-1} x_i^2 + \lambda^2\right) + x_d^2, \quad \text{with } c_0^2 = \frac{\delta_0^2}{2}\min(1, \frac{1}{\delta_0^2 + m_0^2}). \quad (7.40)$$

Proof By Assumption (7.37), $h_d < -\delta_0 < 0$ for any $a \in (0, 1)$,

$$\sum_{i=1}^{d}(x_i - h_i\lambda)^2 \geq \sum_{i=1}^{d-1}\left(x_i^2 - 2ax_i\frac{h_i}{a}\lambda + h_i^2\lambda^2\right) + x_d^2 + h_d^2\lambda^2$$

$$\geq (1 - a^2)\sum_{i=1}^{d-1}x_i^2 - (a^{-2} - 1)m_0^2\lambda^2 + \delta_0^2\lambda^2 + x_d^2.$$

Setting $a^2 = \frac{1}{2}(1 + \frac{m_0^2}{\delta_0^2+m_0^2})$ shows the desired estimate.   □

Now we collect several useful weighted integral bounds on $\widetilde{G}_\alpha$.

Lemma 7.17 *The function* $\widetilde{G}_\alpha(x, t)$ *satisfies the following estimates:*

$$\int_0^\infty |D_x^\ell \widetilde{G}_\alpha(x, t)|dt \leq c|x'|^{2-d-|\ell|}, \quad 1 \leq |\ell| \leq 2, \tag{7.41}$$

$$\int_0^\infty t^{\frac{\alpha}{4}}|\partial_{x_i}\widetilde{G}_\alpha(x, t)|dt \leq c|x'|^{-d+\frac{3}{2}}, \quad i = 1, \ldots, d, \tag{7.42}$$

$$\int_0^\infty t^{\frac{\delta}{2}\alpha}|{}^R\partial_t^\alpha D_x^\ell \widetilde{G}_\alpha(x, t)|dt \leq c|x'|^{\delta-d-|\ell|}, \quad |\ell| = 0, 1, \delta > 0, \tag{7.43}$$

$$\int_{\mathbb{R}^{d-1}} |{}^R\partial_t^{\alpha-1+\zeta}\widetilde{G}_\alpha(x, t)|dx' \leq ct^{-\frac{\alpha}{2}-\zeta}, \quad \zeta \in \mathbb{R}, \tag{7.44}$$

$$\int_{\mathbb{R}^{d-1}} |\partial_t^k \partial_{x_i}\widetilde{G}_\alpha(x, t)|dx' \leq ct^{-1-k}, \quad k = 0, 1, i = 1, \ldots, d, \tag{7.45}$$

$$\int_{\mathbb{R}^{d-1}} |{}^R\partial_t^\nu D_x^\ell \widetilde{G}_\alpha(x, t)|dx' \leq ct^{(1-|\ell|)\frac{\alpha}{2}-\nu-1}, \quad 0 \leq |\ell| \leq 2. \tag{7.46}$$

Proof In the proof, let $\beta = \frac{\alpha}{2}$. By Lemma 7.14, with $r = |x|t^{-\beta}$, there holds

$$|{}^R\partial_t^\nu D_x^\ell \widetilde{G}_\alpha(x, t)| \leq ct^{(2-d-|\ell|)\beta-\nu-1}\gamma_{p(\nu,|\ell|)}(r)e^{-c'r^{\frac{1}{1-\beta}}}, \tag{7.47}$$

with the number $p(\nu, \ell)$ and the function γ_m given in Lemma 7.14. We bound the integrals one by one. The integrals in the estimates are denoted by $I_i, i = 1, \ldots, 6$.
(i) The inequality (7.47) with $\nu = 0$ and (7.38) allows bounding the integral

$$I_1 \leq c\int_0^\infty\int_0^\infty t^{(1-d-|\ell|)\beta-1}\gamma_{d+|\ell|+1}(|x - h\lambda|t^{-\beta})e^{-c'(|x-h\lambda|t^{-\beta})^{\frac{1}{1-\beta}}}d\lambda dt.$$

Due to the difference of γ_m, we distinguish the following two cases (a) $d + |\ell| = 3$ and (b) $d + |\ell| > 3$. In case (a), it follows from (7.29) and Lemma 7.16 and then changing variables $s = |x'|t^{-\beta}$ and $\eta = \lambda t^{-\beta}$ that

$$I_1 \leq c \int_0^\infty \int_0^\infty t^{-2\beta-1}(|x-h\lambda|t^{-\beta})^{-\frac{1}{4}} e^{-\frac{c'}{2}(|x-h\lambda|t^{-\beta})^{\frac{1}{1-\beta}}} \, d\lambda dt$$

$$\leq c \int_0^\infty \int_0^\infty t^{-2\beta-1}(|x'|t^{-\beta})^{-\frac{1}{4}} e^{-c''(|x'|t^{-\beta})^{\frac{1}{1-\beta}}} e^{-c''(\lambda t^{-\beta})^{\frac{1}{1-\beta}}} \, d\lambda dt$$

$$\leq c \left(\int_0^\infty (|x'|s^{-1})^{-1} s^{-\frac{1}{4}} e^{-c''s^{\frac{1}{1-\beta}}} \frac{ds}{s} \right) \left(\int_0^\infty e^{-c''\eta^{\frac{1}{1-\beta}}} \, d\eta \right)$$

$$\leq c|x'|^{-1} \int_0^\infty s^{\frac{1}{4}} e^{-c''s^{\frac{1}{1-\beta}}} \, ds \leq c_1 |x'|^{-1}.$$

Similarly, in case (b), we have

$$I_1 \leq c \int_0^\infty \int_0^\infty t^{(1-d-|\ell|)\beta-1}(|x-h\lambda|t^{-\beta})^{4-d} e^{-c'(|x-h\lambda|t^{-\beta})^{\frac{1}{1-\beta}}} \, d\lambda dt$$

$$\leq c \int_0^\infty \int_0^\infty t^{(2-d-|\ell|)\beta-1}(|x'|t^{-\beta})^{4-d} e^{-c''(|x'|t^{-\beta})^{\frac{1}{1-\beta}}} e^{-c''(\lambda t^{-\beta})^{\frac{1}{1-\beta}}} \, d\lambda dt$$

$$\leq c \int_0^\infty (|x'|s^{-1})^{2-d-|\ell|} s^{4-d} e^{-c''s^{\frac{1}{1-\beta}}} \frac{ds}{s} \leq c|x'|^{2-d-|\ell|}.$$

Combining these two estimates shows the estimate (7.41).

(ii). The estimate (7.47) with $\nu = 0$ and $|\ell| = 2$ yields

$$I_2 \leq c \int_0^\infty \int_0^\infty t^{\frac{\beta}{2}-d\beta-1} \gamma_{d+2}(|x-h\lambda|t^{-\beta}) e^{-c'(|x-h\lambda|t^{-\beta})^{\frac{1}{1-\beta}}} \, d\lambda dt.$$

Like before, due to the different form of the function γ_m, we analyze the cases $d = 2$ and $d \geq 3$ separately. If $d = 2$, by inequality (7.29) and Lemma 7.16, we have

$$I_2 \leq c \int_0^\infty \int_0^\infty t^{\frac{\beta}{2}-2\beta-1}(|x-h\lambda|t^{-\beta})^{-\frac{1}{4}} e^{-\frac{c'}{2}(|x-h\lambda|t^{-\beta})^{\frac{1}{1-\beta}}} \, d\lambda dt$$

$$\leq c \int_0^\infty \int_0^\infty t^{-\frac{3}{2}\beta-1}(|x'|t^{-\beta})^{-\frac{1}{4}} e^{-c''(|x'|t^{-\beta})^{\frac{1}{1-\beta}}} e^{-c''(\lambda t^{-\beta})^{\frac{1}{1-\beta}}} \, d\lambda dt$$

$$\leq c \int_0^\infty (|x'|s^{-1})^{-\frac{1}{2}} s^{-\frac{1}{4}} e^{-c''s^{\frac{1}{1-\beta}}} \frac{ds}{s}$$

$$= c|x'|^{-\frac{1}{2}} \int_0^\infty s^{-\frac{3}{4}} e^{-c''s^{\frac{1}{1-\beta}}} \, ds \leq c|x'|^{-\frac{1}{2}}.$$

Similarly, for $d \geq 3$, we have

$$I_2 \leq c \int_0^\infty \int_0^\infty t^{\frac{\beta}{2}-d\beta-1}(|x-h\lambda|t^{-\beta})^{2-d} e^{-c'(|x-h\lambda|t^{-\beta})^{\frac{1}{1-\beta}}} \, d\lambda dt$$

$$\leq c \int_0^\infty \int_0^\infty t^{\frac{\beta}{2}-d\beta-1}(|x'|t^{-\beta})^{2-d} e^{-c''(|x'|t^{-\beta})^{\frac{1}{1-\beta}}} e^{-c''(\lambda t^{-\beta})^{\frac{1}{1-\beta}}} \, d\lambda dt$$

$$\leq c \int_0^\infty (|x'|s^{-\beta})^{\frac{3}{2}-d} s^{2-d} e^{-c''s^{\frac{1}{1-\beta}}} \frac{ds}{s} \leq c|x'|^{\frac{3}{2}-d}.$$

Combining these two cases shows (7.42).

(iii) The identity (7.47) with $\nu = \alpha$ and (7.38) allows bounding the integral

$$I_3 \le c \int_0^\infty \int_0^\infty t^{(\delta-d-|\ell|-2)\beta-1} \gamma_{d+|\ell|+3}(|x-h\lambda|t^{-\beta})e^{-c'(|x-h\lambda|t^{-\beta})^{\frac{1}{1-\beta}}} \, d\lambda dt.$$

Since $d \ge 2$,

$$\gamma_{d+|\ell|+3}(\zeta) = \zeta^{4-(d+|\ell|+3)} = \zeta^{1-d-|\ell|}.$$

It follows from (7.29) and Lemma 7.16 and changing variables $s = |x'|t^{-\beta}$ and $\eta = \lambda t^{-\beta}$ that

$$I_3 \le c \int_0^\infty \int_0^\infty t^{(\delta-d-|\ell|-2)\beta-1}(|x-h\lambda|t^{-\beta})^{1-d-|\ell|} e^{-c'(|x-h\lambda|t^{-\beta})^{\frac{1}{1-\beta}}} \, d\lambda dt$$

$$\le c \int_0^\infty \int_0^\infty t^{(\delta-d-|\ell|-2)\beta-1}(|x'|t^{-\beta})^{1-d-|\ell|} e^{-c''(|x'|t^{-\beta})^{\frac{1}{1-\beta}}} e^{-c''(\lambda t^{-\beta})^{\frac{1}{1-\beta}}} \, d\lambda dt$$

$$\le c \int_0^\infty (|x'|s^{-1})^{\delta-d-|\ell|} s^{1-d-|\ell|} e^{-c''s^{\frac{1}{1-\beta}}} \frac{ds}{s} \le c|x'|^{\delta-d-|\ell|},$$

which shows the estimate (7.43).

(iv) The definition of $\widetilde{G}_\alpha(x,t)$, the integral representation of $\overline{G}_\alpha(x,t)$, and the differentiation formula (3.28) for Wright function imply

$$^R\partial_t^{\alpha-1+\zeta}\widetilde{G}_\alpha(x,t) = -2\int_0^\infty {}^R\partial_t^{\alpha-1+\zeta}\partial_{x_d}\overline{G}_\alpha(x-h\mu,t)d\mu$$

$$= -2\int_0^\infty \int_0^\infty \partial_{x_d}G_1(x-h\mu,\lambda)^R\partial_t^{\alpha-1+\zeta}\left(t^{-1}W_{-\alpha,0}(-\lambda t^{-\alpha})\right)d\mu d\lambda$$

$$= -2\int_0^\infty \int_0^\infty \partial_{x_d}G_1(x-h\mu,\lambda)t^{-\alpha-\zeta}W_{-\alpha,-\alpha+1-\zeta}(-\lambda t^{-\alpha})d\mu d\lambda.$$

Using the bound (7.30) and Lemma 3.5(iv), and then changing variables $y' = (x'-h'\mu)\lambda^{-\frac{1}{2}}$, we may bound

$$I_4 \le c \int_0^\infty \int_0^\infty \left(\int_{\mathbb{R}^{d-1}} \lambda^{-\frac{d+1}{2}} e^{-c_1\frac{|x-h\mu|^2}{\lambda}} dx'\right) t^{-\alpha-\zeta}|W_{-\alpha,1-\alpha-\zeta}(-\lambda t^{-\alpha})|d\mu d\lambda$$

$$\le c \int_0^\infty \int_0^\infty \lambda^{-1} e^{-c_1\frac{(x_d-h_d\mu)^2}{\lambda}} t^{-\alpha-\zeta}|W_{-\alpha,1-\alpha-\zeta}(-\lambda t^{-\alpha})|d\mu d\lambda.$$

Changing variables $\xi = \mu\lambda^{-\frac{1}{2}}$ and $\eta = \lambda t^{-\alpha}$, the conditions $h_d \le -\delta_0 < 0$ from Assumption (7.37) and Theorem 3.8(iv) give

$$I_4 \le c \int_0^\infty \lambda^{-\frac{1}{2}} t^{-\alpha-\zeta}|W_{-\alpha,1-\alpha-\zeta}(-\lambda t^{-\alpha})|d\lambda \int_0^\infty e^{-c_1\delta_0^2\xi^2} d\xi$$

$$\le ct^{-\beta-\zeta}\int_0^\infty \eta^{-\frac{1}{2}}|W_{-\alpha,1-\alpha-\zeta}(-\eta)|d\eta \le ct^{-\beta-\zeta}, \quad \forall \zeta \in \mathbb{R}.$$

This completes the proof of the estimate (7.44).

(v). The definition of $\widetilde{G}_\alpha(x, t)$ yields

$$\partial_t^k \partial_{x_i}^\ell \widetilde{G}_\alpha(x, t) = -2 \int_0^\infty \int_0^\infty \partial_{x_d x_i}^2 G_1(x - h\mu, \lambda) t^{-1-k} W_{-\alpha,-k}(-\lambda t^{-\alpha}) d\mu d\lambda.$$

Like before, the bound (7.30) and Lemma 3.5(iv) lead to

$$I_5 \leq c \int_0^\infty \int_0^\infty \lambda^{-\frac{3}{2}} e^{-c_1 \frac{(x_d - h_d\mu)^2}{\lambda}} t^{-1-k} |W_{-\alpha,-k}(-\lambda t^{-\alpha})| d\mu d\lambda.$$

Changing variables $\xi = \mu\lambda^{-\frac{1}{2}}$ and $\eta = \lambda t^{-\alpha}$ and Assumption (7.37) give

$$I_5 \leq c \int_0^\infty \lambda^{-1} t^{-1-k} |W_{-\alpha,-k}(-\lambda t^{-\alpha})| d\lambda \int_0^\infty e^{-c_1 \delta_0^2 \xi^2} d\xi$$

$$\leq ct^{-1-k} \int_0^\infty \eta^{-1} |W_{-\alpha,-k}(-\eta)| d\eta \leq ct^{-1-k}, \quad k = 0, 1, \ i = 1, \ldots, d,$$

where the bound on the integral $\int_0^\infty \eta^{-1} |W_{-\alpha,-k}(-\eta)| d\eta$ follows from Theorem 3.8(iv) for large η and the definition of $W_{-\alpha,-k}(-\eta)$ in a neighborhood of $\eta = 0$. This shows the estimate (7.45).

(vi) The definition of $\widetilde{G}_\alpha(x, t)$ and the integral representation of $\overline{G}_\alpha(x, t)$ in (7.4) give

$${}^R\partial_t^\nu D_x^\ell \widetilde{G}_\alpha(x, t) = -2 \int_0^\infty \int_0^\infty D_x^\ell \partial_{x_d} G_1(x - h\mu, \lambda) t^{-\nu-1} W_{-\alpha,-\nu}(-\lambda t^{-\alpha}) d\mu d\lambda.$$

The bound (7.30) yields

$$I_6 \leq c \int_0^\infty \int_0^\infty \lambda^{-\frac{|\ell|}{2}-1} e^{-c_1 \frac{(x_d - h_d\mu)^2}{\lambda}} t^{-\nu-1} |W_{-\alpha,-\nu}(-\lambda t^{-\alpha})| d\mu d\lambda.$$

Changing variables $\xi = \mu\lambda^{-\frac{1}{2}}$ and $\eta = \lambda t^{-\alpha}$, Assumption (7.37) and Theorem 3.8(iv) give

$$I_6 \leq c \int_0^\infty \lambda^{-\frac{|\ell|+1}{2}} t^{-\nu-1} W_{-\alpha,-\nu}(-\lambda t^{-\alpha}) d\lambda \int_0^\infty e^{-c_1 \delta_0^2 \xi^2} d\xi$$

$$\leq ct^{(1-|\ell|)\beta-\nu-1} \int_0^\infty \eta^{\frac{1-|\ell|}{2}} e^{-c'\eta^{\frac{1}{1-\alpha}}} d\eta \leq ct^{(1-|\ell|)\beta-\nu-1}, \quad \text{if } 0 \leq |\ell| \leq 2.$$

This proves the assertion (7.46) and completes the proof of the lemma. □

The next result gives a differentiation formula for the surface potential

$$S_o g(x, t) = \int_0^t \int_{\mathbb{R}^{d-1}} \overline{G}_\alpha(x' - y', x_d, t - s) g(y', s) dy' ds.$$

The notation Q' denotes $Q' = \mathbb{R}^{d-1} \times I$, and accordingly $\overline{Q}' = \mathbb{R}^{d-1} \times \overline{I}$.

Lemma 7.18 *For* $g \in C_0^{0, \frac{1+\theta}{2}\alpha}(\overline{Q}')$, *the following identity holds:*

$$\partial_t^\alpha S_o g(x, t) = \int_0^\infty \int_{\mathbb{R}^{d-1}} {}^R \partial_s^\alpha \widetilde{G}_\alpha(x' - y', x_d, s)(g(y', t - s) - g(y', t))dy'ds. \quad (7.48)$$

Proof Let $v(x, t) = S_o g(x, t)$. The estimate (7.44) implies

$$|v(x, t)| \leq c\|g\|_{C(\overline{Q}')} \int_0^t s^{\frac{\alpha}{2} - 1}ds \leq c\|g\|_{C(\overline{Q}')} t^{\frac{\alpha}{2}},$$

and thus $v(x, t)|_{t=0} = 0$. Further, by assumption, $g(x', t) = 0$ for $t \leq 0$, and thus $\partial_t^\alpha v = {}^R \partial_t^\alpha v$. For any $\epsilon > 0$, let

$$U_\epsilon(x, t) = \int_0^{t-\epsilon} \int_{\mathbb{R}^{d-1}} {}_s I_t^{1-\alpha} \widetilde{G}_\alpha(x' - y', x_d, t - s)g(y', s)dy'ds.$$

It remains to show that the limit $\lim_{\epsilon \to 0^+} \partial_t U_\epsilon$ equals to the right-hand side of (7.48). Indeed, direct computation gives

$$\partial_t U_\epsilon(x, t) = \int_{-\infty}^{t-\epsilon} \int_{\mathbb{R}^{d-1}} {}_s^R \partial_t^\alpha \widetilde{G}_\alpha(x' - y', x_d, t - s)g(y', s)dy'ds$$

$$+ \int_{\mathbb{R}^{d-1}} {}_0 I_t^{1-\alpha} \widetilde{G}_\alpha(x' - y', x_d, \epsilon)g(y', t - \epsilon)dy'$$

$$= \int_\epsilon^\infty \int_{\mathbb{R}^{d-1}} {}^R \partial_s^\alpha \widetilde{G}_\alpha(x' - y', x_d, s)(g(y', t - s) - g(y', t - \epsilon))dy'ds,$$

and

$$\delta_\epsilon(x, t) := \partial_t U_\epsilon(x, t) - \int_0^\infty \int_{\mathbb{R}^{d-1}} {}^R \partial_s^\alpha \widetilde{G}_\alpha(x' - y', x_d, s)(g(y', t - s) - g(y', t))dy'ds$$

$$= \int_\epsilon^\infty \int_{\mathbb{R}^{d-1}} {}^R \partial_s^\alpha \widetilde{G}_\alpha(x' - y', x_d, s)(g(y', t) - g(y', t - \epsilon))dy'ds$$

$$+ \int_0^\epsilon \int_{\mathbb{R}^{d-1}} {}^R \partial_s^\alpha \widetilde{G}_\alpha(x' - y', x_d, s)(g(y', t - s) - g(y', t))dy'ds.$$

Now it follows from the estimate (7.44) that

$$|\delta_\epsilon(x, t)| \leq c[g]_{t, \overline{Q}'}^{(\frac{1+\theta}{2}\alpha)} \left(\epsilon^{\frac{1+\theta}{2}\alpha} \int_\epsilon^\infty s^{-1 - \frac{\alpha}{2}}ds + \int_0^\epsilon s^{\frac{\theta}{2}\alpha - 1}ds\right) \leq c[g]_{t, \overline{Q}'}^{(\frac{1+\theta}{2}\alpha)} \epsilon^{\frac{\theta}{2}\alpha}.$$

Letting $\epsilon \to 0^+$ completes the proof of the lemma. $\square$

The next lemma gives one technical estimate.

Lemma 7.19 *For* $K = \{y' \in \mathbb{R}^{d-1} : |x' - y'| \leq 2|x - \bar{x}|\}$, *and* $h = |x - x'|^{\frac{2}{\alpha}}$,

$$\left|\int_0^h \int_K \partial_{x_i} \widetilde{G}_\alpha(x' - y', x_d, s)dy'ds\right| \leq c, \quad i = 1, \ldots, d.$$

Proof We denote the integral by I, and let $\beta = \frac{\alpha}{2}$. First, we claim the identity

$$\widetilde{G}_\alpha(x,t) = -\frac{2}{h_d}\overline{G}_\alpha(x,t) + \frac{2}{h_d}\sum_{i=1}^{d-1}\int_0^\infty h_i\partial_{x_i}\overline{G}_\alpha(x-h\lambda,t)d\lambda \qquad (7.49)$$

$$:= G_1(x,t) + G_2(x,t).$$

Indeed, since

$$\frac{d}{d\lambda}\overline{G}_\alpha(x-h\lambda,t) = -\sum_{i=1}^{d-1}h_i\partial_{x_i}\overline{G}_\alpha(x-h\lambda,t) - h_d\partial_{x_d}\overline{G}_\alpha(x-h\lambda,t),$$

upon integrating with respect to 0 to ∞, we obtain

$$-\overline{G}_\alpha(x,t) = -\sum_{i=1}^{d-1}\int_0^\infty h_i\partial_{x_i}\overline{G}_\alpha(x-h\lambda,t)d\lambda - h_d\int_0^\infty \partial_{x_d}\overline{G}_\alpha(x-h\lambda,t)d\lambda.$$

This and the identity (7.38) show (7.49). Next we consider the two cases (a) $i \neq d$ and (b) $i = d$ separately. In case (a), by the divergence theorem, we have

$$I = \int_0^h\int_\Sigma \widetilde{G}_\alpha(x'-y',x_d,s)v_i dS_{y'}ds = \int_0^h\int_\Sigma G_2(x'-y',x_d,s)v_i dS_{y'}ds,$$

where $\Sigma = \{y' : |x'-y'| = 2|x-\bar{x}|\}$ and v is the unit outward normal to Σ in $\mathbb{R}^{d-1}$, and $dS_{y'}$ denotes the (infinitesimal) surface area for Σ. By Lemma 7.14, we obtain

$$|I| \leq \int_0^h\int_\Sigma |G_2(x'-y',x_d,s)|dS_{y'}ds$$

$$\leq c\int_0^h\int_0^\infty\int_\Sigma s^{(1-d)\beta-1}\gamma_{d+2}(\xi)e^{-c'\xi^{\frac{1}{1-\beta}}}dS_{y'}d\lambda ds, \qquad (7.50)$$

where $\xi = |x'-y'-h'\lambda|s^{-\beta}$. If $d = 2$, we use (7.40) and (7.29) and obtain (with $\zeta = |x-\bar{x}|s^{-\beta}$)

$$|I| \leq c\int_0^h\int_0^\infty \zeta^{-\frac{1}{4}}e^{-\frac{\sigma_0}{4}\zeta^{\frac{1}{1-\beta}}}e^{-\frac{\sigma_0}{4}(\lambda s^{-\beta})^{\frac{1}{1-\beta}}}s^{-\beta-1}d\lambda ds$$

$$\leq c\int_0^h \zeta^{-\frac{1}{4}}e^{-\frac{\sigma_0}{4}\zeta^{\frac{1}{1-\beta}}}s^{-1}ds \leq c\int_1^\infty \zeta^{-\frac{5}{4}}e^{-\frac{\sigma_0}{4}z^{\frac{1}{1-\beta}}}d\zeta \leq c.$$

Similarly, if $d = 3$, it follows from (7.50) and the identity $|\Sigma| = 2\frac{\pi^{\frac{d-1}{2}}}{\Gamma(d-1)}(2|x-\bar{x}|)^{d-2}$ for the surface area $|\Sigma|$ that

$$|I| \leq c|x-\bar{x}|^{d-2}\int_1^\infty (\zeta|x-\bar{x}|^{-1})^{d-2}\zeta^{2-d}e^{-\frac{\sigma_0}{4}\zeta^{\frac{1}{1-\beta}}}\zeta^{-1}d\zeta \leq c.$$

In case (b), by the definition of $\widetilde{G}_\alpha$ and the decomposition (7.49), we have

$$\partial_{x_d}\widetilde{G}_\alpha(x,t) = -\frac{2}{h_d}\partial_{x_d}\overline{G}_\alpha(x,t) - \frac{1}{h_d}\sum_{i=1}^{d-1}h_i\partial_{x_i}\widetilde{G}_\alpha(x,t). \qquad (7.51)$$

The summation on the right-hand side is already estimated. It remains to prove

$$J := \int_0^h \int_K |\partial_{x_d}\overline{G}_\alpha(x'-y',x_d,s)|dy'ds \le c.$$

By Lemma 7.1(i), $\partial_{x_d}\overline{G}_\alpha(x',x_d,s) \le 0$ and the identity (7.8), we can bound J by

$$J \le c\int_0^\infty \int_{\mathbb{R}^{d-1}} |\partial_{x_d}\overline{G}_\alpha(x',x_d,s)|dx'ds \le c.$$

This completes the proof of lemma. □

The next result collects several estimates on the surface potential $S_o g$. Let $Q = \mathbb{R}_+^d \times (0,T]$ and $Q' = \mathbb{R}^{d-1} \times (0,T]$.

Lemma 7.20 *For the function* $u(x,t) = S_o g$, *the following estimates hold:*

$$[\partial_{x_i}u]_{x,Q}^{(\theta)} + [\partial_{x_i}u]_{t,Q}^{(\frac{\theta}{2}\alpha)} + [u]_{t,Q}^{(\frac{1+\theta}{2}\alpha)} \le c[g]_{C^{\theta,\frac{\theta}{2}\alpha}(\overline{Q}')},$$

$$[\partial_t^\alpha u]_{x,Q}^{(\theta)} + [\partial_t^\alpha u]_{t,Q}^{(\frac{\theta}{2}\alpha)} \le c[g]_{C^{1+\theta,\frac{1+\theta}{2}\alpha}(\overline{Q}')}.$$

Proof The proof follows closely [LSU68, Chapter IV]. To simplify the notation in the proof, we let $\beta = \frac{\alpha}{2}$, and

$$\widetilde{g}(y',t,s) = g(y',t-s) - g(y',t) \quad \text{and} \quad \widehat{g}(y',x',t) = g(y',t) - g(x',t).$$

Then we can split $\partial_{x_i}u(x,t)$ into

$$\partial_{x_i}u(x,t) = \int_0^\infty \int_{\mathbb{R}^{d-1}} \partial_{x_i}\widetilde{G}_\alpha(x'-y',x_d,s)\widetilde{g}(y',t,s)dy'ds$$

$$+ \int_0^\infty \int_{\mathbb{R}^{d-1}} \partial_{x_i}\widetilde{G}_\alpha(x'-y',x_d,s)\widehat{g}(y',x',t)dy'ds$$

$$+ g(x',t)\int_0^\infty \int_{\mathbb{R}^{d-1}} \partial_{x_i}\widetilde{G}_\alpha(x'-y',x_d,s)dy'ds.$$

Fix $x,\bar{x} \in \mathbb{R}_+^d$ with $x \ne \bar{x}$. Then $\partial_{x_i}u(x,t) - \partial_{x_i}u(\bar{x},t)$ can be decomposed into seven terms (with $h = |x - \bar{x}|^{\frac{1}{\beta}}$ and $K = \{y' \in \mathbb{R}^{d-1} : |x' - y'| \le 2|x - \bar{x}|\}$)

$$\partial_{x_i} u(x,t) - \partial_{x_i} u(\bar{x},t) = \int_0^h \int_{\mathbb{R}^{d-1}} \partial_{x_i} \widetilde{G}_\alpha(x' - y', x_d, s) \widetilde{g}(y', t, s) dy' ds$$

$$- \int_0^h \int_{\mathbb{R}^{d-1}} \partial_{\bar{x}_i} \widetilde{G}_\alpha(\bar{x}' - y', \bar{x}_d, s) \widetilde{g}(y', t, s) dy' ds$$

$$+ \int_h^\infty \int_{\mathbb{R}^{d-1}} (\partial_{x_i} \widetilde{G}_\alpha(x' - y', x_d, s) - \partial_{\bar{x}_i} \widetilde{G}_\alpha(\bar{x}' - y', \bar{x}_d, s)) \widetilde{g}(y', t, s) dy' ds$$

$$+ \int_K \widehat{g}(y', x', t) dy' \int_0^\infty \partial_{x_i} \widetilde{G}_\alpha(x' - y', x_d, s) ds$$

$$- \int_K \widehat{g}(y', \bar{x}', t) dy' \int_0^\infty \partial_{\bar{x}_i} \widetilde{G}_\alpha(\bar{x}' - y', \bar{x}_d, s) ds$$

$$+ \int_{\mathbb{R}^{d-1} \setminus K} \int_0^\infty (\partial_{x_i} \widetilde{G}_\alpha(x' - y', x_d, s) - \partial_{\bar{x}_i} \widetilde{G}_\alpha(\bar{x}' - y', \bar{x}_d, s)) \widehat{g}(y', \bar{x}', t) ds dy'$$

$$- \widehat{g}(\bar{x}', x', t) \int_0^\infty \int_K \partial_{x_i} \widetilde{G}_\alpha(x' - y', x_d, s) dx' ds.$$

The seven terms are denoted by I_i, $i = 1, \ldots, 7$. The last term I_7 follows by

$$I_7 = \int_{\mathbb{R}^{d-1} \setminus K} \int_0^\infty \partial_{x_i} \widetilde{G}_\alpha(x' - y', x_d, s)(g(\bar{x}', t) - g(x', t)) ds dy'$$

$$+ g(x', t) \int_{\mathbb{R}^{d-1}} \int_0^\infty \partial_{x_i} \widetilde{G}_\alpha(x' - y', x_d, s) ds dy'$$

$$- g(\bar{x}', t) \int_{\mathbb{R}^{d-1}} \int_0^\infty \partial_{\bar{x}_i} \widetilde{G}_\alpha(\bar{x}' - y', \bar{x}_d, s) ds dy'.$$

Now by Lemma 7.1(i), (7.8), and the identity (7.51), we have

$$\int_{\mathbb{R}^{d-1}} \int_0^\infty \partial_{x_i} \widetilde{G}_\alpha(y', x_d, s) ds dy' = \begin{cases} 0, & \text{if } i \neq d, \\ c(h_d), & \text{if } i = d. \end{cases}$$

Thus

$$\int_{\mathbb{R}^{d-1}} \int_0^\infty \partial_{\bar{x}_i} \widetilde{G}_\alpha(\bar{x}' - y', \bar{x}_d, s) ds dy' = \int_{\mathbb{R}^{d-1}} \int_0^\infty \partial_{\bar{x}_i} \widetilde{G}_\alpha(\bar{x}' - y', x_d, s) ds dy'.$$

Substituting the identity gives the expression for I_7. Next we estimate the terms separately. By the estimate (7.45) with $k = 0$, we bound the terms I_1 and I_2 by

$$|I_1| + |I_2| \leq c[g]_{t,\overline{Q}}^{(\theta\beta)} \int_0^h s^{\theta\beta-1} ds \leq c[g]_{t,\overline{Q}}^{(\theta\beta)} |x - \bar{x}|^\theta.$$

For the term I_3, with $\eta = \bar{x} + \lambda(x - \bar{x})$, we have

$$I_3 = \sum_{k=1}^d \int_h^\infty \int_0^1 \int_{\mathbb{R}^{d-1}} \partial_{\eta_i \eta_k}^2 \widetilde{G}_\alpha(\eta' - y', \eta_d, s)(x_k - \bar{x}_k) \widetilde{g}(y', t, s) dy' d\lambda ds.$$

Then the estimate (7.46) with $|\ell| = 2$ implies

$$|I_3| \leq c[g]_{t,\overline{Q}}^{(\theta\beta)}|x - \bar{x}| \int_h^\infty s^{\theta\beta-1-\beta}ds \leq c[g]_{t,\overline{Q}}^{(\theta\beta)}|x - \bar{x}|^\theta.$$

By the estimate (7.41) with $|\ell| = 1$, we derive

$$|I_4| + |I_5| \leq c[g]_{x,\overline{Q}}^{(\theta)}\left(\int_K |x' - y'|^{\theta-(d-1)}dy'\right.$$

$$\left.+ \int_{|\bar{x}'-y'|\leq 3|\bar{x}-x|} |\bar{x}' - y'|^{\theta-(d-1)}dy'\right) \leq c[g]_{x,\overline{Q}}^{(\theta)}|x - \bar{x}|^\theta.$$

Next again with $\eta = \bar{x} + \lambda(x - \bar{x})$, we can rewrite I_6 as

$$I_6 = \sum_{k=1}^d \int_{\mathbb{R}^{d-1}\backslash K} \int_0^1 \int_0^\infty \partial_{\eta_i \eta_k}^2 \widetilde{G}_\alpha(\eta' - y', \eta_d, s)(x_k - \bar{x}_k)\widehat{g}(y', \bar{x}', t)ds d\lambda dy'.$$

Obviously, $|x - \bar{x}| \leq |\eta' - y'|$, $|\bar{x}' - y'| \leq 2|\eta' - y'|$ for any $\eta = \bar{x} + \lambda(x - \bar{x})$, $y' \in K$, $\lambda \in (0, 1)$. Hence, it follows from (7.41) with $|\ell|=2$ that

$$|I_6| \leq c[g]_{x,\overline{Q}}^{(\theta)}|x - \bar{x}| \int_0^1 \int_{|\eta'-y'|\geq|x-\bar{x}|} |\eta' - y'|^{\theta-d}dy'd\lambda \leq c[g]_{x,\overline{Q}}^{(\theta)}|x - \bar{x}|^\theta.$$

Next we decompose I_7 into the sum

$$I_7 = \widehat{g}(\bar{x}', x', t)\left(\int_h^\infty \int_K \partial_{x_i}\widetilde{G}_\alpha(x' - y', x_d, s)dy'ds\right.$$

$$\left.+ \int_0^h \int_K \partial_{x_i}\widetilde{G}_\alpha(x' - y', x_d, s)dy'ds\right) := \widehat{g}(\bar{x}', x', t)(I_7' + I_7'').$$

By the estimate (7.42), we obtain

$$|I_7'| \leq |x - \bar{x}|^{-\frac{1}{2}} \int_h^\infty \int_K s^{\frac{\beta}{2}}|\partial_{x_i}\widetilde{G}_\alpha(x' - y', x_d, s)|dy'ds$$

$$\leq c|x - \bar{x}|^{-\frac{1}{2}} \int_K |x' - y'|^{\frac{3}{2}-d}dy' \leq c.$$

Now Lemma 7.19 gives the bound $|I_7''| \leq c$. The preceding estimates together imply $[\partial_{x_i} u]_{x,\overline{Q}}^{(\theta)} \leq c[g]_{x,\overline{Q}}^{(\theta)}$. Next we bound $[\partial_t^\alpha u]_{t,\overline{Q}}^{(\theta\beta)}$. Fix any $t > \bar{t}$. By Lemma 7.18, it suffices to we bound

$$\partial_t^\alpha u(x,t) - \partial_t^\alpha u(x,\bar{t})$$

$$= \int_{2\bar{t}-t}^t \int_{\mathbb{R}^{d-1}} {}^R\partial_t^\alpha \widetilde{G}_\alpha(x'-y', x_d, t-s)(g(y',s)-g(y',t))dy'ds$$

$$- \int_{2\bar{t}-t}^{\bar{t}} \int_{\mathbb{R}^{d-1}} {}^R\partial_{\bar{t}}^\alpha \widetilde{G}_\alpha(x'-y', x_d, \bar{t}-s)(g(y',s)-g(y',\bar{t}))dy'ds$$

$$+ \int_{-\infty}^{2\bar{t}-t} \int_{\mathbb{R}^{d-1}} ({}^R\partial_t^\alpha \widetilde{G}_\alpha(x'-y', x_d, t-s) - {}^R\partial_{\bar{t}}^\alpha \widetilde{G}_\alpha(x'-y', x_d, \bar{t}-s))$$

$$\times (g(y',s)-g(y',\bar{t}))dy'ds$$

$$+ \int_{-\infty}^{2\bar{t}-t} \int_{\mathbb{R}^{d-1}} {}^R\partial_t^\alpha \widetilde{G}_\alpha(x'-y', x_d, t-s)(g(y',\bar{t})-g(y',t))dy'ds.$$

We bound the four terms, denoted by II_i, $i = 1,\dots,4$, separately. By the estimate (7.44) in Lemma 7.17, we obtain

$$|\mathrm{II}_1| + |\mathrm{II}_2| \le c[g]_{t,\overline{Q'}}^{(1+\theta)\beta} \left(\int_{2\bar{t}-t}^t (t-s)^{(1+\theta)\beta-\beta-1}ds + \int_{2\bar{t}-t}^{\bar{t}} (\bar{t}-s)^{(1+\theta)\beta-\beta-1}ds \right)$$

$$\le c[g]_{t,\overline{Q'}}^{((1+\theta)\beta)} |t-\bar{t}|^{\theta\beta},$$

and by the mean value theorem,

$$|\mathrm{II}_3|$$

$$\le c[g]_{t,\overline{Q'}}^{((1+\theta)\beta)} \int_{-\infty}^{2\bar{t}-t} \int_{\bar{t}}^t \int_{\mathbb{R}^{d-1}} |{}^R\partial_s^{\alpha+1} \widetilde{G}_\alpha(x'-y', x_d, \xi-s)(\bar{t}-s)^{(1+\theta)\beta}|dy'd\xi ds$$

$$\le c[g]_{t,\overline{Q'}}^{((1+\theta)\beta)} \int_{-\infty}^{2\bar{t}-t} \int_{\bar{t}}^t (\xi-s)^{\beta-2}(\bar{t}-s)^{(1+\theta)\beta}d\xi ds$$

$$\le c[g]_{t,\overline{Q'}}^{((1+\theta)\beta)} \int_{\bar{t}}^t \int_{-\infty}^{2\bar{t}-t} (\bar{t}-s)^{\theta\beta-2}dsd\xi \le c[g]_{t,\overline{Q'}}^{((1+\theta)\beta)} |t-\bar{t}|^{\theta\beta}.$$

Next it follows from the estimate (7.44) and integrating with respect to s that

$$|\mathrm{II}_4| = \left| \int_{\mathbb{R}^{d-1}} (g(y',t)-g(y',\bar{t})){}^R\partial_t^{\alpha-1} \widetilde{G}_\alpha(x'-y', x_d, 2(t-\bar{t}))dy' \right|$$

$$\le c[g]_{t,\overline{Q'}}^{((1+\theta)\beta)} |t-\bar{t}|^{\theta\beta}.$$

The last three estimates imply $[\partial_t^\alpha u]_{t,Q}^{(\frac{\theta}{2}\alpha)} \le c[g]_{t,Q}^{\frac{(1+\theta)}{2}\alpha}$. Next we bound the seminorm $[\partial_t^\alpha u]_{x,\overline{Q}}^{(\theta)}$. Indeed, we may rewrite $\partial_t^\alpha u(x,t)$ as

$$\partial_t^\alpha u(x,t) = \int_{-\infty}^t \int_{\mathbb{R}^{d-1}} {}^R\partial_s^\alpha \widetilde{G}_\alpha(x'-y', x_d, s)(g(y',t-s)-g(y',t))dy'ds.$$

Thus, for any $x \neq \bar{x}$, we split $\partial_t^\alpha u(x,t) - \partial_t^\alpha u(\bar{x},t)$ into (with $h = |x - \bar{x}|^{\frac{2}{\alpha}}$)

$$
\begin{aligned}
&\partial_t^\alpha u(x,t) - \partial_t^\alpha u(\bar{x}',t) \\
&= \int_0^h \int_{\mathbb{R}^{d-1}} {}^R \partial_s^\alpha \widetilde{G}_\alpha(x' - y', x_d, s)(g(y', t - s) - g(y', t)) dy' ds \\
&\quad + \int_0^h \int_{\mathbb{R}^{d-1}} {}^R \partial_s^\alpha \widetilde{G}_\alpha(\bar{x}' - y', \bar{x}_d, s)(g(y', t) - g(y', t - s)) dy' ds \\
&\quad + \int_h^\infty \int_{\mathbb{R}^{d-1}} [{}^R \partial_s^\alpha \widetilde{G}_\alpha(x' - y', x_d, s) - {}^R \partial_s^\alpha \widetilde{G}_\alpha(\bar{x}' - y', \bar{x}_d, s)] \\
&\quad \times (g(y', t - s) - g(y', t)) dy' ds := \sum_{i=1}^3 \mathrm{III}_i.
\end{aligned}
$$

The term III_1 can be bounded by (7.44) from Lemma 7.17 as

$$
\begin{aligned}
|\mathrm{III}_1| &\leq c[g]_{t,\overline{Q}}^{(\frac{1+\theta}{2}\alpha)} \int_0^h \int_{\mathbb{R}^{d-1}} |{}^R \partial_s^\alpha \widetilde{G}_\alpha(x' - y', x_d, s)| s^{\frac{1+\theta}{2}\alpha} dy' ds \\
&\leq c[g]_{t,\overline{Q}}^{(\frac{1+\theta}{2}\alpha)} \int_0^h s^{-\frac{\alpha}{2}-1} s^{\frac{1+\theta}{2}\alpha} ds = c[g]_{t,\overline{Q}}^{(\frac{1+\theta}{2}\alpha)} |x - \bar{x}|^\theta.
\end{aligned}
$$

The term III_2 can be bounded analogously. By the mean value theorem and (7.46) from Lemma 7.17, we have

$$
\begin{aligned}
|\mathrm{III}_3| &\leq c|x - \bar{x}| [g]_{t,\overline{Q}}^{(\frac{1+\theta}{2}\alpha)} \int_0^h \int_{\mathbb{R}^{d-1}} |D {}^R \partial_s^\alpha \widetilde{G}_\alpha(x' - y', x_d, s)| s^{\frac{1+\theta}{2}\alpha} dy' ds \\
&\leq c|x - \bar{x}| [g]_{t,\overline{Q}}^{(\frac{1+\theta}{2}\alpha)} \int_h^\infty s^{-\alpha-1} s^{\frac{1+\theta}{2}\alpha} ds = c[g]_{t,\overline{Q}}^{(\frac{1+\theta}{2}\alpha)} |x - \bar{x}|^\theta.
\end{aligned}
$$

The last two bounds together imply $[\partial_t^\alpha u]_{x,\overline{Q}}^{(\theta)} \leq c[g]_{t,\overline{Q}}^{(\frac{1+\theta}{2}\alpha)} |x - \bar{x}|^\theta$. This shows the second assertion of the lemma. The remaining assertions can be proved similarly, and thus the proof is omitted. $\qquad\square$

With these preliminary estimates, we can state the following existence and uniqueness results for problems (7.34) and (7.35). The proofs are identical with that for Theorem 7.4, and hence they are omitted.

Proposition 7.4 *Let $g \in C_0^{2+\theta, \frac{2+\theta}{2}\alpha}(\overline{Q}')$. Then under condition (7.36), problem (7.34) has a unique solution $v \in C_0^{2+\theta, \frac{2+\theta}{2}\alpha}(\overline{Q})$, and*

$$
\|v\|_{C^{2+\theta, \frac{2+\theta}{2}\alpha}(\overline{Q})} \leq c \|g\|_{C^{2+\theta, \frac{2+\theta}{2}\alpha}(\overline{Q}')}.
$$

Proposition 7.5 *Let $g \in C_0^{1+\theta, \frac{1+\theta}{2}\alpha}(\overline{Q}')$. Then under conditions (7.36) and (7.37), problem (7.35) has a unique solution $w \in C_0^{2+\theta, \frac{2+\theta}{2}\alpha}(\overline{Q})$, and*

$$\|w\|_{C^{2+\theta,\frac{2+\theta}{2}\alpha}(\overline{Q})} \leq c(\delta_0, M_0)\|g\|_{C^{1+\theta,\frac{1+\theta}{2}\alpha}(\overline{Q'})}.$$

7.3.3 Subdiffusion on bounded domains

Now we study subdiffusion on a bounded domain. Let $\Omega \subset \mathbb{R}^d$ be an open bounded domain with a boundary $\partial\Omega$. Let $Q = \Omega \times (0, T]$, and $\Sigma = \partial\Omega \times (0, T]$. We consider the following two initial boundary value problems:

$$\begin{cases} \partial_t^\alpha u + \mathcal{L}(t)u = f, & \text{in } Q, \\ u = g, & \text{on } \Sigma, \\ u(0) = u_0, & \text{in } \Omega, \end{cases} \tag{7.52}$$

and

$$\begin{cases} \partial_t^\alpha u + \mathcal{L}(t)u = f, & \text{in } Q, \\ \sum_{i=1}^d b_i \partial_{x_i} u + b_0 u = g, & \text{on } \Sigma, \\ u(0) = u_0, & \text{in } \Omega. \end{cases} \tag{7.53}$$

The (possibly time dependent) elliptic operator $\mathcal{L}(t)$ is given by

$$\mathcal{L}(t)u = -\sum_{i,j=1}^d a_{ij}(x,t)\partial_{x_i x_j}^2 u + \sum_{i=1}^d a_i(x,t)\partial_{x_i} u + a_0(x,t)u.$$

Throughout, the coefficients a_{ij} and b_i satisfy

$$v|\xi|^2 \leq \sum_{i,j=1}^d a_{ij}\xi_i\xi_j \leq \mu\xi^2, \quad \forall(x,t) \in \overline{Q}, \tag{7.54}$$

$$\sum_{i=1}^d b_i(x,t)n_i(x) \leq -\delta < 0, \quad \forall(x,t) \in \overline{\Sigma}, \tag{7.55}$$

where $n(x)$ denotes the unit outward normal to the boundary $\partial\Omega$ at the point $x \in \partial\Omega$, and $0 < v \leq \mu < \infty$. Then we have the following two regularity results for the Dirichlet and Neumann problems, respectively.

Theorem 7.9 *Let* $\partial\Omega \in C^{2+\theta}$, $a_{ij}, a_i, a_0 \in C^{\theta,\frac{\theta}{2}\alpha}(\overline{Q})$, *for* $i, j = 1, \ldots, d$, *and the following compatibility conditions hold:*

$$g(\cdot, 0) = u_0, \quad \text{on } \partial\Omega,$$
$$\partial_t^\alpha g(x,0) = \mathcal{L}(0)u_0, \quad \text{on } \partial\Omega.$$

Then for $u_0 \in C^{2+\theta}(\overline{\Omega})$, $f \in C^{\theta,\frac{\theta}{2}\alpha}(\overline{Q})$, $g \in C^{2+\theta,\frac{2+\theta}{2}\alpha}(\overline{\Sigma})$, *problem* (7.52) *has a unique solution* $u \in C^{2+\theta,\frac{2+\theta}{2}\alpha}(\overline{Q})$ *satisfying*

$$\|u\|_{C^{2+\theta,\frac{2+\theta}{2}\alpha}(\overline{Q})} \le c\left(\|u_0\|_{C^{2+\theta}(\overline{\Omega})} + \|f\|_{C^{\theta,\frac{\theta}{2}\alpha}(\overline{Q})} + \|g\|_{C^{2+\theta,\frac{2+\theta}{2}\alpha}(\overline{\Sigma})}\right).$$

Theorem 7.10 *Let* $\partial\Omega \in C^{2+\theta}$, a_{ij}, a_i, $a_0 \in C^{\theta,\frac{\theta}{2}\alpha}(\overline{Q})$, *and* b_0, $b_i \in C^{1+\theta,\frac{1+\theta}{2}\alpha}(\overline{\Sigma})$, *for* $i, j = 1, \ldots, d$, *and the following compatibility condition holds:*

$$\sum_{i=1}^{d} b_i(x,0)\partial_{x_i}u_0(x) + b_0(x,0)u_0(x) = g(x,0), \quad on\ \partial\Omega. \tag{7.56}$$

Then for $u_0 \in C^{2+\theta}(\overline{\Omega})$, $f \in C^{\theta,\frac{\theta}{2}\alpha}(\overline{Q})$, $g \in C^{1+\theta,\frac{1+\theta}{2}\alpha}(\overline{\Sigma})$, *problem* (7.53) *has a unique solution* $u \in C^{2+\theta,\frac{2+\theta}{2}\alpha}(\overline{Q})$ *satisfying*

$$\|u\|_{C^{2+\theta,\frac{2+\theta}{2}\alpha}(\overline{Q})} \le c\left(\|u_0\|_{C^{2+\theta}(\overline{\Omega})} + \|f\|_{C^{\theta,\frac{\theta}{2}\alpha}(\overline{Q})} + \|g\|_{C^{1+\theta,\frac{1+\theta}{2}\alpha}(\overline{\Sigma})}\right).$$

The compatibility conditions in Theorems 7.9 and 7.10 consist in the fact that the fractional derivatives $(\partial_t^\alpha)^k u|_{t=0}$, which can be determined for $t = 0$ from the equation and initial condition, must satisfy the corresponding boundary condition. The analysis of problems (7.52) and (7.53) is similar to each other, and thus we discuss only problem (7.53). The overall analysis strategy is identical with that in Section 7.2. First, we consider the case of zero initial data on a small time interval $[0, \tau]$, and show the existence of a solution by means of a regularizer [LSU68, Chapter IV]. We use the following shorthand notation: $Q_\tau = \Omega \times (0, \tau]$, and $\Sigma_\tau = \partial\Omega \times (0, \tau]$.

Lemma 7.21 *Suppose* $u_0 \equiv 0$, $f \in C_0^{\theta,\frac{\theta}{2}\alpha}(\overline{Q})$, $g \in C_0^{1+\theta,\frac{1+\theta}{2}\alpha}(\overline{\Sigma})$ *and the compatibility conditions* (7.56) *are satisfied. Then for sufficiently small* $\tau \in (0, T)$, *problem* (7.53) *has a unique solution* $u \in C_0^{2+\theta,\frac{2+\theta}{2}\alpha}(\overline{Q_\tau}))$ *satisfying the estimate*

$$\|u\|_{C^{2+\theta,\frac{2+\theta}{2}\alpha}(\overline{Q_\tau})} \le c\left(\|f\|_{C^{\theta,\frac{\theta}{2}\alpha}(\overline{Q_\tau})} + \|g\|_{C^{1+\theta,\frac{1+\theta}{2}\alpha}(\overline{\Sigma_\tau})}\right).$$

Proof We prove the lemma by constructing a regularizer, following the standard procedure [LSU68, Chapter IV, Sections 4–7]. We cover the domain Ω with the balls $B_\lambda^{(k)}$ and $B_{2\lambda}^{(k)}$ of radii λ and 2λ, respectively, with a common center $\xi^{(k)} \in \Omega$ for sufficiently small $\lambda > 0$. The index k belongs to one of the following two sets:

$$k \in \begin{cases} \mathbb{I}_1, & \text{if } B_\lambda^{(k)} \cap \partial\Omega \neq \emptyset, \\ \mathbb{I}_2, & \text{if } B_\lambda^{(k)} \cap \partial\Omega = \emptyset. \end{cases}$$

We take

$$\tau = \kappa\lambda^{\frac{2}{\alpha}}, \quad \text{with } \kappa < 1. \tag{7.57}$$

Let $\zeta^{(k)}$ and $\eta^{(k)}$ be the sets of smooth functions subordinated to the indicated overlapping coverings of the domain Ω such that

$$\sum_k \zeta^{(k)}(x)\eta^{(k)}(x) = 1, \quad x \in \Omega,$$

$$|D_x^\ell \zeta^{(k)}| + |D_x^\ell \eta^{(k)}| \le c\lambda^{-|\ell|}, \quad |\ell| \ge 0.$$

By the regularity assumption on the boundary $\partial\Omega$, $\partial\Omega \cap B_{2\lambda}^{(k)}$ can be expressed by $y_d = F(y')$ in a local coordinate y with origin at $\xi^{(k)}$, where the axis y_d is oriented in the direction of the outward normal vector $n(\xi^{(k)})$ to $\partial\Omega$. We straighten the boundary segment $\partial\Omega \cap B_{2\lambda}^{(k)}$ by $z' = y'$ and $z_d = y_d - F(y')$. Let $z = Z_k(x)$ be the transformation of the coordinate x into z. Below we denote

$$\mathcal{L}_0(x, t, \partial_x, \partial_t^\alpha) = \partial_t^\alpha - \sum_{i,j=1}^{d} a_{ij}(x, t)\partial_{x_i}\partial_{x_j} \quad \text{and} \quad \mathcal{B}_0(x, t, \partial_x) = \sum_{i=1}^{d} b_i(x, t)\partial_{x_i},$$

i.e., the leading terms of the operators $\mathcal{L}$ and $\mathcal{B}$, respectively. Let $\mathcal{L}_0^{(k)}$ and $\mathcal{B}_0^{(k)}$ be the operators $\mathcal{L}_0$ and $\mathcal{B}_0$ in the local coordinate y at the origin $(\xi^{(k)}, 0)$:

$$\mathcal{L}_0^{(k)}(\xi^{(k)}, 0, \partial_y, \partial_t^\alpha) = \partial_t^\alpha - \sum_{i,j=1}^{d} a_{ij}^{(k)}\partial_{y_i}\partial_{y_j} \quad \text{and} \quad \mathcal{B}_0^{(k)}(\xi^{(k)}, 0, \partial_y) = \sum_{i=1}^{d} b_i^{(k)}\partial_{y_i},$$

with the constants $a_{ij}^{(k)} = a_{ij}(\xi^{(k)}, 0)$ and $b_i^{(k)} = b_i(\xi^{(k)}, 0)$. Let $\phi = (f, g)$ and define a regularizer $\mathcal{R}$ by

$$\mathcal{R}\phi = \sum_{k \in \mathbb{I}_1 \cup \mathbb{I}_2} \eta^{(k)}(x)u_k(x, t),$$

where the functions u_k, $k \in \mathbb{I}_1 \cup \mathbb{I}_2$, are defined as follows. For $k \in \mathbb{I}_2$, $u_k(x, t)$ solves

$$\begin{cases} \mathcal{L}_0^{(k)}(\xi^{(k)}, 0, \partial_y, \partial_t^\alpha)u_k(x, t) = f_k(x, t), & (x, t) \in \mathbb{R}^d \times I, \\ u_k(x, 0) = 0, & x \in \mathbb{R}^d, \end{cases}$$

with $f_k(x, t) = \zeta_k(x)f(x, t)$. For $k \in \mathbb{I}_1$, let

$$u_k(x, t) = u_k'(z, t)|_{z=Z_k^{-1}(x)},$$

where $u_k'(x, t)$ solves

$$\begin{cases} \mathcal{L}_0^{(k)}(\xi^{(k)}, 0, \partial_z, \partial_t^\alpha)u_k'(z, t) = f_k'(z, t), & (x, t) \in \mathbb{R}_+^d \times I, \\ \mathcal{B}_0^{(k)}(\xi^{(k)}, 0, \partial_z)u_k'(z', 0, t) = g_k'(z', t), & (z', t) \in \mathbb{R}^{d-1} \times I, \\ u_k'(z, 0) = 0, & z \in \mathbb{R}_+^d, \end{cases}$$

with $f_k'(z, t) = \zeta_k(x)f(x, t)|_{x=Z_k(z)}$ and $g_k'(z, t) = \zeta_k(x)g(x, t)|_{x=Z_k(z)}$. According to [LSU68, Chapter IV, Section 6], we can reduce these problems to the case $a_{ij}^{(k)} = \delta_{ij}$, by properly changing the coordinates. Moreover, we can repeat the routine but lengthy calculations of [LSU68, Chapter IV, Sections 6 and 7] to show that the parameter δ_0 in the condition $h_d < -\delta_0 < 0$ depends only on δ and μ from (7.54) and (7.55). Next we rewrite problem (7.53) (with zero initial data) in an operator form:

$$\mathcal{A}u = \phi,$$

where the linear operator $\mathcal{A}$

$$\mathcal{A} : C_0^{2+\theta, \frac{2+\theta}{2}\alpha}(\overline{Q}) \to \mathcal{H}(\overline{Q}) \equiv C_0^{\theta, \frac{\theta}{2}\alpha}(\overline{Q}) \times C_0^{1+\theta, \frac{1+\theta}{2}\alpha}(\overline{\Sigma})$$

is defined by the expressions on the left-hand side of (7.53), and $\mathcal{H}(\overline{Q})$ is the space of function pairs $\phi = (f, g)$, with the norm

$$\|\phi\|_{\mathcal{H}(\overline{Q})} = \|f\|_{C^{\theta, \frac{\theta}{2}\alpha}(\overline{Q})} + \|g\|_{C^{1+\theta, \frac{1+\theta}{2}\alpha}(\overline{\Sigma})}.$$

In view of Propositions 7.2 and 7.4–7.5, we obtain

$$\|\mathcal{R}\phi\|_{C^{2+\theta, \frac{2+\theta}{2}\alpha}(\overline{Q})} \le c\|\phi\|_{\mathcal{H}(\overline{Q})},$$

where the constant c does not depend on λ and τ. Further, for any $\phi \in \mathcal{H}(\overline{Q}_\tau), u \in C^{2+\theta, \frac{2+\theta}{2}\alpha}(\overline{Q}_\tau)$, the following estimates hold:

$$\begin{cases} \|\mathcal{R}\mathcal{A}u - u\|_{C^{2+\theta, \frac{2+\theta}{2}\alpha}(\overline{Q}_\tau)} \le \frac{1}{2}\|u\|_{C^{2+\theta, \frac{2+\theta}{2}\alpha}(\overline{Q}_\tau)}, \\ \|\mathcal{A}\mathcal{R}\phi - \phi\|_{\mathcal{H}(\overline{Q}_\tau)} \le \frac{1}{2}\|\phi\|_{\mathcal{H}(\overline{Q}_\tau)}, \end{cases} \tag{7.58}$$

if τ is sufficiently small, and (7.57) holds. The requisite computations are direct but tedious (cf. [LSU68, Chapter IV, Section 7] for the standard parabolic case). The last two inequalities directly yield the assertion of the lemma, using a basic result in functional analysis. $\qquad\square$

Now we can state the proof of Theorem 7.10.

Proof To remove the restriction on the initial data u_0 in Lemma 7.21 and then to extend the solution from $[0, \tau]$ to $\overline{I}$, we reduce problem (7.53) to the new unknowns with zero initial data (a) at $t = 0$ and (b) at $t = \tau$. In case (a), by the Hestenes-Whitney extension theorem, there exist bounded extensions $\hat{u}_0$ and $\hat{f}$ of u_0 and $f(\cdot, 0) - \mathcal{L}(0)u_0 - \Delta u_0 \in C^\theta(\overline{\Omega})$ from Ω to $\mathbb{R}^d$, with finite support, denoted by $\hat{u}_0 \in C^{2+\theta}(\mathbb{R}^d)$ and $\hat{f} \in C^\theta(\mathbb{R}^d)$. Then let $u^{(0)}$ solve the Cauchy problem

$$\begin{cases} \partial_t^\alpha u^{(0)} - \Delta u^{(0)} = \hat{f}, & \text{in } \mathbb{R}^d \times I, \\ u^{(0)}(0) = \hat{u}_0, & \text{in } \mathbb{R}^d. \end{cases}$$

Then

$$u^{(0)}(\cdot, 0) = u_0, \quad \partial_t^\alpha u^{(0)}(\cdot, 0) = f(\cdot, 0) - \mathcal{L}(0)u_0 \quad \text{in } \Omega.$$

and by Proposition 7.2, $\hat{u}^{(0)} \in C^{2+\theta, \frac{2+\theta}{2}\alpha}(\mathbb{R}^d \times \overline{I})$ with

$$\|u^{(0)}\|_{C^{2+\theta, \frac{2+\theta}{2}\alpha}(\mathbb{R}^d \times [0,T])} \le c(\|\hat{u}_0\|_{C^{2+\theta}(\mathbb{R}^d)} + \|\hat{f}(\cdot, 0)\|_{C^\theta(\mathbb{R}^d)}).$$

Now we look for the solution u of problem (7.53) as $u = v + u^{(0)}$, where v is the solution of a problem of the form (7.53) but with a zero initial condition. Case (b) can be treated exactly as in the proof of Theorem 7.7, and thus the details are omitted.
□

Exercises

Exercise 7.1 Give a direct proof of Lemma 7.1(i) using Bernstein's theorem and complete monotonicity of Mittag-Leffler functions in Theorem 3.5.

Exercise 7.2 Find the moments $\mu_{2n}(t) = \int_{\mathbb{R}^d} |x|^{2n} G_\alpha(x, t) \, dx$.

Exercise 7.3 Give a representation of the solution to the Neumann problem for subdiffusion on the half interval $\Omega = (0, \infty)$:

$$\begin{cases} \partial_t^\alpha u = u_{xx} & \text{in } \Omega \times (0, \infty), \\ u_x(0, t) = 0 & \text{in } (0, \infty), \\ u(\cdot, 0) = u_0 & \text{in } \Omega, \end{cases}$$

using the fundamental solutions, and discuss the solution behavior for incompatible data. (Hint: use an even extension of u).

Exercise 7.4 Prove the asymptotic expansion (7.12).

Exercise 7.5 Let $G(t) = \theta_\alpha(1, t) \equiv \theta_\alpha(-1, t)$. Show that $G(t)$ are C^∞ on $[0, \infty)$, with all finite-order derivatives vanishing at $t = 0$, i.e.,

$$G^{(m)}(0) = 0, \quad m = 0, 1, \dots.$$

Exercise 7.6 Consider the following initial boundary value problem with a mixed boundary condition on the unit interval $\Omega = (0, 1)$ (with $\alpha \in (0, 1)$):

$$\begin{cases} \partial_t^\alpha u - \partial_{xx}^2 u = f, & \text{in } Q, \\ u(0, \cdot) = g_1, & \text{on } I, \\ -\partial_x u(1, \cdot) = g_2, & \text{on } I, \\ u(0) = u_0, & \text{in } \Omega. \end{cases}$$

Find a representation of the solution $u(x, t)$ using fractional θ functions.

Exercise 7.7 Consider the following initial boundary value problem on the unit interval $\Omega = (0, 1)$ with $\alpha \in (0, 1)$

$$\begin{cases} \partial_t^\alpha u = \partial_{xx}^2 u, & \text{in } Q, \\ u(0) = u_0, & \text{in } \Omega, \\ v(0, t) = u(1, t) = 0, & \text{on } I. \end{cases}$$

(i) Prove that the solution u is given by

$$u(x,t) = \sum_{n=1}^{\infty} a_n E_{\alpha,1}(-n^2\pi^2 t^\alpha) \sin n\pi x, \quad \text{with } a_n = 2\int_0^1 u_0(x) \sin n\pi x dx.$$

(ii) Show that the solution u is also given by $u(x,t) = \int_0^1 (\theta_\alpha(x-y,t) - \theta_\alpha(x+y,t))u_0(y)dy.$

(iii) Using the results in (i) and (ii) show that for all x, y and $t > 0$, there holds

$$2\sum_{n=1}^{\infty} E_{\alpha,1}(-n^2\pi^2 t^\alpha) \sin n\pi x \sin n\pi y = \theta_\alpha(x-y,t) - \theta_\alpha(x+y,t).$$

This relation generalizes the well-known result for the classical θ function.

Exercise 7.8 Prove Theorem 7.2.

Exercise 7.9 Derive the representation for problem (7.18), if the elliptic operator has a constant coefficient, i.e., $-a_0\partial^2_{xx}u$, for $a_0 > 0$, instead of $\partial^2_{xx}u$.

Exercise 7.10 Discuss the compatibility condition (7.20) for the Neumann problem (7.19), and determine the nature of the leading singularity term, when the initial condition and Neumann boundary condition are incompatible.

Exercise 7.11 Prove the estimate (7.33).

Exercise 7.12 Does the inequality (7.41) hold for the case $|\ell| = 0$?

Exercise 7.13 Prove the remaining estimates in Lemma 7.20.

Appendix A
Mathematical Preliminaries

In this appendix, we recall various functions spaces (e.g., AC spaces, Hölder spaces and Sobolev spaces), two integral transforms and several fixed point theorems.

A.1 AC Spaces and Hölder Spaces

A.1.1 AC spaces

The space of absolutely continuous (AC) functions is extensively used in the study of fractional calculus. Throughout, $D = (a, b)$, with $a < b$, is an open bounded interval.

Definition A.1 A function $v : D \to \mathbb{R}$ is called absolutely continuous on $\overline{D}$, if for any $\epsilon > 0$, there exists a $\delta > 0$ such that for any finite set of pairwise disjoint intervals $[a_k, b_k] \subset \overline{D}$, $k = 1, \ldots, n$, such that $\sum_{k=1}^{n}(b_k - a_k) < \delta$, there holds the inequality

$$\sum_{k=1}^{n} |v(b_k) - v(a_k)| < \epsilon.$$

The space of these functions is denoted by $AC(\overline{D})$.

It is known that the space $AC(\overline{D})$ coincides with the space of primitives of Lebesgue integrable functions, i.e.,

$$v(x) \in AC(\overline{D}) \quad \Leftrightarrow \quad v(x) = c + \int_a^x \phi(s)\mathrm{d}s, \quad \forall x \in \overline{D}, \quad \text{with} \quad \int_a^b |\phi(s)|\mathrm{d}s < \infty,$$

for some constant c. Hence, AC functions have a summable derivative $v'(x)$ almost everywhere (a.e.). Clearly, $C^1(\overline{D}) \hookrightarrow AC(\overline{D})$, but the converse is not true. For example, for $\alpha \in (0, 1)$, $v(x) = (x - a)^\alpha \in AC(\overline{D})$, but $v(x) \notin C^1(\overline{D})$. However, if v is continuous and $v' \in L^1(D)$ exists a.e., v is not necessary $AC(\overline{D})$. One counterexample is Lebesgue's singular function v (also known as the Cantor-Vitali function or devil's

© The Author(s), under exclusive license to Springer Nature Switzerland AG 2021
B. Jin, *Fractional Differential Equations*, Applied Mathematical Sciences 206,
https://doi.org/10.1007/978-3-030-76043-4

staircase) which is continuous on $[0, 1]$ and has a zero derivative a.e., but is not $AC(\overline{D})$, in fact $v(0) = 0$, $v(1) = 1$ and thus $v(1) - v(0) \neq \int_0^1 v'(s)ds$. The following facts are well-known on a bounded interval: $C^1(\overline{D}) \subset C^{0,1}(\overline{D}) \subset$ differential a.e., and $AC(\overline{D}) \subset$ uniformly continuous $\subset C(\overline{D})$. It is also know that, on a bounded interval, the sum and pointwise product of functions in $AC(\overline{D})$ belongs to $AC(\overline{D})$, and if $v \in AC(\overline{D})$ and $f \in C^{0,1}(\overline{D})$, then $f \circ v \in AC(\overline{D})$. However, the composition of two $AC(\overline{D})$ functions need not be $AC(\overline{D})$.

Definition A.2 $AC^n(\overline{D})$, $n \in \mathbb{N}$, denotes the space of functions $v(x)$ with continuous derivatives up to order $n - 1$ in $\overline{D}$ and the $(n - 1)$th derivative $v^{(n-1)}(x) \in AC(\overline{D})$.

The space $AC^n(\overline{D})$ can be characterized as follows.

Theorem A.1 *The space* $AC^n(\overline{D})$, $n \in \mathbb{N}$ *consists of those and only those functions* $v(x)$ *which are represented in the form* $v(x) = \frac{1}{(n-1)!} \int_a^x (x-s)^{n-1} \varphi(s)ds + \sum_{j=0}^{n-1} c_j(x - a)^j$, *where* $\varphi(t) \in L^1(D)$ *and* $c_j, j = 0, \ldots, n-1$ *are arbitrary constants.*

A.1.2 Hölder spaces

Let $\Omega \subset \mathbb{R}^d$ be an open bounded domain, $\partial\Omega$ be its boundary, and $\overline{\Omega}$ be its closure. Then for any $\gamma \in (0, 1]$, we define the Hölder space $C^{0,\gamma}(\overline{\Omega})$ by

$$C^{0,\gamma}(\overline{\Omega}) := \left\{ v \in C(\overline{\Omega}) : |v(x) - v(y)| \leq L|x - y|^\gamma, \quad \forall x, y \in \overline{\Omega} \right\},$$

where L is a nonnegative constant, equipped with the norm

$$\|v\|_{C^{0,\gamma}(\overline{\Omega})} = \|v\|_{C(\overline{\Omega})} + [v]_{C^\gamma(\overline{\Omega})},$$

with $[\cdot]_{C^\gamma(\overline{\Omega})}$ being the $C^\gamma(\overline{\Omega})$ seminorm, defined by

$$[v]_{C^\gamma(\overline{\Omega})} = \sup_{x,y \in \overline{\Omega}, x \neq y} \frac{|v(x) - v(y)|}{|x - y|^\gamma}.$$

The case $\gamma = 1$ is called Lipschitz, and the corresponding seminorm is the Lipschitz constant. For any multi-index $\ell = (\ell_1, \ldots, \ell_d) \in \mathbb{N}_0^d$, $|\ell| = \sum_{i=1}^d \ell_i$ and $D^\ell = \frac{\partial^{|\ell|}}{\partial x_1^{\ell_1} \ldots \partial x_d^{\ell_d}}$. Then we denote by $C^{k,\gamma}(\overline{\Omega})$, $k \geq 1$, the space

$$C^{k,\gamma}(\overline{\Omega}) := \left\{ v \in C^k(\overline{\Omega}) : |D^\ell v(x) - D^\ell v(y)| \leq L|x - y|^\gamma, \quad \forall x, y \in \overline{\Omega}, \right.$$
$$\left. \forall \ell \in \mathbb{N}_0^d \text{ with } |\ell| = k \right\},$$

with the norm $\| \cdot \|_{C^{k,\gamma}(\overline{\Omega})}$ defined similarly by

$$\|v\|_{C^{k,\gamma}(\overline{\Omega})} = \|v\|_{C^k(\overline{\Omega})} + \sum_{\ell \in \mathbb{N}_0^d, |\ell|=k} [D^\ell v]_{C^{0,\gamma}(\overline{\Omega})}.$$

We often identify $C^\gamma(\overline{\Omega})$, with $\gamma > 0$, with the space $C^{k,\gamma'}(\overline{\Omega})$, where k is an integer, $\gamma' \in (0, 1]$ and $\gamma = k + \gamma'$. The Hölder spaces are Banach spaces. Further, for any $0 < \alpha < \beta \le 1$ and $k \in \mathbb{N}_0$, $C^{k,\beta}(\overline{\Omega}) \hookrightarrow C^{k,\alpha}(\overline{\Omega})$ (i.e., compact embedding). The Hölder spaces $C^{k,\gamma}(\Omega)$ may be defined analogously, if the functions and its derivatives are not continuous up to the boundary $\partial\Omega$. Note that if $v \in C^{0,\gamma}(\Omega)$, $0 < \gamma < 1$, then v is uniformly continuous in Ω, and therefore the Cauchy criterion implies that v can be uniquely extended to $\partial\Omega$ to give a continuous function in $\overline{\Omega}$. This allows us to talk about values on the boundary $\partial\Omega$ of $v \in C^{0,\gamma}(\Omega)$, and also write $C^{0,\gamma}(\Omega) = C^{0,\gamma}(\overline{\Omega})$.

A.2 Sobolev Spaces

Now we review several useful results on Sobolev spaces. The standard references on the topic include [AF03, Gri85]. Let $\Omega \in \mathbb{R}^d$ be an open bounded domain with a boundary $\partial\Omega$, and $\Omega^c = \mathbb{R}^d \setminus \Omega$. The domain Ω is said to be $C^{k,\gamma}$ (respectively, Lipschitz) if $\partial\Omega$ is compact and if locally near each boundary point, Ω can be represented as the set above the graph of a $C^{k,\gamma}$ (respectively, Lipschitz) function.

A.2.1 Lebesgue spaces

The Lebesgue space $L^p(\Omega)$, $1 \le p \le \infty$, consists of functions $v : \Omega \to \mathbb{R}$ with a finite $\|v\|_{L^p(\Omega)}$, where

$$\|v\|_{L^p(\Omega)} = \begin{cases} \left(\int_\Omega |v|^p dx \right)^{\frac{1}{p}}, & \text{if } 1 \le p < \infty, \\ \underset{\Omega}{\text{ess sup}} |v|, & \text{if } p = \infty. \end{cases}$$

A function $v : \Omega \to \mathbb{R}$ is said to be in $L^p_{loc}(\Omega)$, if for any compact subset $\omega \subset\subset \Omega$, there holds $v \in L^p(\omega)$.

In $L^p(\Omega)$ spaces, Hölder inequality is an indispensable tool: if $p, q \in [1, \infty]$ with $p^{-1} + q^{-1} = 1$ (i.e., p, q are conjugate exponents, and often written as $q = p'$), then for all $v \in L^p(\Omega)$ and $w \in L^q(\Omega)$, there holds $vw \in L^1(\Omega)$ and

$$\|vw\|_{L^1(\Omega)} \le \|v\|_{L^p(\Omega)} \|w\|_{L^q(\Omega)}.$$

The equality holds if and only if v and w are linearly dependent. One useful generalization is Young's inequality for the convolution $v * w$ defined on $\mathbb{R}^d$, i.e.,

$$(v * w)(x) = \int_{\mathbb{R}^d} v(x - y)w(y)\mathrm{d}y.$$

One can obtain a version of the inequality on bounded domain by zero extension.

Theorem A.2 *Let $v \in L^p(\mathbb{R}^d)$ and $w \in L^q(\mathbb{R}^d)$, with $1 \le p, q \le \infty$. For any $1 \le r \le \infty$ with $p^{-1} + q^{-1} = r^{-1} + 1$,*

$$\|v * w\|_{L^r(\mathbb{R}^d)} \le \|v\|_{L^p(\mathbb{R}^d)} \|w\|_{L^q(\mathbb{R}^d)}. \tag{A.1}$$

Occasionally we use the weak $L^p(\Omega)$, $1 \le p < \infty$, denoted by $L^{p,\infty}(\Omega)$. For $v : \Omega \to \mathbb{R}$, the distribution λ_v is defined for $t > 0$ by

$$\lambda_v(t) = |\{x \in \Omega : |v(x)| > t\}|,$$

where $|\cdot|$ denotes the Lebesgue measure of a set in $\mathbb{R}^d$. v is said to be in the space $L^{p,\infty}(\Omega)$, if there is $c > 0$ such that, for all $t > 0$, $\lambda_v(t) \le c^p t^{-p}$. The best constant c is the $L^{p,\infty}(\Omega)$-norm of v, and is denoted by $\|v\|_{L^{p,\infty}(\Omega)} = \sup_{t>0} t\lambda_v^{\frac{1}{p}}(t)$. The $L^{p,\infty}(\Omega)$-norm is not a true norm, since the triangle inequality does not hold. For $v \in L^p(\Omega)$, $\|v\|_{L^{p,\infty}(\Omega)} \le \|v\|_{L^p(\Omega)}$, and $L^p(\Omega) \subset L^{p,\infty}(\Omega)$. With the convention that two functions are equal if they are equal a.e., then the spaces $L^{p,\infty}(\Omega)$ are complete, and further, for $p > 1$, $L^{p,\infty}(\Omega)$ are Banach spaces [Gra04].

Example A.1 Let $\omega_\alpha(t) = \frac{t^{\alpha-1}}{\Gamma(\alpha)}$, $\alpha \in (0, 1)$, and $T > 0$ be fixed. Then, $\omega_\alpha \in L^p(0, T)$, for any $p \in [0, (1 - \alpha)^{-1})$, but it does not belong to $L^{\frac{1}{1-\alpha}}(0, T)$. However, $\omega_\alpha \in L^{\frac{1}{1-\alpha},\infty}(0, T)$. Indeed, the distribution λ_{ω_α} is given by $\lambda_{\omega_\alpha}(t) = |\{s \in (0, T) : |\omega_\alpha(s)| > t\}| = (\Gamma(\alpha)t)^{-\frac{1}{1-\alpha}}$. Hence,

$$\|\omega_\alpha\|_{L^{\frac{1}{1-\alpha},\infty}(0,T)} = \sup_{t\in(0,T)} t\lambda_{\omega_\alpha}^{1-\alpha}(t) = \Gamma(\alpha)^{-1} < \infty.$$

This shows the desired assertion.

A.2.2 Sobolev spaces

Let $C_0^\infty(\Omega)$ denote the space of infinitely differentiable functions with compact support in Ω. If $v, w \in L_{loc}^1(\Omega)$ and $\ell = (\ell_1, \ldots, \ell_d) \in \mathbb{N}_0^d$ a multi-index, then w is called the ℓth weak partial derivative of v, denoted by $w = D^\ell v$, if there holds

$$\int_\Omega v D^\ell \phi \, \mathrm{d}x = (-1)^{|\ell|} \int_\Omega w\phi \, \mathrm{d}x, \quad \forall \phi \in C_0^\infty(\Omega),$$

with $|\ell| = \sum_{i=1}^d \ell_i$. Note that a weak partial derivative, if it exists, is unique up to a set of Lebesgue measure zero. We define the Sobolev space $W^{k,p}(\Omega)$, for any $1 \le p \le \infty$ and $k \in \mathbb{N}$. It consists of all $v \in L_{loc}^1(\Omega)$ such that for each $\ell \in \mathbb{N}_0^d$ with

$|\ell| \leq k$, $D^\ell v$ exists in the weak sense and belongs to $L^p(\Omega)$, equipped with the norm

$$
\|v\|_{W^{k,p}(\Omega)} = \begin{cases} \left(\sum_{|\ell| \leq k} \int_\Omega |D^\ell v|^p \, dx \right)^{\frac{1}{p}}, & \text{if } 1 \leq p < \infty, \\ \sum_{|\ell| \leq k} \operatorname{ess\,sup}_\Omega |D^\ell v|, & \text{if } p = \infty. \end{cases}
$$

The space $W^{k,p}(\Omega)$, for any $k \in \mathbb{N}$ and $1 \leq p \leq \infty$, is a Banach space. The subspace $W_0^{k,p}(\Omega)$ denotes the closure of $C_0^\infty(\Omega)$ with respect to the norm $\|\cdot\|_{W^{k,p}(\Omega)}$, i.e., $W_0^{k,p}(\Omega) = \overline{C_0^\infty(\Omega)}^{W^{k,p}(\Omega)}$. The case $p = 2$, i.e., $W^{k,2}(\Omega)$ and $W_0^{k,2}(\Omega)$, is Hilbert space, and often denoted by $H^k(\Omega)$ and $H_0^k(\Omega)$, respectively.

Below we recall a few useful results. The first is Poincaré's inequality. When $p = 2$, the optimal constant $c(\Omega, d)$ in the inequality coincides with the lowest eigenvalue of the negative Dirichlet Laplacian $-\Delta$ on the domain Ω.

Proposition A.1 *If the domain Ω is bounded, then for any $1 \leq p < \infty$*

$$
\|v\|_{L^p(\Omega)} \leq c(\Omega, d) \|\nabla v\|_{L^p(\Omega)}, \quad \forall v \in W_0^{1,p}(\Omega). \tag{A.2}
$$

Sobolev embedding inequalities represent one of the most powerful tools in the analysis. The C^2 regularity assumption on the domain can be relaxed.

Theorem A.3 *Let Ω be a bounded C^1 domain. Then the following statements hold.*

(i) *If $k < \frac{d}{p}$, then the embedding $W^{k,p}(\Omega) \hookrightarrow L^q(\Omega)$ is continuous, with $\frac{1}{q} = \frac{1}{p} - \frac{k}{d}$.*

(ii) *If $k > \frac{d}{p}$, then the embedding $W^{k,p}(\Omega) \hookrightarrow C^{k-\lceil \frac{d}{p} \rceil - 1, \gamma}(\overline{\Omega})$ is continuous, with $\gamma = \lceil \frac{d}{p} \rceil - \frac{d}{p} + 1$ if $\frac{d}{p} \notin \mathbb{N}$, and for any $\gamma \in (0, 1)$ otherwise.*

(iii) *If $k > \ell$ and $\frac{1}{p} - \frac{k}{d} < \frac{1}{q} - \frac{\ell}{d}$, then the embedding $W^{k,p}(\Omega) \hookrightarrow W^{\ell,q}(\Omega)$ is compact.*

The third result is Gagliado-Nirenberg multiplicative embedding.

Theorem A.4 *Let Ω be a bounded Lipschitz domain, and $p \geq 1$. Then for every $s \geq 1$, there holds*

$$
\|v\|_{L^q(\Omega)} \leq c(d, p, s) \|Dv\|_{L^p(\Omega)}^\alpha \|v\|_{L^s(\Omega)}^{1-\alpha}, \quad \forall v \in W_0^{1,p}(\Omega),
$$

where $\alpha \in [0, 1]$, $p, q \geq 1$ are linked by $\alpha = \frac{s^{-1} - q^{-1}}{s^{-1} - p^{-1} + d^{-1}}$, with the admissible range

$$
\begin{cases} q \in [s, \infty], & \alpha \in [0, \frac{p}{p+s(p-1)}], & \text{if } d = 1, \\ q \in [\min(s, \frac{dp}{d-p}), \max(s, \frac{dp}{d-p})], & \alpha \in [0, 1], & \text{if } 1 \leq p < d, \\ q \in [s, \infty), & \alpha \in [0, \frac{dp}{dp+s(p-d)}], & \text{if } p \geq d > 1. \end{cases}
$$

The next result is a corollary of Theorem A.4, with $\alpha = 1$ and $s = 1$.

Corollary A.1 *Let* $1 \leq p < d$, *there holds*

$$\|v\|_{L^{\frac{dp}{d-p}}(\Omega)} \leq c(d, p)\|Dv\|_{L^p(\Omega)}, \quad \forall v \in W_0^{1,p}(\Omega).$$

If the boundary $\partial\Omega$ is piecewise smooth, functions v in $W^{1,p}(\Omega)$ are actually defined up to $\partial\Omega$ via their traces, denoted by γv. The space $W_0^{1,p}(\Omega)$ can be defined equivalently as the set of functions in $W^{1,p}(\Omega)$ whose trace is zero.

Theorem A.5 *If* $\partial\Omega$ *is piecewise smooth, then for* $q \in [1, \frac{(d-1)p}{d-p}]$, *if* $1 < p < d$ *and* $q \in [1, \infty)$, *if* $p = d$, *there holds*

$$\|\gamma v\|_{L^q(\partial\Omega)} \leq c(d, p, \Omega)\|v\|_{W^{1,p}(\Omega)}.$$

A.2.3 Fractional Sobolev spaces

Fractional Sobolev spaces play a central role in the analysis of FDES, see the review [DNPV12] for an overview on fractional Sobolev spaces. For any $s \in (0, 1)$, the Sobolev space $W^{s,p}(\Omega)$ is defined by

$$W^{s,p}(\Omega) = \left\{ v \in L^p(\Omega) \colon |v|_{W^{s,p}(\Omega)} < \infty \right\}, \tag{A.3}$$

where the Sobolev-Slobodeckij seminorm $| \cdot |_{W^{s,p}(\Omega)}$ is defined by

$$|v|_{W^{s,p}(\Omega)} = \left(\iint_{\Omega \times \Omega} \frac{(v(x) - v(y))^p}{|x - y|^{d+ps}} dx \, dy \right)^{\frac{1}{p}}.$$

The space $W^{s,p}(\Omega)$ is a Banach space, when equipped with the norm

$$\|v\|_{W^{s,p}(\Omega)} = (\|v\|_{L^p(\Omega)}^p + |v|_{W^{s,p}(\Omega)}^p)^{\frac{1}{p}}.$$

Equivalently, the space $W^{s,p}(\Omega)$ may be regarded as the restriction of functions in $W^{s,p}(\mathbb{R}^d)$ to Ω. The subspace $W_0^{s,p}(\Omega) \subset W^{s,p}(\Omega)$ is defined as the closure of $C_0^\infty(\Omega)$ with respect to the $W^{s,p}(\Omega)$ norm, i.e.,

$$W_0^{s,p}(\Omega) = \overline{C_0^\infty(\Omega)}^{W^{s,p}(\Omega)}. \tag{A.4}$$

We denote $W^{s,2}(\Omega)$ by $H^s(\Omega)$, and similarly $W_0^{s,2}(\Omega)$ by $H_0^s(\Omega)$. These are Hilbert spaces and frequently used. If the boundary $\partial\Omega$ is smooth, they can be equivalently defined through real interpolation of spaces by the K-method [LM72, Chapter 1]. Specifically, set $H^0(\Omega) = L^2(\Omega)$, then Sobolev spaces with real index $0 \leq s \leq 1$ can be defined as interpolation spaces of index s for the pair $[L^2(\Omega), H^1(\Omega)]$, i.e.,

$$H^s(\Omega) = \left[L^2(\Omega), H^1(\Omega) \right]_s. \tag{A.5}$$

Similarly, for $s \in [0, 1] \setminus \{\frac{1}{2}\}$, the spaces $H_0^s(\Omega)$ are defined as interpolation spaces of index s for the pair $[L^2(\Omega), H_0^1(\Omega)]$, i.e.,

$$H_0^s(\Omega) = \left[L^2(\Omega), H_0^1(\Omega)\right]_s, \quad s \in [0, 1] \setminus \{\tfrac{1}{2}\}. \tag{A.6}$$

The space $[L^2(\Omega), H_0^1(\Omega)]_{\frac{1}{2}}$ is the so-called Lions-Magenes space, $H_{00}^{\frac{1}{2}}(\Omega) = \left[L^2(\Omega), H_0^1(\Omega)\right]_{\frac{1}{2}}$, which can be characterized as [LM72, Theorem 11.7]

$$H_{00}^{\frac{1}{2}}(\Omega) = \left\{ w \in H^{\frac{1}{2}}(\Omega) : \int_\Omega \frac{w^2(x')}{\mathrm{dist}(x', \partial\Omega)} \, dx' < \infty \right\}. \tag{A.7}$$

If $\partial\Omega$ is Lipschitz, then (i) characterization (A.7) is equivalent to the definition via interpolation, and definitions (A.5) and (A.6) are also equivalent to (A.3) and (A.4), respectively; (ii) the space $C_0^\infty(\Omega)$ is dense in $H^s(\Omega)$ if and only if $s \leq \frac{1}{2}$, i.e., $H_0^s(\Omega) = H^s(\Omega)$. If $s > \frac{1}{2}$, $H_0^s(\Omega)$ is strictly contained in $H^s(\Omega)$ [LM72, Theorem 11.1], and the following inclusions hold:

$$H_{00}^{\frac{1}{2}}(\Omega) \subsetneq H_0^{\frac{1}{2}}(\Omega) = H^{\frac{1}{2}}(\Omega),$$

since $1 \in H_0^{\frac{1}{2}}(\Omega)$ but $1 \notin H_{00}^{\frac{1}{2}}(\Omega)$.

Fractional Sobolev spaces of order greater than one are defined similarly. Given $k \in \mathbb{N}_0$ and $0 < s < 1$, the space $H^{k+s}(\Omega)$ is defined by

$$H^{k+s}(\Omega) = \left\{ u \in H^k(\Omega) : |D^\ell u| \in H^s(\Omega) \quad \forall \ell \in \mathbb{N}_0^d \text{ s.t. } |\ell| = k \right\},$$

equipped with the norm

$$\|v\|_{H^{k+s}(\Omega)} = \|v\|_{H^k(\Omega)} + \sum_{|\ell|=k} |D^\ell v|_{H^s(\Omega)}.$$

When $d = 1$, the domain Ω is an open bounded interval $D = (a, b)$. Then the space $H_0^s(D)$ consists of functions in $H^s(\Omega)$ whose extension by zero to $\mathbb{R}$ is in $H^s(\mathbb{R})$. We define $H_{0,L}^s(D)$ (respectively, $H_{0,R}^s(D)$) to be the set of functions whose extension by zero is in $H^s(-\infty, b)$ (respectively, $H^s(a, \infty)$).

A.2.4 $\dot{H}^s(\Omega)$ spaces

We use the spaces $\dot{H}^s(\Omega)$, $s \geq 0$, frequently in Chapter 6. These are spaces with special type of boundary conditions, associated with suitable elliptic operators. Let $A : H^2(\Omega) \cap H_0^1(\Omega) \to L^2(\Omega)$ be a symmetric uniformly coercive second-order elliptic operator with smooth coefficients. The spectrum of A consists entirely of eigenvalues $\{\lambda_j\}_{j=1}^\infty$ (ordered nondecreasingly, with the multiplicities counted). By

$\varphi_j \in H^2(\Omega) \cap H_0^1(\Omega)$, we denote the $L^2(\Omega)$ orthonormal eigenfunctions corresponding to λ_j. For any $s \geq 0$, we define the space $\dot{H}^s(\Omega)$ by

$$\dot{H}^s(\Omega) = \left\{ v \in L^2(\Omega) : \sum_{j=1}^{\infty} \lambda_j^s |(v, \varphi_j)|^2 < \infty \right\},$$

and that $\dot{H}^s(\Omega)$ is a Hilbert space with the norm

$$\|v\|_{\dot{H}^s(\Omega)}^2 = \sum_{j=1}^{\infty} \lambda_j^s |(v, \varphi_j)|^2.$$

By definition, we have the following equivalent form:

$$\|v\|_{\dot{H}^s(\Omega)}^2 = \sum_{j=1}^{\infty} |(v, \lambda_j^{\frac{s}{2}} \varphi_j)|^2 = \sum_{j=1}^{\infty} (v, A^{\frac{s}{2}} \varphi_j)^2 = \sum_{j=1}^{\infty} (A^{\frac{s}{2}} v, \varphi_j)^2 = \|A^{\frac{s}{2}} v\|_{L^2(\Omega)}^2.$$

We have $\dot{H}^s(\Omega) \subset H^s(\Omega)$ for $s > 0$. In particular,

$$\dot{H}^0(\Omega) = L^2(\Omega), \quad \dot{H}^1(\Omega) = H_0^1(\Omega) \quad \text{and} \quad \dot{H}^2(\Omega) = H_0^1(\Omega) \cap H^2(\Omega).$$

Since $\dot{H}^s(\Omega) \subset L^2(\Omega)$, by identifying the dual $(L^2(\Omega))'$ with itself, we have

$$\dot{H}^s(\Omega) \subset L^2(\Omega) \subset (\dot{H}^s(\Omega))'.$$

We set $\dot{H}^{-s}(\Omega) = (\dot{H}^s(\Omega))'$, which consists of all bounded linear functionals on $\dot{H}^s(\Omega)$. It can be characterized by the space of distributions that can be written as

$$v = \sum_{j=1}^{\infty} v_j \varphi_j(x), \quad \text{with} \quad \sum_{j=1}^{\infty} \lambda_j^{-s} v_j^2 < \infty.$$

Next we specify explicitly the space $\dot{H}^s(\Omega)$. If the domain Ω is of class C^∞, then $\dot{H}^s(\Omega) = H^s(\Omega)$ for $0 < s < \frac{1}{2}$, and for $j = 1, 2, \ldots$, such that $2j - \frac{3}{2} < s < 2j + \frac{1}{2}$

$$\dot{H}^s(\Omega) = \left\{ v \in H^s(\Omega) : v = Av = \ldots = A^{j-1} v = 0 \quad \text{on } \partial\Omega \right\}.$$

For the exceptional index $s = 2j - \frac{3}{2}$, the condition $A^{j-1} v = 0$ on $\partial\Omega$ should be replaced by $A^{j-1} v \in H_{00}^{\frac{1}{2}}(\Omega)$. These results can be proved using elliptic regularity theory and interpolation [Tho06, p. 34]. If Ω is not C^∞, then one must restrict s accordingly. For example, if Ω is Lipschitz then the above relations are valid for $s \leq 1$, and if Ω is convex or $C^{1,1}$, then we can allow $s \leq 2$.

A.2.5 Bochner spaces

In order to study the well-posedness and regularity for parabolic-type problems, e.g., the subdiffusion model, Bochner-Sobolev spaces are useful. For any $T > 0$, let $I = (0, T)$. For a Banach space E, we define

$$L^r(I; E) = \{v(t) \in E \text{ for a.e. } t \in I \text{ and } \|v\|_{L^r(I;E)} < \infty\},$$

for any $r \geq 1$, and the norm $\| \cdot \|_{L^r(I;E)}$ is defined by

$$\|v\|_{L^r(I;E)} = \begin{cases} \left(\int_I \|v(t)\|_E^r \, dt \right)^{\frac{1}{r}}, & r \in [1, \infty), \\ \operatorname{esssup}_{t \in I} \|v(t)\|_E, & r = \infty. \end{cases}$$

For any $s \geq 0$ and $1 \leq p < \infty$, we denote by $W^{s,p}(I; E)$ the space of functions $v : I \to E$, with the norm defined by complex interpolation. Equivalently, the space is equipped with the quotient norm

$$\|v\|_{W^{s,p}(I;E)} := \inf_{\widetilde{v}} \|\widetilde{v}\|_{W^{s,p}(\mathbb{R};E)} := \inf_{\widetilde{v}} \|\mathcal{F}^{-1}[(1 + |\xi|^2)^{\frac{s}{2}} \mathcal{F}(\widetilde{v})(\xi)]\|_{L^p(\mathbb{R};E)},$$

where the infimum is taken over all possible $\widetilde{v}$ that extend v from I to $\mathbb{R}$, and $\mathcal{F}$ denotes the Fourier transform. In analogy with the definition in Section A.2.3, one can also define Sobolev-Slobodeckiĭ seminorm $| \cdot |_{W^{s,p}(I;E)}$ by

$$|v|_{W^{s,p}(I;E)} := \left(\int_I \int_I \frac{\|v(t) - v(\xi)\|_E^p}{|t - \xi|^{1+ps}} \, dt d\xi \right)^{\frac{1}{p}},$$

and the full norm $\| \cdot \|_{W^{s,p}(I;E)}$ by

$$\|v\|_{W^{s,p}(I;E)} = \left(\|v\|_{L^p(I;E)}^p + |v|_{W^{s,p}(I;E)}^p \right)^{\frac{1}{p}}.$$

Note that these two norms are not equivalent, and that the former is slightly weaker.

Last, we recall two results for compact sets in $L^p(I; E)$. The following two are classical [Sim86]. The first can be found in [Sim86, Theorem 5], and the second from [Sim86, Lemma 3]. The notation τ_h denotes the shift operator. It is possible to obtain compact embedding into subspaces of E via suitable interpolation [Ama00].

Theorem A.6 *Let E, E_0 and E_1 be three Banach spaces such that $E_1 \hookrightarrow E \hookrightarrow E_0$. If $1 \leq p < \infty$ and*

(i) *W is bounded in $L^p(I; E_1)$;*

(ii) *$\|\tau_h f - f\|_{L^p(0,T-h;E_0)} \to 0$ uniformly as $h \to 0$;*

then W is relatively compact in $L^p(I; E)$.

Theorem A.7 *Let $1 < p_1 \leq \infty$. If W is a bounded set in $L^{p_1}(I; E)$ and relatively compact in $L^1_{loc}(I; E)$, then it is relatively compact in $L^p(I; E)$ for all $1 \leq p < p_1$.*

A.3 Integral Transforms

Now we recall Laplace and Fourier transforms.

A.3.1 Laplace transform

The Laplace transform of a function $f : \mathbb{R}_+ \to \mathbb{R}$, denoted by $\widehat{f}$ or $\mathcal{L}[f]$, is defined by

$$\widehat{f}(z) = \mathcal{L}[f](z) = \int_0^\infty e^{-zt} f(t) dt.$$

Suppose $f \in L^1_{loc}(\mathbb{R}_+)$ and there exists some $\lambda \in \mathbb{R}_+$ such that $|f(t)| \le c e^{\lambda t}$ for large t, then $\widehat{f}(z)$ exists and is analytic for $\Re(z) > \lambda$. The Laplace transform of an entire function of exponential type can be obtained by transforming term by term the Taylor expansion of the original function around the origin [Doe74]. In this case the resulting Laplace transform is analytic and vanishing at infinity.

Example A.2 We compute the Laplace transform of the Riemann-Liouville kernel $\omega_\alpha(t) = \Gamma(\alpha)^{-1} t^{\alpha-1}$ for $t > 0$. Then for $\alpha > 0$, the substitution $s = tz$ gives

$$\mathcal{L}[\omega_\alpha](z) = \frac{1}{\Gamma(\alpha)} \int_0^\infty e^{-zt} t^{\alpha-1} dt = \frac{z^{-\alpha}}{\Gamma(\alpha)} \int_0^\infty e^{-s} s^{\alpha-1} ds = z^{-\alpha},$$

where the last identity follows from the definition of $\Gamma(z)$.

The Laplace transform of the convolution on $\mathbb{R}_+$, i.e.,

$$(f * g)(t) = \int_0^t f(t-s) g(s) ds = \int_0^t f(s) g(t-s) ds$$

of two functions f and g that vanish for $t < 0$ satisfies the convolution rule:

$$\widehat{f * g}(z) = \widehat{f}(z) \widehat{g}(z), \tag{A.8}$$

provided that both $\widehat{f}(z)$ and $\widehat{g}(z)$ exist. This rule is useful for evaluating the Laplace transform of fractional integrals and derivatives.

Example A.3 The Riemann-Liouville fractional integral $_0I_t^\alpha f(t)$ is a convolution with ω_α, i.e., $_0I_t^\alpha f(t) = (\omega_\alpha * f)(t)$. Hence, by (A.8), the Laplace transform of $_0I_t^\alpha f$ is given by $\mathcal{L}[_0I_t^\alpha f](z) = z^{-\alpha} \widehat{f}(z)$.

The Laplace transform of the n-th-order derivative $f^{(n)}$ is given by

$$\mathcal{L}[f^{(n)}](z) = z^n \widehat{f}(z) - \sum_{k=0}^{n-1} z^{n-k-1} f^{(k)}(0^+),$$

which follows from integration by parts, if all the involved integrals make sense.

We also have an inversion formula for the Laplace transform

$$f(t) = \frac{1}{2\pi i} \int_{a-i\infty}^{a+i\infty} e^{zt} \widehat{f}(z)dz,$$

where the integral is along the vertical line $\Re(z) = a$ in $\mathbb{C}$ (also known as Bromwich contour), such that $a > \lambda$, i.e., it is greater than the real part of all singularities of $\widehat{f}$ and $\widehat{f}$ is bounded on the line. The direct evaluation of the inversion formula is often inconvenient. The contour can often be deformed to facilitate the analytic evaluation and numerical computation. For example, it may be deformed to a Hankel contour, i.e., a path in the complex plane $\mathbb{C}$ which extends from $(-\infty, a)$, circling around the origin counterclockwise and back to $(-\infty, a)$.

Laplace transform is an important tool for mathematical analysis. Below we give two examples, i.e., complete monotonicity and asymptotics. Recall that a function $f : \mathbb{R}_+ \to \overline{\mathbb{R}_+}$ is said to be completely monotone if $(-1)^k f^{(k)}(t) \geq 0$, for any $t \in \mathbb{R}_+$, $k = 0, 1, 2, \ldots$. Bernstein's theorem states that a function f is completely monotone if and only if it is the Laplace transform of a nonnegative measure, see [Wid41, Chapter IV] for a proof.

Theorem A.8 *A function $f : \mathbb{R}_+ \to \overline{\mathbb{R}_+}$ is completely monotone if and only if it has the representation $f(t) = \int_0^\infty e^{-ts}d\mu(s)$, where μ is a nonnegative measure on $\mathbb{R}_+$ such that the integral converges for all $t > 0$.*

The asymptotic behavior of a function $f(t)$ as $t \to \infty$ can be determined, under suitable conditions, by looking at the behavior of its Laplace transform $\widehat{f}(z)$ as $z \to 0$, and vice versa. This is described by Karamata-Feller Tauberian theorem, see the [Fel71] for general statements and proofs.

Theorem A.9 *Let $L : \mathbb{R}_+ \to \mathbb{R}_+$ be a function that is slowly varying at ∞, i.e., for every fixed $x > 0$, $\lim_{t\to\infty} L(xt)L(t)^{-1} = 1$. Let $\beta > 0$ and $f : \mathbb{R}_+ \to \mathbb{R}$ be a monotone function whose Laplace transform $\widehat{f}(z)$ exists for all $z \in \mathbb{C}_+$. Then*

$$\widehat{f}(z) \sim z^{-\beta}L(z^{-1}) \quad \text{as } z \to 0 \quad \text{if and only if} \quad f(t) \sim \frac{t^{\beta-1}}{\Gamma(\beta)}L(t) \quad \text{as } t \to \infty,$$

where the approaches are on $\mathbb{R}_+$ and $f(t) \sim g(t)$ as $t \to t_*$ denotes $\lim_{t\to t_*} \frac{f(t)}{g(t)} = 1$.

A.3.2 Fourier transform

The Fourier transform of a function $f \in L^1(\mathbb{R})$, denoted by $\tilde{f}$ or $\mathcal{F}[f]$, is defined by

$$\tilde{f}(\xi) = \mathcal{F}[f](\xi) = \int_{-\infty}^{\infty} e^{-i\xi x} f(x)dx.$$

Note that this choice does not involve the factor 2π that often appears in the literature. It is taken in order to be consistent with several popular textbooks on fractional calculus [SKM93, KST06]. If $f \in L^1(\mathbb{R})$, then $\tilde{f}$ is continuous on $\mathbb{R}$, and $\tilde{f}(\xi) \to 0$ as $|\xi| \to \infty$. The Plancherel theorem states that the Fourier transform extends uniquely to a bounded linear operator $\mathcal{F} : L^2(\mathbb{R}) \to L^2(\mathbb{R})$ satisfying

$$\frac{1}{2\pi} \int_{-\infty}^{\infty} \tilde{f}(\xi)\overline{\tilde{g}(\xi)}\,d\xi = \int_{-\infty}^{\infty} f(x)g(x)\,dx, \tag{A.9}$$

where the overline denotes complex conjugate. The inversion formula for Fourier transform is given by

$$f(x) = \frac{1}{2\pi} \lim_{R \to \infty} \int_{-R}^{R} e^{i\xi x}\tilde{f}(\xi)\,d\xi, \quad x \in \mathbb{R},$$

and the equality holds at every continuous point of f.

Example A.4 The Fourier transform of $f(x) = e^{-\lambda x^2}$, $\lambda > 0$, is given by

$$\tilde{f}(\xi) = \int_{-\infty}^{\infty} e^{-i\xi x}e^{-\lambda x^2}\,dx = \int_{-\infty}^{\infty} \cos(\xi x)e^{-\lambda x^2}\,dx - i\int_{-\infty}^{\infty} \sin(\xi x)e^{-\lambda x^2}\,dx.$$

The second integral vanishes since the integrand is an odd function. Using the identity for any $\lambda > 0$ [GR15, p. 391, 3.896]:

$$\int_{0}^{\infty} e^{-\lambda x^2} \cos(bx)\,dx = \frac{1}{2}\sqrt{\frac{\pi}{\lambda}}e^{-\frac{b^2}{4\lambda}},$$

we deduce

$$\tilde{f}(\xi) = \int_{-\infty}^{\infty} \cos(\xi x)e^{-\lambda x^2}\,dx = \sqrt{\frac{\pi}{\lambda}}e^{-\frac{\xi^2}{4\lambda}}.$$

Similarly, the inverse Fourier transform $\tilde{g}(\xi) = e^{-\lambda \xi^2}$ is given by $\mathcal{F}[\tilde{g}](x) = \frac{1}{\sqrt{4\lambda\pi}}e^{-\frac{x^2}{4\lambda}}$. Note that the Fourier and inverse Fourier transforms differ by a factor of 2π.

A.4 Fixed Point Theorems

Now we describe several fixed point theorems, see [Dei85, Zei86] for complete proofs and extensive applications. These theorems are central to showing various existence and uniqueness results for operator equations in Banach spaces, into which FDES can be reformulated. The most powerful tool is Banach fixed point theorem. There are several versions, with slightly different conditions. The most basic one is concerned with contraction mappings. Let (X, ρ) be a metric space. A mapping $T : X \to X$ is said to be a contraction if there exists a constant $\gamma \in (0, 1)$ such that

$$\rho(Tx_1, Tx_2) \leq \gamma \rho(x_1, x_2), \quad \forall x_1, x_2 \in X.$$

Theorem A.10 *Let (X, ρ) be a complete metric space and T be a contraction. Then T has a unique fixed point, given by $x = \lim_{n \to \infty} T^n x_0$, for any $x_0 \in X$, and further*

$$\rho(x_n, x) \leq (1 - \gamma)^{-1} \gamma^n \rho(x_0, Tx_0).$$

The next result due to Schauder gives the existence of a fixed point, without uniqueness [Zei86, Theorem 2.A, p. 56].

Theorem A.11 *Let K be a nonempty, closed, bounded, convex subset of a Banach space X, and suppose $T : K \to K$ as a compact operator. Then T has a fixed point.*

The next result is an alternative version of Schauder fixed point theorem [Zei86, Corollary 2.13, p. 56].

Theorem A.12 *Let K be a nonempty, compact and convex subset of a Banach space, and let $T : K \to K$ be a continuous mapping. Then T has a fixed point.*

The next result of Krasnoselskii handles completely continuous operators [Kra64, pp. 137, 147–148].

Theorem A.13 *Let X be a Banach space, $C \subset X$ a cone, and B_1, B_2 two bounded open balls of X centered at the origin, with $\overline{B}_1 \subset B_2$. Suppose that $T : C \cap (\overline{B}_2 \setminus B_1) \to C$ is a completely continuous operator such that either*

(i) $\|Tx\| \leq \|x\|$, $x \in C \cap \partial B_1$ *and* $\|Tx\| \geq \|x\|$, $x \in C \cap \partial B_2$, *or*

(ii) $\|Tx\| \geq \|x\|$, $x \in C \cap \partial B_1$ *and* $\|Tx\| \leq \|x\|$, $x \in C \cap \partial B_2$

holds, then T has a fixed point in $C \cap (\overline{B}_2 \setminus B_1)$.

The next result is useful for proving the existence in a cone [Dei85, Chapter 6].

Theorem A.14 *Let X be a Banach space, $C \subset X$ a cone. Let B_R be a ball of radius $R > 0$ centered at the origin. Suppose $T : C \cap B_R \to C$ is a completely continuous operator such that $Tx \neq \lambda x$ for all $x \in \partial(B_R \cap C)$ and $\lambda \geq 1$. Then T has a fixed point in $C \cap B_R$.*

The following nonlinear alternative of Leray and Schauder is powerful [Zei86, pp. 556–557].

Theorem A.15 *Let $K \subset X$ be a convex set of a linear normed space X, and Z be an open subset in K such that $0 \in Z$. Then each continuous compact mapping $T : \overline{Z} \to K$ has at least one of the following properties:*

(i) *T has a fixed point in $\overline{Z}$.*

(ii) *There exist $u \in \partial Z$ and $\mu \in (0, 1)$ such that $u = \mu Tu$.*

References

[AB65] Mohammed Ali Al-Bassam, *Some existence theorems on differential equations of generalized order*, J. Reine Angew. Math. **218** (1965), 70–78. MR 0179405

[Abe26] Niels Henrik Abel, *Auflösung einer mechanischen Aufgabe*, J. Reine Angew. Math. **1** (1826), 153–157. MR 1577605

[Abe81] Niels Henrik Abel, *Solution de quelques problémes à l'aide d'integrales définies*, Gesammelte mathematische Werke 1, Leipzig: Teubner, 1881, (First publ. in Mag. Naturvidenbbeme, Aurgang, 1, no 2, Christiania, 1823), pp. 11–27.

[AC72] Sam M Allen and John W Cahn, *Ground state structures in ordered binary alloys with second neighbor interactions*, Acta Metall. **20** (1972), no. 3, 423–433.

[ACV16] Mark Allen, Luis Caffarelli, and Alexis Vasseur, *A parabolic problem with a fractional time derivative*, Arch. Ration. Mech. Anal. **221** (2016), no. 2, 603–630. MR 3488533

[AF03] Robert A. Adams and John J. F. Fournier, *Sobolev Spaces*, second ed., Elsevier/Academic Press, Amsterdam, 2003. MR 2424078 (2009e:46025)

[Aga53] Ratan Prakash Agarwal, *A propos d'une note de M. Pierre Humbert*, CR Acad. Sci. Paris **236** (1953), no. 21, 2031–2032.

[AK19] Mariam Al-Maskari and Samir Karaa, *Numerical approximation of semilinear subdiffusion equations with nonsmooth initial data*, SIAM J. Numer. Anal. **57** (2019), no. 3, 1524–1544. MR 3975154

[Aka19] Goro Akagi, *Fractional flows driven by subdifferentials in Hilbert spaces*, Israel J. Math. **234** (2019), no. 2, 809–862. MR 4040846

[AKT13] Temirkhan S. Aleroev, Mokhtar Kirane, and Ĭi-Fa Tang, *Boundary value problems for differential equations of fractional order*, Ukr. Mat. Visn. **10** (2013), no. 2, 158–175, 293. MR 3137091

[Al-12] Mohammed Al-Refai, *On the fractional derivatives at extreme points*, Electron. J. Qual. Theory Differ. Equ. (2012), No. 55, 5. MR 2959045

© The Author(s), under exclusive license to Springer Nature Switzerland AG 2021
B. Jin, *Fractional Differential Equations*, Applied Mathematical Sciences 206,
https://doi.org/10.1007/978-3-030-76043-4

[AL14]   Mohammed Al-Refai and Yuri Luchko, *Maximum principle for the fractional diffusion equations with the Riemann-Liouville fractional derivative and its applications*, Fract. Calc. Appl. Anal. **17** (2014), no. 2, 483–498. MR 3181067

[Ali10] Anatoly A. Alikhanov, *A priori estimates for solutions of boundary value problems for equations of fractional order*, Differ. Uravn. **46** (2010), no. 5, 658–664. MR 2797545 (2012e:35258)

[Ali15] Anatoly A. Alikhanov, *A new difference scheme for the time fractional diffusion equation*, J. Comput. Phys. **280** (2015), 424–438. MR 3273144

[Ama00] Herbert Amann, *Compact embeddings of vector-valued Sobolev and Besov spaces*, Glas. Mat. Ser. III **35(55)** (2000), no. 1, 161–177, Dedicated to the memory of Branko Najman. MR 1783238

[AS65]   Milton Abramowitz and Irene A Stegun, *Handbook of Mathematical Functions with Formulas, Graphs, and Mathematical Tables*, Dover, New York, 1965.

[Bai10] Zhanbing Bai, *On positive solutions of a nonlocal fractional boundary value problem*, Nonlinear Anal. **72** (2010), no. 2, 916–924. MR 2579357

[Baj01] Emilia Grigorova Bajlekova, *Fractional Evolution Equations in Banach Spaces*, Ph.D. thesis, Eindhoven University of Technology, Eindhoven, 2001. MR 1868564

[Bar54] John H. Barrett, *Differential equations of non-integer order*, Canadian J. Math. **6** (1954), 529–541. MR 0064936

[BBP15] Leigh C. Becker, Theodore A. Burton, and Ioannis K. Purnaras, *Complementary equations: a fractional differential equation and a Volterra integral equation*, Electron. J. Qual. Theory Differ. Equ. (2015), No. 12, 24. MR 3325915

[BCDS06] Brian Berkowitz, Andrea Cortis, Marco Dentz, and Harvey Scher, *Modeling non-Fickian transport in geological formations as a continuous time random walk*, Rev. Geophys. **44** (2006), no. 2, RG2003, 49 pp.

[Bel43] Richard Bellman, *The stability of solutions of linear differential equations*, Duke Math. J. **10** (1943), 643–647. MR 9408

[Bie56] Adam Bielecki, *Une remarque sur la méthode de Banach-Caccioppoli-Tikhonov dans la théorie des équations différentielles ordinaires*, Bull. Acad. Polon. Sci. Cl. III. **4** (1956), 261–264. MR 0082073

[BKS18] Boris Baeumer, Mihály Kovács, and Harish Sankaranarayanan, *Fractional partial differential equations with boundary conditions*, J. Differential Equations **264** (2018), no. 2, 1377–1410. MR 3720847

[BL76]   Jöran Bergh and Jörgen Löfström, *Interpolation Spaces. An Introduction*, Springer-Verlag, Berlin-New York, 1976. MR 0482275

[BL05]   Zhanbing Bai and Haishen Lü, *Positive solutions for boundary value problem of nonlinear fractional differential equation*, J. Math. Anal. Appl. **311** (2005), no. 2, 495–505. MR 2168413 (2006d:34052)

[BLNT17] Maïtine Bergounioux, Antonio Leaci, Giacomo Nardi, and Franco Tomarelli, *Fractional Sobolev spaces and functions of bounded variation of one variable*, Fract. Calc. Appl. Anal. **20** (2017), no. 4, 936–962. MR 3684877

[BPV99] Hermann Brunner, Arvet Pedas, and Gennadi Vainikko, *The piecewise polynomial collocation method for nonlinear weakly singular Volterra equations*, Math. Comp. **68** (1999), no. 227, 1079–1095. MR 1642797

[BS05] Mário N. Berberan-Santos, *Relation between the inverse Laplace transforms of $I(t^\beta)$ and $I(t)$: application to the Mittag-Leffler and asymptotic inverse power law relaxation functions*, J. Math. Chem. **38** (2005), no. 2, 265–270. MR 2166914

[BWM00] David A Benson, Stephen W. Wheatcraft, and Mark M Meerschaert, *Application of a fractional advection-dispersion equation*, Water Resources Res. **36** (2000), no. 6, 1403–1412.

[Can68] John Rozier Cannon, *Determination of an unknown heat source from overspecified boundary data*, SIAM J. Numer. Anal. **5** (1968), no. 2, 275–286. MR 0231552 (37 #7105)

[Can84] John Rozier Cannon, *The One-Dimensional Heat Equation*, Addison-Wesley, Reading, MA, 1984. MR 747979 (86b:35073)

[Cap67] Michele Caputo, *Linear models of dissipation whose Q is almost frequency independent – II*, Geophys. J. Int. **13** (1967), no. 5, 529–539.

[CD98] John Rozier Cannon and Paul DuChateau, *Structural identification of an unknown source term in a heat equation*, Inverse Problems **14** (1998), no. 3, 535–551. MR 1629991 (99g:35142)

[CDMI19] Fabio Camilli, Raul De Maio, and Elisa Iacomini, *A Hopf-Lax formula for Hamilton-Jacobi equations with Caputo time-fractional derivative*, J. Math. Anal. Appl. **477** (2019), no. 2, 1019–1032. MR 3955007

[CLP06] Eduardo Cuesta, Christian Lubich, and Cesar Palencia, *Convolution quadrature time discretization of fractional diffusion-wave equations*, Math. Comp. **75** (2006), no. 254, 673–696. MR 2196986 (2006j:65404)

[CM71a] Michele Caputo and Francesco Mainardi, *Linear models of dissipation in anelastic solids*, La Rivista del Nuovo Cimento (1971-1977) **1** (1971), no. 2, 161–198.

[CM71b] Michele Caputo and Francesco Mainardi, *A new dissipation model based on memory mechanism*, Pure Appl. Geophys. **91** (1971), no. 1, 134–147.

[CMS76] John M. Chambers, Colin L. Mallows, and B. W. Stuck, *A method for simulating stable random variables*, J. Amer. Stat. Assoc. **71** (1976), no. 354, 340–344. MR 0415982 (54 #4059)

[CMSS18] Yujun Cui, Wenjie Ma, Qiao Sun, and Xinwei Su, *New uniqueness results for boundary value problem of fractional differential equation*, Nonlinear Anal. Model. Control **23** (2018), no. 1, 31–39. MR 3747591

[CN81]   Philippe Clément and John A. Nohel, *Asymptotic behavior of solu-tions of nonlinear Volterra equations with completely positive kernels*, SIAM J. Math. Anal. **12** (1981), no. 4, 514–535. MR 617711

[CNYY09] Jin Cheng, Junichi Nakagawa, Masahiro Yamamoto, and Tomohiro Yamazaki, *Uniqueness in an inverse problem for a one-dimensional fractional diffusion equation*, Inverse Problems **25** (2009), no. 11, 115002, 16. MR 2545997 (2010j:35596)

[CY97]   Mourad Choulli and Masahiro Yamamoto, *An inverse parabolic prob-lem with non-zero initial condition*, Inverse Problems **13** (1997), no. 1, 19–27. MR 1435868 (97j:35160)

[dCNP15] Paulo Mendes de Carvalho-Neto and Gabriela Planas, *Mild solutions to the time fractional Navier-Stokes equations in $\mathbb{R}^N$*, J. Differential Equations **259** (2015), no. 7, 2948–2980. MR 3360662

[Dei85]   Klaus Deimling, *Nonlinear Functional Analysis*, Springer-Verlag, Berlin, 1985. MR 787404 (86j:47001)

[DF02]   Kai Diethelm and Neville J. Ford, *Analysis of fractional differential equations*, J. Math. Anal. Appl. **265** (2002), no. 2, 229–248. MR 1876137

[DF12]   Kai Diethelm and Neville J. Ford, *Volterra integral equations and fractional calculus: do neighboring solutions intersect?*, J. Integral Equations Appl. **24** (2012), no. 1, 25–37. MR 2911089

[Die10]   Kai Diethelm, *The Analysis of Fractional Differential Equations*, Springer-Verlag, Berlin, 2010. MR 2680847 (2011j:34005)

[Die12]   Kai Diethelm, *The mean value theorems and a Nagumo-type unique-ness theorem for Caputo's fractional calculus*, Fract. Calc. Appl. Anal. **15** (2012), no. 2, 304–313. MR 2897781

[Die16]   Kai Diethelm, *Monotonicity of functions and sign changes of their Caputo derivatives*, Fract. Calc. Appl. Anal. **19** (2016), no. 2, 561–566. MR 3513010

[Die17]   Kai Diethelm, *Erratum: The mean value theorems and a Nagumo-type uniqueness theorem for Caputo's fractional calculus [MR2897781]*, Fract. Calc. Appl. Anal. **20** (2017), no. 6, 1567–1570. MR 3764308

[Djr93]   Mkhitar M. Djrbashian, *Harmonic Analysis and Boundary Value Problems in the Complex Domain*, Birkhäuser, Basel, 1993. MR 1249271 (95f:30039)

[DK19]   Hongjie Dong and Doyoon Kim, *L_p-estimates for time fractional parabolic equations with coefficients measurable in time*, Adv. Math. **345** (2019), 289–345. MR 3899965

[DK20]   Hongjie Dong and Doyoon Kim, *L_p-estimates for time fractional parabolic equations in divergence form with measurable coefficients*, J. Funct. Anal. **278** (2020), no. 3, 108338. MR 4030286

[DK21]   Hongjie Dong and Doyoon Kim, *An approach for weighted mixed-norm estimates for parabolic equations with local and non-local time derivatives*, Adv. Math. **377** (2021), 107494, 44. MR 4186022

[DM75] Mkhitar M. Džrbašjan and V. M. Martirosjan, *Theorems of Paley-Wiener and Müntz-Szász type*, Dokl. Akad. Nauk SSSR **225** (1975), no. 5, 1001–1004. MR 0396923

[DM77] Mkhitar M. Džrbašjan and V. M. Martirosjan, *Theorems of Paley-Wiener and Müntz-Szász type*, Izv. Akad. Nauk SSSR Ser. Mat. **41** (1977), no. 4, 868–894, 960. MR 0473707

[DM83] Mkhitar M. Dzhrbashyan and V. M. Martirosyan, *Integral representations and best approximations by generalized polynomials with respect to systems of Mittag-Leffler type*, Izv. Akad. Nauk SSSR Ser. Mat. **47** (1983), no. 6, 1182–1207. MR 727751

[DN61] Mkhitar M. Džrbašjan and Anry B. Nersesjan, *Expansions in certain biorthogonal systems and boundary-value problems for differential equations of fractional order*, Trudy Moskov. Mat. Obšč. **10** (1961), 89–179. MR 0146589

[DN68] Mkhitar M. Djrbashian and Anry B Nersesyan, *Fractional derivatives and the Cauchy problem for differential equations of fractional order*, Izv. Akad. Nauk Armajan. SSR, Ser. Mat. **3** (1968), no. 1, 3–29.

[DNPV12] Eleonora Di Nezza, Giampiero Palatucci, and Enrico Valdinoci, *Hitchhiker's guide to the fractional Sobolev spaces*, Bull. Sci. Math. **136** (2012), no. 5, 521–573. MR 2944369

[Doe74] Gustav Doetsch, *Introduction to the Theory and Application of the Laplace Transformation*, Springer-Verlag, New York-Heidelberg, 1974, Translated from the second German edition by Walter Nader. MR 0344810

[Don20] Hongjie Dong, *Recent progress in the L_p theory for elliptic and parabolic equations with discontinuous coefficients*, Anal. Theory Appl. **36** (2020), no. 2, 161–199. MR 4156495

[DR96] Domenico Delbosco and Luigi Rodino, *Existence and uniqueness for a nonlinear fractional differential equation*, J. Math. Anal. Appl. **204** (1996), no. 2, 609–625. MR 1421467

[DW60] Joaquín Basilio Diaz and Wolfgang L. Walter, *On uniqueness theorems for ordinary differential equations and for partial differential equations of hyperbolic type*, Trans. Amer. Math. Soc. **96** (1960), 90–100. MR 120451

[DYZ20] Qiang Du, Jiang Yang, and Zhi Zhou, *Time-fractional Allen-Cahn equations: analysis and numerical methods*, J. Sci. Comput. **85** (2020), no. 2, Paper No. 42, 30. MR 4170335

[Dzh66] Mkhitar M Dzharbashyan, *Integral Transformations and Representation of Functions in a Complex Domain [in Russian]*, Nauka, Moscow, 1966. MR 0209472

[Džr54a] Mkhitar M. Džrbašyan, *On Abel summability of generalized integral transforms*, Akad. Nauk Armyan. SSR. Izv. Fiz.-Mat. Estest. Tehn. Nauki **7** (1954), no. 6, 1–26 (1955). MR 0069927

[Džr54b] Mkhitar M. Džrbašyan, *On the asymptotic behavior of a function of Mittag-Leffler type*, Akad. Nauk Armyan. SSR. Dokl. **19** (1954), 65–72. MR 0069331

[Džr54c] Mkhitar M. Džrbašyan, *On the integral representation of functions continuous on several rays (generalization of the Fourier integral)*, Izv. Akad. Nauk SSSR. Ser. Mat. **18** (1954), 427–448. MR 0065684

[Džr70] Mkhitar M. M. Džrbašjan, *A boundary value problem for a Sturm-Liouville type differential operator of fractional order*, Izv. Akad. Nauk Armjan. SSR Ser. Mat. **5** (1970), no. 2, 71–96. MR 0414982 (54 #3074)

[Džr74] Mkhitar M. Džrbašjan, *The completeness of a system of Mittage-Leffler type*, Dokl. Akad. Nauk SSSR **219** (1974), 1302–1305. MR 0387940

[EG04] Alexandre Ern and Jean-Luc Guermond, *Theory and Practice of Finite Elements*, Springer-Verlag, New York, 2004. MR 2050138

[EHR18] Vincent John Ervin, Norbert Heuer, and John Paul Roop, *Regularity of the solution to 1-D fractional order diffusion equations*, Math. Comp. **87** (2018), no. 313, 2273–2294. MR 3802435

[Ein05] Albert Einstein, *Über die von der molekularkinetischen Theorie der Wärme geforderte Bewegung von in ruhenden Flüssigkeiten suspendierten Teilchen*, Ann. Phys. **322** (1905), no. 8, 549–560.

[EK04] Samuil D. Eidelman and Anatoly N. Kochubei, *Cauchy problem for fractional diffusion equations*, J. Differential Equations **199** (2004), no. 2, 211–255. MR 2047909 (2005i:26014)

[EMOT55] Arthur Erdélyi, Wilhelm Magnus, Fritz Oberhettinger, and Francesco G. Tricomi, *Higher Transcendental Functions. Vol. III*, McGraw-Hill Book Company, Inc., New York-Toronto-London, 1955, Based, in part, on notes left by Harry Bateman. MR 0066496

[ER06] Vincent J. Ervin and John Paul Roop, *Variational formulation for the stationary fractional advection dispersion equation*, Numer. Methods Partial Differential Equations **22** (2006), no. 3, 558–576. MR 2212226 (2006m:65265)

[Erv21] Vincent J. Ervin, *Regularity of the solution to fractional diffusion, advection, reaction equations in weighted Sobolev spaces*, J. Differential Equations **278** (2021), 294–325. MR 4200757

[Eul29] Leonhard Euler, *Letter to Goldbach, 13 october 1729*, 1729, Euler Archive [E00715], eulerarchive.maa.org.

[Fel71] William Feller, *An Introduction to Probability Theory and its Applications. Vol. II*, Second edition, John Wiley & Sons, Inc., New York-London-Sydney, 1971. MR 0270403

[FLLX18a] Yuanyuan Feng, Lei Li, Jian-Guo Liu, and Xiaoqian Xu, *Continuous and discrete one dimensional autonomous fractional ODEs*, Discrete Contin. Dyn. Syst. Ser. B **23** (2018), no. 8, 3109–3135. MR 3848192

[FLLX18b] Yuanyuan Feng, Lei Li, Jian-Guo Liu, and Xiaoqian Xu, *A note on one-dimensional time fractional ODEs*, Appl. Math. Lett. **83** (2018), 87–94. MR 3795675

[Fri58] Avner Friedman, *Remarks on the maximum principle for parabolic equations and its applications*, Pacific J. Math. **8** (1958), 201–211. MR 0102655 (21 #1444)

[Gal80] R. S. Galojan, *Completeness of a system of eigen- and associated functions*, Izv. Akad. Nauk Armyan. SSR Ser. Mat. **15** (1980), no. 4, 310–322, 334. MR 605051

[Gar15] Roberto Garrappa, *Numerical evaluation of two and three parameter Mittag-Leffler functions*, SIAM J. Numer. Anal. **53** (2015), no. 3, 1350–1369. MR 3350038

[Gar19] Nicola Garofalo, *Fractional thoughts*, New Developments in the Analysis of Nonlocal Operators, Contemp. Math., vol. 723, Amer. Math. Soc., Providence, RI, 2019, pp. 1–135. MR 3916700

[Gau60] Walter Gautschi, *Some elementary inequalities relating to the gamma and incomplete gamma function*, J. Math. and Phys. **38** (1959/60), 77–81. MR 103289

[GK70] Israel C. Gohberg and Mark Grigor'evich Kreĭn, *Theory and Applications of Volterra Operators in Hilbert Space*, AMS, Providence, R.I., 1970. MR 0264447

[GKMR14] Rudolf Gorenflo, Anatoly A. Kilbas, Francesco Mainardi, and Sergei V. Rogosin, *Mittag-Leffler Functions, Related Topics and Applications*, Springer, Heidelberg, 2014.

[GLL02] Rudolf Gorenflo, Joulia Loutchko, and Yuri Luchko, *Computation of the Mittag-Leffler function $E_{\alpha,\beta}(z)$ and its derivative*, Fract. Calc. Appl. Anal. **5** (2002), no. 4, 491–518. MR 1967847 (2004d:33020a)

[GLM99] Rudolf Gorenflo, Yuri Luchko, and Francesco Mainardi, *Analytical properties and applications of the Wright function*, Fract. Calc. Appl. Anal. **2** (1999), no. 4, 383–414. MR 1752379 (2001c:33011)

[GLM00] Rudolf Gorenflo, Yuri Luchko, and Francesco Mainardi, *Wright functions as scale-invariant solutions of the diffusion-wave equation*, J. Comput. Appl. Math. **118** (2000), no. 1-2, 175–191. MR 1765948 (2001e:45007)

[GLY15] Rudolf Gorenflo, Yuri Luchko, and Masahiro Yamamoto, *Time-fractional diffusion equation in the fractional Sobolev spaces*, Fract. Calc. Appl. Anal. **18** (2015), no. 3, 799–820. MR 3351501

[GM98] Rudolf Gorenflo and Francesco Mainardi, *Fractional calculus and stable probability distributions*, Arch. Mech. (Arch. Mech. Stos.) **50** (1998), no. 3, 377–388, Fourth Meeting on Current Ideas in Mechanics and Related Fields (Kraków, 1997). MR 1648257

[GN17] Yoshikazu Giga and Tokinaga Namba, *Well-posedness of Hamilton-Jacobi equations with Caputo's time fractional derivative*, Comm. Partial Differential Equations **42** (2017), no. 7, 1088–1120. MR 3691391

[GR15] Izrail Solomonovich Gradshteyn and Iosif Moiseevich Ryzhik, *Table of Integrals, Series, and Products*, eighth ed., Elsevier/Academic Press, Amsterdam, 2015, Translated from the Russian, Translation edited and with a preface by Daniel Zwillinger and Victor Moll, Revised from the seventh edition [MR2360010]. MR 3307944

[Gra82] Ivan G. Graham, *Singularity expansions for the solutions of second kind Fredholm integral equations with weakly singular convolution kernels*, J. Integral Equations **4** (1982), no. 1, 1–30. MR 640534

[Gra04] Loukas Grafakos, *Classical and Modern Fourier Analysis*, Pearson Education, Inc., Upper Saddle River, NJ, 2004. MR 2449250

[Gri85] Pierre Grisvard, *Elliptic Problems in Nonsmooth Domains*, Pitman, Boston, MA, 1985.

[Gro19] Thomas Hakon Gronwall, *Note on the derivatives with respect to a parameter of the solutions of a system of differential equations*, Ann. of Math. (2) **20** (1919), no. 4, 292–296. MR 1502565

[Grü67] Anton Karl Grünwald, *Uber "begrenzte" Derivationen und deren Anwendung*, Z. angew. Math. und Phys. **12** (1867), 441–480.

[GSZ14] Guang-Hua Gao, Zhi-Zhong Sun, and Hong-Wei Zhang, *A new fractional numerical differentiation formula to approximate the Caputo fractional derivative and its applications*, J. Comput. Phys. **259** (2014), 33–50. MR 3148558

[GW20] Ciprian G. Gal and Mahamadi Warma, *Fractional-in-Time Semilinear Parabolic Equations and Applications*, Springer, Switherland, 2020. MR 4167508

[HA53] Pierre Humbert and Ratan Prakash Agarwal, *Sur la fonction de Mittag-Leffler et quelques-unes de ses généralisations*, Bull. Sci. Math. (2) **77** (1953), 180–185. MR 0060643

[Hač75] I. O. Hačatrjan, *The completeness of families of functions of Mittag-Leffler type under weighted uniform approximation in the complex domain*, Mat. Zametki **18** (1975), no. 5, 675–685. MR 466565

[Han64] Hermann Hankel, *Die Euler'schen Integrale bei unbeschränkter Variabilität des Argumentes*, Zeitschr. Math. Phys. **9** (1864), 1–21.

[Hen81] Daniel Henry, *Geometric Theory of Semilinear Parabolic Equations*, Lecture Notes in Mathematics, vol. 840, Springer-Verlag, Berlin-New York, 1981. MR 610244

[HH98] Yuko Hatano and Naomichi Hatano, *Dispersive transport of ions in column experiments: an explanation of long-tailed profiles*, Water Resour. Res. **34** (1998), no. 5, 1027–1033.

[HHK16] Ma. Elena Hernández-Hernández and Vassili N. Kolokoltsov, *On the solution of two-sided fractional ordinary differential equations of Caputo type*, Fract. Calc. Appl. Anal. **19** (2016), no. 6, 1393–1413. MR 3589357

[HKP20] Beom-Seok Han, Kyeong-Hun Kim, and Daehan Park, *Weighted $L_q(L_p)$-estimate with Muckenhoupt weights for the diffusion-wave equations with time-fractional derivatives*, J. Differential Equations **269** (2020), no. 4, 3515–3550. MR 4097256

[HL28] Godfrey Harold Hardy and John Edensor Littlewood, *Some properties of fractional integrals. I*, Math. Z. **27** (1928), no. 1, 565–606. MR 1544927

[HL32] Godfrey Harold Hardy and John Edensor Littlewood, *Some properties of fractional integrals. II*, Math. Z. **34** (1932), no. 1, 403–439. MR 1545260

[HMS11] Hans Joachim Haubold, Arakaparampil M. Mathai, and Ram Kishore Saxena, *Mittag-Leffler functions and their applications*, J. Appl. Math. (2011), Art. ID 298628, 51. MR 2800586

[HNWY13] Yuko Hatano, Junichi Nakagawa, Shengzhang Wang, and Masahiro Yamamoto, *Determination of order in fractional diffusion equation*, J. Math. Industry **5** (2013), no. A, 51–57.

[HTR+99] Nácere Hayek, Juan J. Trujillo, Margarita Rivero, Blanca Bonilla, and Juan Carlos Moreno Piquero, *An extension of Picard-Lindelöff theorem to fractional differential equations*, Appl. Anal. **70** (1999), no. 3-4, 347–361. MR 1688864

[Hum53] Pierre Humbert, *Quelques résultats relatifs à la fonction de Mittag-Leffler*, C. R. Acad. Sci. Paris **236** (1953), 1467–1468. MR 0054107 (14,872i)

[Isa06] Victor Isakov, *Inverse Problems for Partial Differential Equations*, second ed., Springer, New York, 2006. MR 2193218 (2006h:35279)

[IY98] Oleg Yu. Imanuvilov and Masahiro Yamamoto, *Lipschitz stability in inverse parabolic problems by the Carleman estimate*, Inverse Problems **14** (1998), no. 5, 1229–1245. MR 1654631 (99h:35227)

[JK21] Bangti Jin and Yavar Kian, *Recovery of the order of derivation for fractional diffusion equations in an unknown medium*, Preprint, arXiv:2101.09165, 2021.

[JLLY17] Daijun Jiang, Zhiyuan Li, Yikan Liu, and Masahiro Yamamoto, *Weak unique continuation property and a related inverse source problem for time-fractional diffusion-advection equations*, Inverse Problems **33** (2017), no. 5, 055013, 22. MR 3634443

[JLLZ15] Bangti Jin, Raytcho Lazarov, Yikan Liu, and Zhi Zhou, *The Galerkin finite element method for a multi-term time-fractional diffusion equation*, J. Comput. Phys. **281** (2015), 825–843. MR 3281997

[JLPR15] Bangti Jin, Raytcho Lazarov, Joseph Pasciak, and William Rundell, *Variational formulation of problems involving fractional order differential operators*, Math. Comp. **84** (2015), no. 296, 2665–2700. MR 3378843

[JLPZ14] Bangti Jin, Raytcho Lazarov, Joseph Pasciak, and Zhi Zhou, *Error analysis of a finite element method for the space-fractional parabolic equation*, SIAM J. Numer. Anal. **52** (2014), no. 5, 2272–2294. MR 3259788

[JLZ16a] Bangti Jin, Raytcho Lazarov, and Zhi Zhou, *An analysis of the L1 scheme for the subdiffusion equation with nonsmooth data*, IMA J. Numer. Anal. **36** (2016), no. 1, 197–221. MR 3463438

[JLZ16b] Bangti Jin, Raytcho Lazarov, and Zhi Zhou, *A Petrov-Galerkin finite element method for fractional convection diffusion equation*, SIAM J. Numer. Anal. **54** (2016), no. 1, 481–503. MR 3463699

[JLZ16c] Bangti Jin, Raytcho Lazarov, and Zhi Zhou, *Two fully discrete schemes for fractional diffusion and diffusion-wave equations with nonsmooth data*, SIAM J. Sci. Comput. **38** (2016), no. 1, A146–A170. MR 3449907

[JLZ18] Bangti Jin, Buyang Li, and Zhi Zhou, *Numerical analysis of nonlinear subdiffusion equations*, SIAM J. Numer. Anal. **56** (2018), no. 1, 1–23. MR 3742688

[JLZ19a] Bangti Jin, Raytcho Lazarov, and Zhi Zhou, *Numerical methods for time-fractional evolution equations with nonsmooth data: a concise overview*, Comput. Methods Appl. Mech. Engrg. **346** (2019), 332–358. MR 3894161

[JLZ19b] Bangti Jin, Buyang Li, and Zhi Zhou, *Subdiffusion with a time-dependent coefficient: analysis and numerical solution*, Math. Comp. **88** (2019), no. 319, 2157–2186. MR 3957890

[JLZ20a] Bangti Jin, Buyang Li, and Zhi Zhou, *Pointwise-in-time error estimates for an optimal control problem with subdiffusion constraint*, IMA J. Numer. Anal. **40** (2020), no. 1, 377–404. MR 4050544

[JLZ20b] Bangti Jin, Buyang Li, and Zhi Zhou, *Subdiffusion with time-dependent coefficients: improved regularity and second-order time stepping*, Numer. Math. **145** (2020), no. 4, 883–913. MR 4125980

[JR12] Bangti Jin and William Rundell, *An inverse Sturm-Liouville problem with a fractional derivative*, J. Comput. Phys. **231** (2012), no. 14, 4954–4966. MR 2927980

[JR15] Bangti Jin and William Rundell, *A tutorial on inverse problems for anomalous diffusion processes*, Inverse Problems **31** (2015), no. 3, 035003, 40. MR 3311557

[JZ21] Bangti Jin and Zhi Zhou, *An inverse potential problem for subdiffusion: stability and reconstruction*, Inverse Problems **37** (2021), no. 1, 015006, 26. MR 4191622

[KA13] Mał gorzata Klimek and Om Prakash Agrawal, *Fractional Sturm-Liouville problem*, Comput. Math. Appl. **66** (2013), no. 5, 795–812. MR 3089387

[KBT00] Anatoly A. Kilbas, Blanca Bonilla, and Kh. Trukhillo, *Nonlinear differential equations of fractional order in the space of integrable functions*, Dokl. Akad. Nauk **374** (2000), no. 4, 445–449. MR 1798482

[Kia20] Yavar Kian, *Simultaneous determination of coefficients and internal source of a diffusion equation from a single measurement*, Preprint, arXiv:2007.08947, 2020.

[KKL17] Ildoo Kim, Kyeong-Hun Kim, and Sungbin Lim, *An $L_q(L_p)$-theory for the time fractional evolution equations with variable coefficients*, Adv. Math. **306** (2017), 123–176. MR 3581300

[KM04] Anatoly A. Kilbas and Sergei A. Marzan, *Cauchy problem for differential equation with Caputo derivative*, Fract. Calc. Appl. Anal. **7** (2004), no. 3, 297–321. MR 2252568

[KM05] Anatoly A. Kilbas and Sergei A. Marzan, *Nonlinear differential equations with the Caputo fractional derivative in the space of continuously differentiable functions*, Differ. Uravn. **41** (2005), no. 1, 82–86, 142. MR 2213269

[Koc90] Aantoly N. Kochubeĭ, *Diffusion of fractional order*, Differentsial'nye Uravneniya **26** (1990), no. 4, 660–670, 733–734. MR 1061448 (91j:35133)

[KOM14] Mał gorzata Klimek, Tatiana Odzijewicz, and Agnieszka B. Malinowska, *Variational methods for the fractional Sturm-Liouville problem*, J. Math. Anal. Appl. **416** (2014), no. 1, 402–426. MR 3182768

[KOSY18] Yavar Kian, Lauri Oksanen, Eric Soccorsi, and Masahiro Yamamoto, *Global uniqueness in an inverse problem for time fractional diffusion equations*, J. Differential Equations **264** (2018), no. 2, 1146–1170. MR 3720840

[Kou08] Samuel C. Kou, *Stochastic modeling in nanoscale biophysics: subdiffusion within proteins*, Ann. Appl. Stat. **2** (2008), no. 2, 501–535. MR 2524344

[KR19] Barbara Kaltenbacher and William Rundell, *On an inverse potential problem for a fractional reaction-diffusion equation*, Inverse Problems **35** (2019), no. 6, 065004, 31. MR 3975371

[Kra64] Mark Aleksandrovich Krasnosel'skiĭ, *Positive Solutions of Operator Equations*, P. Noordhoff Ltd. Groningen, 1964. MR 0181881 (31 #6107)

[Kra16] Mykola V. Krasnoschok, *Solvability in Hölder space of an initial boundary value problem for the time-fractional diffusion equation*, Zh. Mat. Fiz. Anal. Geom. **12** (2016), no. 1, 48–77. MR 3477949

[KRY20] Adam Kubica, Katarzyna Ryszewska, and Masahiro Yamamoto, *Time-Fractional Differential Equations*, SpringerBriefs in Mathematics, Springer, Singapore, 2020, A theoretical introduction. MR 4200127

[KST06] Anatoly A. Kilbas, Hari M. Srivastava, and Juan J. Trujillo, *Theory and Applications of Fractional Differential Equations*, Elsevier Science B.V., Amsterdam, 2006. MR 2218073 (2007a:34002)

[KV13] Mykola Krasnoschok and Nataliya Vasylyeva, *On a solvability of a nonlinear fractional reaction-diffusion system in the Hölder spaces*, Nonlinear Stud. **20** (2013), no. 4, 591–621. MR 3154625

[KW04] Peer C. Kunstmann and Lutz Weis, *Maximal L_p-regularity for parabolic equations, Fourier multiplier theorems and H^∞-functional calculus*, Functional Analytic Methods for Evolution Equations, Lecture Notes in Math., vol. 1855, Springer, Berlin, 2004, pp. 65–311. MR 2108959

[KY17] Yavar Kian and Masahiro Yamamoto, *On existence and uniqueness of solutions for semilinear fractional wave equations*, Fract. Calc. Appl. Anal. **20** (2017), no. 1, 117–138. MR 3613323

[KY18] Adam Kubica and Masahiro Yamamoto, *Initial-boundary value problems for fractional diffusion equations with time-dependent coefficients*, Fract. Calc. Appl. Anal. **21** (2018), no. 2, 276–311. MR 3814402

[KY21] Yavar Kian and Masahiro Yamamoto, *Well-posedness for weak and strong solutions of non-homogeneous initial boundary value problems for fractional diffusion equations*, Fract. Calc. Appl. Anal. **24** (2021), no. 1, 168–201. MR 4225517

[Let68] Aleksey Vasilievich Letnikov, *Theory of differentiation with an arbitrary index (in Russian)*, Mat. Sb. **3** (1868), 1–66.

[LHY20] Zhiyuan Li, Xinchi Huang, and Masahiro Yamamoto, *A stability result for the determination of order in time-fractional diffusion equations*, J. Inverse Ill-Posed Probl. **28** (2020), no. 3, 379–388. MR 4104326

[Lio32] Joseph Liouville, *Memoire sue quelques questions de géometrie et de mécanique, et sur un nouveau gentre pour resoudre ces questions*, J. Ecole Polytech. **13** (1832), 1–69.

[Liu08] Jun S. Liu, *Monte Carlo Strategies in Scientific Computing*, Springer Series in Statistics, Springer, New York, 2008. MR 2401592 (2010b:65013)

[LL09] V. Lakshmikantham and S. Leela, *Nagumo-type uniqueness result for fractional differential equations*, Nonlinear Anal. **71** (2009), no. 7-8, 2886–2889. MR 2532815

[LL13] Kunquan Lan and Wei Lin, *Positive solutions of systems of Caputo fractional differential equations*, Commun. Appl. Anal. **17** (2013), no. 1, 61–85. MR 3075769

[LL18a] Lei Li and Jian-Guo Liu, *A generalized definition of Caputo derivatives and its application to fractional ODEs*, SIAM J. Math. Anal. **50** (2018), no. 3, 2867–2900. MR 3809535

[LL18b] Lei Li and Jian-Guo Liu, *Some compactness criteria for weak solutions of time fractional PDEs*, SIAM J. Math. Anal. **50** (2018), no. 4, 3963–3995. MR 3828856

[LLY19a] Zhiyuan Li, Yikan Liu, and Masahiro Yamamoto, *Inverse problems of determining parameters of the fractional partial differential equations*, Handbook of Fractional Calculus with Applications. Vol. 2, De Gruyter, Berlin, 2019, pp. 431–442. MR 3965404

[LLY19b] Yikan Liu, Zhiyuan Li, and Masahiro Yamamoto, *Inverse problems of determining sources of the fractional partial differential equations*, Handbook of fractional calculus with applications. Vol. 2, De Gruyter, Berlin, 2019, pp. 411–429. MR 3965403

[LM72] Jacques-Louis Lions and Enrico Magenes, *Non-homogeneous Boundary Value Problems and Applications. Vol. I*, Springer-Verlag, New York-Heidelberg, 1972, Translated from the French by P. Kenneth,

Die Grundlehren der mathematischen Wissenschaften, Band 181. MR 0350177

[LN16] Ching-Lung Lin and Gen Nakamura, *Unique continuation property for anomalous slow diffusion equation*, Comm. Partial Differential Equations **41** (2016), no. 5, 749–758. MR 3508319

[LRdS18] Wei Liu, Michael Röckner, and José Luís da Silva, *Quasi-linear (stochastic) partial differential equations with time-fractional derivatives*, SIAM J. Math. Anal. **50** (2018), no. 3, 2588–2607. MR 3800228

[LRY16] Yikan Liu, William Rundell, and Masahiro Yamamoto, *Strong maximum principle for fractional diffusion equations and an application to an inverse source problem*, Fract. Calc. Appl. Anal. **19** (2016), no. 4, 888–906. MR 3543685

[LRYZ13] Yuri Luchko, William Rundell, Masahiro Yamamoto, and Lihua Zuo, *Uniqueness and reconstruction of an unknown semilinear term in a time-fractional reaction-diffusion equation*, Inverse Problems **29** (2013), no. 6, 065019, 16. MR 3066395

[LS21] Wenbo Li and Abner J Salgado, *Time fractional gradient flows: Theory and numerics*, Preprint, arXiv:2101.00541, 2021.

[LST96] Christian Lubich, Ian H. Sloan, and Vidar Thomée, *Nonsmooth data error estimates for approximations of an evolution equation with a positive-type memory term*, Math. Comp. **65** (1996), no. 213, 1–17. MR 1322891 (96d:65207)

[LSU68] O. A. Ladyženskaja, V. A. Solonnikov, and N. N. Ural'ceva, *Linear and Quasilinear Equations of Parabolic Type*, AMS, Providence, R.I., 1968. MR 0241822

[Lub83] Christian Lubich, *Runge-Kutta theory for Volterra and Abel integral equations of the second kind*, Math. Comp. **41** (1983), no. 163, 87–102. MR 701626

[Lub86] Christian Lubich, *Discretized fractional calculus*, SIAM J. Math. Anal. **17** (1986), no. 3, 704–719. MR 838249 (87f:26006)

[Lub88] Christian Lubich, *Convolution quadrature and discretized operational calculus. I*, Numer. Math. **52** (1988), no. 2, 129–145. MR 923707 (89g:65018)

[Luc00] Yuri Luchko, *Asymptotics of zeros of the Wright function*, Z. Anal. Anwendungen **19** (2000), no. 2, 583–595. MR 1769012 (2001f:33028)

[Luc08] Yuri Luchko, *Algorithms for evaluation of the Wright function for the real arguments' values*, Fract. Calc. Appl. Anal. **11** (2008), no. 1, 57–75. MR 2379273 (2009a:33024)

[Luc09] Yuri Luchko, *Maximum principle for the generalized time-fractional diffusion equation*, J. Math. Anal. Appl. **351** (2009), no. 1, 218–223. MR 2472935 (2009m:35274)

[LX07] Yumin Lin and Chuanju Xu, *Finite difference/spectral approximations for the time-fractional diffusion equation*, J. Comput. Phys. **225** (2007), no. 2, 1533–1552. MR 2349193

[LY17]   Yuri Luchko and Masahiro Yamamoto, *On the maximum principle for a time-fractional diffusion equation*, Fract. Calc. Appl. Anal. **20** (2017), no. 5, 1131–1145. MR 3721892

[LY19a] Zhiyuan Li and Masahiro Yamamoto, *Inverse problems of determining coefficients of the fractional partial differential equations*, Handbook of Fractional Calculus with Applications. Vol. 2, De Gruyter, Berlin, 2019, pp. 443–464. MR 3965405

[LY19b] Yuri Luchko and Masahiro Yamamoto, *Maximum principle for the time-fractional PDEs*, Handbook of fractional calculus with applications. Vol. 2, De Gruyter, Berlin, 2019, pp. 299–325. MR 3965399

[LY20]   Ping Lin and Jiongmin Yong, *Controlled singular Volterra integral equations and Pontryagin maximum principle*, SIAM J. Control Optim. **58** (2020), no. 1, 136–164. MR 4049386

[LZ14]   Yuri Luchko and Lihua Zuo, *θ-function method for a time-fractional reaction-diffusion equation*, J. Alg. Math. Soc. **1** (2014), 1–15.

[Mai96] Francesco Mainardi, *The fundamental solutions for the fractional diffusion-wave equation*, Appl. Math. Lett. **9** (1996), no. 6, 23–28. MR 1419811 (97h:35132)

[Mai14] Francesco Mainardi, *On some properties of the Mittag-Leffler function $E_\alpha(-t^\alpha)$, completely monotone for $t>0$ with $0<\alpha<1$*, Discrete Contin. Dyn. Syst. Ser. B **19** (2014), no. 7, 2267–2278. MR 3253257

[Mal94] Mark M. Malamud, *Similarity of Volterra operators and related problems in the theory of differential equations of fractional orders*, Trudy Moskov. Mat. Obshch. **55** (1994), 73–148, 365. MR 1468456

[McL10] William McLean, *Regularity of solutions to a time-fractional diffusion equation*, ANZIAM J. **52** (2010), no. 2, 123–138. MR 2832607 (2012g:45018)

[McL21] William McLean, *Numerical evaluation of Mittag-Leffler functions*, Calcolo **58** (2021), no. 1, Paper No. 7. MR 4218357

[MF71]   Richard K. Miller and Alan Feldstein, *Smoothness of solutions of Volterra integral equations with weakly singular kernels*, SIAM J. Math. Anal. **2** (1971), 242–258. MR 0287258

[MG07]   Francesco Mainardi and Rudolf Gorenflo, *Time-fractional derivatives in relaxation processes: a tutorial survey*, Fract. Calc. Appl. Anal. **10** (2007), no. 3, 269–308. MR 2382782

[MK00]   Ralf Metzler and Joseph Klafter, *The random walk's guide to anomalous diffusion: a fractional dynamics approach*, Phys. Rep. **339** (2000), no. 1, 77. MR 1809268

[MK04]   Ralf Metzler and Joseph Klafter, *The restaurant at the end of the random walk: recent developments in the description of anomalous transport by fractional dynamics*, J. Phys. A **37** (2004), no. 31, R161–R208. MR 2090004

[ML03]   Gösta Magnus Mittag-Leffler, *Sur la nouvelle fonction E_a*, C.R. Acad. Sci. Paris **137** (1903), 554–558.

[ML05] Gösta Magnus Mittag-Leffler, *Sur la représentation analytique d'une branche uniforme d'une fonction monogène*, Acta Math. **29** (1905), no. 1, 101–181, cinquième note. MR 1555012

[MLP01] Francesco Mainardi, Yuri Luchko, and Gianni Pagnini, *The fundamental solution of the space-time fractional diffusion equation*, Fract. Calc. Appl. Anal. **4** (2001), no. 2, 153–192. MR 1829592 (2002h:26010)

[MMAK19] William McLean, Kassem Mustapha, Raed Ali, and Omar Knio, *Well-posedness of time-fractional advection-diffusion-reaction equations*, Fract. Calc. Appl. Anal. **22** (2019), no. 4, 918–944. MR 4023101

[MMAK20] William McLean, Kassem Mustapha, Raed Ali, and Omar M. Knio, *Regularity theory for time-fractional advection-diffusion-reaction equations*, Comput. Math. Appl. **79** (2020), no. 4, 947–961. MR 4054212

[MMP10] Francesco Mainardi, Antonio Mura, and Gianni Pagnini, *The M-Wright function in time-fractional diffusion processes: a tutorial survey*, Int. J. Differ. Equ. (2010), Art. ID 104505, 29. MR 2592742 (2011a:60287)

[MO01] Mark M. Malamud and Leonid L. Oridoroga, *Analog of the Birkhoff theorem and the completeness results for fractional order differential equations*, Russ. J. Math. Phys. **8** (2001), no. 3, 287–308. MR 1930376

[MP62] Vladimir Igorevich Macaev and Ju. A. Palant, *On the powers of a bounded dissipative operator*, Ukrain. Mat. Ž. **14** (1962), 329–337. MR 0146664

[MR93] Kenneth S. Miller and Bertram Ross, *An Introduction to the Fractional Calculus and Fractional Differential Equations*, A Wiley-Interscience Publication, John Wiley & Sons, Inc., New York, 1993. MR 1219954

[MS01] Kenneth S. S. Miller and Stefan G. Samko, *Completely monotonic functions*, Integral Transform. Spec. Funct. **12** (2001), no. 4, 389–402. MR 1872377

[MS18a] Khaled Mehrez and Sergei M. Sitnik, *Turán type inequalities for classical and generalized Mittag-Leffler functions*, Anal. Math. **44** (2018), no. 4, 521–541. MR 3877592

[MS18b] Xiangyun Meng and Martin Stynes, *The Green's function and a maximum principle for a Caputo two-point boundary value problem with a convection term*, J. Math. Anal. Appl. **461** (2018), no. 1, 198–218. MR 3759537

[MS98] Kenneth S. Miller and Stefan G. Samko, *A note on the complete monotonicity of the generalized Mittag-Leffler function*, Real Anal. Exchange **23** (1997/98), no. 2, 753–755. MR 1639957

[MT95] Francesco Mainardi and M Tomirotti, *On a special function arising in the time fractional diffusion-wave equation*, Transform Methods & Special Functions, Sofia'94 (P Rusev, I Dimovski, and V Kiryakova, eds.), Science Culture Technology Publ., Singapore, 1995, pp. 171–183.

[MW65] Elliott W Montroll and George H Weiss, *Random walks on lattices. II*, J. Math. Phys. **6** (1965), no. 2, 167–181. MR 172344

[Nag26] Mitio Nagumo, *Eine hinreichende Bedingung für die Unität der Lösung von Differentialgleichungen erster Ordnung*, Jap. J. Math. **3** (1926), 107–112.

[Nak77] Adam Maremovich Nakhushev, *The Sturm-Liouville problem for a second order ordinary differential equation with fractional derivatives in the lower terms*, Dokl. Akad. Nauk SSSR **234** (1977), no. 2, 308–311. MR 0454145 (56 #12396)

[Nig86] Raoul R. Nigmatullin, *The realization of the generalized transfer equation in a medium with fractal geometry*, Physica Status Solidi (b) **133** (1986), no. 1, 425–430.

[Nol97] John P. Nolan, *Numerical calculation of stable densities and distribution functions*, Comm. Statist. Stochastic Models **13** (1997), no. 4, 759–774. MR 1482292 (98f:60024)

[Nol20] John P. Nolan, *Univariate Stable Distributions*, Springer Nature, Switzerland, 2020, Models for Heavy Tailed Data.

[NR20] Tokinaga Namba and Piotr Rybka, *On viscosity solutions of space-fractional diffusion equations of Caputo type*, SIAM J. Math. Anal. **52** (2020), no. 1, 653–681. MR 4062803

[NSY10] Junichi Nakagawa, Kenichi Sakamoto, and Masahiro Yamamoto, *Overview to mathematical analysis for fractional diffusion equations—new mathematical aspects motivated by industrial collaboration*, J. Math-for-Ind. **2A** (2010), 99–108. MR 2639369

[OP97] Iosif Vladimirovich Ostrovskiĭ and I. N. Peresyolkova, *Nonasymptotic results on distribution of zeros of the function $E_\rho(z, \mu)$*, Anal. Math. **23** (1997), no. 4, 283–296. MR 1629981 (99g:30037)

[OS74] Keith B. Oldham and Jerome Spanier, *The Fractional Calculus*, Academic Press, New York-London, 1974, Theory and applications of differentiation and integration to arbitrary order, With an annotated chronological bibliography by Bertram Ross. MR 0361633 (50 #14078)

[Osl70] Thomas J. Osler, *Leibniz rule for fractional derivatives generalized and an application to infinite series*, SIAM J. Appl. Math. **18** (1970), 658–674. MR 0260942

[Par02] Richard Bruce Paris, *Exponential asymptotics of the Mittag-Leffler function*, R. Soc. Lond. Proc. Ser. A **458** (2002), no. 2028, 3041–3052. MR 1987525 (2004f:33041)

[Par20] Richard B. Paris, *Asymptotics of the Mittag-Leffler Function $E_a(z)$ on the negative real axis when $a \to 1$*, Preprint, 2005.05737v1, 2020.

[PBM86] Anatoliĭ Platonovich Prudnikov, Yurii Aleksandrovich Brychkov, and Oleg Igorevich Marichev, *Integrals and Series. Vol. 1*, Gordon & Breach Science Publishers, New York, 1986. MR 874986

[Pod99] Igor Podlubny, *Fractional Differential Equations*, Academic Press, Inc., San Diego, CA, 1999. MR 1658022 (99m:26009)

[Pól18] Georg Pólya, *Über die Nullstellen gewisser ganzer Funktionen*, Math.
         Z. **2** (1918), no. 3-4, 352–383. MR 1544326

[Pol46] Harry Pollard, *The representation of e^{-x^λ} as a Laplace integral*, Bull.
         Amer. Math. Soc. **52** (1946), 908–910. MR 0018286 (8,269a)

[Pol48] Harry Pollard, *The completely monotonic character of the Mittag-
         Leffler function $E_a(-x)$*, Bull. Amer. Math. Soc. **54** (1948), 1115–
         1116. MR 0027375 (10,295e)

[Pop06] Anton Yu. Popov, *On the number of real eigenvalues of a boundary
         value problem for a second-order equation with a fractional derivative*,
         Fundam. Prikl. Mat. **12** (2006), no. 6, 137–155. MR 2314136

[Prü93] Jan Prüss, *Evolutionary Integral Equations and Applications*, Mono-
         graphs in Mathematics, vol. 87, Birkhäuser Verlag, Basel, 1993. MR
         1238939

[PS38]   Everett Pitcher and W. E. Sewell, *Existence theorems for solutions of
         differential equations of non-integral order*, Bull. Amer. Math. Soc.
         44 (1938), no. 2, 100–107. MR 1563690

[PS11]   Anton Yur'evich Popov and A. M. Sedletskiĭ, *Distribution of roots of
         Mittag-Leffler functions*, Sovrem. Mat. Fundam. Napravl. **40** (2011),
         3–171. MR 2883249 (2012k:33049)

[PS12]   Pierre Patie and Thomas Simon, *Intertwining certain fractional
         derivatives*, Potential Anal. **36** (2012), no. 4, 569–587. MR 2904634

[Psk05a] Arsen V. Pskhu, *On the real zeros of a function of Mittag-Leffler type*,
         Mat. Zametki **77** (2005), no. 4, 592–599. MR 2178026

[Psk05b] Arsen V. Pskhu, *Partial Differential Equations of Fractional Order
         (in Russian)*, "Nauka", Moscow, 2005. MR 2330051

[Psk09] Arsen V. Pskhu, *The fundamental solution of a diffusion-wave equa-
         tion of fractional order*, Izv. Ross. Akad. Nauk Ser. Mat. **73** (2009),
         no. 2, 141–182. MR 2532450

[PV94]   Arvet Pedas and Gennadi M. Vainikko, *The smoothness of solutions to
         nonlinear weakly singular integral equations*, Z. Anal. Anwendungen
         13 (1994), no. 3, 463–476. MR 1290314

[PV06]   Arvet Pedas and Gennadi Vainikko, *Integral equations with diagonal
         and boundary singularities of the kernel*, Z. Anal. Anwend. **25** (2006),
         no. 4, 487–516. MR 2285098

[Qi10]   Feng Qi, *Bounds for the ratio of two gamma functions*, J. Inequal.
         Appl. (2010), Art. ID 493058, 84. MR 2611044

[RFMM+04] Gabriel Ramos-Fernández, José L Mateos, Octavio Miramontes, Ger-
         minal Cocho, Hernán Larralde, and Bárbara Ayala-Orozco, *Lévy walk
         patterns in the foraging movements of spider monkeys (Ateles geof-
         froyi)*, Behav. Ecol. Sociol. **55** (2004), no. 3, 223–230.

[Rie76] Bernhard Riemann, *Versuch einer allgemeinen Auffassung der Inte-
         gration und Differentiation*, Gesammelte Mathematische Werke und
         Wissenschaftlicler, Teubner, Leipzig, 1876, pp. 331–344.

[Ros16] Sabrina D. Roscani, *Hopf lemma for the fractional diffusion operator and its application to a fractional free-boundary problem*, J. Math. Anal. Appl. **434** (2016), no. 1, 125–135. MR 3404551

[RSL95] Bertram Ross, Stefan G. Samko, and E. Russel Love, *Functions that have no first order derivative might have fractional derivatives of all orders less than one*, Real Anal. Exchange **20** (1994/95), no. 1, 140–157. MR 1313679

[RV12] J. Diego Ramírez and Aghalaya S. Vatsala, *Generalized monotone iterative technique for Caputo fractional differential equation with periodic boundary condition via initial value problem*, Int. J. Differ. Equ. (2012), Art. ID 842813, 17. MR 2975374

[Sch96] Walter Rudolf Schneider, *Completely monotone generalized Mittag-Leffler functions*, Exposition. Math. **14** (1996), no. 1, 3–16. MR 1382012 (97a:33041)

[Sed94] A. M. Sedletskiĭ, *Asymptotic formulas for zeros of functions of Mittag-Leffler type*, Anal. Math. **20** (1994), no. 2, 117–132. MR 1311296 (95m:30038)

[Sed00] A. M. Sedletskiĭ, *On the zeros of a function of Mittag-Leffler type*, Mat. Zametki **68** (2000), no. 5, 710–724. MR 1835453 (2002m:33028)

[Sed04] A. M. Sedletskiĭ, *Nonasymptotic properties of the roots of a function of Mittag-Leffler type*, Mat. Zametki **75** (2004), no. 3, 405–420. MR 2068803 (2005m:33037)

[SH09] Hansjörg Seybold and Rudolf Hilfer, *Numerical algorithm for calculating the generalized Mittag-Leffler function*, SIAM J. Numer. Anal. **47** (2008/09), no. 1, 69–88. MR 2452852 (2009i:30024)

[Sim86] Jacques Simon, *Compact sets in the space $L^p(0, T; B)$*, Ann. Mat. Pura Appl. (4) **146** (1986), 65–96. MR 916688

[Sim14] Thomas Simon, *Comparing Fréchet and positive stable laws*, Electron. J. Probab. **19** (2014), no. 16, 25. MR 3164769

[Sim15] Thomas Simon, *Mittag-Leffler functions and complete monotonicity*, Integral Transforms Spec. Funct. **26** (2015), no. 1, 36–50. MR 3275447

[SKM93] Stefan G. Samko, Anatoly A. Kilbas, and Oleg I. Marichev, *Fractional Integrals and Derivatives*, Gordon and Breach Science Publishers, Yverdon, 1993. MR 1347689 (96d:26012)

[Sol76] Vsevolod Alekseevich Solonnikov, *Estimates of the solution of a certain initial-boundary value problem for a linear nonstationary system of Navier-Stokes equations*, Zap. Naučn. Sem. Leningrad. Otdel Mat. Inst. Steklov. (LOMI) **59** (1976), 178–254, 257, Boundary value problems of mathematical physics and related questions in the theory of functions, 9. MR 0460931

[ST07] Thomas Schmelzer and Lloyd N. Trefethen, *Computing the gamma function using contour integrals and rational approximations*, SIAM J. Numer. Anal. **45** (2007), no. 2, 558–571. MR 2300287

[Sta70] Bogoljub Stanković, *On the function of E. M. Wright*, Publ. Inst. Math. (Beograd) (N.S.) **10 (24)** (1970), 113–124. MR 0280762

[Sty16] Martin Stynes, *Too much regularity may force too much uniqueness*, Fract. Calc. Appl. Anal. **19** (2016), no. 6, 1554–1562. MR 3589365

[SW78] E. J. P. Georg Schmidt and Norbert Weck, *On the boundary behavior of solutions to elliptic and parabolic equations—with applications to boundary control for parabolic equations*, SIAM J. Control Optim. **16** (1978), no. 4, 593–598. MR 497464

[SW89] Walter Rudolf Schneider and Walter Wyss, *Fractional diffusion and wave equations*, J. Math. Phys. **30** (1989), no. 1, 134–144. MR 974464

[SW06] Zhi-Zhong Sun and Xiaonan Wu, *A fully discrete difference scheme for a diffusion-wave system*, Appl. Numer. Math. **56** (2006), no. 2, 193–209. MR 2200938 (2006k:65227)

[SY11] Kenichi Sakamoto and Masahiro Yamamoto, *Initial value/boundary value problems for fractional diffusion-wave equations and applications to some inverse problems*, J. Math. Anal. Appl. **382** (2011), no. 1, 426–447. MR 2805524

[Tam30] Jacob David Tamarkin, *On integrable solutions of Abel's integral equation*, Ann. of Math. (2) **31** (1930), no. 2, 219–229. MR 1502936

[Tho06] Vidar Thomée, *Galerkin Finite Element Methods for Parabolic Problems*, second ed., Springer-Verlag, Berlin, 2006. MR 2249024 (2007b:65003)

[TT16] Niyaz Tokmagambetov and Berikbol T. Torebek, *Fractional analogue of Sturm-Liouville operator*, Doc. Math. **21** (2016), 1503–1514. MR 3603928

[TY17] Erwin Topp and Miguel Yangari, *Existence and uniqueness for parabolic problems with Caputo time derivative*, J. Differential Equations **262** (2017), no. 12, 6018–6046. MR 3624548

[TYZ19] Tao Tang, Haijun Yu, and Tao Zhou, *On energy dissipation theory and numerical stability for time-fractional phase-field equations*, SIAM J. Sci. Comput. **41** (2019), no. 6, A3757–A3778. MR 4036095

[US07] Sabir R. Umarov and È. M. Saĭdamatov, *Generalization of the Duhamel principle for fractional-order differential equations*, Dokl. Akad. Nauk **412** (2007), no. 4, 463–465. MR 2451337

[Vai89] Gennadi M. Vainikko, *On the smoothness of the solution of multidimensional weakly singular integral equations*, Mat. Sb. **180** (1989), no. 12, 1709–1723, 1728. MR 1038224

[VZ10] Vicente Vergara and Rico Zacher, *A priori bounds for degenerate and singular evolutionary partial integro-differential equations*, Nonlinear Anal. **73** (2010), no. 11, 3572–3585. MR 2718161

[VZ15] Vicente Vergara and Rico Zacher, *Optimal decay estimates for time-fractional and other nonlocal subdiffusion equations via energy methods*, SIAM J. Math. Anal. **47** (2015), no. 1, 210–239. MR 3296607

[Web19a] Jeffrey R. L. Webb, *Initial value problems for Caputo fractional equations with singular nonlinearities*, Electron. J. Differential Equations (2019), Paper No. 117, 32. MR 4028821

[Web19b] Jeffrey R. L. Webb, *Weakly singular Gronwall inequalities and applications to fractional differential equations*, J. Math. Anal. Appl. **471** (2019), no. 1-2, 692–711. MR 3906348

[Wei75] Dennis G. Weis, *Asymptotic behavior of some nonlinear Volterra integral equations*, J. Math. Anal. Appl. **49** (1975), 59–87. MR 367596

[Wei01] Lutz Weis, *Operator-valued Fourier multiplier theorems and maximal L_p-regularity*, Math. Ann. **319** (2001), no. 4, 735–758. MR 1825406

[Wen48] James G. Wendel, *Note on the Gamma function*, Amer. Math. Monthly **55** (1948), 563–564. MR 29448

[Wey12] Hermann Weyl, *Das asymptotische Verteilungsgesetz der Eigenwerte linearer partieller Differentialgleichungen (mit einer Anwendung auf die Theorie der Hohlraumstrahlung)*, Math. Ann. **71** (1912), no. 4, 441–479. MR 1511670

[Wid41] David Vernon Widder, *The Laplace Transform*, Princeton University Press, Princeton, N. J., 1941. MR 0005923

[Wim05] A. Wiman, *Über die Nullstellen der Funktionen $E^a(x)$*, Acta Math. **29** (1905), no. 1, 217–234. MR 1555016

[Wri33] Edward M. Wright, *On the coefficients of power series having exponential singularities*, J. London Math. Soc **8** (1933), 71–79.

[Wri35a] Edward M. Wright, *The asymptotic expansion of the generalized Bessel function*, Proc. London Math. Soc. **38** (1935), 257–270.

[Wri35b] Edward M. Wright, *The asymptotic expansion of the generalized hypergeometric function*, J. London Math. Soc. **10** (1935), 287–293.

[Wri40] Edward M. Wright, *The generalized Bessel function of order greater than one*, Quart. J. Math., Oxford Ser. **11** (1940), 36–48. MR 0003875 (2,285e)

[WW95] Aleksander Weron and Rafał Weron, *Computer simulation of Lévy α-stable variables and processes*, Chaos—the interplay between stochastic and deterministic behaviour (Karpacz, 1995), Lecture Notes in Phys., vol. 457, Springer, Berlin, 1995, pp. 379–392. MR 1452625 (98b:60029)

[WW96] Edmund Taylor Whittaker and George Neville Watson, *A Course of Modern Analysis*, Cambridge University Press, Cambridge, 1996. MR 1424469

[WY13] Hong Wang and Danping Yang, *Wellposedness of variable-coefficient conservative fractional elliptic differential equations*, SIAM J. Numer. Anal. **51** (2013), no. 2, 1088–1107. MR 3036999

[WZ99a] Roderick Wong and Yu-Qiu Zhao, *Smoothing of Stokes's discontinuity for the generalized Bessel function*, R. Soc. Lond. Proc. Ser. A Math. Phys. Eng. Sci. **455** (1999), no. 1984, 1381–1400. MR 1701756 (2000c:41046)

[WZ99b] Roderick Wong and Yu-Qiu Zhao, *Smoothing of Stokes's discontinuity for the generalized Bessel function. II*, R. Soc. Lond. Proc. Ser. A Math. Phys. Eng. Sci. **455** (1999), no. 1988, 3065–3084. MR 1807056 (2001k:33007)

[WZ02] Roderick Wong and Yu-Qiu Zhao, *Exponential asymptotics of the Mittag-Leffler function*, Constr. Approx. **18** (2002), no. 3, 355–385. MR 1906764 (2003c:41042)

[WZ20] Kai Wang and Zhi Zhou, *High-order time stepping schemes for semilinear subdiffusion equations*, SIAM J. Numer. Anal. **58** (2020), no. 6, 3226–3250. MR 4172731

[Yam18] Masahiro Yamamoto, *Weak solutions to non-homogeneous boundary value problems for time-fractional diffusion equations*, J. Math. Anal. Appl. **460** (2018), no. 1, 365–381. MR 3739910

[YGD07] Haiping Ye, Jianming Gao, and Yongsheng Ding, *A generalized Gronwall inequality and its application to a fractional differential equation*, J. Math. Anal. Appl. **328** (2007), no. 2, 1075–1081. MR 2290034

[YKF18] Yubin Yan, Monzorul Khan, and Neville J. Ford, *An analysis of the modified L1 scheme for time-fractional partial differential equations with nonsmooth data*, SIAM J. Numer. Anal. **56** (2018), no. 1, 210–227. MR 3744997

[Zac05] Rico Zacher, *Maximal regularity of type L_p for abstract parabolic Volterra equations*, J. Evol. Equ. **5** (2005), no. 1, 79–103. MR 2125407

[Zac08] Rico Zacher, *Boundedness of weak solutions to evolutionary partial integro-differential equations with discontinuous coefficients*, J. Math. Anal. Appl. **348** (2008), no. 1, 137–149. MR 2449333

[Zac09] Rico Zacher, *Weak solutions of abstract evolutionary integro-differential equations in Hilbert spaces*, Funkcial. Ekvac. **52** (2009), no. 1, 1–18. MR 2538276

[Zac13a] Rico Zacher, *A De Giorgi–Nash type theorem for time fractional diffusion equations*, Math. Ann. **356** (2013), no. 1, 99–146. MR 3038123

[Zac13b] Rico Zacher, *A weak Harnack inequality for fractional evolution equations with discontinuous coefficients*, Ann. Sc. Norm. Super. Pisa Cl. Sci. (5) **12** (2013), no. 4, 903–940. MR 3184573

[ZDK15] Vasily Zaburdaev, Sergey Denisov, and Joseph Klafter, *Lévy walks*, Rev. Modern Phys. **87** (2015), no. 2, 483–530.

[Zei86] Eberhard Zeidler, *Nonlinear Functional Analysis and its Applications. I*, Springer-Verlag, New York, 1986, Fixed-point theorems, Translated from the German by Peter R. Wadsack. MR 816732

[Zha00] Shuqin Zhang, *The existence of a positive solution for a nonlinear fractional differential equation*, J. Math. Anal. Appl. **252** (2000), no. 2, 804–812. MR 1800180

[Zha06a] Shuqin Zhang, *Existence of solution for a boundary value problem of fractional order*, Acta Math. Sci. Ser. B (Engl. Ed.) **26** (2006), no. 2, 220–228. MR 2218359

[Zha06b] Shuqin Zhang, *Positive solutions for boundary-value problems of non-linear fractional differential equations*, Electron. J. Differential Equations (2006), No. 36, 12 pp. MR 2213580 (2006k:34061)

[ZK13] Mohsen Zayernouri and George Em Karniadakis, *Fractional Sturm-Liouville eigen-problems: theory and numerical approximation*, J. Comput. Phys. **252** (2013), 495–517. MR 3101519

[ZX11] Ying Zhang and Xiang Xu, *Inverse source problem for a fractional diffusion equation*, Inverse Problems **27** (2011), no. 3, 035010, 12. MR 2772529 (2011j:65203)

[ZZ17] Zhidong Zhang and Zhi Zhou, *Recovering the potential term in a fractional diffusion equation*, IMA J. Appl. Math. **82** (2017), no. 3, 579–600. MR 3671483

Index

© The Author(s), under exclusive license to Springer Nature Switzerland AG 2021
B. Jin, *Fractional Differential Equations*, Applied Mathematical Sciences 206,
https://doi.org/10.1007/978-3-030-76043-4

Printed in the United States
by Baker & Taylor Publisher Services